USEFUL PLANTS OF GHANA

Useful Plants Of Ghana

West African uses of wild and cultivated plants

DANIEL K. ABBIW

INTERMEDIATE TECHNOLOGY PUBLICATIONS and
THE ROYAL BOTANIC GARDENS, KEW

Published by Intermediate Technology Publications Ltd.,
103/105 Southampton Row, London WC1B 4HH, UK, and
Royal Botanic Gardens, Kew, Richmond, Surrey TW9 3AB, UK.

© Daniel Abbiw, 1990

Reprinted 1995

ISBN 1 85339 043 7

Printed by SRP, Exeter, UK

CONTENTS

Map of Ghana..vi
Preface vii
Acknowledgements ix
Introduction xi

1. Forests and Conservation 1
2. Food and Fodder.................................. 22
3. Industrial or Cash Crops 59
4. Building and Construction 78
5. Furnishings 90
6. Fuel 94
7. Tools and Crafts 103
8. Potions and Medicines 118
9. Poisons, Tannin, Dyes, etc....................... 206
10. Amenity Landscaping and Gardening............. 232
11. Weeds 232
12. Plants and Soil Nutrients 252

Bibliography and References........................ 259

Index
   Local Names (Ghanaian)........................ 277
   Common Names (English) 283
   Scientific Names................................ 288
   General 313

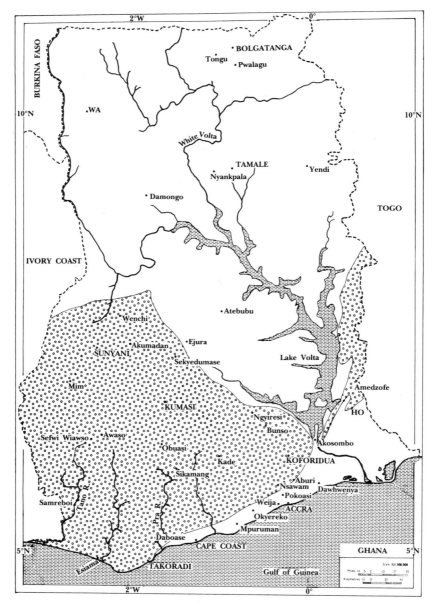

Sketch map of Ghana showing regional capitals in block letters, also selected towns and villages mentioned for their productive or distributive importance. Original forest zone is shaded.

Preface

It is encouraging to know that the author is a Ghanaian. He is a botanist and Curator of the University of Ghana Herbarium at Legon. He writes as a leading expert on the forests of his country and is familiar with the uses of trees and herbaceous plants, many of which have a wide distribution in Africa far beyond Ghana's frontiers. For this reason I enjoy visiting forests with him!

He has brought together a mass of fascinating information for the librarian, student, forester, agriculturalist, horticulturist, herbalist and a host of other specialists, as well as the general public. Indeed it is aimed at the average reader who will find the text conveniently divided into subjects such as food, fuel, medicines and weeds. The plants are listed by their scientific and Ghanaian vernacular names, also with English ones if available, making this of international application founded on scientific research.

While industrial crops and plants of farms and gardens are included, many of the uses are of local occurrence and of minor importance but of great interest to the ethnobotanist and biochemist searching for little known remedies as an indication of active chemicals for further research. In an age of rapid change when environmental issues have assumed significant popular and political standing, it is important to document as much as possible and make it available to a wide public. Hence it is appropriate that Ghana and Great Britain should collaborate over the production of this book, and that the Royal Botanic Gardens, Kew, Oxford Forestry Institute and Intermediate Technology Publications also co-operate in its publication. Kew botanists have studied African plants for some two centuries and many publications, such as the *Flora of West Tropical Africa*, have been produced there. In recent years Intermediate Technology of London has pioneered the sharing of useful knowledge with countries of the Third World, helping people to locate and use technologies appropriate to the social, cultural, economic and environmental needs of the communities throughout the developing world. Intermediate Technology Publications make relevant literature available at affordable prices, hence the present work is substantially subsidised.

An attractive and informative feature of this book is the inclusion of colour photographs of plant products which has been made possible by a generous grant from a Quaker trust. Many of the photographs have been taken in Ghana by Daniel Abbiw himself, a man of many parts.

I am sure that the contents of this book will be of use to Africans and a constant source of interest and reference throughout the world, especially across the Atlantic Ocean where many residents trace their roots back to Africa and wish to know about African plant uses. Further information on folklore and religion beyond the scope of the current volume is being gathered into the forthcoming *Traditional Plants of West Africa* — those uses and practices based on beliefs, symbols, signs and values — by the same author.

In conclusion, I think it is right to repeat the author's warning (on p. 125) concerning the use of medicinal plants that 'self-medication is dangerous' and 'it is advisable to consult practising herbalists or qualified elders in all cases of herbal treatment'.

F. Nigel Hepper
The Herbarium
Royal Botanic Gardens
Kew, England

Acknowledgements

I wish to express my sincere thanks to all those who helped to make the successful completion of this book possible. I am grateful to Mr O.B. Dokosi of the Herbs of Ghana Project and Mr Albert A. Enti of Forestry Enterprises Limited in Ghana for their fatherly advice and constant encouragement; and to Mr Barry Hughes, then Senior Lecturer at the Zoology Department, University of Ghana, Legon, for his interest, the provision of relevant references, his regular enquiries as to the progress of the manuscript, and information on plants in relation to snakes. I thank Mr Joseph K. Osei, Research Officer of the Agricultural Research Station, Kade, Ghana, for information on the varieties of Sweet Orange, Cocoyam, and Plantain; and on the weed problem at the Station. The Curators of the Botanical Gardens and the Grounds at Legon were equally helpful on the weed situation there and its control; and Mr C.W. Agyakwah of IITA in Nigeria furnished useful information on the weed problem in developing countries generally.

I am grateful to Mr K. Amoe-Abban, Managing Director of Tema Textiles Limited, and Mr J. Ofosu-Amaah, Operations Manager of Ghana Textiles Manufacturing Co. Limited, for information on cotton importation and cultivation and on the production figures at these factories. The information on the production and export of both processed and unprocessed cocoa was made available to me by Mr Isaac K. Abbiw, my brother and Factory Manager of Cocoa Processing Company Limited (G.C.M.B.), Takoradi. I thank Mr S.P. Yawson, General Manager of Fibre Products Company Limited, and Mr Yaw Nkansah-Gyamfi of the Bast Fibres Development Board, both at Kumasi in Ghana, for the information on cultivation of fibre plants and production of jute bags; and the Chief Conservator of Forests in Ghana, Dr Johnny François, for the list of sites and areas of the major fuel-wood plantations in Ghana.

I am indebted to Miss Amanda Jones, then a student at Forestry and Land management at the Oxford Forestry Institute, for a review of the synopsis and first draft; and to Miss Soo Tasker and Mr David Menzies of the Orchid Unit at the Royal Botanic Gardens, Kew, for reading through the paragraphs on the orchids, and for their corrections, additions, and information on decorative and endangered species and West African orchids introduced to Kew. Mr H.M. Burkill, also of Kew and author of *Useful Plants of West Africa* Vol. 1 (A–D), made useful comments and criticisms on the manuscript and kindly put at my disposal his list of revised names of the West African flora for which I am most grateful. My greatest debt is to Moro Ibrahim, a herbalist at the

University of Ghana, Legon, for first-hand information on the curative properties of the Hairy Indigo and many other herbs. My sincere thanks also go to the many herbalists, craftmen, carpenters, carvers, weavers, farmers, firewood contractors, charcoal burners, palm-wine tappers, and so on — too many to single out by name — who kindly supplied information on the usefulness of plants and plant material.

I acknowledge the very useful role which existing literature played in contributing to the writing of this book — particularly *The Useful Plants of West Africa* by J.M. Dalziel and *Woody Plants of Ghana* by F.R. Irvine. I express my appreciation for the facilities at the Ghana Herbarium (GC) and other Herbaria in Ghana; the Balme Library and the Faculty of Agriculture Library, University of Ghana; the Herbarium and Library (UCJ), Centre National de Floristique, Université d'Abidjan, and Office de la Recherche Scientifique et Techniques Outre-Mer (ABI), la Côte d'Ivoire; the Herbarium and Library, Université du Benin, Lomé, Togo; the Herbarium and Library of the Missouri Botanical Gardens (MO); the Smithsonian Institution and the United States National Arboretum and Herbarium (US), Washington DC; the Agricultural Research Centre, Beltsville (BARC), Maryland; the Shore Acres State Park and Library, Coos Bay, Oregon; the Forest Herbarium (FHO), Oxford; and the Gardens, Museum, Herbarium (K), and Library of the Royal Botanic Gardens, Kew

I am grateful to the United States National Science Foundation for providing an airline ticket and expenses for me to attend the AETFAT 1985 Congress in St Louis, Missouri; and the then Secretary General, Dr Peter Goldblatt, for kindly consenting to reschedule the return flight with a stop-over in Britain to enable me to confer with my supervisors. Subsistence for this period of study attachment at Oxford and Kew was sponsored by a generous grant from the Commonwealth Foundation in London. The Foundation also paid the extra expenses involved to air-freight the necessary stationery for the final typing of the manuscript in Ghana. Funds for extra travelling expenses were kindly paid for by Mr and Mrs Emil Zivney, the veteran seed collectors of Lincoln City, Oregon and Miss Renate Orf of Centralia, Washington State.

The final typing of the manuscript was ably done by Mr Francis C. Essandoh of the Institute of African Studies, University of Ghana, Legon —assisted by Messrs Frederick K. Abredu and Emmanuel A. Gyabaah. All the indices were kindly typed on the word processor free of charge by Cynthia Styles of Oxford, England; and programmed for alphabetical sorting by Dr Matthew Jebb, Dr Kit Goodwin-Bailey, and Mrs Julie Smith, all of the Oxford Forestry Institute.

Last but not least, I am grateful to Mr Neal Burton of Intermediate Technology Publications for his interest, and to the Director of the Royal Botanic Gardens, Kew, for allowing his staff to spend time advising on, editing, typesetting and checking this work, especially Mr F.N. Hepper who also sent many of his photographs for reproduction.

Introduction

Since prehistoric times man has used plants for various purposes and he will continue to do so as long as life continues on this planet. Even in an age of substitute man-made materials, plants and plant products are still in great demand. The living world depends on plant life. Plants purify the air we breathe and serve as food for both man and beast; they are a source of fuel for cooking, lighting, heating, and provide materials for building and construction. Trees are used for carving and other household articles. Their extraction yields sweet juices to drink, and potions and medicines to cure diseases. Cash crops provide our industries with much-needed raw materials. We use wood for our buildings, chairs, tables, and other furnishings. The forest houses and protects game, stabilizes the environment, and prevents soil erosion. There is no substitute for the aesthetic values of parks and the beauty of decorative plants. Even noxious weeds may be utilized to advantage as manure.

The English saying 'touch wood' presupposes the presence of wood wherever one might be, even in this age of modern materials such as plastics. No one would want the world to remain unchanged but, as new materials replace the old, it is good to see that wood at least will never be dispensable; it has been used by too many craftsmen over too long a period to produce too great a variety of durable objects, to be lightly set aside. There is hardly any aspect of human activity which does not in one way or another depend on or require the use, either directly or indirectly, of a plant or plant material. Some of these potential sources have probably not yet been realized, while others are regrettably underused. The Chief Conservator of Forests in Ghana, Dr Johnny François, observed that, in the view of those most closely associated with the story of these forests, what is known is but a minuscule portion of what there is to know; the products and services now utilized are perhaps only an infinitesimal part of the full potential.

According to the World Wide Fund for Nature only 15 per cent of tropical moist forest species have been catalogued and only one per cent intensively screened for possible benefits to humanity. Of the many timber trees in West Africa, less than 5 per cent are of importance economically in the world market. In Ghana most of the Class 3 and 4 species (Table 1.2) — over 95 per cent — are of little or no value outside the boundaries. Research has been conducted by the Utilization Branch of the Department of Forestry and the Forest Products Research Institute of the Centre for Scientific and Industrial

Research (CSIR) on the suitability of these timbers for furniture, with promising results. Unfortunately, it is another matter convincing local carpenters to use them.

There is untold potential in our wild plants. Now that the Centre for Scientific Research into Plant Medicine (CSRPM) is established in Ghana, systematic analysis may well resolve and unravel the hidden secrets of the curative properties of plants. Eventually it should be possible to reduce dosages from, say, a calabashful of unrefined juices to a spoonful, or even a few tablets.

The entire dependence of man on plants and plant products, directly for his basic needs as food, clothing, and shelter and indirectly for their beneficial influence on the climate and maintenance of his immediate and remote environment, makes plants vital to his survival and the basis of his continued existence. Our survival and continued existence in turn depend on the efficiency with which man, with all the resources and technology available to him, harnesses, develops, and utilizes plants and plant products.

The uses to which plants in our area may be put are listed in *The Useful Plants of Nigeria*, J.H. Holland, 1922; *Plants of the Gold Coast*, F.R. Irvine, 1930; *The Useful Plants of West Tropical Africa*, J.M. Dalziel, 1948 (vol. 1 revised by H.M. Burkill, 1985); and *Woody Plants of Ghana*, F.R. Irvine, 1961. Some uses of our plants are contained in *Synecology and Silviculture*, C.J. Taylor, 1960, and *Medicinal Plants of West Africa*, Edward Ayensu, 1978, gives the medicinal uses of members of selected families of plants. The present work not only attempts to combine the material in *The Useful Plants of West Tropical Africa* and *Woody Plants of Ghana*, but also includes material from sources other than these. Unlike its predecessors, where the uses are given under plant families arranged alphabetically or phylogenetically, the uses here are grouped into chapters by subject.

The binomial system of naming plants is used. The nomenclature is as followed in the *Flora of West Tropical Africa* by J. Hutchinson and J.M. Dalziel (revised by R.W.J. Keay and F.N. Hepper 1954–72). Name changes and revisions since the appearance of this work are given in brackets. Common English names are given where available, and reliable Ghanaian vernacular names are also given in brackets. It is desirable that the usage of vernacular names be standardized and for this reason no species is given more than one name. The use of one vernacular name for two or more species is also generally avoided. The plants mentioned occur mainly in tropical West Africa and to some extent tropical Africa generally, but the emphasis is particularly on West Africa with special reference to Ghana. Uses in other tropical and subtropical countries where our plants grow, or temperate countries where they have been introduced, are included.

ONE
Forests and Conservation

The paradox of the forest is that where it occurs man's tendency is to destroy it, and where it does not occur the tendency is to create it. Directly and indirectly, it serves a multiplicity of purposes. Hall and Swaine (1981) observed that

> 'many reasons may be adduced for the conservation of representative intact forest, some cultural and some economic. The forest environment has set the scene for the historical development of most of the peoples of southern Ghana. It is our duty to ensure that such an important part of the country's heritage is preserved for future generations. Experience elsewhere in the world has shown that urbanised, industrialised man needs wilderness areas into which, from time to time, he can escape from his artificial working environment. It can also be argued that forest animals —elephants, monkeys, antelopes, birds — have a right to continued existence'.

Beside the protection of game and wildlife, the forest provides fertile land and atmosphere suitable for the cultivation of food and cash crops; and is a source of the finest timber, curls, and figured wood of commerce. The forest is a collecting site for new, undescribed species and plants of unknown potential and a source of curative herbs, seeds, roots, and stem-bark. It is a place for collecting fuel-wood, snails, and mushrooms. It is believed to be the abode of dwarfs and a sacred grove where our ancestors were buried. Folklore has it that the forest is the domain of the mythical *Sasabonsam* – a legendary figure responsible for all the woes of mankind and to which mishaps and everything evil are attributable. The benefits (or otherwise) are many; but, unlike the story of the elephant and the three blind men, where the individual different descriptions together make up the whole, the benefits it offers add up to a fraction only of the total potential of the forest.

Tropical rain forest (the 'Guinea-Congolian' rain forest) in West Africa covers Sierra Leone, Liberia, and parts of Guinea, the Ivory Coast, Ghana, Nigeria, and Cameroun. The area of the forest zone in Ghana is about 82,000 km^2, representing 34 per cent of the total area of the country. There have been three main classifications of the forest zone. Those of Taylor in 1960 and of Mooney in 1961 are based on the inventory of dominant, emerging trees only. A more recent survey by Hall and Swaine (1976) introduces three basic criteria. The study involved those species that occur four or more times in a total enumeration of 155 sample plots measuring 25 m x 25 m (0.0625 ha).

These must be located in a truly representative climax and in a homogenous close-canopy forest.

There are some two hundred Forest Reserves in the forest zone (see Table 1.1), and these together represent just over 20 per cent of the total area of this zone. Through constant clearing for habitation, and especially farming, the forest zone is diminishing with alarming speed. Dr Vietmeyer of the Smithsonian Institution in Washington lamented that nowhere is the loss greater or more serious than in the tropics, where vast areas are being deforested by spreading agriculture and human settlement, by commercial timbering, and by the gathering of firewood, which has been spurred to an unprecedented scale by the rising cost of kerosene. The fact that the proportion of forest outside reserves in Ghana declined from 20 per cent in 1953 to 5 per cent in 1972, according to estimates in the annual reports of the Forestry Department for those years, supports this statement. This rape of the forest is actually going on in Forest Reserves as well. Despite heavy fines for offenders, illegal farming in reserves continues. On the whole, however, the Department has been quite successful in checking illegal farming and there were 525 successful prosecutions in the courts during 1972. It is not only the forest which is at stake but also the much-needed fertile topsoil as well. Dr Vietmeyer observes that removing the tropical cover leaves the soil deprived of nutrients (because in a tropical forest the bulk of nutrients is in the trees rather than in the soil, as in temperate zones); deforestation also leaves the ground unprotected from rain that in one storm can wash away fertile topsoil that took centuries to accumulate.

The forest serves two main purposes — first, the influence that the association of trees as a whole has on the immediate locality or the area generally, and secondly, the forest produce that may be obtained from it. While there is an inbuilt mechanism that automatically and naturally checks and balances the state of the forest by regulating energy loss and ensures a steady, uniform development to the climatic climax in the former, it is the way and manner in which the latter is exploited that results in the gradual decline or complete destruction of the forest. Products or by-products of the forest are termed direct benefits; as opposed to indirect benefits.

Table 1.1 Forest Reserves in the forest zone of Ghana

Reserve	Area (km)	Reserve	Area (km)
Abasumba	1.0	Abrimasu	26.2
Abisu	9.1	Afao Hills	34.7
Aboben Hill	7.3	Afia Shelterbelt	21.0
Aboma	45.6	Afram Headwaters	201.2
Aboniyere Shelterbelt	41.2	Afrensu-Brohuma	72.5

1. FORESTS AND CONSERVATION

Reserve	Area (km)	Reserve	Area (km)
Ahirasu	1.0	Bia Tributaries North	356.1
Aiyaola	34.7	Bia Tributaries South	305.6
Ajenjua Bepo	5.7	Bimpong	104.1
Ajuesu	9.6	Birim	39.1
Akrobong	2.6	Birim Extension	21.8
Amama Shelterbelt	44.0	Bobiri	54.6
Angoben Shelterbelt	34.7	Bodi	175.3
Anhwiaso East	124.3	Boi Tano	128.5
Anhwiaso North	3.6	Boin River	277.6
Anhwiaso South	22.3	Bomfoum	294.7
Ankaful Fuelwood	2.1	Bonkoni	75.1
Ankasa River	518.0	Bonsa Ben	155.4
Anum Su	43.8	Bonsa River	160.6
Anum Su Southern Sect.	12.7	Bonsam Bepo	124.3
Apamprama	34.7	Bosomoa	170.9
Aparapi Shelterbelt	19.2	Bosumkese	138.3
Apedwa	4.1	Bosumtwi Range	78.7
Apepesu	60.6	Boti Falls	1.3
Asenanyo River	227.9	Bowiye Range	120.2
Asin Apimanim	11.4	Brimso	10.6
Asin Atandaso	153.6	Bura River	103.1
Asonari	1.6	Cape Three Points	51.0
Asubima	78.7	Chai River	182.3
Asufu Shelterbelt East	11.4	Chiremoasi Bepo	6.0
Asufu Shelterbelt West	13.5	Chirimfa	114.0
Asukese	265.0	Dadiaso	171.2
Asuokoko River	116.0	Dampia Range	80.3
Atewa Range	232.3	Dede	51.0
Atewa Range Extension	26.4	Denyau Shelterbelt	12.4
Auro River	8.5	Desiri	151.0
Awura	133.9	Disue River	23.6
Ayum	112.9	Dome River	80.5
Baku	13.0	Draw River	235.4
Bandai Hills	160.8	Ebi River Shelterbelt	25.9
Bediako	7.0	Esen Epam	46.1
Bemu	43.8	Esuboni	28.5
Ben East	25.4	Esukawkaw	122.2
Ben West	55.7	Fum Headwaters	72.5
Bia Shelterbelt	29.5	Fure	158.2
Bia Tano	193.3	Fure Headwaters	169.6
Bia Tawya	678.6	Gianima	17.1

Reserve	Area (km)	Reserve	Area (km)
Goa Shelterbelt	23.8	Northern Scarp East	49.2
Jade Bepo	5.2	Northern Scarp West	64.8
Jade Bepo Extension	0.8	Nsemre	18.1
Jema Assamkrom	66.0	Nsuansa	62.7
Jeni River	21.5	Numia	50.2
Jimira	62.9	Nyamibe Bepo	22.3
Kabakaba Hills East	7.3	Obotumfo	1.6
Kabakaba Hills North	1.3	Oboyow	63.7
Kabakaba Hills West	5.4	Ochi Headwaters I	3.4
Kabo River	136.0	Ochi Headwaters II	9.3
Kade Bepo	16.8	Oda River	164.2
Kajeasi	26.7	Odomi River	16.1
Kakum	212.4	Ofin Headwaters	13.0
Klemu Headwaters	10.9	Ofin Shelterbelt	60.3
Kokotintin Shelterbelt	9.1	Ongwam I, II & III	31.3
Komenda Fuelwood	2.1	Onuem Bepo	34.4
Kpandu Range Dayi Block	30.3	Onuem Nyamibe Shelterbelt	24.9
Kpandu Range West	35.5	Onyimsu	8.5
Krochua	10.6	Opimbo I & II	1.0
Krokosua Hills	481.7	Opon Mansi	116.6
Kronwam	5.7	Opro River	129.2
Kunda	0.5	Pamu Berekum	189.1
Kunsimua Bepo	10.1	Pompo Headwaters	12.2
Kwamisa	82.9	Pra Anum	123.3
Kwekaru	11.7	Pra Birim	23.6
Kwesi Anyinima	1.8	Pra Suhien I	82.1
Mamang River	54.4	Pra Suhien II	104.1
Mamiri	45.3	Prakaw	9.8
Mankrang	85.5	Santomang	21.2
Manzan	305.6	Sapawsu	15.3
Minta	21.8	Sawsaw	62.9
Mirasa Hills	67.3	Sekondi Waterworks II & III	10.1
Mpameso	322.5	South Fomangsu	41.4
Muro	63.5	Southern Scarp	154.6
Ndumfri	72.5	Subin	238.3
Neung	157.7	Subin Shelterbelt	22.5
Nkawanda	8.0	Subri River	587.9
Nkonto Ben	14.5	Suhuma	358.5
Nkrabia	100.2	Sukusuku	147.6
North Bandai Hills	72.8	Sui River	333.9
North Fomangsu	42.7	Sumtwitwi	3.6

Reserve	Area (km)	Reserve	Area (km)
Supong	35.7	Tonton	146.3
Supuma Shelterbelt	25.0	Totua Shelterbelt	63.5
Tain Tributaries I	30.6	Upper Wasaw	100.8
Tain Tributaries II	509.2	Volta River	50.5
Tano Anwia	153.1	Wawahi	138.9
Tano Ehuro	176.1	Worobong Kwahu	55.2
Tano Nimri	205.9	Worobong North	13.2
Tano Ofin	402.2	Worobong South	106.2
Tano Suhien	84.4	Yaya	51.3
Tano Suraw	28.5	Yenku	21.2
Tanp Suraw Extension	75.1	Yogaga	0.8
Tinte Bepo	115.5	Yongwa	7.8
Togo Plateau	150.0	Yoyo	235.7

Direct Benefits

MAJOR FOREST PRODUCE

Generally, the woody material is classified as being major forest produce. This consists first of timber which is mostly obtained from the boles of trees, and second fuel-wood, obtained from the branches of big trees or from small trees, shrubs, and dead wood. In Ghana the Forestry Department classifies timber according to its international value. Irvine (1961) lists about 400 timbers suitable for heavy construction, general carpentry and joinery, windows, doors, plywood, and veneers (see Chapters 4 and 5). Of these only some 50 species are classified as merchantable timbers (see Table 1.2), and 23 species (mostly those in Classes 1 and 2) as commercially important for export, as logs or sawn timber, processed into veneers and plywood, or as self-assembly furniture. In the Ivory Coast only 49 of the 300 main tropical species in the forests are felled and marketed. The actual export of timber trees is concentrated around only a few species of timbers. *Khaya anthotheca* White Mahogany (Krumben); *K. ivorensis* African Mahogany (Dubini); *Entandrophragma angolense* Gedu Nohor (Edinam); *E. cylindricum* Sapele (Apenkwa); *E. utile*, Utile; *Pericopsis elata* (*Afromosia elata*) (Kokrodua), *Tieghemella heckelii* (Baku), and *Triplochiton scleroxylon* (Wawa) account for about 84 per cent of all exportable timber. The rest of the merchantable timbers and the remaining classified ones are not being exported in quantities commensurate with their quality and abundance.

Hall and Swaine (1981) put the total annual timber production in Ghana (including local use) as averaging about 1.85 million m³ since 1958, between 50 per cent and 70 per cent being exported. Timber used to be the second most important export after cocoa — the two accounting for about 75 per cent of the country's income. It is now third, after cocoa and minerals. The manufacture of veneers and plywood both for local use and for export is an important part of the timber industry. In Ghana, there are three main veneer and plywood mills: African Timber and Plywood Ltd at Samreboi, Gliksten (W.A.) Ltd at Sefwi Wiawso, and Takoradi Veneer and Lumber Ltd (formerly F. Hills Ghana Ltd) at Takoradi — all in the Western Region.

Table 1.2 Forest timber trees

	Local Name		Local Name
Class I(a)		*Guarea cedrata*	(Kwabohoro)
		G. thompsonii	(Kwadwuma)
Milicia (Chlorophora) excelsa	(Odum)	*Lophira alata*	(Kaku)
M. regia	(Odum-nua)	*Piptadeniastrum africanum*	(Danhoma)
Entandrophragma angolense	(Edinam)		
E. cylindricum	(Penkwa)	*Class II(b)*	
E. utile	Utile		
Khaya anthotheca	(Krumben)	*Antiaris toxicaria (africana*	
K. grandifoliola	(Kruba)	*and welwitschii)*	(Kyenkyen)
K. ivorensis	(Dubini)	*Guibourtia ehie*	(Anokye-hyedua)
Tieghemella heckelii	(Baku)	*Mansonia altissima*	(Oprono)
Nauclea diderrichii	(Kusia)	*Mitragyna ciliata*	(Subaha)
		M. stipulosa	(Subaha-akoa)
Class I(b)		*Nesogordonia papaverifera*	(Danta)
Pericopsis elata (Afrormosia elata)	(Nyankom)	*Turraeanthus africanus*	(Apapaye)
		Class III	
Lovoa trichilioides	(Dubinibiri)		
Terminalia ivorensis	(Emire)	*Afzelia africana*	(Papao)
Triplochiton scleroxylon	(Wawa)	*A. bella* var. *glacior*	(Papaonua)
		Albizia adianthifolia	(Pampena)
Class I(c)		*A. ferruginea*	(Awiemfosamina)
Heritiera utilis (Tarrietia utilis)	(Nyankom)	*A. zygia*	(Okoro)
		Allanblackia parviflora	
		(floribunda)	(Sonkyi)
Class II(a)		*Anopyxis klaineana*	(Kokote)
		Canarium schweinfurthii	(Bediwonua)
Entandrophragma candollei	Candollei	*Celtis adolfi-friderici*	(Esakosua)

1. FORESTS AND CONSERVATION

	Local Name		Local Name
C. mildbraedii	(Esa)	*Petersianthus macrocarpus*	
C. zenkeri	(Esakokoo)	(*Combretodendron*	
Cylicodiscus gabunensis	(Denya)	*macrocarpum*)	(Esia)
Cynometra ananta	(Ananta)	*Pycnanthus angolensis*	(Otie)
Diospyros sanza-minika	(Sanzamulike)	*Scottellia klaineana* (*chevalieri*	
Distemonanthus benthamianus	(Bonsamdua)	and *coriacea*)	Odoko
Erythrophleum ivorense	(Odom)	*Sterculia rhinopetala*	(Wawabima)
E. suaveolens (*guineense*)	(Potrodom)	*Strombosia glaucescens*	
Holoptelea grandis	(Nakwa)	var. *lucida*	(Afena)
Mammea africana	(Bompagya)	*Terminalia superba*	(Ofram)
Morus mesozygia	(Wonton)		

Where the forest is degraded and poorly stocked, economic trees are introduced or helped to regenerate. One such system of improving the stock of the forest is the Tropical Shelterwood, where the forest is regenerated naturally under the canopy of existing trees. Another system, the taungya, is an agro-forestry operation, a combination of planting food crops and trees together in the initial stage. It operates better in areas where there is a genuine shortage of farming land, and serves the double purpose of artificially regenerating economic trees and simultaneously growing food crops. The period of association is about three years and the farmers care for the economic trees as they tend their crops. *Terminalia ivorensis* (Emire), *Khaya* species Mahogany, and *Entandrophragma* species Cedar are economic trees usually planted in taungyas in Ghana. Combined operations consist of the following measures:

(a) stock survey — an inventory of the economic trees of exploitable girth;
(b) enumeration survey — an assessment of all the economic trees below exploitable girth; and
(c) improvement thinning — the cutting of climbers and removal by killing with the arboricide sodium arsenite ($Na_2HAs_2O_3$).

In this way uneconomic trees were removed from the vicinity of useful ones. Improvement thinning has since been abandoned because many timber trees of future economic importance were being lost: as the dead trees fell they destroyed some of the very trees for which they were killed. The system was also too costly.

Since fuel-wood is a consumable item, the need for it exceeds by far the volume of timber removed from the forest. In 1962, Foggie and Piasecki estimated the annual consumption in Ghana as 4.5 million m^3 used directly as firewood and 1.5 million m^3 converted to charcoal. According to Hall and

Swaine (1981) the population of Ghana has almost doubled since Foggie and Piasecki reported their estimates, and if fuel-wood consumption has increased proportionally it seems likely that at least twice as much wood is now burnt as is used for timber. Recent figures from the Department of Forestry indicate that as much as ten times more fuel-wood is removed from the forests and the savanna woodland together than timber, and Dr Alain Lacroix (1982), editor of *The Courier*, observed that in Thirld World countries 90 per cent of the wood cut is used for fuel. The inefficient and wasteful method of traditional open-fire cooking accounts principally for the astronomically high proportion of fuel-wood consumed. At the World Wide Fund for Nature (WWF) International Council Meeting in Divonne, France, in 1985, fuel-saving stoves designed to combat the serious deforestation facing many Third World countries went on practical display. These stoves, built from locally found materials, can result in fuel savings of up to 90 per cent; by using a stove instead of cooking on an open fire, villagers cannot only dramatically reduce their consumption of wood, but also the amount of time they spend cooking or searching for fuel.

Wood may be crushed, ground, and processed for other purposes; woodpulp is the raw material for the manufacture of paper, for example. In Ghana a 100 km^2 plantation of *Gmelina arborea* with *Leucaena leucocephala* Leucaena and *Pinus* Pines is being established in the Subri Forest Reserve near Daboase on the Pra river to feed the proposed 60,000-ton capacity paper pulp mill. Large quantities of wood, usually of low grade, are also utilized in chip form to manufacture chipboard, fibreboard, and hardboard. There is a chipboard factory near Nkawkaw in Ghana.

MINOR FOREST PRODUCE

Other products that are obtained from the forest besides timber and firewood are collectively termed minor forest produce. The term 'minor' is only relative, for some of these produces, for example rubber and fibre, may be very important economically.

Chewing sponges, chewing sticks, and teeth cleaners

Chewing sponges (*sawe*) are prepared from the stems of the forest climbers *Acacia pentagona* and *A. kamerunensis* (Oguaben). The fruits of *Cnestis ferruginea* (Akitase), *Agelaea obliqua* (Apose), *A. trifolia* and the seeds of *Soyauxia grandifolia* (Abotesima) serve as teeth cleaners, as do the sap of *Jatropha curcas* Physic Nut, with salt, and the calyx of *Mussaenda afzelii* and *Datura innoxia*. The twigs or split stems of many plants are used as chewing sticks. Adu-Tutu and others (1979) estimated a total of 28 plants in their survey 'Chewing Stick Usage in Southern Ghana' and Irvine (1961) estimates about a hundred such plants for the whole country. Roots used as chewing sticks include *Anogeissus leiocarpus* (Sakanee), *Cassia sieberiana* Africana Laburnum, *Guiera senegalensis*,

1. FORESTS AND CONSERVATION

Pentadesma butyraceum (*butyracea*) Tallow Tree, (Abotoasebie); *Paullinia pinnata* (Toa-ntini), *Terminalia glaucescens* (Ongo), and *Waltheria indica* (Sawai), a common savanna weed. The roots of *Penianthus zenkeri* and *Sphenocentrum jollyanum* (Krakoo) are chewed more as an aphrodisiac than as teeth cleaners. *Garcinia* species are the most popular chewing sticks, in forest areas especially and urban centres generally; and together with *Teclea verdoorniana* (Owebiribi), a dry forest plant, account for about 86 per cent of all chewing sticks in Ghana. These two species are essentially the only ones sold as such. *G. kola* (Tweapea) is a high forest plant, slow-growing and fairly rare. It is worth developing into a crop to prevent extinction owing to the rate at which it is cut and used. *G. afzelii* and *G. epunctata* (Nsokodua) occur in the forest–savanna boundary and wet forests respectively.

Other plants used as chewing sticks are *Acacia nilotica* var. *tomentosa*, *Breonadia salicina* (*Adina microcephala*), *Allophylus africanus* (Hokple), *Paropsia adenostegia* (*Androsiphonia adenostegia*) (Nkatie), *Aulacocalyx jasminiflora* (Sanbrim), *Azadirachta indica* Neem Tree, *Baphia nitida* Camwood, (Odwen); *Burkea africana* (Pinimo), *Carpolobia alba* (Afiafia), *C. lutea* (Otwewa), and *Casearia barteri* (Punum). Also used are *Castanola paradoxa* (Abokodidua), *Cleidion gabonicum* (Mpawu), *Cola laurifolia* Laurel-Leaved Cola, *Craterispermum caudatum* (Duade), *C. laurinum*, *Dialium guineense* Velvet Tamarind, *Dichapetalum madagascariense* (*guineense*) (Antro), *Diospyros barteri* (Aheneba-nsatea), *D. elliotii*, *D. heudelotii*, *D. tricolor* (Ako), *Ehretia cymosa* (Okosua), *Eugenia coronata* (Kraku), *Gongronema latifolium* (Nsurogya), and *Griffonia simplicifolia* (Kagya). The rest are *Hibiscus rosa-sinensis* Shoe Flower, *Lasiodiscus mannii* var. *chevalieri* (*chevalieri*) (Dafa), *Maerua crassifolia*, *Mallotus oppositifolius* (Satadua), *Microdesmis puberula* (Fema), *Napoleonaea leonensis* (*Napoleona leonensis*), *N. vogelii* (Obua), *Olax gambecola*, *Oxyanthus speciosus* (Korantema), *O. tenuis*, *Margaritaria discoidea* (*Phyllanthus discoideus*) (Pepea), *P. muellerianus* (Potopoleboblo), *Psychotria subobliqua* (Aposin), *Rinorea subintegrifolia* (Atobe-gyaso), *Sorindeia juglandifolia*, and *Tiliacora dielsiana* (Kadze). *Drypetes floribunda* (Bedibesa) and *Tetrorchidium didymostemon* (Aboagyedua) are chewed for the sweet stems and sap respectively and *Strychnos afzelii* (Duapepere) for the aromatic scent. The flowers of *Nicotiana rustica*, *N. tabacum* Tobacco, *Solanum dasyphyllum*, and *S. macrocarpon* (Ntoropo) are used to stain the teeth.

Plants for basketry

The stem of some climbing plants are suitable for making baskets (see Chapter 7).

Plants as binding material

The stems of a number of plants are useful as rope, twine, or tietie for loads, firewood, building frames, rafters, platforms, and other purposes. Many plants in the Celastraceae family are used as tietie. These include *Loeseneriella africana* (*Hippocratea africana*) (Noto), *H. apocynoides* subsp. *guineensis* (*guineensis*) (Gwodei), *Reissantia indica* (*H. indica*), *L. rowlandii* (*H. rowlandii*) (Ntwea), and *Helictonema velutina* (*Hippocratea velutina*). *Adenia cissampeloides* (Akpeka), *A. rumicifolia* var. *miegei* (*lobata*) (Peteha), *Adenium somalense*, *Sherbournea bignoniiflora* (Kyerebeteni), and *Grewia barteri* have flexible and strong stems suitable for tying things. All the climbing palms are suitable for use as rope. The stems of *Stachyanthus occidentalis* (*Neostachyanthus occidentalis*) (Mutuo), *Neuropeltis acuminata*, *Smilax kraussiana* (Kokora), *Tetracera potatoria* (Twihama), *Tiliacora dielsiana* (Kadze), *T. dinklagei* (Susanfo), and *Ziziphus mauritiana* Indian Jujube are also suitable for tying (See also Chapter 4).

Fibre plants

Bast fibre plants include *Adansonia digitata* Baobab, *Alafia barteri* (Momorehemo), *Bauhinia rufescens* (Jinkiliza), *Calotropis procera* Sodom Apple, *Celtis integrifolia* Nettle Tree (Samparanga), *Christiana africana* (Sese-dua), and the stems of *Clappertonia ficifolia* (Nwohwea), which yield a valuable fibre comparable with jute and suitable for paper pulp. Other bast fibre plants are *Cola gigantea* var. *glabrescens* (Watapuo), *C. laurifolia* Laurel-Leaved Cola, *Cordia myxa* Sapistan Plum, *C. rothii*, *Cryptostegia grandiflora*, *Duparquetia orchidacea* (Pikeabo), and the root-bark as well as the stem of *Entada africana* (Kaboya). Fibre-yielding plants include *Ficus ingens* var. *ingens* (Kunkwiya), *Grewia bicolor*, *G. mollis* (Yualega), and *G. villosa*. The list also includes *Harungana madagascariensis* (Okosoa), *Hildegardia barteri* (Akyere), *Lannea microcarpa*, *L. kerstingii* (Kobewu), *Lonchocarpus sericeus* Senegal Lilac, *Ostryocarpus riparius*, *Piliostigma reticulatum*, *P. thonningii* (Opitipata), *Rauvolfia vomitoria* (Kakapenpen), *Sclerocarya birrea* (Nanogba), *Sesbania pachycarpa* (*bispinosa*), *Sterculia tragacantha* African Tragacanth (Sofo), and *Tabernaemontana* species. The rest are *Tamarindus indica* Indian Tamarind, *Thespesia populnea* (Frefi), *Trema orientalis* (*guineensis*) (Sesea), *Urera cameroonensis*, *Voacanga africana* (Ofuruma), *Xylopia quintasii* Elo (Asimba), and *Xylopia staudtii* (Alari). A number of climbing plants and lianas yield durable cordage and the fruit of *Cocos nucifera* Coconut Palm provides a coir fibre (see Plants as binding material, above, and Chapters 4 and 7). Other sources of plant fibre are *Kosteletzkya stellata*, *Wissadula amplissima*, *Dracaena fragrans*, *D. mannii* (Kesene), *Thalia welwitschii*, and the stems and roots of *Cissus quadrangularis* Edible-Stemmed Vine (Kotokoli). Commercially, *Hibiscus cannabinus* Kenaf, *H. sabdariffa* Roselle, *Corchorus* species Jute, and *Urena lobata* Congo Jute are the sources of bast fibre (see Chapter 3).

Poles for huts and sheds

Poles and forked sticks for the erection of temporary huts and sheds are obtainable from the forest (see Chapter 4).

1. FORESTS AND CONSERVATION

Pestles

Celtis species, especially *C. mildbraedii* (Esa), *C. zenkeri* (Esakokoo), and *C. wightii* (*brownii*) (Esa-fufuo) are the poles traditionally used as pestles in the forest zone. Others are *Albizia zygia* (Okoro), *Blighia unijugata* (Akyebiri), *B. welwitschii* (Akyekobiri), *Cola chlamydantha* (Tanamfre), *C. buntingii*, and *Morus mesozygia* (Wonton). The rest are *Diospyros kemerunensis* (Omenewa), *Necepsia afzelii*, *Pachystela brevipes* (Aframsua), *Swartzia fistuloides* (Asomanini), and *Xylopia quintasii* Elo (Asimba). In the savanna areas *Dialium guineense* Velvet Tamarind is the favourite pole for pestles. *Acacia* species, *Pericopsis laxiflora* (*Afrormosia laxiflora*) Satinwood, and *Afraegle paniculata* (Obuobi) are also used. Other poles cut for pestles are *Balanites aegyptiaca* Desert Date, *Vitellaria paradoxa* (*Butyrospermum paradoxum* subsp. *parkii*) Shea Nut Tree, *Faurea speciosa* (Se ngo se bari), *Hymenocardia acida* (Sabrakyie), *Lecaniodiscus cupanioides* (Dwindwera), *Manilkara multinervis* subsp. *lacera* (*lacera*) African Pearwood, *Piliostigma reticulatum*, *P. thonningii* (Opitipata), *Prosopis africana* (Sanga), and *Tamarindus indica* Indian Tamarind.

Wild fruits

The forest is a source of a wide variety of wild fruits. Some species of *Diospyros*, for example, have edible fruits. They are *D. barteri* (Aheneba-nsatea), *D. canaliculata* Flint Bark (Otwabere); *D. heudelotii* (Omenewabere), *D. kamerunensis* (Omenewa), *D. viridicans* (*kekemi*) Gaboon Ebony (Atwea), *D. mespiliformis* West African Ebony, *D. sanza-minika* (Sanzamulike), *D. soubreana* (Otweto), and *D. tricolor* (Ako). Several species of *Salacia* also have edible fruits. These include *S. staudtiana* var. *leonensis* (*callei* and *tshopoensis*) (Kpleng), *S. ituriensis*, *S. stuhlmanniana* (*lomensis*) (Abontore), *S. chlorantha* (*senegalensis*), *S. owabiensis* (*pyriformis*), *S. whytei* (*nitida*) (Mumue), and *S. togoica*. Certain wild fruits are often sold in the markets and sometimes cultivated. They include *Dialium guineense* Velvet Tamarind (Asenamba), *Spondias mombin* Hog Plum (Ataaba) (probably introduced); *Vitex doniana* Black Plum (Afua); *Chrysophyllum delevoyi* (*albidum*) White Star Apple, *C. africanum* Star Apple, *Dacryodes klaineana* (Adwea), and *Vitellaria paradoxa* (*Butyrospermum paradoxum* subsp. *parkii*) Shea Nut Tree. Other wild fruits sold in the markets are *Uvaria chamae* (Akotompotsen), *U. angolensis* subsp. *guineensis*, *U. doeringii* (Agbana), *U. ovata* (Akotompo), *Landolphia calabarica*, *L. heudelotii* (Pempen), *L. owariensis* White Rubber Vine (Obowe); *Saba florida* (Akontoma), *Coula edulis* African Walnut (Bodwue), and *Treculia africana* var. *africana* African Bread Fruit (Brebretim).

The list of wild edible fruits includes *Grewia carpinifolia* (Ntanta), *Dichapetalum madagascariensis* (*guineense*) (Antro), *Deinbollia pinnata* (Woteegbogbo), *Flacourtia flavescens* Niger Plum (Amugui); *Ancylobotrys amoena*, *Myrianthus arboreus* (Anyankoma), *Tieghemella heckelii* (Baku). *Adansonia digitata* Baobab, *Napoleonaea leonensis* (*Napoleona leonensis*), *Parinari curatellifolia* (Atena), *P. excelsa* Guinea Plum (Ofam), and *Maranthes glabra* (*P. glabra*) (Punini). The rest are *Chrysobalanus icaco* (*ellipticus* and *orbicularis*) (Abeble), *C. icaco* subsp. *atacorensis* Coco Plum, *Balanites aegyptiaca* Desert Date, *Carpolobia lutea* (Otwewa), *Pentaclethra macrophylla* Oil Bean Tree, *Lecaniodiscus cupanioides* (Dwindwera), *Ziziphus abyssinica* (Laruklukor), and *Z. mauritiana* Indian Jujube.

A group of wild fruits of interest as possible sources for sweeteners are *Synsepalum dulcificum* Miraculous Berry (Asaa); *Thaumatococcus daniellii* (Katemfe), and *Dioscoreophyllum cumminsii* Guinea Potato, West African Serendipity Berry. The sweetness of *Synsepalum* is due to a glycoprotein called miraculin. When the ripe fruits are eaten it causes sour fruits taken afterwards to taste sweet. Small plantations have been established at Agricultural Research Stations and Botanical Gardens in Ghana. The sweetening effect of *Thaumatococcus* is due to thaumatin, a protein. Weight for weight it is 1,600 times as sweet as sucrose. It grows in semi-deciduous and deciduous forests in Ghana where the rainfall does not exceed 200 cm annually. A plantation was established at the Agricultural Station at Bunso but has since been abandoned. Mr A.A. Enti, of Forestry Enterprises, a plant exporting company in Ghana, estimates that if truly exploited about 350 tons dry fruits annually should be possible. In addition the leaves serve as food wrappers. *Dioscoreophyllum* contains a substance called monellin which makes the fruits extremely sweet. Unlike the rest, the fruits are stable and keep for weeks at room temperature. It is a forest plant and the tubers, like yam, are reported to be eaten by the Binis in southern Nigeria and the Central African Republic. The leaves of *Gymnema sylvestre*, on the contrary, destroy the taste for sweetness for several hours. (See also Chapter 2 for more wild edible fruits.)

Wild vegetables and mushrooms (see Chapter 2).

Plants as sources of drinking water

Water may be obtained from the stems or bark of some plants. The stem of *Adansonia digitata* Baobab, *Cleistopholis patens* Salt and Oil Tree, *Costus afer* Ginger Lily, *Phyllanthus muellerianus* (Potopoleboblo), and *Sterculia setigera* (Pumpungo) when cut yield clear drinking water. The sap from the stilt roots of *Musanga cecropioides* Umbrella Tree (Dwumma); *Myrianthus arboreus* (Anyankoma), and the leafy tips of *Cnestis ferruginea* (Akitase) yield water. Water is obtainable from the cut stems of *Artabotrys thomsonii, Ampelocissus gracilipes, Cissus populnea* (Agyako), *Byttneria catalpifolia* subsp. *africana* (Sukuruwa), *Tetracera alnifolia* (Akotopa), *T. potatoria* (Twihama), *T. leiocarpa, T. podotricha*, and *Jaundea pubescens. Calotropis procera* Sodom Apple (Burkill 1985) and *Daniellia oliveri* African Copaiba Balsam Tree (Sanya) are indicators of underground water. *Ravenala madagascariensis* Travellers' Palm, an introduced palm from Madagascar, is so called because the water that

1. FORESTS AND CONSERVATION

accumulates in the leaf bases has been used for drinking in cases of need (Willis, 1973).

Plants as sources of honey

Honey is made by bees from nectar. Plants visited by bees for their nectar include *Acacia farnesiana, A. gourmaensis* (Gowuraga), *A. sieberiana* var. *villosa* (Kulgo), *Albizia lebbeck* East Indian Walnut, Lebbeck; *Anacardium occidentale* Cashew Nut, *Asclepias curassavica* Blood Flower, Red Head; *Burkea africana* (Pinimo), *Vitellaria paradoxa* (*Butyrospermum paradoxum* subsp. *parkii*) Shea Butter Tree, *Cassia alata* Ringworm Shrub, *Ceiba pentandra* Silk Cotton Tree, *Cocos nucifera* Coconut Palm, and *Crescentia cujete* Calabash Tree (Dwereba). Others are *Diospyros mespiliformis* West African Ebony (Okisibiri); *Eugenia jambos* Rose Apple, *Haematoxylon campechianum* Logwood (the honey is of fine quality and said to be the best in the world), *Lantana camara* Wild Sage, *Mangifera indica* Mango (the honey has a beautiful flavour), *Manihot glaziovii* Ceara Rubber, *Melia azedarach* Persian Lilac, *Mitragyna inermis* (Kukyafie), *Moringa oleifera* Horse-Radish Tree, Oil of Ben Tree; *Persea americana* Avocado Pear, and *Pithecellobium dulce* Madras Thorn. The rest are *Tamarindus indica* Indian Tamarind, *Terminalia glaucescens* (Ongo) (the honey is said to be particularly good), *Vernonia colorata*, and *Vitex doniana* Black Plum (Afua). The stem-juice of *Cissus populnea* (Agyako) is used as a honey adulterant. Honey is useful as a sweetening agent in food and medicine. It is also prescribed for coughs in children.

Plant beverages and wine (see Chapter 2).

Medicinal and poisonous plants

The forest is the source of plants with medicinal properties, antidotes, poisons, dyes, etc. (See Chapters 8 and 9).

Latex-producing plants

These are mostly in the families Apocynaceae, Euphorbiaceae, Moraceae and Sapotaceae. Those in the Apocynaceae include *Alafia scandens* (Momonimo), *Ancylobotrys scandens* (Bomene), *Aphanostylis mannii, Dictyophleba macrophylla, Landolphia dulcis* (also for birdlime), *L. hirsuta* (Pumpune) (also for birdlime), *L. owariensis* White Rubber Vine (Obowe); *Ancylobotrys amoena* (*L. amoena*), *Orthopichonia barteri, Parquetina nigrescens* (Aba-kamo) (also an adulterant: see below), and *Saba thompsonii*. The tubers of *Adenium obesum* yield latex. The best indigenous true rubber is obtained from *Funtumia elastica* West African Rubber Tree. The latex coagulates readily to a solid, sticky mass, about one-third of which is pure rubber of excellent quality and comparable to that of the best Para Rubber. Latex-producing plants in the Euphorbiaceae include *Hevea brasiliensis* Para Rubber (see Chapter 3), *Manihot dichotoma* Jequé Rubber Tree, and *M. glaziovii* Ceara Rubber. Those in the Moraceae include *Ficus elastica* Rubber Plant, *F. leprieuri* (Amangyedua) also for birdlime), and *F. dekdekena* (*thonningii*) (Gamperoga). *Gluema ivorensis* in the family Sapotaceae produces latex useful for repairing broken calabashes.

Adulterants

These are substances derived from latex-yielding plants added to good rubber to increase the quantity. Adulteration also lowers the quality. Plants producing adulterants include *Alafia barteri* (Momorohemo), *Alstonia boonei* (Sinduro) (also for birdlime), *Antiaris toxicaria* (*africana* and *welwitschii*) Bark Cloth Tree (Kyankyen); *Trilepisium madagascariense* (*Bosqueia angolensis*) (Okure) (also for birdlime), *Milicia* (*Chlorophora*) *excelsa* Iroko (Odum); *Tabernaemontana chippii*, *Ficus ingens* var. *ingens* (Kunkwiya) (also for birdlime), and *F. lutea* (*vogelii*) (Fonto). Others are *Funtumia africana* False Rubber Tree, *Holarrhena floribunda* (Sese), *Morus mesozygia* (Wonton), and *Tabernaemontana crassa* (Pepae).

Birdlime

This is a sticky latex used to catch birds; it is usually boiled and wound on to sticks or fruit trees. Sources of birdlime include the latex of *Afraegle paniculata* (Obuobi), *Chrysophyllum delevoyi* White Star Apple, *Ficus glumosa* var. *glaberrima* (Galinziela), *F. ovata*, *F. platyphylla* Gutta-Percha Tree, *F. dekdekena* (*thonningii*) (Gamperoga), *Dictyophleba leonensis*, *Voacanga africana* (Ofuruma), and the mucilaginous fruit pulp of *Cordia myxa* Sapistan Plum (Tungbo) and *Rhigiocarya racemifera*.

Coagulants

These help to change the fluid rubber to a more or less solid state. Plant sources of coagulants include *Acacia nilotica* var. *tomentosa*, *Adansonia digitata* Baobab, *Alafia* species (Homafuntum), *Daniellia thurifera* Frankincense Tree (Kwanga); *Hymenodictyon floribundum* (Amandidua), *Omphalocarpum ahia* (Duapompo), the flower juice of *Panda oleosa* (Kokroboba), *Pavetta crassipes* (Nyenyanke), *Piliostigma reticulatum*, and *P. thonningii* (Opitipata). Others are *Strophanthus barteri*, *S. preussii* (Dietwa), *S. sarmentosus* (Adwokuma), and *Tamarindus indica* Indian Tamarind.

Gum-yielding trees

The Acacias are generally good quality gum-yielding trees. Among these are *Acacia albida* (Gozanga), *A. dudgeoni* (Gosei) (source of gum arabic), *A. farnesiana*, *A. hockii* Shittim Wood (edible when fresh), *A. polycantha* subsp. *campylacantha* African Catechu (Gorpila); *A. nilotica* var. *tomentosa*, and *A. sieberiana* var. *villosa* (Kulgo) (edible when fresh). The Gum Arabic of

1. FORESTS AND CONSERVATION

commerce is obtained from *A. senegal* Gum Arabic which grows from Mauritania to northern parts of the Ivory Coast, northern Nigeria, and extending to the Red Sea and eastern India (see Chapter 3).

Other good quality gum-yielding trees in West Africa are *Burkea africana* (Pinimo), *Albizia lebbeck* East Indian Walnut, *A. zygia* (Okoro), *Anacardium occidentale* Cashew Nut, *Azadirachta indica* Neem Tree, *Combretum nigricans* var. *elliotii* (edible), *Mangifera indica* Mango, *Samanea saman* Rain Tree, *Spondias mombin* Hog Plum, Ashanti Plum (Ataaba); *Sterculia setigera* (Pumpungo) (edible), and *S. tragacantha* African Tragacanth (Sofo). Gums of lesser quality are obtained from *Adenanthera pavonina* Bead Tree, *Afraegle paniculata* (Obuobi), *Albizia adianthifolia* (Pampena), *Anogeissus leiocarpus* (Sakanee), *Antidesma venosum* (Mpepea), *Balanites aegyptiaca* Desert Date (edible when fresh), *Bauhinia purpurea* (edible), and *Borassus aethiopum* Fan Palm. Other poorer quality gum-yielding trees are *Cassia fistula* Golden Shower, *Cola gigantea* var. *glabrescens* (Watapuo), *Combretum collinum* subsp. *binderianum* (*binderianum*) (Domapowa), *C. fragrans* (*ghasalense*) (Kwaginyanga) (edible), *C. collinum* subsp. *hypopilinum* (*hypopilinum*) (Chinchapula), *C. sericeum* (Peytuba), *C. molle* (Gburega), *Cordia myxa* Sapistan Plum, (Tungbo), and *Crateva adansonii* (*religiosa*) (Chelum Punga).

Other gum-yielding trees are *Cussonia arborea* (*barteri*) (Saaborofere), *Cynometra vogelii* (Boboe), *Delonix regia* Flamboyante, Flame Tree; *Desplatzia subericarpa* (Esonowisamfie-bere) (fruits), *Dialium guineense* Velvet Tamarind (Asenamba) (also a resin); *Dichrostachys cinerea* (*glomerata*) Marabou Thorn, *Diospyros mespiliformis* West African Ebony (Okisibiri); *Drypetes afzelii* (Opahanini) (scented and used for anointing), *Entada africana* (Kaboya), *Erythrophleum africanum* Africanum Black Wood (Bupunga), and *Haematostaphis barteri* Blood Plum. The rest are *Hildegardia barteri* (Akyere), *Khaya senegalensis* Dry-Zone Mahogany (also a resin), *Lannea acida* (Kuntunkuri) (edible), *L. kerstingii* (Kobewu), *L. microcarpa* (edible), *Macaranga heterophylla* (Opamkokoo), *Anthonotha fragrans* (Totoronini), *Melia azedarach* Persian Lilac, *Moringa oleifera* Horse-Radish Tree, *Mammea africana* African Mammy Apple (Bompagya); *Opuntia* species Prickly Pear, *Piliostigma reticulatum*, *P. thonningii* (Opitipata) (both for caulking canoes), *Piptadeniastrum africanum* (Danhoma), *Pithecellobium dulce* Madras Thorn, *Prosopis africana* (Sanga), *Symphonia globulifera* (Ehureke) (for caulking canoes), *Tamarindus indica* Indian Tamarind, *Terminalia catappa* Indian Almond, and the fruit of *Paullinia pinnata* (Toa-ntini)

Resin-yielding trees

These include *Allanblackia parviflora* (*floribunda*) Tallow Tree (Sonkyi); *Amphimas pterocarpoides* (Yaya), *Berlinia grandiflora* (Tetekono), *Copaifera salikounda* Bubinga (Entedua); *Daniellia thurifera* Frankincense Tree, *Dichapetalum toxicarium* West African Ratbane, *Dioclea reflexa* Marble Vine, *Garcinia kola* Bitter Cola (Tweapea); *Zanthoxylum xanthoxyloides* (*Fagara* zanthoxyloides) Candle Wood, *Nauclea latifolia* African Peach (Sukisia); *Pseudospondias microcarpa* var. *microcarpa* (Katawani), *Psorospermum corymbiferum* var. *corymbiferum*, and *Pterocarpus erinaceus* Senegal Rose Wood Tree. The shell-oil of *Anacardium occidentale* Cashew Nut is an important source of phenolic resin for commerce, and the seeds of *Caesalpinia bonduc* Bonduc contain bonducin resin. Gum resin is found in *Bombax buonopozense* Red-Flowered Silk Cotton Tree, *Boswellia dalzielii* Frankincense Tree (Kabona); *Commiphora pedunculata* and other members of the family Burseraceae generally; *Carapa procera* Crabwood, Monkey Cola; *Ceiba pentandra* Silk Cotton Tree, *Detarium senegalense* Tallow Tree, *Erythrophleum suaveolens* (*guineense*) Ordeal Tree (Potrodom), and *Tetrapleura tetraptera* (Prekese). Gum resins are burned as incense and for fumigating garments and repelling insects.

Gum copal trees

These are *Balanites wilsoniana* (Kurobow), *Boswellia dalzielii* Frankincense Tree (Kabona), usually burned with the resin of *Lannea kerstingii* (Kobewu), *Canarium schweinfurthii* Incense Tree, *Daniellia ogea* Gum Copal Tree (Hyedua) (one of the first exports of Ghana with 500 tons in 1876); *D. oliveri* African Copaiba Balsam Tree, *D. thurifera* Frankincense Tree, Niger Copal Tree; *Guibourtia copallifera*, *G. ehie* (Anokye-hyedua), *Pellegriniodendron diphyllum* (Fetefele), and *Trachylobium verrucosum* East African Copal, an introduced tree from East Africa. Gum copal is burned to fumigate rooms and clothing and to drive away evil spirits. It is used in religious ceremonies and ground to perfume the body, alone or mixed with pomade. Small lumps of copal are hung around the necks of children — again, to drive away evil spirits causing disease.

Gutta-percha

This is a resinous, rubbery but inelastic plant exudate. Commercially gutta is useful in electrical insulations and the manufacture of golf balls. Sources include *Vitellaria paradoxa* (*Butyrospermum paradoxum* subsp. *parkii*) Shea Nut Tree, *Chrysophyllum perpulchrum* (Atabene), *Ficus sycomorus* (*gnaphalocarpa*) (Kankanga), *F. platyphylla* Gutta-Percha Tree, *Simirestis welwitschii* (*Hippocratea welwitschii*) (Akladepka), *Manilkara obovata* (*multinervis*) (Berekankum), *Omphalocarpum ahia* (Duapompo) (used also as rubber coagulant), *Salacia staudtiana* var. *leonensis* (*callei*) (Kpleng), *S. debilis* (Hama-kyereben), *S. stuhlmaniana* (*lomensis*) (Abontore), and *S. togoica*.

Decorative beads

Trees and climbers yielding decorative beads for ornaments, rosaries, and musical instruments are some of the minor products of the forest (see Chapter 7).

1. FORESTS AND CONSERVATION

Animal products

This group includes game for food and sports; hides, horns, ivory, wax, honey (see above), snails, tortoise, crabs, and freshwater fish. Pet animals, like monkeys, and birds, like parrots, are products of the forest.

Minerals

From the forest soil mineral products such as stones, sand, and gravels for building and construction are obtainable. In Ghana bauxite is excavated at Awaso in the Western Region and there are large deposits at Nyinahin in Ashanti and the Atewa Range Forest Reserve in the Eastern Region. There are diamond mines at Akwatia in the Birim valley and gold mines at Obuase in Ashanti. In Ghana minerals are the second most important export after cocoa. (See Table 1.3 for complete list.)

Table 1.3 Main mineral deposits in Ghana

GOLD	DIAMOND	MANGANESE	BAUXITE
Abontiakoon	Akwatia	Dawtiem	Atewa Range — Kibi
Aboso	Bandaye	Himakrom	Awaso — Kanaiyerebo Hill
Bandaye	Bonsa	Hotopo	Nkawkaw — Mt. Ejuanema
Bibiani	Huniso	Nsuta	Nyinahin
Bogoso	Kanyankaw		Sefwi Bekwai
Bremang	Nsuta		
Konongo	Oda		
Obuasi			
Preastea			
Tamso			
Tarkwa			

Indirect Benefits

AMELIORATION OF LOCAL CLIMATE

The forest serves as a shield against the rays of the sun — the thick canopy reduces the degree and intensity of penetration by sunlight. As a result the forest atmosphere is comparatively cooler than that in the open country on sunny days. But the warming effect of the sun during the day is still noticeable hours later in the forest, since air movement is restricted. It is thus warmer at night than the open country. The cool temperatures in the day and the warm ones at night make conditions in the forest more favourable and equitable than the hot days and cold nights in the open country.

Rainfall

The influence of forest on rainfall is debatable. There is a story of a colonial forest officer giving this forest–rainfall relationship to persuade an Akwapim chief in Ghana to give up a portion of his tribal land known as 'stool land' for reservation. The chief wanted to know whether it rained on the open sea where there were no forest reserves. Generally, the rainfall in forest areas exceeds that of non-forested areas — although other factors may be involved. As the wind blows over large expanses of forest country it collects moisture in the form of water vapour transpired from the leaves of the trees. When this collection of moisture in the air is forced upwards by a hill mass, it cools, condenses, and falls as rain locally. The rainfall makes more water available for absorption and transpiration to repeat the process.

Relative humidity

As a result of water vapour (mainly from the transpiration of plants), the relative humidity in the forest environment is comparatively higher than it is in the open country. Within tropical rain-forest it may not fall below 80 per cent: indeed, for considerable periods it may be near saturation point. The high relative humidity in the forest environment is beneficial to agriculture generally. Food and cash crops can successfully be grown in the protective environment of the forest. In contrast, the open savanna country may register extremes of 15 per cent or less in the dry season to a maximum of 80 per cent in the rainy season.

Wind

Physically, the forest is a barrier against the wind and its desiccating effects on the weather. The effects of the dry north-east trade winds (harmattan) which blow in West Africa between November and March are not as severe in the forest zones to the south as they are in the northern savanna. The primary cause of the reduction in the severity of the dry harmattan is the forest over which it blows on its journey from the north to the south. To ensure a continuous band of forest, Shelterbelt Forest Reserves are created. Examples in Ghana are the Amama Shelterbelt Forest Reserve, between the Asukese and Bosomkese Forest Reserves, and the Bia Shelterbelt Forest Reserve, between the Mpameso and the Bia Tano Forest Reserves. Other Shelterbelt Reserves include Aboniyere, Angoben, Aparapi, Pru, and Totua (see Table 1.1). Shelterbelt Reserves are purely protective in function.

PROTECTION

Watersheds and Catchment Areas

The establishment and maintenance of forests on catchment areas regulates

1. FORESTS AND CONSERVATION

the flow of water and reduces excessive evaporation and run-off to a minimum — this ensures a perennial flow of water and prevents seasonal dry-up. The forest cover allows the water to penetrate deeply into the soil to reach the water table. Reserves created at such sources are called Headwaters Forest Reserves. Examples in Ghana are at the sources of the rivers Afram, Bia, Fum, Fure, Klemu, Ochi, Offin, Pompo, and Tain (see Table 1.1).

Animals

Forests, by nature, provide protection and a home for animals. When the forest is set aside as a Game Reserve, it is exclusively intended for the complete protection of game (see Table 1.4). Game Reserves protect the animals from indiscriminate hunting by poachers while offering them shelter, shade, and home. Game Reserves attract tourism. They are also a source of meat and other animal products such as tusks for ivory and hides for rugs. Hunting in other reserves is allowed in accordance with the hunting permit regulations.

Table 1.4 Game Reserves in Ghana

NAME	AREA IN HECTARES	LOCATION
Ankasa Game Production Reserve	20 740	Near Axim, Western Region
Bia Game Production Reserve	22 810	Near Sefwi Wiawso, Western Region
Bia National Park	7780	Near Sefwi Wiawso, Western Region
Bomfobiri Wildlife Sanctuary	5180	Near Kumawu, Ashanti Region
Bui National Park	207 360	Near Wenchi, Brong Ahafo — extends to Northern Region
Digya National Park	312 600	Near Atebubu, Brong Ahafo — extends to Ashanti, Ashanti Region
Gbele Game Production Reserve	54 690	Near Tumu, Upper West Region
Kalakpa Game Production Reserve	32 400	Near Ho, Volta Region
Kogyae Strict Nature Reserve	32 400	Near Ejura, Ashanti Region
Mini Suhien National Park	10 630	Near Axim, Western Region
Mole National Park	491 440	Near Damongo, Northern Region — extends to Upper East Region
Owabi Wildlife Sanctuary	7260	Near Kumasi, Ashanti Region
Shai Hills Game Production Reserve	5440	Near Tema, Greater Accra Region
	1 210 730	

Crops

The protection provided by the forest for crops needs to be emphasized. Forests provide the environment most suitable for the proper growth of some crops. Cocoa, Coffee, Oil Palm, Plantain, Cocoyam, are examples — and generally most of the cash crops and staple foods (except cereals) are better adapted to growing in the forest zone than the savanna country. Besides the high rainfall they require, the forest zone provides the most suitable climate.

Soil

Without the forest cover (or vegetative cover) the soil would be blown away by the wind, washed away by the rain, or parched dry by the scorching sun. The presence of the forest builds up and maintains the soil texture and nutrients. A number of physiological and chemical processes together contribute to this process. These are the physical break-up of the soil as the roots penetrate it; the binding of the soil as the roots ramify through it; the processes involved in the absorption of water and mineral salts from the soil and air from the atmosphere for the manufacture of the plant food in the presence of sunlight (photosynthesis); the return to the soil of organic matter in the form of dead leaves and vegetable remains and the decomposition of these materials by bacteria and fungi to form humus. These decomposing agents (bacteria) work faster in favourable temperatures and, as a result, in tropical temperatures they completely oxidize the organic matter, causing it to disappear.

Forest soils are rich in humus, spongy, and porous, with high water retentive properties. The forest canopy further checks evaporation and conserves water. Channels of various sizes created by the ramifying roots open the soil to aeration. Rain water first penetrates the forest canopy to reach the forest floor. The thicker the canopy, the longer this takes. The intercepted rain eventually finds its way to the forest floor in drips which may continue hours after the rain has stopped. The gradual dripping enables the water to sink deeper into the soil, and it is thus made available to plant roots and springs. There is hardly any run-off water in the forest, unlike the open country where rain water runs off into gutters, rivers, and eventually the sea.

PREVENTION OF EROSION

Erosion, caused by run-off water, is prevented by introducing an appropriate vegetative cover. The following plants are useful sand binders and therefore help to prevent erosion: *Indigofera spicata*, *Philoxerus vermicularis* (Koklotade), *Passiflora glabra (foetida)* Stinking Passion Flower, and a number of plants in the family Convolvulaceae. These include *Ipomoea pes-caprae* subsp. *brasiliensis* Beach Convolvulus, *I. asarifolia*, *I. stolonifera*, *Merremia umbellata* subsp. *umbellata*, and *Evolvulus nummularius*. The sedge *Remirea maritima* is an effective sand binder along the sea shore. The following grasses are equally good sand binders: *Cymbopogon citratus* Lemon Grass, *Cynodon dactylon*

1. FORESTS AND CONSERVATION

Bahama Grass, *Eleusine indica, Panicum repens, Saccharum spontaneum* var. *aegyptiacum, Sacciolepis africana, Sporobolus virginicus, S. robustus,* and *Stenotaphrum secundatum* Buffalo Grass. Other sand binders are *Alternanthera maritima* and *Dodonaea viscosa* Switch Sorrel (Fomitsi)

DRAINAGE

The forest may be effectively employed to drain swampy areas. The method is based on the principle of the absorption of water by the roots and the transpiration of this water through the leaves to the atmosphere. *Gliricidia sepium* Mother of Cocoa is often used in this respect. Others are *Sesbania sesban* Egyptian Sesban (Tingkwanga) and the grass *Vossia cuspidata.*

TWO
Food and Fodder

The need to produce more food in developing countries has been stressed by successive governments before and since independence. 'Back to the Land', 'Grow More Food', 'Operation Feed Yourself', 'Green Revolution', and so on — such slogans are aimed at increased food productivity and self-sufficiency. Despite these appeals and the fact that many of these countries are potentially agricultural, food shortages are common, food prices are often prohibitive, and large quantities of food, mainly grains, are imported to supplement local production. In 1983 alone over 55 million dollars-worth of food was imported by Ghana, besides the continuous supply of various food relief aids.

Low productivity is attributable to several factors. Among the important ones are:

1. the land tenure system in which our chiefs, other landowners, and our customary laws feature prominently;
2. commercial establishments engaged in the manufacture and/or importation and sale of agricultural chemicals, machinery, and implements;
3. the prices that local and overseas markets are prepared to offer for our agricultural commodities;
4. the state of the economy and development in many Third World countries, with special respect to the level of taxation, industries, communications, transportation, and the availability of credit facilities for the development of agriculture;
5. the social and educational background of the actual farmers, as well as the level of scientific and technological training available to them.

There is also the problem of storage and preservation. An estimated 30 per cent of post-harvest crops are lost through lack of suitable storage facilities.

Until recently, when commercial farming was started in West Africa, the main source of foodstuffs was from the peasant farms. Professor Benneh of the University of Ghana noted that the traditional farmer accounts for well over 90 per cent of the agricultural output of the country. Since these farms depend entirely on human labour, only limited acreages are cultivated. With the increase in population and the corresponding increase in demand for food, a combination of technical and mechanical assistance as well as scientific research, into the improvement of seeds, for example, is needed to enable the peasant farmer to increase the present production levels. Unfavourable and unpredictable weather has been the cause of low yields

2. FOOD AND FODDER

and poor harvest. Irrigation, as practised at Dahwenya, Weija, Okyereko, and Tongu in Ghana, would be a further step in helping peasant farmers increase their yield.

A third source of food, especially of vegetables, is backyard farming, which supplies the family with its primary requirements or supplements them — little or none being sold. Since many city and urban dwellers are denied backyard farming, the provision of farming plots within easy walking distance will enable free weekends and holidays to be used for a more productive purpose. A fourth source of food is direct importation.

Cereals

Cultivated cereals in West Africa include *Zea mays* Maize, *Oryza sativa* Asiatic Rice, *O. glaberrima* Upland Rice, *Digitaria exilis* Hunry Rice (Kabuga), *Sorghum bicolor* Guinea Corn, and *Pennisetum americanum* Bulrush Millet, also called Pearl Millet or Spiked Millet. Some *Triticum aestivum* Wheat is being recently cultivated at Depali, near Tamale in Northern Ghana, and Obuoho, near Begoro in the Eastern Region, on an experimental basis. Some wheat is cultivated in Nigeria and in East Africa. The entire wheat requirements of many West African countries are however imported — mainly from the United States.

ZEA MAYS — Maize

Maize, Corn is the third most important cereal crop in the world. It is widely cultivated in many West African countries, both in the forest areas and the savanna zones. In Ghana, the Sekyedumase and Techiman areas in the Brong Ahafo Region, areas around Sefwi Bekwai in the Western Region, and the Gomua area — especially around Mpuruman — are some of the main growing areas. There are several local varieties as to colour — pure white, red, and red with white on the same cob. There are also varieties in height and in period of maturation — 60 days and 90 days after planting. An estimated 550,000 tonnes of maize is consumed in Ghana annually, most of it produced locally. Maize is more important as a staple among the ethnic groups in the southern savanna than in the forest zone and the northern savanna. The fresh grain is roasted or boiled on the cob as a side dish. Roasted maize with groundnuts is a popular dessert. The bulk of the crop is dried and milled into a dough of cornflour used to prepare solid foods like *banku* and *kenkey*, or lighter foods like porridge and *agidi* or *kaafa*. Mixed with cassava dough it is used to prepare *akple*. Many other dishes like *boodoo, bodongo, kakro*, bread, and various pastries are prepared from corn dough. *Ablemamu* or 'Tom-Brown' (a porridge) and *aprampransa* are prepared from roasted and milled corn. *Kpekpoi* is a special corn dish used at the Homowo Festival of the Gas in Ghana. *Ahei*, or *ngmeda*, and *tuei* are beverages prepared from corn. Like Guinea corn, maize can be brewed into the native beer *pito*.

Oil can be extracted from the germ of the grain and is used for cooking and the manufacture of soap and glycerine. The leaves serve as feed for livestock and the grains are used in poultry feed preparations. The sheaths of corn cobs are used as wrappers for *Ga-kenkey* in Ghana, and for mat-making, bags, and doormats (see Chapter 7). The dried stems serve as fuel. Nutritionally, maize has a low protein value but plenty of vitamin B; there is also a little vitamin A — yellow corn containing much more than the white. (For medicinal properties see Chapter 8.)

ORYZA SATIVA — Rice

Rice is one of the most common cereal crops produced in tropical areas. Producing countries in West Africa, in order of volume, are Nigeria, Sierra Leone, Guinea, the Ivory Coast, Senegal, Liberia, Mali, and Ghana. In Ghana rice is cultivated on swampy or irrigated land along the coast in the Western Region, Accra Plains around Dawhenya and Kpong, and in the northern parts of the country. Rice is also cultivated on suitable sites in the forest region. *O. glaberrima* Upland Rice is grown only in West Africa in a triangle from Senegal to Lake Chad to the Nigerian coast (Portères, 1976). In Ghana it is cultivated around Hani in the Brong Ahafo Region and surrounding areas. Rice cultivation is reported to have started in the country in 1880. Quantities of the grain were supplied as part salary to the forces before the Second World War. There are several varieties, according to size of grain — long and short; colour of grain — red (*kaya-mo*) and white; and period of maturation — early, medium, and late. It is a major food product throughout the world, and a staple, together with Guinea Corn and Millet, in the northern parts of Ghana. It is the staple food in Senegal, parts of the Ivory Coast, Gambia, Sierra Leone, and Liberia, and popular in the other countries in West Africa generally. Estimates, such as those prepared by the World Indicative Plan, suggest that the consumption of rice within West Africa will exceed 5 million tons by 1985 (Leakey and Wills, 1977). Rice is also an important item of diet in institutions. Where it becomes a staple diet it is important that the whole grain is eaten to prevent the occurrence of beriberi. An estimated 60,000 to 95,000 metric tonnes are produced annually in Ghana (though this could be improved upon, and even doubled) and about ₵10 million imported to supplement local production. Rice is eaten boiled with soup, sauce, or stew; or prepared as jollof rice. Rice is also used for preparing porridge (rice water), and when milled as flour it is used with wheat flour to prepare bread. The stem is good fodder for stock, containing 4.72 per cent crude protein, 32.21 per cent carbohydrate, and 1.87 per cent fat. The chaff is useful as fuel.

DIGITARIA EXILIS — Hungry Rice

Hungry Rice (Kabuga) is cultivated mainly in the Guinea savanna woodland or Sudanian regional centre of endemism, and used as porridge or added to

2. FOOD AND FODDER

other cereals as meal. It is the staple food of many semi-Bantu tribes of northern Nigeria.

SORGHUM BICOLOR — Guinea corn

Guinea Corn (Atoko) is the fifth most important cereal grain in the world — exceeded only by wheat, maize, rice, and barley. In West Africa cultivation is mainly in the Sudan and Guinea savanna woodland countries. Nigeria and Burkina Faso (Upper Volta) are leading producing countries. There are two main types — the common loose, open panicle and the compact, erect one (usually in drier areas). Both white and red grain varieties occur. It is a staple food in an area where both rainfall and other conditions have proved hostile to Bulrush Millet. Guinea Corn is taken as porridge, *banku*, or *kenkey*; or brewed into *pito* — a nutritious alcoholic beverage with a definite vitamin value. The grains are rich in vitamin B when taken with the bran which, with the grains, are fodder, for livestock and poultry respectively. As fodder, young plants may be poisonous to stock as a result of the presence of prussic acid developed by enzyme action on the glucoside dhurrine during germination. Crops affected by drought are especially liable to suspicion; and tiller shoots after harvesting the main crops are also often poisonous to stock. The stems are useful as fuel. (For other uses of the stem see Chapter 4.)

PENNISETUM AMERICANUM — Millet

Bulrush, Pearl, or Spiked Millet (Ewio) is a short season crop which matures during the brief rainy season, and is thus suitable for the Sahel, Sudan, and Guinea savanna woodland countries in West Africa. Together with Guinea Corn, it forms about 80 per cent of the food requirements of the states in these vegetational zones. Producing countries in order are Nigeria, Niger, Mali, Chad, and Senegal. There are several varieties — big and small grain; smooth and bristly grain; quick-maturing, early and late forms. There are varieties of colour, too. It is used to prepare solid meals or porridge; or milled into flour for pastries — alone or mixed with wheat flour. Millet is used for the brewing of a local beer and other beverages. It is also grown as fodder. The millets include six genera of grasses — *Panicum, Setaria, Echinochloa, Pennisetum, Paspalum,* and *Eleusine.*

WILD EDIBLE GRAINS

Many wild grasses have edible grains, some of which are worth developing into crops. They include *Brachiaria distichophylla, B. xantholeuca, B. jubata, B. falcifera,* and *B. stigmatista; Cenchrus biflorus* (either eaten or for preparing a drink), *Dactyloctenium aegyptium* (for porridge), *Digitaria debilis, D. horizontalis, D. ciliaris, Echinochloa colona* Jungle Rice, Shama Millet (formerly cultivated in Egypt), *E. pyramidalis* Antelope Grass, and *E. stagnina; Eleusine indica* (with high protein content — nitrogenous), *Eragrostis cilianensis* Stink Grass,

E. ciliaris, E. gangetica, E. pilosa, and *E. tremula.* Others are *Eriochloa fatmensis (nubica), Leptothrium senegalense, Oryza barthii* Wild Rice, and *O. longistaminata.* Wild Rice is a semi-arid zone species found across the Sahel and savanna zones from Sudan to the West Coast. It is the closest wild relative to *O. glaberrima* Upland Rice. *Panicum pansum, P. fluviicola, P. subalbidum, P. laetum* (best of the wild cereals, sometimes sold for cakes and porridge and appreciated by Europeans), and *P. turgidum* also have edible grains. The rest are *Paspalum scrobiculatum (orbiculare)* Bastard or Ditch Millet (best ones from dry land rather than swamp), *Pennisetum unisetum (Beckeropsis uniseta)* (also for brewing beer), *Pseudobrachiaria deflexa, Sorghum arundinaceum* Kamerun Grass, *Setaria verticillata* Rough Bristle Grass, *S. pallide-fusca* Bristly Foxtail Grass, Cat's Tail Grass; *Sacciolepis africana, Sporobolus pyramidalis* Rat's Tail Grass, and *Oxytenanthera abyssinica* Bamboo (used as rice substitute by the Acholis of Uganda and in Sudan).

Roots, Tubers and Plantain

MANIHOT ESCULENTA — Cassava

Cassava (Bankye) is a native of South America. Zaire is the leading producing country in tropical Africa, followed by Nigeria. Brazil is by far the world's greatest producer, producing more than double the amount from any other country (Onwueme, 1978). Professor Doku, formerly Dean of the Faculty of Agriculture, University of Ghana, reports that it used to be grown only in Southern Ghana, particularly in the non-forested area which still produces about 50 per cent of the total; the crop is now being increasingly cultivated in the forest and savanna areas in other parts of the country. It is the most widely cultivated of all food crops in Ghana with an estimated 3½–4 million tonnes annual production; it is also the most widely consumed. There are numerous varieties. Professor Doku gives about 90 local ones, based mainly on the branching habit, and petiole colour. Cassava is a staple among the ethnic groups of southern Ghana particularly — the basic food of the poor and middle class, since it is comparatively the cheapest of the roots and tubers. It is usually boiled and eaten as *ampesi* or pounded into *fufu* and eaten with soup. It may also be eaten roasted or fried. A large proportion of cassava is eaten as *gari* (roasted, previously fermented cassava dough) or *kokonte* (sun-dried cassava chips milled or ground into flour and cooked). Both *gari* and *kokonte* store well and are conveniently transported. They form the major part of food parcels by air from Ghana to friends and relatives in Europe, Britain, and the United States. Fresh cassava dough (*agbelima*) is mixed with corn dough to prepare a meal — or steamed alone as *yakayaka*; or fried with other ingredients as *agbelekakro*. Cassava flour with wheat flour is used to bake cassava bread. The young leaves are useful as spinach. Cassava starch is an important raw material in textile manufacture. The starch is used to prepare

2. FOOD AND FODDER

tapioca. The dried or fresh root peelings are feed for sheep and goats. Cassava is affected by dry rot caused by *Rhizopus* species, soft rot by *Bacillus* species, and wet rot caused by *Fomes* species.

DIOSCOREA SPECIES — Yams

West Africa has the following cultivated species: *D. alata* Water Yam, Winged Yam, Greater or Ten-month Yam; *D. cayenensis* Guinea Yam, *D. esculenta* Lesser Yam, *D. rotundata* White Yam, and *D. bulbifera* Potato Yam; and the following wild but edible species: *D. praehensilis* Wild Yam (high forest type) and *D. lecardii* Wild Yam (savanna type). *D. dumetorum* is a poisonous wild yam with prickly bulbils that twines clockwise as opposed to anticlockwise, a characteristic of the edible species of *Dioscorea*. The tubers are poisonous, and cases of such yam poisoning occur. It is deliberately planted at the edge of farms to punish thieves and monkeys.

Within West Africa, yam production is confined to the northern forest and southern savanna zones, stretching from the Ivory Coast to Cameroun (Onwueme, 1978), the major producing countries in order of importance being Nigeria, the Ivory Coast, Ghana, Togo, Benin Republic, and Guinea. Nigeria alone accounts for 78 per cent of world production. Professor Doku observes that the cultivation of yams in Ghana is centred on northern Ashanti, Brong Ahafo, Krachi, and along the main roads leading to Wa and Tamale and Yendi in the north. Yams are the most costly of the roots and tubers and, as such, only the rich and upper class generally can afford them. Ayensu and Coursey (1972) note that in much of West Africa, they are still the preferred staple food among most of the inhabitants of the forest zone and southern part of the savanna. Yams are eaten boiled (*ampesi* or *pusa*); pounded into *fufu*; fried as chips; prepared as a thick porridge; or mashed into *eto*. Ayensu and Coursey observed that the quantity of the yam commonly consumed in West Africa by those using this staple is sufficient to supply almost a third of the basic protein requirements of an adult male, and it may well be that much of the malnutrition reported in recent years especially in urban populations, has been associated with the replacement of yams by cassava in the diet. The report adds that yams are also nutritionally useful in that they supply substantial amounts of minerals and vitamin C to the diet —scurvy (avitaminosis C) is virtually unknown in the yam zone of West Africa. The antiscorbutic properties of yams, being appreciated by mariners of the sailing-ship era, probably contributed to the wide dissemination of the crop in the tropics. Smaller but still useful amounts of some of the B vitamins and mineral elements are also present. The steroidal sapogenins contained in *Dioscorea* can be effectively utilized as a means of contraception. Yam peelings serve as feed for livestock, especially pigs. *D. praehensilis* Wild Yam is a delicacy for elephants and bush pigs. Yams are affected by two main diseases: internal brown spot, probably caused by a virus, and tuber rot/soft rot caused by various organisms including *Aspergillus niger* and *Rhizopus nodosus*.

Dioscorea alata. Water Yam is a native of eastern India, introduced to East Africa but now distributed throughout West Africa and cultivated in the tropics as an important food crop. The tubers are variously shaped — straight, curved, coiled, fan-shaped, palm-shaped, and branched or unbranched; and may be variously coloured — white, pink, or purple according to variety. The aerial tubers are rounded and may be lobed. The tubers mature earlier — ten months — than those of Guinea yam. It stores well.

D. cayenensis subsp. ***cayenensis*** Guinea Yam. This includes some of the commonly cultivated yams in West Africa, from where it was introduced to America during the slave trade. It grows throughout the year and prefers moist conditions. It takes a much longer time to mature and does not store as well as the Water yam.

D. esculenta Lesser Yam is an introduced crop, probably of eastern Malayan origin. The tubers are comparatively small but abundant. They are elongate-oval, radiating, and mature near the surface of the soil. They are quite sweet and much preferred.

D. cayenensis subsp. ***rotundata*** White Guinea Yam is indigenous to Africa. It is popular because of its quality and taste and one of the most commonly cultivated. It stores better than most of the other yams, and since it is harvested in the eighth month it is better suited to the savanna, with a shorter growing season.

D. bulbifera Potato Yam is cultivated for the edible bulbils (aerial tubers). These are cooked like potatoes. Some forms may be eaten raw. The subterranean tubers are simple, round or oblong, and are too acrid to be used as food. They may sometimes be absent in cultivated plants. Wild varieties with toxic, angular, greyish bulbils, or with small, purplish bulbils occur. They may be cultivated among the edible ones deliberately, to prevent thieving by predators.

In addition to the wild yams mentioned above are *D. hirtiflora* which grows in the savanna and *D. smilacifolia*, a forest plant. The tubers are rarely eaten.

Xanthosoma mafaffa Cocoyam, Tannia (Mankani) is a native of tropical America but now naturalized in West Africa and extensively cultivated in the forest regions. It was introduced from the West Indies in 1843. Africa accounts for well over 75 per cent of the world production of cocoyams, most of it from West Africa (Onwueme, 1978). Important producing countries are Nigeria, Ghana, and the Ivory Coast. There are about four varieties based on the colour of the cormels. In *mankani-nkontia* both the suckers (corms) and tubers (cormels) are eaten. In the other varieties — *mankani-pa, mankani-fitaa* or *-fufuo*, and *serwaa* — only the cormels are normally eaten. Cocoyams are a staple, particularly in the forest areas. They are either boiled and eaten as *ampesi* or pounded into *fufu*. Boiled tubers may also be mashed with palm oil as *eto* — or cut into pieces and cooked as porridge (*mpihu* or *mankani-potowee*). They may also be roasted or fried in oil (*koliko*); or used to prepare chips. The

2. FOOD AND FODDER

peeled, dried tubers are ground or milled into flour for *kokonte*. The flour is used for biscuits and pastries. Cocoyam leaves (*kontomire*) are a favourite spinach when young and tender. *Xanthosoma brasiliense* Tahitian Taro or Belembe from tropical America is cultivated for the leaves only. Cocoyams are attacked by collar and root rot disease caused by *Sclerotium rolfsii* which causes the plants to wilt; the plant is also affected by nematode disease, the agents being *Helicotylenchus multinctus* and *Pratylenchus brachyrus*.

Colocasia esculenta Eddoes, Dasheen, Taro (Kookoo or Ntwibo) is originally of tropical Asia and Malaya from where it was introduced. It is now naturalized by streams and damp places in the forest — sometimes growing wild. It is cultivated, but not on as wide a scale as cocoyams. There are varieties in size of corms, colour, and degree of acridity. The corms are eaten boiled or fried after repeated washing, and are wholesome but generally believed to be inferior to cocoyams. Like cocoyams, the young leaves may be eaten as a potherb, after cooking with soda. The leaves can also be ensiled for animal feed. Taro is attacked by Taro leaf blight, caused by *Phytophthora colocasiae*; corm rot, caused by *Phythium* species and *Phytophthora* species; sclerotium rot, caused by *Corticium rolfsii*; and soft rot caused by *Fusarium oxysporum*.

Ipomoea batatas Sweet Potato (Santom) is probably of American origin. There are many varieties and cultivars showing differences in leaf shape, colour, and size of tubers and yield. The three distinct forms recognized locally are the orange-fleshed and white-fleshed varieties. Varieties or cultivars called Centennial, Nemagold, Jewel, Gem, and Eland are orange-fleshed varieties Miguela and Brondal are white-fleshed. A third variety has red tubers with red flesh. A high yielding (over 20 tonnes/ha), early maturing (3 or 4 months) variety with no pest and disease problems, ranging from very sweet to non-sweet has been developed. Sweet potatoes are cultivated in the interior and coastal savanna zones throughout West Africa, all on small scale. Compared with the others it is not such an important root crop. The tubers are eaten boiled or fried in oil and may also be roasted. The Home Science Department of the University of Ghana has developed techniques for making excellent bread, doughnuts, etc. by substituting up to 30 per cent sweet potato flour for wheat flour. It is sometimes used as sugar substitute in the preparation of *boodoo* (baked corn dough) or *kakro* (fried corn dough mixed with red plantain). The tubers are useful as feed for livestock and for processing into starch. The leaves are used as spinach and as feed for fish in domestic and commercial fish ponds. Sweet potato is attacked by the following diseases: black rot caused by *Caratocystis fimbriata*; blue mould rot caused by *Penicillium* species; charcoal rot caused by *Macrophomina phaseolina*; and dry rot caused by *Diaporthe batatis*.

Solanum tuberosum Irish Potato. In tropical Africa potatoes are produced at high altitudes in the eastern part of the continent. They have been cultivated on small-holdings around Amedzofe in Ghana. They could also be grown along the Atewa Range (altitude 600 m plus) and at similar altitudes in the country. The bulk of Irish Potatoes eaten in West Africa are imported.

Musa paradisiaca Plantain is a native of tropical Asia and was introduced probably through Egypt. It is cultivated in the forest region especially and in similar situations in the southern savanna of West Africa, in which regions it is a staple. Plantains are not cultivated in East Africa although bananas grow well in the area. There are about 21 varieties divided into three main groups: the *apantu*, the *apem* (thousand), and the intermediate between the two. The varieties are based on the length of the fruit peduncle and the form, size, number, and colour of fruit, and their arrangement on the peduncle. Some of the *apem* varieties are *apempa, osabum, onniaba, brodekokoa-apem, brodehene*, and *nyiretia-apem*. The *apantu* varieties consist mainly of *brodewio*, double-bunch, *nyiretia, kaamenko*, and *abomianu*. The intermediate variety consists of *osakoro, osameanu, osameansa*, and *adoso*. Plantains are boiled and eaten when unripe as *ampesi* or pounded into *fufu*, sometimes mixed with cassava. They are also eaten roasted (with groundnuts), either unripe or more usually ripe. Fried, ripe plantain (sometimes wrongly called *tatare*) is a favourite dish with bean stew of *aboboe*, Vigna subterranea. Unripe plantains are fried as chips or dried and powdered as *kokonte*. The very ripe ones are pounded with corn dough and other ingredients and fried as *tatare* or *kakro*, or used alone with porridge or pap as sugar substitute. The stem yields a fibre for fishing tackle and for a sponge and towel used by elderly women. The peduncle also yields good fibre. The burned peelings of the fruit yield potash for local soap-making. (For the medicinal properties see Chapter 8.) The plantain is attacked by many diseases; some of these are crown rot caused by *Botryodiplodia theobromae*, which also causes the fruit rot; pitting caused by *Pyricularia grisea*; speckle caused by *Deightoniella torulosa* and *Thielaviopsis*; and stalk rot caused by *Ceratocystis paradoxa*.

Cyperus esculentus Tiger Nut, Rush Nut, Earth Almond (Atadwe) is a native of the Mediterranean and western Asia and naturalized in many warm countries. It is a savanna plant and grows widely in West Africa. In Ghana it is cultivated along the coastal savanna, particularly around Senya Bereku in the Central Region, in the Volta Region, along the Kwahu Scarp in the Eastern Region, and on small-holdings throughout the country. The nuts are usually chewed raw as a side dish or used to prepare a thick dessert, *atadwe-milk* (finely ground fresh nuts, strained and boiled with wheat flour and sugar). The nuts are said to cure constipation when chewed and swallowed whole. Tiger Nuts are widely acclaimed as an aphrodisiac alone or with other ingredients.

WILD, EDIBLE ROOTS, TUBERS, CORMS AND RHIZOMES

Some of these are wholesome after some treatment, while others can be eaten in emergency situations like famine. It is worthwhile promoting their growth as crops. Plants with edible underground parts include the rhizomes of *Nymphaea lotus* Water Lily; the tuberous stems of *Ipomoea aquatica* are eaten in scarcity, also by wart-hogs; the tubers of *Asparagus flagellaris*, *Amorphophallus aphyllus* and *A. dracontioides* after drying and boiling to remove the acrid property; the tubers of *Anchomanes difformis* after washing with water and ashes and boiling for a long time, or cut up and soaked in water, dried in the sun and stored for use — these tubers are acrid owing to raphides and saponin; the rhizomes of *Stylochiton lancifolius* in famine, after repeated washing in lye of ashes before cooking, and the rhizomes of *Typha domingensis* (*australis*) Bulrush, also in famine. The roots of *Dioscoreophyllum cumminsii* Guinea Potato are occasionally eaten boiled, like potatoes, especially in Gabon. The corms of *Gladiolus daleni*, *G. klattianus*, *G. gregarius*, and *G. unguiculata* collectively called Sword Lily or Corn-flag are edible, as are the corms of *Zygotritonia crocea*. *Solenostemon rotundifolius* Hausa, Frafra, or Salaga Potato is semi-cultivated and propagated by cuttings or tubers, maturing in 5–6 months. There are four varieties. Others are the tubers of *Smilax kraussiana* (Kokora), *Mariscus sumatrensis* (*alternifolius*), *Icacina olivaeformis* (*senegalensis*) False Yam after treatment, and the tubers of *Tacca leontopetaloides* South Sea Arrowroot as famine food after much washing — otherwise they are indigestible. The rest are the roots of *Pachycarpus lineolatus*, *Jatropha curcas* Physic Nut and *J. multifida* Coral Plant, Spanish Physic Nut, which is like cassava, and the rhizome of *Cyperus distans* is used in sauce.

Food Legumes

Legumes are the cheapest source of proteins for the rural communities and the low income group, in view of the high cost of eggs, fish, and meat. An estimated five million children under five years old die yearly in the developing countries as a result of malnutrition — basically in the form of protein deficiency. Legumes, in West Africa, consist of *Arachis hypogaea* Groundnut, Peanut; *Phaseolus lunatus* Lima Bean, Butter Bean; *Cajanus cajan* Pigeonpea, *Vigna unguiculata* Cowpea, *V. subterranea* (*Voandzeia subterranea*) Bambara Groundnut, *Canavalia ensiformis* Sword Bean, and more recently *Psophocarpus tetragonolobus* Winged Bean. *Sphenostylis stenocarpa* African Yam Bean is cultivated on a limited scale by some tribes such as the Ewes of Ghana, Togo, and Benin for the seeds and, sometimes, the small tubers. The underground beans of *Kerstingiella geocarpa* are used as food and the plant is sometimes cultivated for the purpose.

Arachis hypogaea Groundnut, Peanut, probably introduced from tropical South America, is a predominantly savanna crop grown in West Africa mostly in the Sudan and Guinea savanna woodland countries. Major producing countries in Africa are Nigeria, Senegal, and Sudan. In Ghana cultivation is centred around the Northern and Upper Regions and in Brong Ahafo Region which produces about a fifth of the country's supply. There are several varieties. The nuts may be chewed uncooked, but they are usually eaten boiled or roasted as a side dish — preferably with boiled or roasted corn. Groundnut soup is prepared from ground, previously roasted nuts or groundnut paste. The paste also serves as peanut butter, a useful addition to a meatless diet. A variety of sweets may be prepared from the roasted nuts or the paste. Groundnuts contain up to 38 per cent protein, and some high-protein baby foods are manufactured in parts of tropical Africa from peanut flour, skim milk powder, and added vitamins and minerals. Groundnut contains about 40–50 per cent oil — and is among the most costly of the cooking oils. It yields a solid fat when hydrogenated and may be used in the preparation of margarine and cheese. The residue (about 55 per cent of kernels) after oil extraction is an excellent livestock feed in the form of cakes, and in specially prepared form is used for human consumption as groundnut flour, a protein-rich food supplement. Groundnut tops are a valuable stock feed. In drier areas groundnuts are attacked by *Striga*, a parasitic weed (see Chapter 11). Groundnuts are also attacked by nematode diseases caused by *Helicotylenchus multinctus*, *Rotylenchus reniformis* and *Tylenchus* species among others.

Phaseolus lunatus Lima Bean, Butter Bean is grown as a backyard crop or on small-holdings throughout West Africa. The beans are boiled for food or used to thicken soups. The dry bean has a high nutritional value — it is rich in protein, calcium, and iron. Fresh seeds are rich in calcium, iron, and vitamin C, but may contain some hydrogen cyanide which can be removed sufficiently by cooking (see Chapter 9). The other species of edible *Phaseolus* are *P. aureus* Green Gram and *P. mungo* Black Gram which have recently been introduced into West Africa — probably from India and South-East Asia, where they are highly commercialized. *Macroptilium atropurpureus* (*P. atropurpureus*) and *M. lathyroides* (*P. lathyroides*), cover crops or weeds, have edible beans and are worth cultivating. The leaves and immature pods of Lima Bean are sometimes used as a pot-herb. The crop is attacked by web blight of bean in which brown fungal threads of *Corticium solani* grow up the stem and on to the pods and leaves which rapidly rot and hang. Clerk (1974) reports that the disease is very common and very severe during wet weather.

Cajanus cajan Pigeonpea is cultivated around homes, as a hedge, and among cultivated crops because of its ability to fix nitrogen. It is more frequent in the savanna zone. The immature seeds serve as green peas, the young pods as vegetables, and the leaves as fodder. The plant is more of value as green manure and cover crop than as a food crop. The mature seeds are eaten, but are not a popular food. They seem to be of little importance for human consumption; otherwise one would certainly find them on the market. The leaves and seeds are medicinal (see Chapter 8).

Vigna unguiculata Cowpea is cultivated extensively, mostly in the savanna areas of West Africa. It is the most widely used leguminous crop. Major producing countries in order are Niger, Nigeria, Burkina Faso, and Senegal. There are many varieties in habit — prostrate, climbing, dwarf, etc., and colour and size of seeds. White varieties are generally preferred to the red ones. It is cooked with rice (rice and beans), or boiled and eaten with *gari* (*gari*

2. FOOD AND FODDER

and beans), or used to prepare bean stew or to thicken soup. *Koose* or *akla*, a popular doughnut, is prepared from milled cowpeas. The unripe, tender pods are used as vegetables and the young leaves as spinach. *Striga gesnerioides* Witchweed is reported to parasitize cowpeas in Sudan. The crop is also attacked by the nematode *Meloidogyne* species.

Vigna subterranea (*Voandzeia subterranea*) Bambara Groundnut grows well in the Sudan and Guinea savanna woodland countries of West Africa (Heppei, 1970). Major producing countries in order are Nigeria, Niger, Ghana, Togo, and the Ivory Coast. It is relatively tolerant of poor soils, and features prominently in many traditional farming systems in Africa as an intercrop of cereals and root crops. Martin (1984) notes that it is widely distributed as a poor man's crop throughout tropical Africa. In Ghana, Doku and Karikari (1971) report that it ranks next only to *Vigna unguiculata* Cowpea in production and consumption. The seeds vary in colour and size. It is used in preparing *aboboi*, a popular local dish eaten usually with fried, ripe plantain or *tatare* (a preparation of ripe plantain and corn dough). The nuts may also be boiled, salted, and eaten as a side dish. *Koose* or *akla* is prepared from the milled beans in northern Ghana. These beans in gravy are now being canned, and Doku and Karikari further report that the Nsawam Cannery in southern Ghana cans over 40,000 tins of various sizes annually. Young (1978) observes that the seeds contain about 6 per cent oil, 60 per cent carbohydrates, and 18 per cent protein and are highly nutritious. The introduction of Bambara groundnut into food shops in predominantly African communities in Britain is being considered. It may also be used as animal feed, although the foliage is unsuitable as fodder. The crop is affected by leaf spot caused by *Cercospora* species and powder mildews caused probably by *Oidium* species.

Canavalia ensiformis. Sword Bean is planted around homes and on smallholdings. It is useful as both green manure and cover crop. The beans are white, but some varieties have wine-red beans. They are edible and the young pods may be used as vegetables. The cultivation of sword bean on a wider scale is recommended to supplement protein shortage.

Psophocarpus tetragonolobus Winged Bean, a native of India and South-east Asia, was recently introduced into West Africa — at the Agricultural Research Station at Kade in Ghana and ORSTOM in Abidjan, Ivory Coast. *CERES*, an FAO periodical, observes that this legume is extremely rich in protein — the leaves, flowers, and pods contain 10–15 per cent, the dry beans 30–37 per cent, and the tubers (also edible) 12–15 per cent. *CERES* adds that if a mere 20 per cent of winged bean flour is added to the cassava flour normally eaten in developing countries, the protein content per portion rises from 9 g to 41 g. The young, tender pods are prepared and eaten like vegetables. Hymowitz and Boyd (1977) report that the immature pods are also eaten raw and in some areas the young leaves and shoots are eaten both raw and cooked. The report adds that the use of the tuberous roots for food (eaten both raw and cooked) appears to be exclusive to Burma. Winged bean is also useful as green manure and cover crop, and the dry stalks as fodder. The crop is reported to be plagued by the fungus *Woroninella psophocarpi* in Java. The indigenous *P. palustris* and *P. monophyllus* are inedible.

Mucuna. Some species of *Mucuna* are reported to be edible. Cultivated plants bear pods lacking the irritating hairs found in the wild ones.

Leaf Protein

In addition to legumes, fruits, and seeds, protein may be obtained from leaves. The process involves grinding fresh green leaves to a pulp and pressing out the juice. The proteins and other nutrients are then separated from the juice by heating and filtering through cloth bags. The water is pressed from the protein concentrate, which may be preserved by drying it to a powder or pickling with salt. It is eaten with conventional foods by mixing it in during cooking, or it can be made into a tonic drink or sweets. Any leaf which is not poisonous can be used. Leaf protein or leaf nutrient is rich in protein, lipids, and vitamins E and A (as B-carotene) and also contains useful amounts of the minerals iron, calcium, and magnesium (see Table 2.1). The villagers of Kpong Bawaleshie, Nanomang, and Abobya in Ghana together support the production of leaf-protein at Kpone Bawaleshie on the Accra–Dodowa road. The project would be worth extending to many more villages in view of the protein-deficient local diet. The project is funded by a group of aid agencies and a trust fund, including Find Your Feet and the André Simon Memorial Fund in London and OANIDA in Denmark. Leaf protein is currently being produced in the USA, France, India, Mexico, Pakistan, and Sri Lanka. There are plans to start projects in Thailand, Egypt, Sudan, and Zambia.

Table 2.1 Nutritive value of leaf protein

	Provided by 20 g (2 teaspoonsful) of leaf protein	*% of Daily requirement (FAO/WHO)*
Protein	12 g	52%
Energy	70 kcal	5%
Vitamin A (B-carotene)	17600 IU	880%
Vitamin E	5.6 IU	80%
Niacin	5.4 mg	60%
Magnesium	100 mg	80%
Calcium	400 mg	50%
Iron	18 mg	120%

2. FOOD AND FODDER

Another source of plant protein is *Spirulina* species — blue-green algae which grow in water containing up to 14,000 mg/litre of chlorine and can withstand an alkaline level up to pH 11. It has been used as a source of protein by the inhabitants of the Lake Chad area where the algae grows, and in Mexico, for many generations. Up to 60-70 per cent of *Spirulina* is good-quality protein. It is also rich in vitamins. *Spirulina* grows in the Volta Lake in Ghana, but this rich source of cheap protein is not yet utilized. The algae is easily recovered by filtering through fine sieve or muslin, a great advantage. The filtrate may be spray-dried in the sun, cut into blocks, and cooked as a green vegetable. *Spirulina* is already being sold in Mexico as a high-protein, high-carotene additive for chicken feed. It would be worth encouraging its use as human food in West Africa in view of the shortage of protein in the average diet and the high cost of fish, meat, and eggs.

Fleshy Vegetables

These are a source of vitamins, minerals, and other nutrients.

Lycopersicon esculentum (*lycopersicum*). Tomato is cultivated as a backyard crop, on small-holdings, and on large farms, mostly in the Sudan and Guinea savanna woodland countries of West Africa. Major producing and marketing centres in Ghana are the Tongu Irrigation Site in the Upper Region, Akumadan in northern Ashanti, Amanfrom Kwahu at the Afram Plains, and Akorwu-Bana in the Eastern Region. These are several varieties. Tomatoes are easily perishable and as such are fairly cheap when in season, but costly during the lean season. They are eaten raw in salads, but the bulk is used as flavouring in sauce, stew, and soup. IBPGR observes that if consumption is high tomatoes contribute to the value of the meal in respect of carotene, thiamin, niacin, and vitamin C. The Nsawam Cannery and the Pwalagu Tomato Factory in Ghana process tomatoes into paste for canning. The leaves are used medicinally for earache and the fruits as a remedy for diseases of the urinary passage. The crop is attacked by several diseases including nematode disease caused by *Meloidogyne* species, bacterial soft rot caused by *Erwinia carotovora* and others, and fruit rot caused by *Didymella lycopersici*.

Capsicum annuum Pepper is of South American origin and cultivated as a backyard or farm crop on small and large-scale holdings. It is a predominant favourite and widely-used ingredient in cooking in Ghana — many cannot enjoy a meal (especially fresh fish) without pepper. It is either eaten fresh (as in *kenkey* and hot pepper) or in sauce, stew, and soup. Pepper is sold fresh, dried or powdered. It has the highest average nutritional value (ANV) (see Table 2.2). Since it can be dried and stored it has the advantage of maintaining a more or less uniform price level irrespective of season. The Nsawam Cannery in Ghana cans ground pepper as paste. The leaves are sometimes eaten as a vegetable. Pepper is an ingredient in many local medicinal prescriptions. *C. frutescens*, which is propagated by birds, is also

Table 2.2 Average nutritive value of vegetables. Data from: Food composition table for use in East Asia (FAO, 1972)

Type of produce	Waste %	DM g	Energy Kcal	per 100 g edible portion									ANV per 100g. dry matter
				protein g	fibre g	Ca mg	iron mg	carotene mg	thiamine mg	ribo-flavine mg	niacin mg	vit.C mg	ANV	
fleshy vegetables														
tomato	6	6.2	20	1.2	0.7	7	0.6	0.5	0.06	0.04	0.6	23	2.39	38.5
egg plant	4	8.0	26	1.6	1.0	22	0.9	0.0	0.08	0.07	0.7	6	2.14	26.8
sweet pepper	13	8.0	26	1.3	1.4	12	0.9	1.8	0.07	0.08	0.8	103	6.61	82.6
hot pepper	13	34.6	116	6.3	15.0	86	3.6	6.6	0.37	0.51	2.5	96	27.92	36.7
okra	10	10.4	31	1.8	0.9	90	1.0	0.1	0.07	0.08	0.8	18	3.21	30.9
cucumber	20	3.8	12	0.6	0.5	21	0.4	0.1	0.03	0.04	0.2	11	1.69	44.5
pumpkin	17	8.1	27	0.7	0.8	24	0.7	0.8	0.03	0.04	0.5	14	2.68	33.1
watermelon	37	6.8	21	0.6	0.2	8	0.2	0.1	0.03	0.03	0.2	6	0.90	13.2
melon (white-green)	22	7.6	26	1.0	0.5	16	0.5	0.0	0.05	0.02	0.4	25	2.33	30.7
bitter gourd	20	6.0	19	0.8	1.0	26	2.3	0.1	0.06	0.04	0.3	57	4.10	68.3
leafy vegetables														
amaranth	40	10.7	26	3.6	1.3	154	1.9	6.5	0.01	0.22	0.7	23	11.32	105.8
kangkong	28	10.0	30	2.7	1.1	60	2.5	2.9	0.09	0.16	1.1	47	7.57	75.7
Chinese cabbage, leaf type	14	5.8	17	1.7	0.7	102	2.6	2.3	0.07	0.13	0.8	63	6.99	120.6
lettuce	26	6.4	20	1.4	0.6	56	2.1	2.0	0.06	0.12	0.5	17	5.35	83.6
white cabbage	15	7.0	22	1.6	0.8	55	0.8	0.3	0.06	0.06	0.3	46	3.52	50.3
cassava leaves	13	19.0	60	6.9	2.1	144	2.8	8.3	0.16	0.32	1.8	82	16.67	87.7

Table 2.2 continued

Type of produce	Waste %	DM g	Energy Kcal	protein g	fibre g	Ca mg	iron mg	carotene mg	thiamine mg	ribo-flavine mg	niacin mg	vit.C mg	ANV	ANV per 100g. dry matter
leguminous vegetables														
hyacinth bean (dry)	0	87.9	334	21.5	6.3	93	3.9	0.0	0.40	0.12	1.8	0	14.03	16.0
asparagus bean (pods)	12	11.7	37	3.0	1.6	14	0.2	0.2	0.12	0.11	1.0	22	3.74	36.7
Lima bean (fresh)	43	31.5	119	8.4	1.0	25	2.2	0.1	0.16	0.16	1.5	30	1.88	17.9
mung bean (sprouted)	7	9.9	30	4.2	0.0	15	1.2	0.0	0.11	0.10	0.8	18	2.94	31.2
sprouts, bulbs, tubers, etc.														
onion (dry)	6	11.4	38	1.6	0.7	30	1.0	0.0	0.06	0.04	0.2	9	2.05	20.0
carrot	17	10.4	37	1.1	0.9	36	1.2	4.2	0.06	0.05	0.7	9	6.48	64.2
bamboo shoots	44	9.0	28	2.5	1.2	17	0.9	0.0	0.11	0.09	0.6	9	2.55	30.9
mushroom	9	11.3	37	2.7	0.9	8	1.0	0.0	0.10	0.42	4.8	3	2.10	19.3
turnip	21	7.2	23	1.2	0.3	43	0.9	0.0	0.04	0.04	0.5	22	2.03	28.2
taro (as vegetable)	16	24.6	34	1.1	0.3	14	1.2	0.0	0.12	0.04	1.0	8	2.28	9.7

used as a spice and has medicinal properties. The crop is attacked by bacterial spot caused by *Xanthomonas vesicatoria*, grey mould caused by *Botrytis* species, and anthracnose caused by *Colletotrichum* species among other diseases.

Solanum Egg Plants. There are many species in Ghana. (*Solanum aethiopicum* L. Gilo group) Garden Egg or Scarlet Eggplant (Awororo) is a savanna crop and has the widest cultivation in Ghana, as a backyard crop throughout the country, but cultivation is more concerned in the southern savanna and along the Cape Coast–Kumasi road. The fruits may be chewed uncooked, but they are usually boiled to thicken light soup (*nkakra*) or for stew (*ntoroba foroye*) or mashed with pepper, onions, and roasted *korbi* (*abomu*). Garden eggs are nutritious, and weight for weight more garden eggs are probably consumed than any of the food legumes (beans, etc.). They are tinned by the Nsawam Cannery in Ghana. Other edible species used in Ghana are *S. anguivi* (*indicum*) (Amponimpo etc.) for flavouring and medicine; *S. aethiopicum* Kumba group and *S. macrocarpon* (Atropo) with edible leaves as well as fruits; *S. melongena* Aubergine (Ntorobabanyin) with edible fruits; *S. scabrum*, *S. americanum* (*nigrum*) and even *S. torvum* (Samanntoroba). Many different vernacular names are applied to eggplants in Ghana, (see a more detailed account of *Solanum* in Africa by K. Pipe-Wolferstan in *Traditional Food Plants* Food & Nutrition paper 42: 450–66 (1988) FAO, Rome).

Diseases of the garden egg include cottony leak caused by *Pythium aphanidermatum*, grey mould caused by *Botrytis cinerea*, and *Rhizopus* rot caused by *R. stolonifer*.

Allium ascalonicum Shallot is of Asian origin and cultivated in the savanna areas of West Africa. The Awunas in the Volta Region of Ghana are, by tradition, growers of shallot. The crop is also cultivated in the Afram Plains and the Kwahu area. It is a common backyard crop in homes. Shallots are used in the preparation of sauce, stew, and soup as a flavouring. They are eaten raw in salad and more often in preparing hot pepper for *kenkey*. Shallots dry and store easily — this reduces transport costs and ensures uniform availability and consequently stable prices all the year. The nutritional value is comparatively low. According to IBPGR, if immature bulbs and tops are consumed it improved the average nutritional value scores 3.96 with a high vitamin C and calcium content. When shallots are used in salads, therefore, the nutritional value is higher. The medical properties of the onion family are often stressed. Garlic, shallot, onion, and other species have a disinfecting action in the intestines and purify the blood. *A. cepa* onion is also cultivated in West Africa. There are about six varieties based on size of bulb, colour, and degree of scent. Some onions are imported from Nigeria to supplement production in Ghana. Shallot is affected by several diseases including black rot and white rot caused by *Aspergillus niger* and *Sclerotium cepivorum* respectively.

Abelmoschus esculentus (*Hibiscus esculentus*) Okro, Okra is mostly cultivated in the Sudan and Guinea savanna woodland countries of West Africa and as a backyard crop. There are several varieties based on size of plant — tall or

2. FOOD AND FODDER

short, perennial or annual; size of fruit — long or short; and colour — reddish or greenish. Okro is eaten as a vegetable in soup or stew, or it may form the main part of these dishes as in okro soup. It may be sliced, dried, and stored. It is rich nutritionally with a high calcium content. This gelatinous vegetable helps to relieve constipation when eaten in quantity. The leaves as well are eaten as a vegetable — both fresh and dried. The crop is attacked by nematode disease caused by *Helicotylenchus multinctus* and *Pratylenchus brachyurus* among others; it is also affected by parasitic sooty moulds caused by *Irenopsis aciculosa*.

Sechium edule Chayote is cultivated in Ghana around Awutu on the Accra-Winneba road; around Eisam, on the Winneba-Cape Coast road and around Aburi. The vegetable is used to prepare stew and soup in the same way as egg plant. Besides the fruits and young leaves the fleshy roots are also edible. Its cultivation on a wider scale than at present is to be encouraged. In Africa only Sierra Leone seems to be a place where it is commonly grown.

Colocynthis lanatus (*vulgaris*) Water Melon is cultivated in the southern savanna and, to some extent, in the north, but is available in almost all urban markets when in season. There are varieties according to size of fruit — large or small, and shape of fruit — spherical or elongated-oval. The fruits are refreshing and filling. Some varieties with inedible pulp are cultivated for the edible seeds only, which are prepared in the same way as Agushie (below). The food value of the seed is given as fat 45 per cent and protein 54 per cent. In Ghana the seeds are generally discarded. They may be roasted as a coffee substitute.

Cucurbita pepo Pumpkin, Vegetable Marrow is cultivated for the fruits which are dried to make gourd vessels for a variety of purposes. However, the immature fruits and leafy shoots serve as vegetable. There are about three varieties. Locally, however, they are of minor importance. *C. maxima* Squash Gourd, Melon Pumpkin is also cultivated. The inside of the fruit is generally discarded, only the outside being used as a vegetable.

Cucumeropsis edulis (Agushie) is cultivated on small-holdings and around villages for the edible seeds. They may be roasted and eaten; but more often the seeds are ground and used to thicken soup or to prepare stew alone (*aketewa-foroye*), or with other vegetables. The seeds yield oil. Agushie, like mushrooms, is a favourite vegetable among vegetarians. Cucurbits are generally affected by leaf spot caused by *Carcospora citrullina*.

Telfairia occidentalis Fluted Pumpkin grows wild in the forest zone. It is often cultivated for the edible seeds (also eaten by animals), which are prepared in the same way as Agushie. The young, green shoots of *Telfairia*, locally called Krobonko, make an excellent vegetable. The seeds yield oil. The stems are useful as sponge.

The young fruits of *Luffa cylindrica* Vegetable Sponge, *Momordica charantia* African Cucumber, and *Lagenaria siceraria* Bottle Gourd, all members of the family Cucurbitaceae, are used as vegetables.

Mushrooms

The mushroom is a fungus in the group basidiomycetes, and on the whole grows wild in the region, mostly in the forest zone. Cultivation is practised on a small scale only by individuals for domestic use. A large proportion of the mushrooms sold in the markets and on the roadside are harvested during the season from those growing wild. *Termitomyces* species and *Volvariella* species are generally preferred and popular. Other edible ones are *Auricularia* species, *Pleurotus* species, and *Calvatia* species. Zoberi (1973) observes that, having high-quality protein and containing most of the vitamins, they could in fact become an important source of food for the increasing world population. Mushrooms are an indispensable part of a vegetarian diet. Eating poisonous mushrooms is however inadvisable and might be fatal (see Chapter 9). The fungi are not only important for edible purposes, but also in fermentation, baking, and in the production of antibodies and for their role in disease, decay, and other ecologically important processes.

Spinach or Leaf Vegetables

Cultivated spinach include *Basella alba* Indian or Ceylon Spinach, *Amaranthus hybridus* subsp. *incurvatus* (sometimes growing wild), and *Celosia argentea* (Nkyewodue). The young shoots and leaves of the bottle gourd and the young leaves of pumpkin and squash gourd are used as spinach. In Ghana the leaves of *Xanthosoma mafaffa* Cocoyam (Kontomire) are the most popular spinach in the forest zone and the southern savanna. The leaves of *Colocasia esculenta* Eddoes (Kokoo) are occasionally used as spinach. IBPGR warns that older leaves may contain raphids of calcium oxalate which do not dissolve during cooking. High levels of oxalates are also reported in *Talinum, Celosia, Corchorus*, and *Amaranthus* species. Of recent introduction is *Cnidoscolus aconitifolius* Chaya, a fast-growing shrub the cooked leaves and young shoots of which, says Professor Ayensu of the Smithsonian Institution, are apparently rich in various nutrients including protein. Newton (1979) observes that since Chaya grows as a branching shrub it can also be used to make an effective and attractive hedge. A hedge with edible clippings is a valuable asset.

Semi-cultivated leaf vegetables include *Talinum triangulare* Water Leaf, Wild Spinach (Fan), *Sesamum indicum* Sesame, Beniseed, *Corchorus aestuans*, and *Launaea taraxacifolia* (*Lactuca taraxacifolia*) Wild Lettuce which serves both as spinach and salad leaves. *Vernonia amygdalina* Bitter Leaf (Bonwen) and *V. colorata*, though bitter, are used as spinach. The bitterness is reduced by soaking in water or by boiling. The leaves and young shoots of the Fluted Pumpkin and the leaves of *Cleome gynandra* (*Gynandropsis gynandra*) and *Piper guineense* West African Pepper (Soro-wisa) serve as spinach. Other leaf vegetables are *Aerva tomentosa* (Bameha), *Amaranthus viridis* Wild or Green Amaranth, *Cardiospermum halicacabum* Balloon Vine, *Laportea aestuans*

(*Fleurya aestuans*) (Hunhon), *Pentodon pentandrus* (Buburanya), *Portulaca oleracea* Purslane, Pigweed; young leaves of *Bidens pilosa* Spanish Needles, *Crassocephalum rubens* (Banfa-banfa), *C. crepidioides*, and *Sonchus oleraceus* Sow-thistle. Others are *Emilia coccinea*. *E. sonchifolia* (both for salads as well), *Feretia apodanthera* (Bitinamusa), *Struchium sparganophora*, many wild species of *Solanum* including *S. anomalum* (Nsusoa), *S. dasyphyllum*, and the cultivated *S. aethiopicum* and *S. macrocarpon* (Atoropo). *S. nigrum* (Nsusuabiri) is only edible after cooking and may be more or less poisonous. Others are *Senecio biafrae* (*Crassocephalum biafrae*), *Ipomoea alba* Moon Flower, *I. eriocarpa*, the young shoots of *I. aquatica*, *I. batatas* Sweet Potato, *Jacquemontia tamnifolia* (Boeboe), *Ceratotheca sesamoides* (Bungu), *Sesamum alatum*, *S. radiatum* (Zinzam), *Justicia insularis* (Asipiriwa) (which is sometimes cultivated), *Aeolanthes pubescens*, *Urera cameroonensis*, *Hyptis pectinata* (Peaba), and *Solenostemon monostachys* (Sisiworodo). The young, leafy shoots of *Vitex doniana* Black Plum (Afua) with groundnuts, salt, pepper, etc., form a Hausa foodstuff, the plant often being grown in villages for its fruits and leaves. Others are *Balanites aegyptiaca* Desert Date, *Piliostigma thonningii* (Opitipata), and the palm cabbage of *Phoenix reclinata* Wild Date Palm, *Cocos nucifera* Coconut Palm, *Raphia hookeri* Wine Palm, *Laccosperma secundiflorum* (*Ancistrophyllum secundiflorum*), and the young buds of *Calamus deeratus*. Young shoots of *Oxytenanthera abyssinica* Bamboo are eaten in Japan and the Far East.

Introduced vegetables include *Brassica oleracea* var. *capitata* White Cabbage, *B. oleracea* var. *botrytis* Cauliflower, *B. oleracea* var. *italica* Broccoli, *B. campestris* var. *rapifera* Turnip, *Daucus carota* Carrot, and *Launaea sativa* (*Lactuca sativa*) Lettuce. The cultivation of these vegetables is concentrated mainly in the urban centres on small-holdings and backyards, mostly for use by Europeans and the higher class.

Wild vegetables include the leaves of *Haumaniastrum lilacinum* (Taga), *Adansonia digitata* Baobab, the young leaves of *Afzelia africana* (Papao), *Albizia chevalieri*, *A. zygia* (Okoro), *Carica papaya* Pawpaw and *Anacardium occidentale* Cashew Nut, eaten raw with rice as flavouring in Java and Malaya; the leaves of *Annona senegalensis* var. *senegalensis* Wild Custard Apple, *Antidesma venosum* (Mpepea), *Barleria opaca* (Mu), *Bauhinia purpurea* in India, *Rhodognaphalon brevicuspe* (*Bombax brevicuspe*) (Onyinakoben), *B. buonopozense* Red-flowered Silk Cotton Tree and the sepals, *Boscia salicifolia* in emergency, and leaves and twigs of *Cadaba farinosa* in Sudan and northern Nigeria. The leaves of *Capparis corymbosa* and *Cassia nodosa* Pink Cassia, though purgative, are sometimes eaten in South-East Asia. Others are the young leaves of *C. obtusifolia* (*tora*) Foetid Cassia; the leaves of *Castanola paradoxa* (Abokodidua) by the Binis of southern Nigeria, *Chenopodium murale*, *Ceiba pentandra* Silk Cotton Tree, *Celtis integrifolia* Nettle Tree (Samparanga), young leaves of *Milicia* (*Chlorophora*) *excelsa* Iroko (Odum); *Christiana africana* (Sesedua) mixed with corn to make a meal in Nigeria, and the leaves of *Cleistopholis patens* Salt and Oil Tree eaten with cola nuts in Sierra Leone; leaves of *Combretum paniculatum* (Omeha), *C. platypterum* (Owhirem), *Corchorus olitorius* Jew's Mallow, Long-fruited Jute; *C. tridens* (Sanvoa), *Crateva adansonii* (*religiosa*) (Chelum Punga), sometimes sold in the markets as such, and the leaves and flowers of *Crotalaria ochroleuca*, eaten in Central African Republic and Congo respectively; young leaves of *Daniellia oliveri* African Copaiba Balsam Tree (Sanya) in times of famine, *Dendrocalamus strictus* Male Bamboo in India, *Dracaena mannii* (Kesene), sometimes used as asparagus by Europeans, young leaves of *Monanthotaxis foliosa* (*Enneastemon foliosus*) (Ntetekon), *Euadenia trifoliolata*, *Eriosema glomeratum*, *Fadogia cienkowskii*, *Ficus glumosa* var. *glaberrima* (Gilinziela), *F. sycomorus* (*gnaphalocarpa*) (Kankanga), *F. vallis-choudae* (Aloma-Bli), *Globimetula braunii*, young leaves of *Grewia carpinifolia* (Ntanta) in soup, and *Heinsia crinita* Bush Apple sometimes sold as such; the leaves of *Hibiscus lunariifolius* as pot-herb, *H. rosa-sinensis* Garden Hibsicus, Shoe flower in northern Ghana, *H. rostellatus*, and *H. tiliaceus* (Nwohwea); *Isoberlinia tomentosa* (*dalzielii*) (Kangkalaga), *Lannea acida* (Kuntunkuri) (soluble gum also edible), *L. microcarpa*, *Laccosperma secundiflorum* (*Ancistrophyllum secundiflorum*) after boiling to remove the bitterness, *Isonema smeathmannii* as pot-herb in Sierra Leone, *Jatropha curcas* Physic Nut in Java, *Justicia extensa* in Central African Republic, *Leucaena leucocephala* Leucaena, Horse or Wild Tamarind (young pods as vegetable), *Lippia multiflora* Gambian Tea Bush (Saa-nunum) boiled with palm nuts, *Maerua angolensis* (Pugodigo), *M. crassifolia*, young leaves of *Manihot esculenta* Cassava, *M. glaziovii* Ceara Rubber as spinach in East Africa, *Microdesmis puberula* (Fema) in Central African Republic, *Mikania chevalieri* (*cordata*) Climbing Hemp Weed, and *Moringa oleifera* Horse-Radish Tree, Oil of Ben Tree. The flowers and young pods of *Moringa* are also boiled as vegetable. The young leaves of *Myrianthus arboreus* (Anyankoma) and *M. libericus* (Anyankom-nini) are edible.

Others are the young leaves of *Neuropeltis acuminata*, *Ouratea affinis* (Ananse Don), *O. calophylla* (Opunini), and *Rhabdophyllum affine* subsp. *myrioneurum* (*O. myrioneura*) (Duabogo) as spinach in Central African Republic; the leaves of *Parkia clappertoniana* West African Locust Bean mixed with cereal flour, *Pavetta crassipes* (Nyenyanke), *Phyllanthus muellerianus* (Potopoleboblo) in food or soup in Sierra Leone and southern Nigeria; the leaves of *Piper umbellatum* (Amuaha), *Pseuderanthemum tunicatum*, *Leptadenia hastata*, young shoots of *Quisqualis indica* Rangoon Creeper in Indonesia (the seeds are also edible and taste like those of Indian Almond), *Rangia grandis*, the young leaves of *Sesbania grandiflora* together with the fresh flowers and immature fruits, *S. sesban* Egyptian Sesban, *Sphenostylis stenocarpa* African Yam Beans, *Sterculia tragacantha* African Tragacanth (Sofo); *Tamarindus indica* India Tamarind in soup by the Nankanis and other tribes in northern Ghana, *Tetracera alnifolia* (Akotopa), *T. potatoria* (Twihama), *Thespesia populnea* (Frefi), *Trema orientalis* (*guineensis*) (Sesea) and the fruits, *Trichosanthes cucumerina* var. *anguina* Snake Gourd, *Triumfetta cordifolia*, and *T. rhomboidea* Burweed.

Fruits

With the exception of a few semi-cultivated fruits like Wild Mango, Star Apple, Velvet Tamarind, Black Plum, and the Miraculous Berry, the cultivated fruits have all been introduced from other tropical or sub-tropical countries.

CITRUS SPECIES

Those cultivated in West Africa include *C. sinensis* Sweet Orange, *C. aurantiifolia* Lime, *C. reticulata* Tangerine, Mandarin, *C. paradisi* Grapefruit, and *C. limon* Lemon. Also cultivated, but on a smaller scale, are *C. aurantium* Sour Oranges and *C. medica* Citron.

Citrus sinensis Sweet Oranges were introduced by the Portuguese and are extensively cultivated in the forest regions of West Africa for the sweet, refreshing fruits. The juice is canned or bottled as orange juice or syrup by Nsawam Cannery, Nkulenu Industries, and Astek Industries in Ghana. Several varieties occur — including Washington, Late Valencia, Sekkan, Ovelleto, Sanguino, Shama, Kwesi Nyarko, and Otumi. Hybrids between Sweet Orange and Grapefruit include Lake Tangelo, Ortenique, King de Semis, and Minneola Tangelo. Orange peel yields orange oil for flavouring and pectin. The essential oil from the flowers and leaves are used in perfumery.

Citrus reticulata Tangerine, Mandarin — a native of China — is not as widely cultivated as Sweet Oranges; probably due to the difficulty in transporting the easily perishable fruits. Varieties include Satsuma, Ponkan, and Clementine.

Citrus paradisi Grapefruit is cultivated around rest houses and private homes. Except on Agricultural Research Stations, the fruit is not cultivated on commercial scale in West Africa. *C. grandis* Shaddock, Pommelo has fruits as large as a football — spherical or sometimes pear-shaped.

Citrus aurantiifolia Lime is a native of South-East Asia. In Ghana it is cultivated extensively in the Central Region to feed the lime-juice factory at Abakrampa (near Cape Coast). The juice is used for lime-juice cordial, limeade, marmalade, citric acid, and as flavouring.

Citrus limon Lemon is used commercially for lemon squash, as a flavouring, in cosmetics, and in the production of lemon oil, citric acid, and pectin. Like *C. aurantium* Sour or Seville Orange, it is used as stock on which oranges and mandarins are budded or grafted. The possibility of using indigenous trees in the family Rutaceae, like *Aeglopsis chevalieri* (Kokoro), *Afraegle paniculata* (Obuobi), and *Citropsis articulata* African Cherry Orange, still needs investigation.

Citrus species are attacked by a number of diseases including green mould or fruit rot caused by the fungus *Penicillium digitatum* which enters the fruit through various types of wounds. *Citrus* is also attacked by foot rot caused by the fungus *Phytophthora parasitica*. Sour Orange and Lemon are fairly resistant, though. *Phragmanthera* species Mistletoe often attack *Citrus* and reduce yield considerably.

OTHER CULTIVATED FRUITS

Mangifera indica Mango, a native of the East Indies and Burma, is now naturalized in tropical West Africa. In Ghana it grows better along the coastal savanna and the savanna areas of Wenchi and Kintampo. There are several varieties in size of fruit, colour, texture, and flavour of ripe fruit. The tree is freely cultivated for the deliciously refreshing fruits which are extensively sold in the markets when in season. They are rich in vitamins A and C. The starchy kernels are edible when roasted. The fruits may be eaten green as a vegetable. The ripe fruits are processed into jam. The leaves are used as fodder. The tree yields a gum, some tannin, and a yellow dye. The seeds, leaves, bark, and root-bark have various medicinal uses (see Chapter 8).

Ananas comosus Pineapple is a native of South America, but now half-naturalized in tropical Africa and extensively cultivated in West Africa. Wild varieties grow in the forest, although the crop is basically a savanna one. It is cultivated in the southern savanna area of Ghana especially, by both peasant farmers and commercial companies for local consumption and for export. Pineapples have been exported since independence, mainly by the private sector, chiefly to Britain and Switzerland. The exporting companies in Ghana include Combined Farms, Worac Farms, Koranc Farms, John Lawrence Farms, Gaffic Export and Trading, Akramang Farms, Ghana Federation of Agricultural Co-operatives, E.A. Ackom Farms, Mankoadze Farms, Danakof Farms, and Canbod Farms and Industries. Over ₵6 million ($120,000) annual export is estimated. The crop is also cultivated for export in the Ivory Coast. Pineapples contain vitamins A, B, and C. The juice and the edible head are both canned by the Nsawam Cannery and Astek Industries in Ghana. The plant has medicinal properties. The crop is attacked by root rot caused by the fungus *Ceratostomella paradoxa* and black rot caused by the fungus *Thielaviopsis paradoxa*. The crop is also affected by nematodes.

Musa paradisiaca var. ***sapientum*** Sweet Banana was probably introduced by the Portuguese or the missionaries. There are varieties according to colour, size of fruit, and flavour. It is cultivated in the forest region and to some extent in the southern savanna areas of West Africa for the finger-like fruits. These are sold throughout the year and eaten as a side dish or dessert. The fermented fruits are used in East Africa for preparing an alcoholic beverage. Bananas contain vitamins A and C and, when fully ripe, form a complete meal with milk for infants. They are a potential export. Nigeria and the Ivory Coast are the major exporting countries in West Africa. The leaves are good fodder and both the stem and peduncle yield fibre. The crop is attacked by several

2. FOOD AND FODDER

diseases including cigar-end rot caused by *Stachylidium theobromae*; Sigatoka disease or leaf spot caused by *Cercospora musae*, and Panama disease or banana wilt caused by the fungus *Fusarium oxysporum* f. *cubense*.

Persea americana Avocado Pear is a native of tropical America and widely cultivated in the forest region and similar conditions in West Africa. There is more than one variety. Trees with quality fruits are better propagated by grafting, since seedlings mostly prove to be inferior in quality to the parent tree. The ripe fruits serve as butter substitute and as dessert. Avocado contains vitamins A, B, C, and E; about 2 per cent proteins and 17—27 per cent fat mainly of the glycerol esters of palmitic, stearic, and oleic acids. The oil is used in the cosmetics industry. The crop is affected by fruit rot caused by the fungus *Trachysphaera fructigena*.

Annona muricata Sour Sop is a native of tropical America and is grown on small-holdings and private gardens for the refreshing fruits. They are also used in preparing a drink and for jelly. The bark contains tannin. The leaves are medicinal, while the seed oil is used in killing lice. *A. squamosa* Sweet Sop is also a native of tropical America and cultivated in and around homes and on small-holdings. The ripe fruit is sweeter than that of Sour Sop but inferior in flavour. Other introduced plants in the genus are *A. cherimola* Cherimola and *A. reticulata* Custard Apple; both with edible fruits. Indigenous *Annona* species with edible fruits include *A. senegalensis* subsp. *onlotricha* (*arenaria*) (Aboboma) and *A. senegalensis* subsp. *senegalensis* Wild Custard Apple. The fruit of *A. glabra* Marsh Cockwood, Monkey Apple (Adadima) is sometimes eaten, though unpalatable.

Carica papaya Pawpaw is a native of tropical America. It is cultivated mainly for the fruits — usually eaten ripe as a dessert or used to prepare mixed fruit salad. There are varieties according to height of the tree, size of fruit, or fruit colour. It contains vitamins A and C. The immature fruits serve as a vegetable when cooked. The dried latex (papain), like the leaves, is used to tenderize meat. The young leaves and the pith may be eaten. The root acts as a salt substitute while the seeds, leaves, and roots are medicinal (see Chapter 8). The plant is attacked by leaf spot caused by *Cercospora caricae*; powdery mildews caused by *Phyllactinia corylea*; and nematode disease caused by *Meloidogyne javanica*.

Anacardium occidentale Cashew Tree was introduced from tropical America. In Ghana cultivation is mainly on the Accra plains, particularly on the south-eastern savanna and around Dodowa. The ripe fruits (apple) are eaten fresh or used for preserves. They are a source of an alcoholic drink. The roasted kernels are a delicacy and sometimes used as nuts in the manufacture of chocolate, or sold as one of the important nuts of the world. The nut shells contain anacardic acid, which is a poison. The shell-oil is a source of phenolic resin of commerce. The bark and leaves contain tannin and the wood is fairly termite-proof and suitable for household constructional work. It is a good fuel-wood. The tree has various medicinal uses (see Chapter 8).

Other introduced fruit trees to West Africa include *Nephelium lappaceum* Rambutan, *Passiflora edulis* Passion Fruit, *P. quadrangularis* Giant Granadilla, *Bertholetia excelsa* Brazil Nut, *Artocarpus communis* Breadfruit (some varieties are seedless), *A. heterophyllus* Jack Fruit and *Chrysophyllum cainito* Star Apple; indigenous edible species are *C. pelpulchrum* (Atabene) and *C. pruniforme* (Duatadwe). Other introduced fruits trees are *Terminalia catappa* Indian Almond, *Averrhoa carambola* Carambola, *A. bilimbi* Bilimbing, and *Ziziphus jujuba* Indian Jujuba; indigenous edible species are *Z. abyssinica* (Larukluror), *Z. mauritiana* Indian Jujube, *Z. mucronata* Buffalo Thorn, and *Z. spina-christi* var. *microphylla*. Also introduced are *Flacourtia indica* Governor's Plum, *Spondias mombin* Hog Plum, *S. cytherea* English Plum, *Lecythis zabucajo* Monkey Pot, *Coccoloba uvifera* Seaside Grape (sold in Togo markets), *Psidium guajava* Guava, *P. cattleianum* Strawberry Guava, *Eugenia jambos* Rose Apple, *E. malaccensis* Malay Apple, *E. uniflora* Pitanga Cherry, *Dillenia indica* Honda Para, *Punica granatum* Pomegranate, *Garcinia xanthochymus* Egg Plant, and *G. mangostana* Mangosteen (one of the best tropical fruits); edible indigenous species are *G. afzelii* (Nsokodua), *G. kola* Bitter Cola (Tweapea), and *G. smeathmannii (polyantha)* False Chewstick Tree. The rest are *Achras zapota* Sapodilla Plum, *Melicoccus bijugatus* Honey Berry, *Vitis vinifera* Vine Plant, *Rubus* species West Indian Raspberry, *Elaeocarpus serratus* Ceylon Olive, *Prosopis chilensis* Cashaw Bean, *Malpighia glabra* Barbados Cherry, *Tamarindus indica* Indian Tamarind, *Samanea saman* Rain Tree, and *Canarium luzonicum* Java Almond — the indigenous *C. schweinfurthii* Incense Tree has edible seeds, but is only eaten cooked.

WILD FRUITS

Most wild fruits in our forests and woodland are of local importance only and are not known beyond the growing areas. They are eaten by hunters, farmers, field workers, and animals. A few popular ones are sold in the markets. Wild fruits include *Acacia albida* (Gozanga) and seeds, *Afrosersalisia afzelii* (Bakunini), *Alsodeiopsis staudtii* (Bonsa-dua), *Sherbournia bignoniiflora* (Kyere Beteni), *S. calycina*, *Ampelocissus gracilipes*, *Ancylobotrys scandens* (Obomene), *Aningeria robusta* (Samfena), *Anonidium mannii* (Asumpa), *Antidesma venosum* (Mpepea), *Antrocaryon micraster* (Aprokuma), *Lepisanthes senegalensis* (*Aphania senegalensis*) (Akisibaka), *Atroxima afzeliana*, *Bauhinia rufescens* (Jinkiliza), *Beilschmiedia mannii* Spicy Cedar and the seeds, *Bequaertiodendron megalismontanum* (Nufu-nufu), *B. oblanceolatum* (Nnanfuro), *Boscia senegalensis* (Dila), *Trilepisium madagascariense* (*Bosqueia angolensis*) (Okure) and the roasted seeds, *Bridelia stenocarpa* (*micrantha*) (Opam), *B. scleroneura* (Ba-Udiga) eaten by the Nankanis in Ghana, *Buchholzia coriacea* (Esono-bise), *Caloncoba gilgiana* (Kotowhiri), and *C. glauca*; the fruits of *Capparis corymbosa*, *C. erythrocarpos* (Apana), *Carissa edulis* (Botsu), *Landolphia dulcis*, *L. hirsuta* (Pumpune), *L. landolphioides* Cameroons Mountain Rubber Vine sometimes

2. FOOD AND FODDER

sold in the market, *Dictyophleba leonensis, Carpoloba alba* (Afiafia), *Cassia kirkii* var. *guineensis, Cathormion altissimum* (Abobonkakyere) and the seeds, *Celtis integrifolia* Nettle Tree (Samparanga); *Milicia (Chlorophora) regia* (Odum-nua), *Cissus aralioides* (Asirimu), *C. arguta, C. cornifolia* (Sintanatora), *Commiphora pedunculata, Cordia africana, C. myxa* Sapistan Plum (Tungbo) often cultivated, *C. rothii, Crateva adansonii (religiosa)* (Chelum Punga), *Deinbollia grandifolia* (Potoke) and the seeds, *Desplatzia chrysochlamys, D. dewevrei* (Wisamfia), *D. subericarpa* (Esonowisamfie-bere), *Detarium microcarpum, D. senegalense* Tallow Tree often sold in the markets, *Dichrostachys cinerea (glomerata)* Marabou Thorn and seeds, *Dinophora spenneroides, Machaerium lunatum (Drepanocarpus lunatus)* (Nkako), *Drypetes floribunda* (Bedibesa), *D. gilgiana* (Katrikanini), *D. ivorensis, Ectadiopsis oblongifolia, Ehretia cymosa* (Okosua), *Monanthotaxis foliosa (Enneastemon foliosus)* (Ntetekon), *E. vogelii* (Anmada), *Eudenia eminens* (Dinsinkoro) and the seeds, and *Euclinia longiflora* (Anyofon-bokowa).

Many *Ficus* species have edible figs — these are *F. barteri, F. capensis* (Nwadua), (the present name, which is conserved, has been used instead of *F. sur* the revised one), *F. capreifolia, F. congensis, F. elegans, F. eriobotryoides, F. glumosa* var. *glaberrima* (Galinziela), *F. sycomorus (gnaphalocarpa)* (Kankanga), *F. ingens* var. *igens* (Kunkwiya), *F. iteophylla* (Kwonkwia), *F. platyphylla* Gutta-Perch Tree, *F. polita* (Blohunyi), *F. umbellata* (Gyedua), *F. urceolaris, F. vallis-choudae* (Aloma-bli), *F. vogeliana* (Opanto), and *F. lutea (vogelii)* (Ofonto). *Grewia* species with edible fruits include *G. barteri* eaten with *Parkia* fruits (Daudawa), *G. bicolor, G. carpinifolia* (Ntanta) (fruits of all three varieties, *carpinifolia, rowlandii,* and *hierniana,* are edible), *G. venusta (mollis), G. mollis (pubescens)* (Yualega), and *G. villosa;* fruits of *Haematostaphis barteri* Blood Plum and the kernels, *Quassia undulata (Hannoa undulata)* (Kunmuni), *Harungana madagascariensis* (Okosoa), *Heinsia crinata* Bush Apple, *Heisteria parvifolia,* (Sikakyia), *Hexalobus crispiflorus* (Etwa prada), *H. monopetalus* var. *monopetalus, H. monopetalus* var. *parvifolius, Hoslundia opposita* (Asifuaka), *Hunteria eburnea* (Kanwenakoa), the young fruits of *Hymenocardia acida* (Sabrakyie), *Irvingia gabonensis* Wild Mango (Abesebuo) sometimes cultivated for the seed-oil, *Ixora brachypoda, Jasminum dichotomum* (Krampa), *Ancylobotrys amoena* eaten in Sudan, *Lannea acida* (Kuntunkuri) eaten in northern Ghana, *L. kerstingii* (Kobewu), *L. microcarpa, L. velutina* (Sinsa), *L. welwitschii* (Kumanini), *Lantana trifolia, Lecaniodiscus cupanioides* (Dwindwera), *Leea guineensis* (Okatakyi), *Macaranga heterophylla* (Opamkokoo), *Maerua crassifolia* eaten in Mauritania, *Maesobotrya barteri* var. *sparsiflora* (Apotrewa) also used for fruit tarts and jam, *Maesopsis eminii* (Onwam-dua) also contains an oil, *Mammea africana* African Mammy Apple (Bompagya) and the oily seeds, *Maytenus senegalensis* (Kumakuafo), *Microdesmis puberula* (Fema), *Monodora tenuifolia* (Motokuradua) eaten by children, *Morelia senegalensis* eaten in Nigeria, young pods of *Moringa oleifera* Horse-Radish Tree, Oil of Ben Tree and the mature, roasted seeds; *Morus mesogygia* (Wonton) which taste like white grapes, *Musanga cecropioides* Umbrella Tree (Dwumma); *Mussaenda elegans* (Damaram), *Myrianthus libericus* (Anyankom-nini), *M. serratus* (Bangama), *Nauclea diderrichii* (Kusia), *N. latifolia* African Peach (Sukisia); *N. pobeguinii* (Sukusia), *Ochna afzelii* (Okoli Awotso), *Olax subscorpioides* (Ahoohenedua), *Omphalocarpum procerum* (Gyatofo-Akongua) used in palm soup and also eaten by elephants and other mammals, *Oncoba spinosa* Snuffbox Tree (Asratoa); *Ongokea gore* (Bodwe) and the seeds which are eaten in small numbers in Central African Republic for their purging property, *Opuntia* species, and *Campylospermum flavum* (*Ouratea flava*) (Epebegai); the fruits of *Oxyanthus tubiflorus* eaten in Sierra Leone and *Pachystela brevipes* (Aframsua)

Parinari species with edible fruits include *P. congensis* (Tulingi) eaten in Congo, *Neocarya macrophylla* (*P. macrophylla*) (Nya), and *Maranthes polyandra* (*P. polyandra*) (Abrabesi) which is scarcely edible. The *Parkia* is an important fruit tree in Sudan and Guinea savanna woodland countries of West Africa, the species are *P. bicolor* (Asoma), *P. biglobosa* (Duaga), and *P. clappertoniana* West African Locust Bean (Daudawa). Other edible fruits are *Paullinia pinnata* (Toa-ntini) pulp sometimes eaten, *Peddiea fischeri* eaten in Guinea, *Margaritaria discoidea* (*Phyllanthus discoideus*) (Pepea) eaten raw in West Africa but cooked with coconuts in East Africa, *P. muellerianus* (Potopoleboblo) and the leaves, *P. reticulatus* var. *reticulatus* and *P. reticulatus* var. *glaber* (Awobe) eaten in scarcity and also by stock with the roots; the pods of *Piliostigma reticulatum* and *P. thonningii* (Opitipata) eaten or pounded to make a drink in Chad; the fruits of *Pseudospondias microcarpa* var. *microcarpa* (Katawani) eaten, with the seeds sometimes, in Congo, *Rhoicissus revoilii*, *Ritchiea reflexa* (Aayerebi), *Rutidea glabra* (Nserewedua), *Saba senegalensis* (Sono-nantin) sometimes sold in markets, *Sabicea africana* (Adwokule), *S. vogelii* eaten in Sierra Leone, *Sacoglottis gabonensis* (Tiabutuo), *Santaloides afzelii* (Humatarakwa) also eaten by birds, *Santiria trimera* sometimes sold in markets in Sierra Leone and oily seeds eaten in Liberia, *Scaphopetalum amoenum* (Nsoto) eaten in Liberia, *Sclerocarya birrea* (Nanogba): the fruit-juice makes an agreeable drink and oily kernels also edible; *Scytopetalum tieghemii* (Aprim) also eaten by monkeys, *Securinega virosa* (Nkanaa) eaten by children and birds, the young fruits of *Sesbania grandiflora* and *S. sesban* Egyptian Sesban, (Tingkwanga) eaten in Somaliland, *Smeathmannia pubescens* (Turunnua), *Sorindeia juglandifolia* eaten in Sierra Leone but rather astringent, *S. warneckei* (Akpokpoe), the fruit pulp of *Strychnos spinosa* Kaffir Orange (Akankoa); *S. innocua* subsp. *innocua* var. *pubescens* (Kampoye), and of the introduced *S. nux-vomica* Nux Vomica, Poison Nut. The nuts are, however, poisonous as the name suggests.

Syzygium species with edible fruits include *S. cumini* Java Plum, Jambolan; *S. guineense* var. *guineense* (Sunya) also eaten by monkeys and birds, *S. guineense* var. *littorale* (Avunle) dry ground varieties said to be larger, more pulpy and edible, and suitable for making beverages while other varieties are inedible, *S. guineense* var. *macrocarpum* (Kultia), and *S. owariensis*; fruits of *Teclea afzelii*,

2. FOOD AND FODDER

Tiliacora dielsiana (Kadze), *Trema orientalis* (*guineensis*) (Sesea) eaten in Congo, *Trichilia emetica* subsp. *suberosa* (*roka*) (Kisiga), *Trichoscypha arborea* (Anaku) sweet — even eaten when green in Liberia, *T. chevalieri*, and *T. oba* which is probably edible, *Tristemma hirtum*, *T. incompletum* (Anidan), *Uapaca corbisieri* (*esculenta*) (Kuntammiri), *U. guineensis* Sugar Plum (Kuntan) the pulp of the closed forest form said to be sweeter and sold in the markets; *U. heudelotii* (Kuntan-akoa) and *U. togoensis*; (Dzogbedzro); the other wild fruits are *Vitex ferruginea*, *V. fosteri* (Otwentorowa), *V. grandifolia* (Supowa), *V. micrantha* (Nyamele-buruma), *V. rivularis* (Bli), *V. simplicifolia* (Abisa), *Ximenia americana* Wild Olive, *Zanha golungensis*, *Tacca leontopetaloides*, *Zygotritonia crocea* eaten in Guinea, *Asparagus racemosus*, the fruit pulp of *Aframomum daniellii*, *A. sulcatum*, *Sarcophrynium brachystachys* (*Koto-haban*), and the seed pulp of *Trachyphrynium braunianum*; the seeds of *Sterculia rhinopetala* Sterculia Brown (Wawabimma); *Manniophyton fulvum* (Hunhun), *Pterocarpus santalinoides* (Hote), *Balanites wilsoniana* (Kurobow), *Panda oleosa* (Kokroboba), *Prosopis africana* (Sanga) used in northern Nigeria like *Parkia* (Daudawa), *Ricinodendron heudelotii* (Wamma) roasted for food in the Ivory Coast, *Heritiera utilis* (*Tarrietia utilis*) (Nyankom), *Crescentia cujete* Calabash Tree, and the fruits and kernels of *Icacina olivaeformis* (*senegalensis*) False Yam in scarcity. (See Chapter 1 for more wild fruits.)

In addition to the Oil Palm and the Coconut Palm, *Borassus aethiopum* Fan Palm, *Hyphaene thebaica* Dum Palm, and *Phoenix reclinata* Wild Date Palm have edible fruits or nuts. *P. dactylifera* Date Palm, an exotic, grows in some botanical gardens in our area.

Salt Substitutes

In the absence of common salt the following have been substituted: the wood-ashes (impure carbonate of soda) or potash of *Acacia polyacantha* subsp. *campylacantha* African Catechu, *Ceiba pentandra* Silk Cotton Tree, *Ficus mucuso* (Bambra), *Macaranga heterophylla* (Opamkokoo), *Maytenus senegalensis* (Kumakuafo), *Mimosa pigra* (Kwedi), *Maranthes polyandra* (*Parinari polyandra*) (Abrabesi), *Phyllanthus muellerianus* (Potopoleboblo), *Margaritaria discoidea* (*P. discoideus*) (Pepea), *Ricinodendron heudelotii* (Wamma), *Sterculia rhinopetala* Sterculia Brown (Wawabimma); *S. tragacantha* African Tragacanth (Sofo); *Vernonia colorata*, *V. conferta* (Flakwa), *Voacanga africana* (Ofuruma), and *V. thouarsii*. The root ashes of *Calamus deeratus* (Demmere), *Carica papaya* Pawpaw, *Grewia mollis* (Yualega), and *Lippia multiflora* Gambian Tea Bush (Saa-nunum). Leaf-ashes of *Avicennia africana* Mangrove with roots, *Cocos nucifera* Coconut Palm, *Elaeis guineensis* Oil Palm with the flowering stalks and *Jatropha curcas* Physic Nut with the branches; also *Psychotria obscura* with the branches. The pod-ashes of *Calpocalyx brevibracteatus* (Atrotre), *Pentaclethra macrophylla* Oil Bean Tree, *Piliostigma thonningii* (Opitipata), *Pterocarpus santalinoides* (Hote), and *Tetrapleura tetraptera* (Prekese). Ashes of the whole plant of *Sabicea africana* (Adwokule); the pericarp ashes of *Caloncoba echinata* (Gorli); the ashes from the stipes of *Hyphaene thebaica* Dum Palm; the ashes of freshly felled trees of *Musanga cecropioides* Umbrella Tree (Dwumma), and the ashes of the fronds of *Raphia hookeri* Wine Palm. The ashes of water plants like *Limnophyton obtusifolium*, *Pistia stratiotes* Water Lettuce, *Hygrophila spinosa*, *H. auriculata* (Eyitro) (sometimes cultivated for the purpose), and *Typha domingensis* (*australis*) Bulrush are used as salt substitutes. Stem ashes of *Cyrtosperma senegalense* and *Sesamum indicum* Sesame, Beniseed are a salt substitute, as are the ashes of the sedges *Cyperus haspan* and *Fuirena umbellata*, and the grass *Echinochloa pyramidalis* Antelope Grass. *Nelsonia canescens* is slightly acidic to the taste and is used when fresh as a salt substitute.

Beverages

Plant beverages include tea, coffee, wine, alcoholic drinks, intoxicants, and sweet beverages.

TEA SUBSTITUTES

Camellia sinensis Tea is grown in Kenya, Uganda, Tanzania, Zambia, and Malawi in East Africa and in Zaire. In Ghana it is cultivated only on Mt Gemi, near Amedzofe in the Volta Region. Tea has been grown on Mt Tonkui in the Ivory Coast experimentally to ascertain its suitability to the climate. It is also cultivated on the Mambilla Plateau in Nigeria and in Cameroun. The bulk of the requirement for West Africa is imported. As a substitute the following plants are used: the young leaves of *Cassia mimosoides* (Langrinduo) *C. nigricans*, *Pulicaria crispa*, *Senecio biafrae* (*Crassocephalum biafrae*), *Stachytarpheta cayennensis* Brazilian Tea, *Hyptis suaveolens* Bush Tea-Bush, *Ocimum* species, and the leaves of *Citrus aurantiifolia* Lime or any of the *Citrus* species. The leaves of *Lippia multiflora* Gambian Tea Bush (Saa-nunum); *Erythroxylum coca* Cocaine Plant; leaves or bark of *Cinnamomum zeylanicum* Cinnamon for scenting tea, also the flowers of *Jasminum sambac*; the leaves of *Leonotis nepetifolia* var. *africana* and *L. nepetifolia* var. *nepetifolia*, *Cymbopogon citratus* Lemon Grass, and the dried leaves and roots of *Launaea taraxacifolia* (*Lactuca taraxacifolia*) Wild Lettuce.

COFFEE SUBSTITUTES (see Chapter 3)

WINE

The following palm trees are tapped for wine: *Borassus aethiopum* Fan Palm, *Cocos nucifera* Coconut Palm, *Hyphaene thebaica* Dum Palm, *Phoenix reclinata* Wild Date Palm, *Raphia* species Wine Palm, (Adoka), and *Elaeis guineensis* Oil Palm. The last two are the most important source of palm wine. Wine is also brewed from *Zea mays* Maize, *Sorghum bicolor* Guinea Corn, and *Pennisetum americanum* Millet.

2. FOOD AND FODDER

ALCOHOLIC BEVERAGES AND LIQUEURS

The following plants are used in the preparation of alcoholic drinks: the fruits of *Anacardium occidentale* Cashew Nut by the Krobos in Ghana, the fruits of *Annona muricata* Sour Sop, *Antrocaryon micraster* (Aprokuma), *Balanites aegyptiaca* Desert Date, *Ficus sycomorus* (*gnaphalocarpa*) (Kankanga), *Landolphia owariensis* White Rubber Vine, *Lannea acida* (Kuntunkuri), *Ongokea gore* (Bodwe), *Parinari excelsa* Guinea Plum, (Ofam), and the fruits of *Parkia biglobosa* (Duaga). The rest are the fruits of *Aframomum melegueta* Melegueta, Guinea Grains; *Sclerocarya birrea* (Nanogba), *Strychnos spinosa* Kaffir Orange (Akankoa); *Vitex grandifolia* (Supowa), *Ximenia americana* Wild Olive for brewing beer, and the fruits of *Ziziphus mauritiana* Indian Jujube.

INTOXICANTS

The following plant parts are used as intoxicants: the bark of *Corynanthe pachyceras* (Pamprama), *Cremaspora triflora* (Otu), *Musanga cecropioides* Umbrella Tree (Dwumma), and the whole plant of *Pseudospondias microcarpa* var. *microcarpa* (Katawani); the rest are the leaves of *Morinda morindoides*, the root-bark of *Alchornea floribunda*, and the stems of *Triclisia patens* added to palm wine to increase the potency. The local gin, *akpeteshi*, is normally distilled from palm wine but may also be distilled from sugar-cane juice or pineapple juice. Triple-distilled *akpeteshi* is bottled under various trade names in Ghana (see also Chapter 8).

SWEET BEVERAGES

The fruits of *Anacardium occidentale* Cashew Tree as a refreshing, thirst-quenching drink; the fruit-pulp of *Adansonia digitata* Baobab stirred up in water, boiled, and allowed to cool; the fruit-juice of *Annona muricata* Sour Sop with water and sugar; fruits of *Balanites aegyptiaca* Desert Date macerated in water; immature fruits of *Bombax buonopozense* Red-Flowered Silk Cotton Tree used as beverage; seeds of *Borassus aethiopum* Fan Palm contain refreshing liquid; the fruits of *Caloncoba echinata* (Gorli) used to prepare a sweet drink by the Mendes of Sierra Leone; the nuts of *Cola nitida* Bitter Cola (Bese) for non-alcoholic, soft drinks like Coca-Cola; macerated fruit pulp of *Dialium guineense* Velvet Tamarind in cold water as a refreshing drink; the fruit infusion of *Gardenia erubescens* (Dasuli) used to prepare a non-fermented drink; the swollen calyx of *Hibiscus subdariffa* Red Sorrel or Roselle; fruit juice of *Opuntia* species with sugar as a drink; the fruit-pulp of *Parkia clappertoniana* West African Locust Bean macerated in water as drink; pods of *Piliostigma thonningii* (Opitipata) and *P. reticulata* pounded and boiled as drink; the fruit-juice of *Sclerocarya birrea* (Nanogba) as a drink; the fresh fruit of *Spondias mombin* Hog Plum, Ashanti Plum as drink; the berries of *Syzygium guineense* var. *littorale* (Avunle) as beverage; the seeds of *Treculia africana* var. *africana* African Bread Fruit as drink; the seed pulp of *Vitex doniana* Black Plum, (Afua) as beverage; the fruit of *Ziziphus mauritiana* Indian Jujube in water; the roots of *Combretum ghasalense* (Atena) used in making a beverage and the fruit-juice of *Citrus* species (see FRUITS above).

Spices and Flavourings

These season, preserve, and give aroma to food and render it more tasty.

SPICES AND SEASONERS

The spices normally used in everyday cooking are onions, shallots, peppers, ginger, etc. Melegueta, West African Black Pepper, Black Pepper, Ethiopia Pepper (Hwenetia), and *Xylopia parviflora* (Gyambobre) are also used. Others are *Combretum racemosum* leaves for seasoning soup; the fruit pulp of *Parkia biglobosa* (Duaga) and *P. clappertoniana* West African Locust Bean; the rhizomes of *Cucurma domestica* Turmeric and *Kaempferia aethiopica*; the seeds of *Monodora myristica* (Ayerewamba), *M. brevipes* (Abotokuradua), and *M. tenuifolia* (Motokuradua); *Myristica fragrans* Nutmeg; also the seeds of *Zanthoxylum xanthoxyloides* (*Fagara zanthoxyloides*) Candle Wood and the bark extract or the fleshy, dried roots of *Moringa oleifera* Horse-Radish Tree, Oil of Ben Tree. *Eugenia caryophyllata* Clove, a native of Indonesia, has been introduced to Benin and some other parts of West Africa.

FLAVOURING

Food flavourings include *Ocimum basilicum*, *O. canum*, and *Cucurma domestica* Turmeric, the main constituent of curry powder. *Ocimum* species are usually cultivated near homes for the purpose. Common food flavourings in northern Ghana are *Parkia* species (also for seasoning; see above) and *Tetrapleura tetraptera* (Prekese) in the forest regions, particularly among the Ashantis of Ghana. Others are *Aframomum melegueta* Melegueta, Guinea Grains (also as spice); *Hyptis suaveolens* Bush Tea-Bush, *Lantana trifolia* (aromatic leaves); the leaves and bark of *Cinnamomum zeylanicum* Cinnamon; the roots of *Cochlospermum planchonii* and *C. tinctorium* (Kokrosabia) to flavour soup in Nigeria; the juice of the sedges *Kylinga erecta*, *K. pumila*, *K. squamulata*, and *K. tenuifolia* for flavouring food and medicine; *Cymbopogon citratus* Lemon Grass and *C. giganteus* var. *giganteus*; the roots of *Carissa edulis* (Botsu) for food and drink; the pericarp of *Raphia hookeri* Wine Palm; the seeds of *Beilschmiedia mannii* Spicy Cedar; the young leaves of *Anacardium occidentale* Cashew Nut; the leaves of *Blighia welwitschii* (Akyekobiri) for soups, *Tephrosia linearis* and *T. purpurea* for milk and Guinea-corn pap, and *Cordyline fruticosa* leaves, a native of Java, for rice.

Edible Oils

In addition to palm oil, palm kernel oil, coconut oil, groundnut oil, and shea

2. FOOD AND FODDER

butter, edible oils are obtainable from *Aeglopsis chevalieri* (Kokoro), *Afraegle paniculata* (Obuobi), *Balanites aegyptiaca* Desert Date, *Blighia sapida* (Akye), *Tieghemella heckelii* (Baku), *Treculia africana* var. *africana* African Bread Fruit, *Trichilia emetica* subsp. *suberosa* (*roka*) (Kisiga), *Moringa oleifera* Horse-Radish Tree, Oil of Ben Tree, *Sterculia foetida* (both introduced), *Adansonia digitata* Baobab, *Anacardium occidentale* Cashew Nut, *Antrocaryon micraster* (Aprokuma), *Beilschmiedia mannii* Spicy Cedar, and *Canarium schweinfurthii* Incense Tree. Others are *Ceiba pentandra* Silk Cotton Tree, *Chrysophyllum delevoyi* White Star Apple, *Coula edulis* African Walnut (Bodwue); *Gossypium arboreum* Cotton, *Irvingia gabonensis* Wild Mango (Abesebuo); *Klainedoxa gabonensis* var. *oblongifolia* (Kroma), *Lophira alata* Red Ironwood (Kaku); *L. lanceolata, Maesopsis eminii* (Onwamdua), *Mammea africana* African Mammy Apple (Bompagya); *Phyllocosmus africanus* (*Ochthocosmus africanus*), and *Panda oleosa* (Kokroboba). The rest are *Lagenaria siceraria* Calabash, Bottle Gourd, White Pumpkin; *Colocynthis lanatus* (*vulgaris*) Watermelon, *Pentaclethra macrophylla* Oil Bean Tree (Ataa), *Pentadesma butyraceum* (Abotoasebie), *Telfairea occidentalis* Fluted Pumpkin, (Krobonko); *Terminalia catappa* Indian Almond, *Cucumeropsis edulis* (Agushie), *Cucumis melo* var. *agrestis* Cucumber, *Calophyllum inophyllum* Alexandra Laurel, *Ximenia americana* Wild Olive, and *Raphia hookeri* Wine Palm. The seed of *Sesamum indicum* Sesame, Beniseed contains 50–57 per cent oil suitable for cooking and margarine manufacture. (For essential oils and oils for soap manufacture see Chapter 9.)

Fodder

Apart from roots and tubers, fodder consists mainly of the leaves of small trees, shrubs, and grasses. McGinnies, Goldman, and Paylore (1971) noted that shrubs are of a higher quality than grass and generally contribute a greater amount to the grazing animal's diet because of greater abundance. They also noted that browse species are higher in protein, phosphorus, and carotene (vitamin A). In contrast, grasses are superior to shrubs only in energy-yielding qualities (metabolizable energy). Suitable plants for fodder are the leaves of *Acacia albida* (Gozanga) (general), *A. dudgeoni* (Gosei) (general), *A. farnesiana* (sheep), *A. gourmaensis* (Gowuraga), *A. hockii* Shittim Wood, *A. nilotica* var. *adansonii* Egyptian Mimosa (general), *A. nilotica* var. *tomentosa, A. kamerunensis* (*pennata*) (Oguaben), *A. senegal* Gum Arabic, *A. sieberiana* var. *villosa* (Kulgo) and the pods, *Adansonia digitata* Baobab (horses), *Afzelia africana* (Papao) (general), *Albizia lebbeck* East Indian Walnut, *Annona senegalensis* var. *senegalensis* Wild Custard Apple, *Anogeissus leiocarpus* (Sakanee) (generally considered inferior), *Artocarpus communis* Bread Fruit and the fruits, *Azadirachta indica* Neem Tree (cattle), *Balanites aegyptiaca* Desert Date (general), *Bambusa vulgaris* Green and Yellow Striped Bamboo, *Bauhinia rufescens* (Jinkiliza), *Bombax buonopozense* Red-Flowered Silk Cotton Tree (goats), *Boscia senegalensis* (Dila) and the fruits, *Burkea africana* (Pinimo),

Cadaba farinosa, Cajanus cajan Pigeonpea, *Capparis polymorpha* (Sansangwa) (camels), and *Carissa edulis* (Botsu); the leaves of *Cassia mimosoides* (Langrinduo), *C. siamea* (poisonous to pigs), *C. obtusifolia* (*tora*) Foetid Cassia (ostriches), *Ceiba pentandra* Silk Cotton Tree (goats), *Celtis africana* (Yisa), *C. integrifolia* Nettle Tree (Samparanga) (cattle); *Chrozophora senegalensis* (camels), *Cochlospermum tinctorium* (Kokrosabia), *Cola laurifolia* Laurel-Leaved Cola, *Commicarpus plumbagineus, Corchorus olitorius* Jew's Mallow, Long-Fruited Jute; *C. tridens* (Sanvoa), *Cordia rothii, Crateva adansonii* (*religiosa*) (Chelum Punga), *Cynometra vogelii* (Boboe), *Daniellia oliveri* African Copaiba Balsam Tree (Sanya), *Dendrocalamus strictus* Male Bamboo (horses, buffaloes, and elephants). *Desmodium velutinum* (Koheni-koko) (horses), *Detarium senegalense* Tallow Tree (kernels beaten into cakes as cattle feed), *Entada abyssinica* (Sankasaa), *E. africana* (Kaboya), *Erythrina senegalensis* Coral Flowers, *Dichrostachys cinerea* (*glomerata*) Marabou Thorn and the pods, *Diospyros mespiliformis* West African Ebony, *Dodonaea viscosa* Switch Sorrel (Fomitsi) (cattle and camels, and the roots for increasing milk production in cattle); the fallen leaves of *Euphorbia balsamifera* Balsam Spurge (Aguwa) (browsed by sheep), and the leaves of *Zanthoxylum xanthoxyloides* (*Fagara zanthoxyloides*) Candle Wood (sheep).

The list of plants suitable for fodder includes the leaves of *Ficus sycomorus* (*gnaphalocarpa*) (Kankanga) and the fruits (sheep and goats), *F. ingens* var. *ingens* (Kunkwiya), *F. iteophylla* (Kwonkwia) (goats, especially young leaves), *Gliricidia sepium* Mother of Cocoa (goats), *Gmelina arborea, Griffonia simplicifolia* (Kagya) (said to promote multiple birth — sheep and goats), *Guiera senegalensis* (camels), *Hibiscus tiliaceus* (Mwohwea) (contains 51 per cent starch and 12 per cent nitrogenous matter dry weight — cattle) and the leaves of *Hoslundia opposita* (Asifuaka); the leaves of *Hyphaene thebaica* Dum Palm (cattle; outer fruit-husks, donkeys; shavings of seeds, horses and cattle; and roasted kernels, sheep), *Clitoria ternatea* Blue Pea (sheep and goats), *Indigofera spicata, Khaya senegalensis* Dry-Zone Mahogany (cattle and camels), *Laguncularia racemosa* White Button Wood (Abin); *Leucaena leucocephala* (*glauca*) Leucaena, Wild Tamarind (general; boiled seeds, cattle) and the seeds of *Irvingia gabonensis* (Abesebuo) as cattle-cake; the leaves of *Lonchocarpus laxiflorus* (Nalenga) (goats), *Maerua angolensis* (Pugodigo) (sheep and goats), *Manihot esculenta* Cassava and root peelings, *Maytenus senegalensis* (Kumakuafo), *Melia azedarach* Persian Lilac, *Mikania chevalieri* Climbing Hemp Weed (cattle), *Millettia thonningii* (Sante) (sheep), *Millingtonia hortensis* (an introduced tree, goats), *Mitragyna inermis* (Kukyafie), *Moringa oleifera* Horse-Radish Tree, Oil of Ben Tree (general); *Oxytenanthera abyssinica* Bamboo (especially the young leaves, also the seedlings), *Parkia clappertoniana* West African Locust Bean and the fruits, *Peltophorum pterocarpum* Copper Pod (the pods relished by cattle), and the roots and fruits of *Phyllanthus reticulatus* var. *reticulatus* and *P. reticulatus* var. *glaber* (Awobe). The list of fodder plants includes the pods of *Piliostigma reticulatum* and *P. thonningii* (Opitipata);

2. FOOD AND FODDER

leaves of *Pithecellobium dulce* Madras Thorn (general), *Prosopis africana* (Sanga), *Pterocarpus erinaceus* Senegal Rose Wood Tree, African Kino; *Samanea saman* Rain Tree and the pods, *Sclerocarya birrea* (Nanogba), *Securidaca longepedunculata* (Kpaliga), *Sesbania pachycarpa* (*bispinosa*), *S. grandiflora*, *S. sesban* Egyptian Sesban (Tingkwanga), *Spondias mombin* Hog Plum, Ashanti Plum; *Stylosanthes fruticosa* (*mucronata*), *Tecomaria capensis* Red Tecoma, *Tephrosia bracteolata* (horses), *T. linearis* (horses), *T. platycarpa* (Buruguni), *Trichilia emetica* subsp. *suberosa* (*roka*) (Kisiga), *Urena lobata* Congo Jute, *Vitex doniana* Black Plum, (Afua); *Ziziphus abyssinica* (Laruklukor) (relished by sheep and goats, next after *Acacia seyal*), *Z. mauritiana* Indian Jujube, *Z. mucronata* Buffalo Thorn (cattle), and *Z. spina-christi* var. *microphylla* (camels and cattle — fruits, sheep and goats).

Herbaceous fodder plants include *Amaranthus spinosus* Spiny or Prickly Amaranth, *A. viridis* Wild or Green Amaranth, *Celosia argentea* (Nkyewodue), *Blepharis linariifolia* (all stock), *Boerhavia diffusa* and *B. repens* Hogweed, *Alysicarpus rugosus* (best harvested after fruiting and dried as hay as young ones cause mucous diarrhoea), *Desmodium triflorum*, *Lablab niger* Lablab Bean and the fruit, *Struchium sparganophora* (cattle, in Java), *Spilanthes filicaulis* Para or Brazil Cress (cattle and horses in Java), *Ipomoea mauritiana* (Dinsinkoro) (cattle), *Nelsonia canescens*, *Stachytarpheta cayennensis* Brazilian Tea (cattle, in Java), *Hyptis suaveolens* Bush Tea-Bush, and *Commelina* species; the following plants in the family Compositae are relished by hares and rabbits: *Melanthera scandens*, *M. elliptica*, *Aspilia africana* Haemorrhage Plant, and *Synedrella nodiflora*. A few sedges serve as fodder — these are *Cyperus rotundus* Nut Grass, *C. sphacelatus* (sheep and goats), *Fuirena umbellata*, *Kyllinga erecta*, *K. pumila*, *K. squamulata*, *K. tenuifolia*, and *K. umbellata*.

Grasses are an important source of fodder and hay. The following provide good pastures for grazing: *Brachiaria brizantha* and hay, *B. distichophylla*, *B. xantholeuca*, *Axonopus compressus*, and *Cynodon dactylon* Bermuda or Bahama Grass; others are *B. jubata*, *B. falcifera*, *Digitaria debilis*, *D. horizontalis*, and *D. ciliaris* — all three suitable for horses and cattle and for hay; *Eleusine indica* suitable for all stock and for hay; *Eriochloa fatmensis* (*nubica*), *Hackelochloa granularis* (in Bauchi, horses), *Leersia hexandra* Rice Grass (in damp places), *Melinis minutiflora* var. *minutiflora* Stink Grass and for hay, *Pennisetum polystachion* Mission Grass and for hay, *Setaria barbata* and *S. pallide-fusca* Bristle Foxtail Grass, Cat's Tail Grass (both drought resistant); *Rottboellia exaltata* and for hay (sometimes cultivated), *Stenotaphrum secundatum* Baffalo Grass, Pimento Grass (often cultivated), and *S. dimidiatum*.

The list of useful grasses for fodder include *Andropogon gayanus* varieties *gayanus*, *tridentatus*, *squamulatus*, and *bisquamulatus* when young and up to flowering; *A. schrensis* when young; *A. pseudapricus*; *A. tectorum* when young, especially for horses; *A. curvifolius*; *A. incanellus* and *A. fastigiatus* when young; *Acroceras amplectens* and *A. zizanioides* and for hay; *Allopteropsis paniculata* and *A. semialata*; *Aristida adscensionis* (13 per cent protein and only 2 per cent free silica ashes on analysis); *A. mutabilis* and *A. sieberiana* for camels; *A. hordeacea, A. kerstingii,* and *A. recta*; *Axonopus flexuosus* (often cultivated); *Pennisetum unisetum* (*Beckeropsis uniseta*) when young and *P. laxior* (*B. laxior*); *Pseudobrachiaria deflexa* (*Brachiaria deflexa*) (drought resistant and suitable for dry regions); *Brachiaria brachylopha, B. distachyoides, B. plantaginea, B. stigmatisata,* and *Urochloa mutica* (*B. mutica*) Para, Mauritius, or Water Grass (often cultivated); *Cenchrus biflorus* and *C. ciliaris* (by all stock); *C. echinatus, C. setigerus, Centotheca lappacea, Chasmopodium caudatum* (horses and cattle, but is reported toxic to stock in Futa-Jallon), *Chloris barbata, C. gayana, C. pilosa, C. pycnotrix* and hay, *C. prieuri, Coix lacryma-jobi* Job's Tears, and *Cymbopogon giganteus* var. *giganteus* when young; *Dactyloctenium aegyptium* and hay, *Digitaria exilis* Hungry Rice, *D. gayana* (up to seedling time), *Diheteropogon amplectens* var. *catangensis,* and *D. hagerupii*; *Echinochloa colona* Jungle Rice, Shama Millet; *E. pyramidalis* Antelope Grass (excellent, especially when young — cultivated as hay), *E. stagnina* and hay (cattle and horses), and *E. crus-pavonis*; *Elionurus elegans, E. hirtifolius,* and the following species of *Eragrostis*: *E. cilianensis* Stink Grass (not given to horses in northern Nigeria), *E. ciliaris, E. gangetica, E. pilosa,* and *E. tremula* (all four species make good hay), *E. aspera, E. atrovirens, E. barteri, E. blepharostachya, E. chalarothyrsos, E. cylindriflora, E. domingensis, E. egregia, E. namaquensis* var. *diplachnoides, E. namaquensis* var. *namaquensis, E. pobeguinii, E. scotelliana, E. squamata, E. turgida,* and *E. welwitschii*; others are *Eriochloa meyerana, Heteropogon contortus* Spear Grass (up to flowering only), *Hyperthelia dissoluta* (when young), *Hyparrhenia rufa* (when young or kept low by grazing), and the young stages of *H. cyanescens, H. familiaris, H. glabriuscula, H. involucrata* var. *breviseta, H. involucrata* var. *involucrata, H. mutica, H. nyassae, H. rudis, H. subplumosa,* and *H. welwitschii.*

Other useful grasses for fodder are *Jardinea congoensis* (all stock), *Leutothrium senegalense, Leersia drepanotrix, Loudetia annua,* and *L. hordeiformis* (both grazed by cattle when young), *L. arundinacea, L. kagerensis,* and *L. simplex* (when young and tender in all three), *Melinis effusa* and *M. tenuissima* (when young — also for hay); *Microchloa indica, M. kunthii, Monocymbium ceresiiforme* (when young), *Oplismenus burmannii,* and *O. hirtellus* (both when young); *Oryza barthii, O. longistaminata,* and *O. punctata* (when young in all three). *Panicum* species are a good source of fodder — *P. maximum* Guinea Grass being one of the best. It is sometimes cultivated both as green fodder and hay, and sold in the markets. Others are *P. anabaptistum, P. baumannii, P. brevifolium, P. comorense, P. congoense, P. dinklagei, P. griffonii, P. hochstetteri, P. laetum, P. laxum, P. lindleyanum, P. pansum, P. paucinode, P. parvifolium, P. phragmitoides, P. porphyrrhizos, P. praealtum, P. pubiglume, P. repens, P. subalbidum* (when young), *P. trichoides,* and *P. turgidum.*

Other species of grass useful as fodder are *Paspalidium geminatum, Paspalum conjugatum* Sour Grass (withstands drought, remaining green in the dry season), *P. vaginatum, P. scrobiculatum* (*orbiculare*) Bastard or Ditch Millet,

2. FOOD AND FODDER

P. dilatatum (introduced from America), *Pennisetum hordeoides* (when young), *P. pedicellatum* (cattle and horses, also useful as hay when cut before flowering), *P. purpureum* Elephant Grass (resists drought), and *P. subangustum*; *Setaria verticillata* Rough Bristle Grass (when cut before flowering), *S. chevalieri* Buffel Grass, *S. aurea*, and *S. sphacelata* Golden Timothy Grass, Rhodesian Timothy (when young); *Perotis patens*, *P. hildebrandtii*, *Rhynchelytrum villosum*, *R. repens*, *Rhytachne triaristata*, *Sacciolepis africana*, and *S. micrococca*. Several species of *Schizachyrium* are good fodder when young — especially for cattle; they include *S. brevifolium*, var. *brevifolium*, *S. exile*, *S. delicatum*, *S. maclaudii*, *S. nodulosum*, *S. platyphyllum*, *S. pulchellum*, *S. ruderale*, *S. sanguineum*, *S. schweinfurthii*, *S. rupestre*, and *S. urceolatum*. The following are also useful as fodder: *Schoenefeldia gracilis* (when young), *Sorghum arundinaceum* Kamerun Grass, *S. lanceolatum*, and *S. vogelianum*; *Sporobolus* species used as fodder are *S. africanus*, *S. festivus*, *S. microprotus*, *S. pyramidalis* Rat's Tail Grass (when young), *S. robustus*, and *S. sanguineus*. The rest are *Thelepogon elegans* (for coarse grazing), *Themeda triandra*, *Vetiveria nigritana* (when fresh, roots eaten by wart-hogs), *V. fulvibarbis*, *V. zizanioides* Vetiver (cattle, when young), and *Vossia cuspidata* (when young).

The following plants are good feed for fowl: *Croton lobatus*, *Boerhavia diffusa* and *B. repens* Hogweed (to stimulate egg laying), *Commelina forskalaei*, *C. erecta* subsp. *erecta*, *Sesbania pachycarpa* (*bispinosa*), and the seeds of *Hibiscus lunariifolius*. *Pistia stratiotes* Water Lettuce is used to feed ostriches in northern Nigeria and as pig feed in China.

Water plants are a source of food for fish, water snails, and other aquatic animals which are, in turn, eaten by man. Such plants are, therefore, a useful link in the food chain. In addition, aquatic vegetation provides shelter and breeding grounds for fish as well as oxygenating the water and absorbing compounds from the water (John, 1986). Water plants are thus economically useful. Aquatic plants eaten by fish are duckweeds; *Lemna perpusilla* (*paucicostata*), *Spirodela polyrhiza*, and *Wolffia arrhiza*. Other aquatic plants eaten by fish are *Azolla pinnata* var. *africana* (*africana*), *Ottelia ulvifolia*, *Salvinia nymphellula*, *Lagarosiphon hydrilloides*, *Potamogeton schweinfurthii*, *P. octandrus*, *Utricularia inflexa* var. *inflexa*, *U. inflexa* var. *stellaris*, *U. reflexa*, *U. subulata*, *U. baoulensis*, *U. spiralis* var. *tortilis*, and *U. gibba* subsp. *exoleta*. Useful for both decoration and food in domestic aquariums are *Ceratophyllum demersum*, *Vallisneria aethiopica*, *Anubias* species such as *A. minima*, and *Elodea canadensis* American Water Weed. Algal blooms occur in the Volta Lake in Ghana — the largest man-made lake S. of Sahara, constructed to generate electricity, but inland fishing is now an important industry there. Species recorded from the Afram arm alone include *Eudorina elegans*, *Volvox tertius*, *V. aureus*, *Ankistrodesmus falcatus*, *Closterium kutzingii*, *Staurastrum paradoxum*, *Scenedesmus serratus*, *Pediastrum duplex*, *Mongeotia scalaris*, *Spirogyra* species, and *Oedogonium* species. Others are *Melosira granulata*, *Stephanodiscus lantzschii*, *Synedra acus*, *S. falciculata*, *Fragillaria* species, *Microcystis aeruginosa*, *Oscillatoria princeps*,

O. curviceps, O. pimosa, O. tenuis, Lyngbya limnetica, Anabaena limnetia, and *Peradinium cinctum.* Seaweeds are also a source of food for fish, and for man in China. Common species include *Gracilaria dentata, Hypnea musciformis, H. flagelliformis, Gigartina acicularis,* and *Grateloupia filicina.*

THREE
Industrial or Cash Crops

The success of agriculture-dependent industries relies on continuous production at full capacity, which in turn depends on uninterrupted supplies of raw materials and efficient performance of the machinery. Cash crops as raw materials are the backbone of these industries. Several setbacks have contributed to a reduction in production of these raw materials and consquently of optimum production capacity of the industries concerned. These setbacks are due to technical, economic, logistical, or human failings; or a combination of these. In the case of *Theobroma cacao* Cocoa they have been caused in Ghana by:

1. the Aliens Compliance Order, which expelled many of farm-hands in the early 1970s, coupled with the steady drift of the youth to urban centres for a taste of city life, leaving the ageing fathers to cope with the farm work alone;
2. the depreciation of the currency and the unattractive government price;
3. cumbersome and difficult marketing procedures such as the abrogated chit system of payment;
4. the practice of cultivating food crops instead of cash crops for faster returns and spot cash sales;
5. the practice of smuggling produce across the borders to neighbouring countries;
6. the lack of transport from the hinterlands to the ports;
7. the problem of pests, the shortage of insecticides and spraying machines, or the lack of effective distribution and sale of these inputs directly to genuine farmers at approved prices;
8. the question of the land tenure system, the landless farmer, and permanent crops;
9. the incidence of widespread bush fires which extensively destroy large areas of cash crops, food farms, and forests.

A nation-wide rehabilitation programme to replant an estimated 120,000 hectares of burned cocoa farms has been launched, while the Akuafo Cheque System has replaced the chit system. The hardships consequent on this system were the main cause of the threat to cultivate food crops exclusively or to cut down on cash crops in favour of food crops. In practice, every cash crop farmer also cultivates food crops. A Chief Farmer said, 'We use the money from the cocoa to educate our children, and live on the income from plantain and melegueta.' Furthermore, cocoa is harvested only for a short season in the year, and payment for the crop was then considerably delayed; by growing several crops instead of one, the farmer can spread his income over the year. The landless farmer and the land tenure system varies slightly from place to place, but is usually centred on the tripartite system: the yield is divided into three parts — a part each to the landless farmer, the land and the land-owner. The land-owner also owns the crops. The farm work is done by the landless farmer alone.

Table 3.1 Cocoa, coffee and sheanut producer prices. Source: Policy Planning, Monitoring and Research Department of the Cocoa Board.

PRODUCER PRICES
(CEDIS/TONNE)

YEAR	COCOA	COFFEE	SHEANUT
1969/70	293.86	373.33	50.40
1970/71	293.86	373.33	50.40
1971/72	293.86	373.33	50.40
1972/73	367.32	373.33	56.00
1973/74	440.79	485.33	168.00
1974/75	550.98	634.67	196.00
1975/76	587.71	634.67	196.00
1976/77	734.64	933.33	193.60
1977/78	1333.33	1530.00	193.60
1978/79	2666.67	3570.00	800.00
1979/80	4000.00	3570.00	800.00
1980/81	4000.00	3570.00	1760.00
1981/82	12 000.00	8500.00	6400.00
1982/83	12 000.00	8500.00	6400.00
1983/84	20 000.00	20 400.00	11 520.00
1984/85	30 000.00	30 600.00	17 280.00
1985/86	56 600.00	62 260.00	27 200.00
1986/87	87 000.00	94 095.00	28 000.00
1987/88	150 000.00	156 800.00	30 800.00
1988/89	165 000.00	186 000.00	31 000.00

3. INDUSTRIAL OR CASH CROPS 61

Theobroma cacao

Cocoa, a native of Central America, is a tropical rainforest crop. With the introduction of the crop into West Africa at the end of the nineteenth century, a tremendous expansion in production occurred to reach 1.5 million tonnes per year. Over 70 per cent of world production now comes from the Ivory Coast, Ghana, Nigeria, and Cameroun, most of the remainder from the Americas and the Caribbean. Cocoa was introduced into Akyim in Ghana (the south-east corner of the country in Akwapim) in 1878 from Bioko (Fernando Po) by Opanyin Tetteh Quarshie, a blacksmith. In 1885 the first export of 54 kg (121 lb), worth £6.05, was made. The crop later spread to Ashanti, Brong Ahafo, and the remaining forest areas to give rise to the prosperous West African industry. Hall and Swaine (1981) report that it was adopted eagerly by farmers in the eastern part of the forest zone, and by 1911, Ghana was the leading producer of the crop — a position it held until 1978, thus remaining the ascendant cocoa-producing nation for exactly one hundred years. Between 1909 and 1923 annual cocoa exports rose from 20,000 to 200,000 tons. By 1965 production had increased to 557,000 tons; then there was a gradual decline to about 300,000 tons in 1978. Ghana is now the third major producing country, after Brazil and the neighbouring Ivory Coast, with an average production of 400,000 tons per annum, representing about 30 per cent of total world output. For the producer price per tonne see Table 3.1. The world market price for the beans (as at Sept. 1988) is $1,823 per tonne and for cocoa butter it is $3,207 per tonne.

The main producing areas in Ghana are the Eastern, Ashanti, Brong Ahafo, Central, Western, and the Volta Regions. It is estimated that there are between four and six million acres of farms under cultivation by approximately 480,000 peasant farmers throughout the growing areas. In addition, the Ghana Cocoa Board has developed on a scientific basis plantations of at least 2,000 acres each with high-yielding varieties at Maabang in Tepa District of Ashanti Region, Bokabo-Tumantu in the Sefwi Wiawso District of Western Region, and at Assin Nsuta in Assin Apemanim District of the Central Region. A total of 50,000 acres of cocoa plantations is initially planned to be established in all the growing regions of the country. These plantations are expected to produce an average of 1,000 lb an acre as against the 300 lb an acre produced by peasant farmers. The Amelonado cocoa used to supply about 98 per cent of the West African cocoa crop; but since it is the most susceptible variety to the Cocoa Swollen Shoot Virus (CSSV) disease, it has since been replaced by the F3 Amazon — a more resistant type but with similar flavour.

The cocoa industry is the backbone of Ghana's economy and it appears that it will be so in the forseeable future — accounting for about 70 per cent of foreign exchange earnings with the export of both processed and unprocessed beans to the United States, Britain, West Germany, the

Netherlands, the USSR, and East Germany. There are three processing factories in the country — the Cocoa Products Factory and West African Mills Limited, both at Takoradi, and the Cocoa Products Factory at Tema. The factories produce TAKSI, WAM, and PORTEM brands of cocoa butter, cocoa cake, cocoa liquor, cocoa powder, and two other by-products: cocoa residue and shells (see Table 3.2). These factories export 99 per cent of their products to the countries above.

The cocoa plant can be put to a number of uses, but the seed is economically the most important part. The sap around the seed is edible and sweet. The fermented beans when roasted are cracked to remove the shells and ground into cocoa mass from which cocoa powder and cocoa butter are produced. Chocolate is made by grinding sugar with the cocoa mass and adding extra cocoa butter. West African Amelonado and F3 Amazon types are the most suitable for chocolate making because of their mild flavour. The Cocoa Products Factory at Tema manufactures various kinds of Golden Tree Chocolates like bars, pebbles, and beads mainly for local consumption. Cocoa butter is useful in the cosmetic and pharmaceutical industries, and the powder is used to make a variety of drinks and beverages —hot or cold. In combination with milk, cocoa beverages are ideal for convalescence and give strength and energy. Cocoa is also used in the manufacture of wines and spirits. The Cocoa Research Institute of Nigeria (CRIN) has researched into the manufacture of cocoa wine, whisky, brandy, and vinegar using both powdered cocoa beans with sugar and water and also the juice or mucilage of the cocoa pod. The by-products include floor tiles, ceiling boards, and shoe polish. Are and Gwynne-Jones (1974) observe that such uses of cocoa could prove of considerable economic importance, especially in times of over-production and low world prices. Ainslie (1937) reported that as the seeds contain theobromine, cocoa is a stimulant to heart, kidney, and muscle, and a diuretic. Theobromine has been proved to be poisonous to cattle, sheep, pigs, and chickens. Locally, the dried pods are burned and the ashes, which contain about 30–40 per cent potassium, are used for manufacturing soft soap. Plantain peels, which are otherwise used for the same purpose, are thus available as fodder. Cocoa pod husks are used as a substitute for or with maize as the main ingredients for the preparation of animal feed, in effect making more maize available for human consumption. Cocoa pod meal compares favourably with that of cassava or groundnuts. Experiments by Alba and others with dairy cows fed with a ration containing 50 per cent cassava or groundnut meal demonstrated no difference in the total milk production. The digestibility of cocoa husk meal is, however, comparatively poor.

Recently the Cocoa Research Institute of Ghana (CRIG) has successfully researched into the use of cocoa husks for the manufacture of fertilizer. Cocoa husks contain about 40 per cent potassium and this could be used to cultivate both food crops and cocoa. Dr Adomako of the CRIG estimates that from the annual output of 400,000 tons of dry cocoa beans 636, 157 tons of dry

TABLE 3.2 Cocoa processing in Ghana (1984)

Factory	Grade of Cocoa Processed	Annual Tonnage Cocoa Processed	Annual Tonnage Cocoa Powder	Annual Tonnage Cocoa Butter	Finished Products				
					Butter	Liquor	Cake	Powder	Residue
TAKSI	I & II	18 000	180	9 000	35%	30%	30%	5%	-
PORTEN	I & II	10 000	2 000	5 000	35%	-	30%	2%	-
WAM	III	20 000	-	10 000	80%	-	-	-	20%
TOTAL		48 000	2 180	24 000					

cocoa husks is available to produce about 60,000 tons of ash for the manufacture of potash fertilizer. Cocoa pod husks also contain polysaccharides, among others pectin (6 to 12 per cent) which is used in large quantities for the manufacture of jams and jellies. Pectin enables the jam to set easily. Although the methoxyl group in cocoa husk pectin is low, and therefore inferior to the pectins of commerce, Dr Adomako states that low methoxyl pectins are being increasingly used in the presence of calcium salts to prepare jellies, jams, and other confectionery which have low sugar and acid content. The fresh juice or sweatings may be utilized for the production of wine or alcohol, but the traditional way of fermentation, the heap method, does not easily allow the collection of sweatings.

The cocoa crop is affected by several diseases. Swollen shoot of cocoa is a virus disease transmitted by mealybugs, the most important being *Planococcus njalensis*. The removal of diseased trees has proved an effective form of control. Blackpod of cocoa, or pod rot, is a serious fungal disease caused by *Phytophthora palmivora*. It is controlled by spraying with fungicides. Capsid bug disease, or *akate*, attacks the shoots of the crop and it is controlled by spraying affected trees with *Kum-akate*, Gammalin 20.

Coffea Species

Coffea arabica Arabian, Brazilian or Abyssinian Coffee, the coffee plant of commerce, is a native of the mountains of Abyssinia (now Ethiopia), thriving best at altitudes of 2,000–4,000 ft (600–1200 m). In tropical Africa coffee is an important foreign exchange earner in Uganda, Ethiopia, Kenya, Burundi, and Rwanda (all in East AFrica); Angola and Equatorial Guinea (in South-West Africa), and the Central African Republic. Important producing countries in West Africa are the Ivory Coast, Cameroun, Togo, and Guinea. The largest producing country, however, is Brazil which accounts for about half the world's production. It is not certain when Abyssinian coffee was introduced into West Africa; however it is the first introduced export crop of importance to be grown in the forest zone. Hall and Swaine (1981) report that annual exports from Ghana peaked at about 70 tons in the 1890s.

Coffee is grown in all the six forest regions of Ghana — Eastern, Ashanti, Brong Ahafo, Central, Western, and the Volta Regions — mainly by peasant farmers on small-scale farms averaging only about three acres. The Ghana Cocoa Board, which also develops, purchases, and exports coffee, has established plantations at Adomfe in the Ashanti Akim District of Ashanti Region; at Dormaa Akwamu in the Dormaa District of Brong Ahafo Region; and at Suhuma in the Sefwi Wiawso District of the Western Region. The plantations are the first stage of a 25,000 acre project to establish plantations in all coffee-growing regions of the country. Annual exports of 6,527 tons and 6,953 tons were recorded for 1964 and 1970 respectively. (For the producer price per tonne see Table 3.1). The dried, roasted, and ground seeds of

3. INDUSTRIAL OR CASH CROPS

the coffee plant make a stimulating, refreshing, and popular beverage. Local processing factories produce brands like Excelsior Coffee, Bongo Coffee, Flamingo Coffee, and St Louis Coffee mainly for home consumption. Nescafé, an instant coffee, is imported to supplement these.

The most important constituent of the seeds is caffeine, which acts on the central nervous system, the muscular tissues, and the kidneys. Ainslie (1937) observed that its action on the brain produces alertness and increased mental activity, facilitating physical performance of all kinds by increasing the total amount of work obtainable from the muscle. The pulse is accelerated and raises the blood pressure slightly. It is an important diuretic. Coffee is used as a cardiac stimulant in chronic nephrites. It is useful in cases of nervous headache and migraine and is often effective in relieving asthma.

Irvine (1961) reported that about ten indigenous tropical African coffee species have seeds which can be brewed and of these only three are cultivated.

Coffea canephora Rio Nunez or Robusta Coffee grows well at altitudes below 2,000 ft (600 m) in the forest area and is, therefore, better suited for cultivation in West Africa than *C. arabica,* and it is the species mainly cultivated in Ghana.

Coffea liberica var. **liberica** Liberian or Monrovia Coffee is indigenous around Monrovia and the Ivory Coast and is grown widely in the tropics. Though it requires less shade than other coffee plants, the beans are inferior to those of *C. arabica, C. canephora,* and *C. stenophylla.* Var. *dewevrei,* which is the grower's 'excelsa', is now more widely planted than 'liberica'.

Coffea stenophylla Sierra Leone (Upland) or Narrow-Leaves Coffee is a native of Lower Guinea and Sierra Leone where it has been cultivated for many years. It is best suited to hilly areas up to 2,000 ft (600 m) and requires fairly good shade and good rainfall.

'Arabusta' coffee is a new hybrid *arabica* × *canephora* developed in Ivory Coast.

It is likely that packets of roasted, ground coffee displayed for sale locally may be any of the above three species and not necessarily *C. arabica.* The coffee crop is attacked by several diseases including fruit-rot caused by the fungus *Trachysphaera fructigena* and leaf spot caused by *Cercospora coffeicola* which destroys the leaf tissue.

WILD COFFEE

The beans of the following wild coffee plants are roasted and brewed as a beverage — *Argocoffeopsis (Coffea) afzelii, A. (C.) rupestris* (Aduba), *A. (C.) subcordata, Calycosiphonia (Coffea) macrochlamys,* and *C. (Coffea) spathicalyx.* The seeds of *Cassia occidentalis* Negro Coffee, Coffee Senna, *C. obtusifolia (tora)* Foetid Cassia, *Boscia senegalensis* (Dila), and the fruit of *Ziziphus mucronata* Buffalo Thorn, are roasted, ground, and substituted for coffee although these contain no caffeine. The seeds of *Colocynthis lanatus (vulgaris)* Water Melon, *Abelmoschus esculentus (Hibiscus esculentus)* Okro or Okra,

Ipomoea turbinata (*muricata*), *Entada pursaetha* Sea Bean, *Parkia biglobosa* (Duaga), *Pavetta corymbosa* (Kronko), *Psilanths mannii* (Gbomete), and *Sericanthe chevalieri* (*Tricalysia chevalieri*) may also be roasted, ground, and substituted for coffee. The nut of *Cyperus esculentus* Tiger Nut is also used as substitute for coffee.

Vitellaria paradoxa (*Butyrospermum paradoxum* subsp. *parkii*)

Shea Butter Tree is indigenous to the Guinea savanna woodland. In Ghana it virtually covers about two-thirds of the country, mostly in the wild state. There is variation in this species: trees on cultivated ground, in farms, and around villages are often the finest. The *Ghana Cocoa Marketing Board News* of December 1980 estimates the existing trees in the country at about 9.4 million, with a possible yield of 100,000 tons of shea nuts per annum. At the current price of $990 per ton this amounts to $99 million per annum. Unfortunately the industry is not being fully tapped for the national benefit and still remains a rural, at best a regional, one; as a result the actual quantity of nuts collected is between 4,000 and 8,000 tons per annum — a meagre fraction of the potential figure. Several factors contribute to the vast discrepancy. Perhaps the most important are that

1. trees scorched by bush fires when flowering do not fruit that season; and judging from the prevalence of bush fires in the northern savanna a fair proportion of shea butter trees probably do not fruit at all;
2. by tradition the fruits are collected exclusively by women and girls;
3. the nuts from distant and inaccessible trees are most likely never collected — only trees around villages, settlements, camps, and farms are reached; that is, only trees within a day's walking distance (approximately fifteen miles in and out or seven miles' radius) are ever harvested.

The net result is that shea butter tree is an under-exploited resource. Shea butter is a fat that is increasingly assuming great importance because of its incorporation into margarine and chocolate formulations, and is thus a well-priced commodity on the international market. Unfortunately, the shea nut is left to rot year after year. The establishment of a Shea Nuts Marketing Board at the centre of the producing area to co-ordinate production and marketing would be economically sound.

The pulp around the seeds is edible and extensively eaten in the Guinea savanna woodland since the trees yield in May–June when there is general shortage of food in the area. The fruit-pulp is also eaten by animals. Shea butter is obtained from the seeds. The extraction, which is an important local industry in shea nut growing areas, is by tradition also done by women. The oil content of the kernels is about 45 to 55 per cent by weight. The roasted kernels are pounded, ground, and boiled with water. The juice of *Ceratotheca sesamoides* (Bungu) is added to the boiling pot to accelerate separation of the

3. INDUSTRIAL OR CASH CROPS

fat. The residual meal after the oil is skimmed off is used by people in northern Ghana, particularly Dagombas, as a waterproof material on the walls of their huts.

The butter is one of the cheapest and most important cooking oils for the ethnic groups in shea nut growing areas in West Africa. It also serves as a pomade (in other parts of the region as well) and an illuminant. Shea butter is commercially used in soap and candle making, in cocoa butter substitutes, and is processed into baking fat or incorporated into margarine formulations. It is used for cosmetics, skin and hair creams, and for the production of lubricants. For the producer price per tonne see Table 3.1. The branchlets serve as fuel-wood, for making charcoal, and the timber is useful for various items like stools, mortars, bowls, and furniture. Being termite-proof it is also useful for fencing and rafters. (For medicinal properties see Chapter 8.) *Globimetula braunii*, a semi-parasite, grows on the shea butter tree.

Palms

Cocos nucifera Coconut Palm is cultivated along the coast throughout the tropics. It requires a humid climate and plenty of sunlight, and withstands high concentrations of salt in the soil. Two varieties are recognized — the tall type and the dwarf type. Child (1974) estimates that the total area covered by the crop in West Africa is of the order of 70,000 hectares, the main producing countries in order of importance being Benin Republic, the Ivory Coast, Ghana, Nigeria, and Togo. In Ghana, cultivation is centred in the Western Region in the Nzema area generally, particularly at Awiebo, Esiama, Half Assini, and Tikobo No. 1, where large plantations have been established, and in the Keta area. There are also plantations at Asuansi, Ejumako, Essarkyir, and along the Winneba-Cape Coast road in the Central Region.

The country has the potential to produce 60,000 tonnes of copra (dried coconut fruits) worth almost $43 million annually — copra selling at $715 per tonne. The local mills have a consumption capacity of 20,000 tonnes. The remaining 40,000 tonnes would fetch $28.4 million annually if exported. Actual production is below this target and as a result the GIHOC Vegetable Oil Mills at Esiama and Atuabo have been producing below capacity. (There are two other mills, at Atebubu and Tamale, for processing ground-nut oil.) The lower level of production is also due partly to competition from local coconut oil manufacturers, private oil mills like the Crystal and Esiama mills, and to the high incidence of large quantities of copra consumed as famine food. Some 2,000 tonnes of copra are exported from Ghana to Germany alone and some to other European countries.

The fresh fruit of the coconut is eaten — the milk being very refreshing as it contains dissolved CO_2 from the respiration of the internal tissues. Another drink from the coconut is the juice or wine obtained by tapping the unopened flowers. In Ghana coconut wine is not as popular and common as palm wine (that tapped from *Elaeis guineensis* Oil Palm). The dried fruit or copra may also be eaten, but is usually crushed for the extraction of oil. Coconut oil is used for cooking and the manufacture of soap, detergents, and cosmetics. The residue — copra cake — has high protein content and is useful as animal feed, particularly for pigs. Coconut flour is prepared from the desiccated fruit and is rich in protein and fat. The coir fibre is useful for doormats, mats, ropes, and twine, and the midrib for brooms. The fronds are used for fences, thatching, temporary buildings, mats, and baskets (see Chapters 4 and 7). The crop is attacked by Cape St Paul Wilt which in 1960 killed over 100,000 plants in the Keta area in Ghana. Weaver birds (*Quelea quelea*) also represent a very serious pest by defoliating the trees. The major beetles that attack the coconut are the Rhinoceros Beetle (*Oryctes monoceros*) and the Hope Beetle (*Xylotripes gideon*).

Elaeis guineensis Oil Palm is indigenous to tropical African forest and its fringes. The major producing countries in West Africa are Nigeria, Ivory Coast, Ghana, Benin Republic, and Togo. There are about ten varieties. The variety supplied by Kusi in Ghana is Tenera — a cross between Dura and Pisifera. The Oil Palm Research Centre at Kusi supplies about two million seedlings annually to farmers throughout the growing areas, but it has the potential to produce up to five million annually. Suitable regions for its cultivation in Ghana are the Eastern, Central, and Western Regions and parts of Ashanti. There are about twenty-six State Oil Palm Plantations in the growing areas, the major ones being those at Juaso, Okumanin, Assin Foso, Akwanseram, Prestea, Sese, Kwamoso, and Huhunya. There are World Bank Project Plantations at Benso, Kwae, and Twifu. There are, in addition, numerous private oil palm plantations ranging from a few acres in size to about 200 acres per farmer. The area of planted oil palm in 1986 was estimated at 84,729 hectares. Palm oil production for the same period was estimated at between 51,300 and 53,500 tonnes, with equivalent palm kernel oil of 25,000 tonnes. The annual production of palm oil increased to 60,000 tonnes in 1988. Ghana's annual palm oil requirement is estimated to be between 80,000 and 100,000 tonnes. Thus about 40,000 tonnes palm oil are imported annually to supply Lever Brothers Limited, the soap manufacturing company at Tema.

The fruits of the oil Palm are used in the preparation of palm soup (*abenkwan*). Two types of oil are produced from the fruit — from the pulp and from the seeds or kernels, each containing about 50 per cent oil. High-yielding varieties of oil palm yield about 2,376 litres (twelve 44-gallon drums) of oil per tonne of fruits. Palm oil extraction is practised more in the growing areas and palm kernel oil extraction generally outside the growing areas, and in the urban centres. The red colour of palm oil is due to carotenes, the precursors of vitamin A. This is important to the diet, since the average intake of vitamin A would otherwise be inadequate. The high cost of palm oil and edible oil generally is attributable to the dual purpose that palm oil and

3. INDUSTRIAL OR CASH CROPS

the other edible oils serve, as food and as raw materials for manufactured products. Despite large plantations, palm oil is still imported into Ghana to feed soap manufacturing industries. The use of inedible oils or less popular oils like *Tieghemella heckelii* (Baku), *Adansonia digitata* Baobab, and *Pentadesma butyraceum* (*butyracea*) (Abotoasebie) would relieve the high demand for palm oil.

Lower grades of both palm oil and kernel oil are used locally for the manufacture of soap. The burnt tree ashes, K_2O, are also used in soap manufacture. The higher grades of both oils are utilized industrially for the manufacture of margarine and compound cooking fat. Palm oil is used as a substitute for cocoa butter in the manufacture of chocolate. The kernel oil, like palm oil, is used in the manufacture of confectionery, ice cream, and baking fat. Palm oil is also used as cosmetic and lamp oil. Zeven (1967) observes that the use of palm oil for embrocation and lighting has considerably decreased since the introduction of soap, medicinal oils, and kerosene, although in many areas it is still used for these purposes.

Palm wine — a popular alcoholic drink — is obtained by tapping the male flowers. Usually the tree must be felled before it is tapped (a very wasteful practice); some can be tapped standing, like *Raphia hookeri* Wine Palm — nevertheless the tree still dies. Wine from newly felled trees is sweeter than that from older trees. Palm wine contains a high proportion of natural yeast, so the residue may be used for baking. The central shoot or palm cabbage is edible. The nut shells and the residue of the pericarp are useful as fuel. Like the coconut palm, the midrib is used for household sweeping brooms, the fronds woven as fences or for thatching and temporary shelters, baskets and matting (see Chapters 4 and 7.) The crop is attacked by collar rot caused by the fungus *Ganoderma lucidum* and by leaf spot caused by *Cercospora elaedis*. It is also attacked by nematodes.

Gossypium species

Cotton is indigenous to the tropics, but the origin of the cultivated cotton of commerce is uncertain. There are four important cultivated species: *G. herbaceum, G. arboreum, G. hirsutum,* and *G. barbadense*. The crop has been the centre of commerce in the region since the middle of the last century. Penzer (1920) observed that by 1860, Lagos had become the largest cotton port of West Africa and Kano a great cotton market; prices paid at Kano were often double what was paid for cotton at Liverpool. Producing countries in order of importance are Nigeria, Cameroun, Ghana, and Sierra Leone. Its cultivation in Ghana preceded the introduction of cocoa, although commercial cotton production did not become important until the middle of the nineteenth century. The crop is cultivated in the Upper West, Upper East, Northern, Brong Ahafo, and Volta Regions by both peasant farmers and some of the textile companies mainly for the lint which is the leading vegetable fibre in the world, and accounts for about 56 per cent of natural fibre used by man.

Peasant farmers intercrop cotton with food crops like maize, groundnut, and cassava. An estimated 7,000 acres were cultivated by such farmers in 1981-2 and 2,000 acres by the Ghana Textile Manufacturing Company Ltd, leading manufacturers of wax print, *Dumas*, in the country. The Ghana Cotton Company, formerly Cotton Development Board, distributes seeds of improved varieties to farmers for planting. These are Allen 333 and BJA 592 known to be resistent to bacteria. Others are Allen 26J, Allen 444, and HAR 444. Average yields range between 600 and 800 lb per acre. Koli (1973) records that trials by the Crops Research Institute of Ghana yielded 2,500 lb per acre in Nyankpala, Tamale, Damongo, and Yendi areas, provided that the crop was planted at the optimum time, weeded early, encouraged by animal manure and fertilizer, and protected against pests. Where the rainfall distribution is less favourable during the growing season, as it is in the Volta Region and Brong Ahafo, yields of about 1,000 to 1,500 lb per acre were obtained.

In addition to locally produced cotton, large quantities are imported from the United States as supplement by the textile companies (see Table 3.3). There are four main textile companies in Ghana — Akosombo Textiles, Juapong Textiles, the Ghana Textile Manufacturing Company at Tema, and Tema Textiles — and several other smaller ones. The last two companies have a capacity to produce 43.2 million and 20 million linear metres of fabric per annum respectively; actual production is about 7 million linear metres worth ₡420 million ($84 million) and 2.7 million linear metres worth ₡28 million ($560,000) respectively — the shortfall being the result of under-supply of raw materials.

Table 3.3 Cotton produced and imported by Ghana.

Year	Bales		
	Domestic Supply	Imported	Total
1970	2 500	52 500	55 000
1975	15 000	60 000	75 000
1980	50 000	40 000	90 000
1985*	16 000	17 900	33 900

* The drop in domestic cotton supply in 1985 with the corresponding drop in imported cotton worth only $6 million due to economic constraints, could explain the large importation of secondhand clothing (popularly called *broniwanu*) available in markets.

3. INDUSTRIAL OR CASH CROPS

The lint of the cotton is the most important part of the fruit and source of the most widely used natural fibre. It is spun into yarn and woven into fabrics for various purposes — clothing, bedding, curtains, and so on. Cotton is also used in the manufacture of cordage and twine and processed into absorbent cotton wool used for medical and surgical purposes. Low-grade fibres are used for felt in mattresses, bedding products, and furniture upholstery. Still lower grades may be utilized for high-grade writing paper and for the manufacture of rayon. In the chemical industry cotton is used for photographic and X-ray films and for the production of explosives. The seeds yield about 25 per cent cotton-seed oil used for cooking, salads, and the manufacture of soap. The residual seed-cake is used as manure or cattle-feed; it is unsuitable for pigs and chickens, the seeds containing gossypol, a polyphenolic substance toxic to these animals. The crop is attacked by black-arm caused by the bacterium *Xanthomonas malvacearum* which also causes angular leaf spot, a disease that affects all above-ground parts of the plant.

Bast Fibres

There are many bast fibre plants in West Africa (see Chapter 1 — Direct Benefits), but four main ones are being developed as crops by the Bast Fibres Development Board in Ghana (see Table 3.4). Congo Jute is a common weed around villages; Kenaf appears to be indigenous to tropical and sub-tropical Africa, where it grows wild; both Jute and Roselle are widely cultivated in the tropics. According to recent research by Dr Obeng Asamoah of the Volta Basin Research Project in Ghana, other wild plants suitable for bast fibre are *Triumfetta rhomboidea* Burweed, *T. cordifolia*, *Sida cordifolia*, *Wissadula amplissima*, and *Waltheria indica* (Sawai).

TABLE 3.4 Plants yielding bast fibre

Type	*Varieties*	
Hibiscus cannabinus Kenaf	a. Cuba 108	b. Cuba 2032
	c. G 45	d. A 63-440
	e. GT 3	f. GT -7
Corchorus species Jute	a. B 7-1	b. B 7-2
	c. B 7-2-2	d. B 7-1-3
Hibiscus sabdariffa Roselle	a. Thai Red	b. Nyankpala White
Urena lobata Congo Jute	Ex-Mokwa	

The main areas where bast fibre is cultivated in Ghana are Eastern Region — Kwahu Tafo; Ashanti Region — Mampong and Kumasi Districts; Brong Ahafo Region — Tekyiman, Nkoransa and Atebubu Districts; Northern Region — around Yapei; Upper West Region — Wa District; and Upper East Region — around Zebella. Among peasant farmers the crop is interplanted with food crops, so estimated total acreage cultivated by these farmers is not readily available. However, the estimated total area cultivated by the Board in 1975 was 428.9 hectares and total tonnage harvested in the same year was 382,094. There are plantations in the Upper West, Upper East, and Northern Regions. There is also a 4,000-hectare farm at Aframso in Ashanti of which 800 hectares are under cultivation.

The products from these farms feed the Fibre Products Manufacturing Company Limited (Jute Factory) in Kumasi. The factory is capable of producing 12 million sacks yearly from 12,000 tons of raw material to meet local demand. Actual production, however, is only 6 million sacks.

Cola species

Cola acuminata Commercial Cola Nut Tree and *C. nitida* Cola Nut, Bitter Cola (Bese) are indigenous in the forest but are also cultivated mainly in Sierra Leone, the forest area in Ghana, and the Niger estuary in Nigeria. The nuts of *C. nitida* are richest in caffeine and are the most appreciated. The principal areas of kola nut production in Ghana are Brong Ahafo — Chiraa, Tichiri, Adrubum, Tanoso, Afroni, Bechem, Techimantia, Yamfo, Kenyasi, and Techiman; Ashanti — Akumadan, Nkenkasu, Teppa, Sekodumasi, Barikese, Mampong West, and Effiduasi; Central Region — Twifo Praso, Assin Foso, and Assin Manso; and Eastern Region — Akroso, Asamankese, Kade Asuom, Abomoso, Anyinam, Tafo, Suhum, and Kibi. The best trees are reported to grow around Sunyani and Goaso in the Brong Ahafo Region. The largest production comes from individual peasant farmers. The State Farms Corporation had 760 acres (304 hectares) of plantation under cultivation in Ashanti and Eastern Regions, but only 272 acres (109 hectares) at present exists; the University of Ghana Agricultural Research Station at Kade has 35 acres (14 hectares) under cultivation for research purposes.

In Ghana, kola nuts have been exported since prehistoric times to the northern savanna, and in recent times to Nigeria and Europe. Annual production levelled at 14,000 tons in 1921. D. T. Adams (1940) reported 10 million lb production in 1928 valued at £169,184; and peak exports of £282,773 in 1925 declining to £138,322 in 1930. Between 1972 and 1975, 251,000 tons of nuts worth $447,800 were exported. Kola nut was an ingredient in the Coca-Cola formulations which started off the fashion for cola drinks, although the flavourings and stimulants are now obtained from other sources. The nut is chewed fresh, or more usually dried, as a stimulant to

3. INDUSTRIAL OR CASH CROPS

prevent drowsiness and sleep. It is also extensively chewed to control hunger and thirst due to the presence of the alkaloids caffeine (1.25 per cent) and theobromine. The nuts also contain fat, cola red, enzymes, starch and sugar. Other edible kolas are *C. caricifolia* Fig-leaved Cola, *C. chlamydantha* (Tanamfre), *C. gigantea* var. *glabrescens* (Watapuo), *C. heterophylla* (We-ana), and *C. lateritia* var. *maclaudi* (Watapuobere); the seeds of *Garcinia kola* Bitter Cola (Tweapea) called *michingoro* are chewed as kola. The edible seeds of *Strephonema pseudocola* (Awuruku), which resemble kola nuts, are sometimes used to adulterate the true kola nuts.

Nicotiana tabacum

Tobacco is a native of South America from where it was introduced by the early Portuguese traders. Major producing countries in West Africa are Nigeria, Cameroun, the Ivory Coast, Ghana, and Sierra Leone. The crop is also extensively cultivated in East Africa, and it is an important foreign exchange earner in Malawi and Zimbabwe. The local species, *N. rustica* Wild or yellow-flowered Tobacco from northern Ghana, is also cultivated. In Ghana tobacco is cultivated by peasant farmers in derived savanna areas like northern Ashanti, Brong Ahafo, and Volta Region; in the transitional forest zones around Nsawam, Abura Dunkwa, Assin Foso, Koforidua, Ashanti Bekwai, and Agona Swedru; and in the savanna areas of Northern, Upper East, and Upper West Regions. The largest growing area is around Wenchi and Ejura — both in the Brong Ahafo Region. Farmers alternate the cultivation of tobacco with food crops like maize, groundnuts, or yams. Cultivation was on small-holdings and at backyards until the establishment of the Pioneer Tobacco Company (PTC) in 1952. The company, based in Takoradi in the Western Region, promotes the cultivation of tobacco and its processing into cigarettes and cigars. Some brands of cigarettes manufactured by the company are Embassy, Rothmans, 555, and Hollywood.

The two largest growing areas, Wenchi and Ejura, together cultivated an area of about 5,000 acres (2,000 hectares) in 1968 with a total yield of about 1,200 tons. Productivity is easily the highest in Africa, with a 1976 figure of 726 lb per acre (813 kg/ha). Ghana still imports both raw and processed tobacco.

Tobacco leaves are dried in the air (air-cured) in Bolgatanga and Tamale areas; or by fire (fire-cured) in Akatsi, Koforidua, and Bekwai areas; or smoked (flue-cured) in Wenchi, Ejura, and Damongo areas. The dried, smoked, or cured leaves are used as a narcotic for smoking, for chewing, or ground into a snuff. Tobacco provides a source of revenue for the government since it is heavily taxed. As the specific name suggests, tobacco contains nicotine, a drug which causes cardiovascular damage and lung, mouth, and throat cancer when used in excess. Most of the nicotine, however, is burnt during smoking. The Surgeon General warns that cigarette smoking is injurious to health and a contributary cause of lung cancer. The crop is affected by foot rot, referred to as Black shank and caused by *Pythium sphanidermatum* and *Phytophthora parasitica* var. *nicotianae*, leaf curl disease, and leaf spot or frog eye caused by *Cercospora nicotianae* among others.

Saccharum officinarum

Sugar-cane is cultivated in low, damp locations, along streams and watercourses or on swamp areas and land liable to flood. Irvine (1969) notes that there are many world varieties with varying characteristics such as thickness of stem, height, colour, hardness and sweetness of stem, and shape of joints. Red and white-stemmed varieties are recognized. The stems of the white varieties are used as chewing cane — more as a dessert than a meal. This variety is also useful in the distillation of local gin, *akpeteshie*, but not in the extraction of sugar. The stem of the red variety is crushed and the juice used in the manufacture of sugar, an important item of food in the world. Molasses (the liquid remaining after the sugar crystals are separated) is distilled into gin. The bagasse (crushed, dried stems of sugar-cane) is used as fuel. Mixed with molasses, it has been fed to livestock. It is a good source of cellulose, and is now recycled into plastics, paper, and compressed fibreboard manufacture. Ethanol is a by-product of the sugar industry. Of all the energy crops, sugar-cane has the highest yield of ethanol per hectare.

There are two sugar factories belonging to the Ghana Sugar Estates Limited (GHASEL), at Komenda in the Central Region and Asutsuare in the Eastern Region. About 43 per cent of the plant capacity of both factories is satisfied from local sugar-cane production — both from the factory's own farms and from peasant farmers. Sugar-cane growing under irrigation is encouraged to double the yield from 23 to 45 tons per acre. In addition, a further 2,400 acres (960 hectares) for Asutsuare and 3,000 acres (1,200 hectares) for Komenda is planned, utilizing a sprinkler irrigation system. Also an irrigation project of 30,000 acres (12,000 hectares) is planned for Akuse/Kpong of which 16,000 acres (6,400 hectares) is to be developed for sugar-cane to enable the Asutsuare Factory to produce at full capacity. The country's annual sugar needs — far exceeding 200,000 tons — are mostly imported. Studies are being carried out into the Aveyime, Havi, Agogo, and Juapong sugar-cane projects. Small cottage sugar factories are also envisaged for Mankesem, Tsito Avenorpeme/Torve, and Pokoasi/Media. The crop is affected by a number of diseases including gummosis caused by *Xanthomonas vasculorum*, leaf scald caused by *X. albilineans*, and mosaic disease caused by a virus.

Hevea brasiliensis

Para Rubber Tree, a native of tropical South America, is cultivated in plantations in the forest areas of West Africa. Producing countries in the region are Liberia, Nigeria, Ghana, and the Ivory Coast. Malaya is the largest crude rubber producing country, supplying about half the world's requirement. Supplementing the 5,600 acres (2,240 hectares) already established in the Western Region of Ghana, an additional 45,000 acres (18,000 hectares) is being planted in the Western Region — 80 per cent by the State Farms Corporation and Ghana Rubber Estates and 20 per cent by peasant farmers. The State Farm is establishing a 25,000-acre (10,000 hectares) plantation at Enchi. Other rubber plantations are Offin Rubber Estates, Holland Rubber Estates near Bunso, Ghana Community Farms Estates, and Rubber Growers' Association Estates. There are also plantations at Kumasi, Avrebo, Nsuaem, and the University of Ghana Agricultural Research Station at Kade. The produce of these plantations feeds the Bonsa Tyre Factory (formerly Firestone Ghana Ltd) at Bonsaso near Tarkwa in the Western Region. Before the establishment of the factory the rubber was exported. D. T. Adams (1940) reports that exports peaked at £53,283 in 1926, declining to £21,987 in 1930. There are also tyre manufacturing factories in Nigeria and Liberia.

Rubber is also obtainable from wild rubber-producing plants like *Landolphia owariensis* White Rubber Vine and *Funtumia elastica* West African Rubber Tree, which has been extensively tapped for the purpose. The demand for wild rubber, especially during the war, was a result of high prices of Para Rubber. Knorr (1945) noted that high prices stimulated increased collection, low prices tended to contract output. The annual production for export reached 3,000 tons by the end of the nineteenth century. Scars from the tapping marks are still visible on the trees in the forests today. Irvine (1961) reports that the latex coagulates readily to a solid, sticky mass, about one-third of which is pure rubber of excellent quality, and comparable to that of the best Para Rubber. *F. africana* False Rubber Tree was used as a rubber adulterant. (For other latex-producing plants, see Chapter 1.)

Spices

Aframomum melegueta. Grains of Paradise, Guinea-grains, Melegueta (Famu-wisa) (to distinguish it from (Soro-wisa) *Piper guineense)* is indigenous to West Africa and cultivated in Ghana, Guinea, the Ivory Coast, Sierra Leone, Benin Republic, and other West African countries. It was first described from the western side of the Atlantic in South America — probably carried across with the slave trade. Between 1865 and 1920 it was exported, to the United Kingdom and Europe mainly, as a spice and flavouring. The earliest record of Melegueta was at a festival held at Treviso in 1214. Physicians as far apart as Nicosia, Rome, Lyons, and Wales included it in their prescriptions. The spice was then carried by Mandingo traders from West Africa across the Sahara to the port of Mundibarca on the coast of Tripoli, but later brought by the Portuguese by sea from the Guinea coast, which was also known as the Grain

Coast. It is now of little importance in the export trade, the bulk of production being used locally. In Ghana it is cultivated mostly around the Obuase area in the villages of Sikamang, Mensonso, and Brofoyedru in Adanse Ashanti. Other parts of the country noted for the cultivation of Melegueta are Mim-Goaso area in Brong Ahafo Region and Dunkwa in the Central Region. It is cultivated as a subsidiary or secondary crop to cash crops like *Theobroma cacao* Cocoa or food crops like *Musa paradisiaca* Plantain. Melegueta has occasionally been found in the Atewa Range Forest Reserve and elsewhere growing wild at forest edges and in clearings. The fruits of the wild plants and the vegetative parts are often much smaller than the cultivated ones. According to a survey by Lock, Hall, and Abbiw (1977), the yield of about 100 bags from Sikamang and neighbouring villages together weigh about 3000 kilograms. This being only one of the several cultivation areas, it is safe to estimate that the gross national amount cultivated stands at about 600 to 800 bags weighing between 21.6 and 28.2 tonnes. Melegueta was one of the first exports from Ghana with 86,719 kg in 1871, 281,567 kg in 1872, and 68,909 kg in 1875. The fruit pulp around the seeds is edible, especially before it is fully ripened, and it is chewed as a stimulant. As a spice Grains of Paradise is used for flavouring food and strengthening drink; medicinally alone or with other plant parts to relieve various ailments (see Chapter 8).

Zingiber officinale Ginger (Kakaduro) was introduced from Asia and has been widely cultivated for the rhizomes in the forest zone from ancient times. The principal producing countries in West Africa are Nigeria and Sierra Leone. Experimental trials have been carried on in Ghana in recent years. Ashanti and Brong Ahafo in Ghana are the main ginger cultivation regions. Adding lime to the soil or manuring it greatly increases the yield where the soil is acid. Ginger is used as a spice for seasoning and flavouring food and for a variety of medicinal purposes both externally and internally (see Chapter 8). *Kakaduro* means it relieves and cures toothache. Ginger is used commercially in the preparation of ginger ale and ginger beer, and also in the preparation of Hausa beer, an effervescent, refreshing beverage; ginger biscuits and gingerbread. Ginger peelings are valued for the essential oils which are obtained by distillation. The oil is used to cure rheumatic pain affecting joints and muscles, boils or carbuncles, and other ailments. The annual yield is estimated to be about 500 tonnes. The world price fluctuates between £45 and £150 per tonne. *Z. zerumbet*, which resembles the true ginger vegetatively, has been cultivated in error in parts of Ghana, especially the Brong Ahafo Region.

Other cash crops

Sesamum indicum and ***S. radiatum***. Sesame, Beniseed is an important crop in Nigeria. It is also grown in Ghana, Guinea, Mali, and the Ivory Coast. It contains 50 to 57 per cent sesame oil used overseas for margarine and

3. INDUSTRIAL OR CASH CROPS

fine-grade machine oil; also for cooking, soap manufacture, lighting, and fuel. The leaves serve as spinach and the seeds are roasted and ground for meal.

Ceratotheca sesamoides (Bungu) resembles beniseed and contains 37 per cent oil. It may be used as adulterant of sesame. The seeds are roasted, ground, and added to soup. The leaves serve as a vegetable.

Agave sisalana. Sisal Hemp is cultivated throughout the region for the hemp. The main centre of production in tropical Africa is Tanzania in East Africa.

Citrus aurantiifolia. Lime is cultivated on small-holdings and in plantations for the juice which is exported for lime-juice cordial, limeade, marmalade, citric acid, and as flavouring. In Ghana cultivation is centred around Abakrampa (near Cape Coast), in the Central Region.

Simmondsia chinensis. Jojoba is a native of the semi-arid region of the Sonora desert in northern Mexico and the south-western USA. It is thus suitable for cultivation profitably in the arid and semi-arid zones of developing countries like the Sahel and Sudan savanna in West Africa. The plant is important for the seeds which contain an unusual liquid vegetable wax, jojoba oil, which has been suggested as a substitute for sperm whale oil.

Acacia senegal. Gum Arabic grows from Mauritania to northern parts of the Ivory Coast, northern Nigeria and extending to the Red Sea and eastern India. It is commercially important for the gum. Sudan produces about 85 per cent of the world's demand. *A. seyal* and *Prosopis chilensis*, an introduction from tropical and sub-tropical America, also produce gums of commercial quality. (See also Chapter 1.)

Chrysanthemum cinerariifolium. Pyrethrum is cultivated mainly in Kenya in East Africa for the flowers which are used to make the insecticide Pyrethrum. Bisacre and others (1984) observe that, in the concentrations used, the insecticide is non-toxic to both plants and higher animals and is widely used on livestock and edible plants. Pyrethrin sprays have also been used successfully for herbarium fumigation at the Missouri Botanical Garden (Gereau, 1985). (A herbarium is a collection of dried and preserved plant materials systematically arranged for scientific and other purposes.)

Vernonia galamensis. This promising new industrial crop for semi-arid areas is a superior natural source of seed oil which yields 32 per cent epoxy acid suitable for the manufacture of plastic formulations and protective coating. The plant grows from Senegal and Guinea to Ethiopia, south to Zimbabwe and Mozambique. Dr Robert Perdue (1985) reports that preliminary trial plantings in Zimbabwe in 1984 provided yields up to 2,000 kg/ha.

Calotropis procera Sodom Apple and *Euphorbia* species which can grow on marginal lands are promising sources of liquid hydrocarbons from their milky latex.

FOUR
Building and Construction

Perhaps the two primary requirements that materials for construction purposes must satisfy are strength and durability. Other properties to be considered are availability, relative cost, and how easily they may be worked. As a material for building and constructional work, wood predates burnt bricks, landcrete, concrete, slate, metal, and fibreglass. It was, historically, probably the first building material after mud.

In many parts of West Africa trees are ubiquitous; therefore wood is easily available. It has the advantage of being produced on the spot or near the locality of ultimate use. This saves or minimizes transportation and importation costs and serves to reduce the overall cost of the material. Poles may be felled just at the right size for the work required — an example is *Strombosia glaucescens* var. *lucida* (Afena) for telegraph poles — saving the additional labour and cost of cutting or reshaping them. Wood is easy to work with; it saws well and takes nails. Timbers for building and construction satisfy the minimum standard required in strength and durability. For light construction timbers of medium hardness and weight, 35–40 lb/cu ft air dry, are used. Examples are *Terminalia ivorensis* (Emire), *T. superba* (Ofram), and *Mansonia altissima* (Oprono). Heavy construction demands timbers which are very durable, hard, and heavy, 40–70 lb/cu ft air dry. *Piptadeniastrum africanum* (Danhoma), *Milicia* (*Chlorophora*) *excelsa* Iroko (Odum), and *Lophira alata* Red Ironwood (Kaku) are examples. Quite a number of timbers are termite-proof like *Afzelia africana* (Papao), *Bridelia stenocarpa* (*micrantha*) (Opam), *Anopyxis klaineana* (Kokote), and *Malacantha alnifolia* (Fafaraha), to mention a few; *Craterispermum laurinum*, *Lovoa trichilioides* African Walnut, *Carapa procera* Crabwood, and Iroko are fairly resistant to fire, while *Borassus aethiopum* Fan Palm and *Distemonanthus benthamianus* African Satinwood (Bonsamdua) are resistant to water. These qualities make wood a first-class choice of material for building and construction.

Buildings

TEMPORARY STRUCTURES

The erection of these structures for various purposes is quite common, especially in rural areas. On farms, sheds are used to store foodstuffs and farm equipment, or to serve as resting and dwelling places. Foresters,

4. BUILDING AND CONSTRUCTION

surveyors, and hunters often erect sheds as camps. Poles for sheds are selected from understorey and small trees or branches of large trees. These include *Acacia sieberiana* var. *villosa* (Kulgo), *Adenodolichos paniculatus*, *Bridelia stenocarpa (micrantha)* (Opam), *Carpolobia alba* (Afiafia), *C. lutea* (Otwewa), *Cassia kirkii* var. *guineensis*, *Cola acuminata* Commercial Cola Nut Tree, *Celtis* species, *Combretum fragrans (ghasalense)* (Kwaginyanga), *C. glutinosum* (Nkunga), *Faurea speciosa* (Se. ngo se bari), *Hexalobus monopetalus* var. *monopetalus*; and the petioles of *Hyphaene thebaica* Dum Palm, *Raphia* species Wine Palm, *Cocos nucifera* Coconut Palm, *Elaeis guineensis* Oil Palm, and *Phoenix reclinata* Wild Date Palm. Other trees are *Maesa lanceolata*, *Parinari curatellifolia* (Atena), *Sacoglottis gabonensis* (Tiabutuo), *Scottellia klaineana* (Odoko), *Soyauxia grandifolia* (Abotesima), *Sterculia tragacantha* African Tragacanth (Sofo); *Syzygium guineense* var. *guineense* (Sunya), *Terminalia glaucescens* (Ongo), *Tetrorchidium didymostemon* (Aboagyedua), and *Trichilia tessmannii (lanata)* (Tanuronini). The rest are *Uapaca heudelotii* (Kuntanakoa), *Vismia guineensis* (Kosowanini), *Voacanga thouarsii* (Foba), *Xylopia quintasii* Elo (Asimba); *X. staudtii* (Alari), *Ziziphus abyssinica* (Larukluror), *Z. mauritiana* Indian Jujube, and *Z. mucronata* Buffalo Thorn. *Oxytenanthera abyssinica* and *Bambusa vulgaris* Bamboo is used for practically all purposes for which poles or posts are required, and is quite suitable for the erection of sheds — either split or in the round. Besides the bamboo other grasses like *Phragmites karka* Common Reed, *Pennisetum purpureum* Elephant Grass, and *Gynerium saccharum* are used for the erection of huts.

Sawn timber is also used to erect temporary sheds in the cities, small towns, and villages. These sheds are used by fitters, bricklayers, proprietors of roadside restaurants, *fufu* and drinking bars, and other commercial enterprises. Sheds also serve as bus stops. In the cities and urban centres wooden kiosks for selling various merchandise are a common sight. The frames and floors of these kiosks are constructed from beams and planks of wood respectively and the sides are made of plywood. Kiosks have the advantage of being portable.

PERMANENT STRUCTURES, MATERIALS, ETC.

These are built of mud (the *atakpame* type) with skeleton wooden frames; or built completely of wood; or of bricks, cement blocks, or concrete with wooden parts.

Mud Houses

The *atakpame* type of mud house is the popular model in camps, hamlets, and villages. They are simple, practical, and cheap — all the materials being obtainable from the surroundings. In addition to the trees listed above, the following are useful for house posts or for frames of permanent structures: *Acacia nilotica* var. *tomentosa*, *Pericopsis laxiflora* (*Aframomum laxiflora*)

Satinwood, *Anogeissus leiocarpus* (Sakanee), *Antidesma venosum* (Mpepea), *Bersama abyssinica* subsp. *paullinioides* (Duantu), *Aporrhiza urophylla*, *Borassus aethiopum* Fan Palm, *Bridelia ferruginea* (Opam-fufuo), *Caloncoba gilgiana* (Kotowhiri), *Calpocalyx brevibracteatus* (Atrotre), *Cola buntingii*, and *C. digitata*. Others are *Combretum molle* (Gburega), *Coula edulis* Gaboon Nut (Bodwue); *Croton zambesicus* (Dodwatu), *Deinbollia grandifolia* (Potoke), *Diospyros abyssinica* (Gblitso), *D. gabunensis* Flint Bark Tree (Kusibiri); *D. kamerunensis* (Omenewa), *Dodonaea viscosa* Switch Sorrel (Fomitsi); *Drypetes afzelii* (Opahanini), *Eriocoelum pungens* (Onibona), and *E. racemosum* (Onibonakokoo). Also suitable for house posts are *Harungana madagascariensis* (Okosoa), *Homalium letestui* (Esononankoroma), *Hunteria eburnea* (Kanwene-akoa), *H. umbellata*, *Irvingia gabonensis* Wild Mango (Abesebuo); *Isoberlinia tomentosa* (*dalzielii*) (Kangkalaga), *I. doka* (Sapelaga), *Leptaulus daphnoides* (Afena-akoa), *Macaranga hurifolia* (Kpazina), *Malacantha alnifolia* (Fafaraha), *Massularia acuminata* (Pobe), *Memecylon afzelii* (Otwe-ani), *Monodora tenuifolia* (Motokura-dua), and *Morinda lucida* Brimstone Tree (Kankroma). The rest are *Phyllocosmus africanus* (*Ochthocosmus africanus*), *Octoknema borealis* (Wisuboni), *Pachypodanthium staudtii* (Fale), *Parinari congensis* (Tulingi), *Maranthes robusta* (*P. robusta*) (Kwinabuka), *Pausinystalia lane-poolei*, *Placodiscus pseudostipularis* (Ankye-Wobiri), *Greenwayodendron oliveri* (*Polyalthia oliveri*) (Duabiri), *Pterocarpus erinaceus* Senegal Rose Wood Tree, African Kino; *Scaphopetalum amoenum* (Nsoto), *Securidaca longepedunculata* (Kpaliga), *Talbotiella gentii* (Kpeteple), and *Vitex grandifolia* (Supuwa).

Like iron rods in concrete structures, the wooden framework is a structural reinforcement for the mud walls. It makes the walls stronger (particularly laterally) and more durable. The framework also helps to prevent the cracking of the walls by the heat of the sun.

Plants used as tietie to bind the skeleton framework in place are the split stems of the climbing palms *Laccosperma opacum* (*Ancistrophyllum opacum*) Rattan Palm, *L. secundiflorum* (*A. secundiflorum*) (Ayike), *Calamus deeratus* (Demmere), and that of *Eremospatha hookeri*. Also useful as binding material are stem strips of *Cissus populnea* (Agyako), *Flabellaria paniculata* (Okpoi), *Asparagus africanus* (Adedende), *A. flagellaris*, *Cercestis afzelii* (Batatwene), *Cochlospermum tinctorium* (Kokrosabia), *Ipomoea asarifolia*, *Merremia aegyptiaca*, *Sechium edule* Chayote, and *Entada pursaetha* Sea Bean. Many members of the family Celastraceae and split stems of the members of Marantaceae are suitable as binding material. The stems of some grasses are woven as cordage and used in binding. These include *Eragrostis ciliaris*, *E. gangetica*, *E. pilosa*, *E. tremula*, and *Schoenefeldia gracilis*.

The skeleton frame is filled in on both sides with mud or clay. A slimy solution of the macerated root-bark or stem of *Cissus populnea* (Agyako) is sometimes mixed with the mud to provide a smooth surface. Dried grass is usually cut up and mixed with the clay for building — a practice which further

4. BUILDING AND CONSTRUCTION

holds the clay together to prevent cracking in the sun. The grasses used include *Andropogon pseudapricus*, *Hyparrhenia rufa*, *Monocymbium ceresiiforme*, *Oryza sativa* Rice, *Pennisetum pedicellatum*, *Schizachyrium exile*, and *Digitaria exilis* Hungry Rice. The leaves of *Ceratotheca sesamoides* (Bungu) mixed with chopped grass and clay make it adhesive. The leaves of *Aspilia africana* Haemorrhage Plant are added to mud and the resultant material is used for plastering floors.

Roofing Materials

Plant materials useful for roofing include grass or palm fronds as thatch, split wood as shingles, and tree bark or leaves as a covering sheet. The leaves of *Cola gigantea* var. *glabrescens* (Watapuo), *Indigofera pulchra*, *Mitragyna stipulosa* (Subaha-akoa), *Pterygota macrocarpa* Pterygota (Kyereye), and *Theobroma cacao* Cocoa are used as roofing material for temporary huts. Leaves of *Carapa procera* Crabwood are termite-proof and suitable for thatching. The bark of Crabwood, *Milicia excelsa* (*Chlorophora excelsa*) Iroko (Odum), *Bombax buonopozense* Red-Flowered Silk Cotton Tree (Okuo), and the stems of *Agelaea trifolia* and *Calotropis procera* Sodom Apple are used in roofing. A decoction of the bark of *Bridelia ferruginea* (Opam-fufuo) with clay forms a cement used for flat roofs of huts. Palm fronds used as thatch are *Cocos nucifera* Coconut Palm and the climbing palms *Laccosperma opacum* (*Ancistrophyllum opacum*) Rattan Palm, *L. secundiflorum* (*A. secundiflorum*) (Ayike), *Calamus deeratus* (Demmere), and *Eremospatha macrocarpa*; and in the rainforest, the slender palm *Sclerosperma mannii*. Other palms useful as thatch are *Borassus aethiopum* Fan Palm, *Hyphaene thebaica* Dum Palm, *Phoenix reclinata* Wild Date Palm, *Elaeis guineensis* Oil Palm, and *Raphia hookeri* Wine Palm. Leaves of *Ataenidia conferta* (Konkon sibere), *Halopegia azurea*, *Marantochloa cuspidata* (Ntentrema), *M. leucantha* (Sibere), *M. mannii*, *M. purpurea* (Sugugwa), *Megaphrynium macrostachyum*, *Sarcophrynium brachystachys* (Koto-haban), *S. prionogonium*, *Thaumatococcus daniellii* Katemfe (Awuram-asie); *Thalia welwitschii*, and *Trachyphrynium braunianum* — all members of the family Marantaceae — are used for thatching temporary huts. Burger (1967) reports that the leaves of *Typha domingensis* (*australis*) Bulrush are used for thatch in Ethiopia.

Terminalia ivorensis (Emire) is preferred for shingles because it splits easily and lasts about fifteen years. Other timbers used for shingles are *T. superba* Afara (Ofram); *Milicia excelsa* (*Chlorophora excelsa*) Iroko (Odum); *Craterispermum laurinum*, *Cordia millenii* Drum Tree, *C. platythyrsa* (Tweneboa); *Distemomanthus benthamianus* African Satinwood (Bonsamdua); *Hildegardia barteri* (Akyere), *Mansonia altissima* (Oprono), *Musanga cecropioides* Umbrella Tree, *Pericopsis laxiflora* (*Afrormosia laxiflora*) Satinwood, *Margaritaria discoidea* (*Phyllanthus discoideus*) (Pepea), *Pycnanthus angolensis* African Nutmeg (Otie); *Heritiera utilis* (*Tarrietia utilis*) (Nyankom), *Tieghemella heckelii* (Baku), and *Triplochiton scleroxylon* (Wawa). The life of shingles depends on the type of tree used. Iroko shingles last about ten years, those of the Umbrella Tree only two to three years.

By far the most popular roofing material for native houses is grass thatch. Grasses used for thatching include *Anadelphia afzeliana, A. trispiculata, A. leptocoma,* and *A. pumila. Andropogon* species are popular as thatch. These are *A. gayanus* varieties *gayanus, tridentatus, squamulatus,* and *bisquamulatus* — used for the circular bands of conical thatch roofing. *A. pseudapricus, A. tectorum, A. tenuiberbis, A. curvifolius, A. africanus, A. incanellus, A. fastigiatus, A. ascinodis, A. perligulatus, A. canaliculatus, A. macrophyllus,* and *A. pteropholis* are equally used as thatch. Other grasses used are *Aristida adscensionis, A. mutabilis, A. sieberiana, Axonopus flexuosus, Pennisetum unisetum (Beckeropsis uniseta), Brachiaria lata, Chloris robusta, Ctenium elegans, C. canescens, C. newtonii, C. villosum, Cymbopogon giganteus* var. *giganteus, Echinochloa pyramidalis* Antelope Grass, *E. stagnina,* and *E. crus-pavonis. Eragrostis* species are equally popular as thatching material. These are *E. cilianensis* Stink Grass, *E. ciliaris, E. gangetica, E. pilosa, E. aspera, E. barteri, E. chalarothyrsos, E. namaquensis* var. *diplachnoides, E. namaquensis* var. *namaquensis, E. egregia, E. domingensis, E. squamata, E. tremula,* and *E. turgida.* Other species of grass used for thatching are *Heteropogon contortus* Spear Grass, *Diheteropogon amplectens* var. *amplectens, D. hagerupii, Hyperthelia dissoluta, Hyparrhenia rufa, H. subplumosa, H. glabriuscula, H. nyassae, H. cyanescens, H. familiaris, H. rudis, H. smithiana* var. *major, H. involucrata* var. *brevisata,* and *H. welwitschii. Imperata cylindrica* var. *africana* Lalang, a notorious weed (see Chapter 11), is usefully employed as thatch. Thatching material include *Loudetia arundinacea, L. simplex, L. flavida, Monocymbium ceresiiforme, Oryza barthii* Wild Rice, *O. longistaminata, O. punctata, Panicum anabaptistum, P. phragmitoides, P. maximum* Guinea Grass, *P. turgidum, P. dinklagei, Paspalum polystachyum, Pennisetum americanum* Bulrush Millet, *P. pedicellatum, P. purpureum* Elephant Grass, and *Phragmites karka* Common Reed. The rest are *Setaria chevalieri* Buffel Grass, *S. longiseta, S. anceps, S. pallide-fusca* Bristle Foxtail Grass, Cat's Tail Grass; *S. megaphylla, Saccharum spontaneum* var. *aegyptiacum, Schizachyrium exile, S. sanguineum, S. platyphyllum, S. ruderale, Schoenefeldia gracilis, Sorghum* species Guinea Corn, *Sporobolus festivus, Urelytrum annuum, U. muricatum, U. pallidum, Vetiveria fulvibarbis,* and *V. nigritana.* Some sedges, like *Scleria depressa* and *Mariscus alternifolius,* are useful as thatch.

Material for thatching is usually readily available and is obtainable inexpensively or at no cost at all. Thatched roofs promote comfort. They are cool during the hot afternoon sun and warm on chilly harmattan nights. Grass thatch lasts up to ten years or more; but the ease with which thatch catches fire and blazes is a serious disadvantage, particularly where the houses are close to each other — as is usually the case.

Wooden Houses

In West Africa wooden buildings are fairly rare, although wood is relatively

4. BUILDING AND CONSTRUCTION

cheaper than other building materials. In Ghana, the Timber Marketing Board, the Utilization Branch of the Department of Forestry, and the Forest Products Research Institute — a branch of the Centre for Scientific and Industrial Research (CSIR) — have researched into the suitability of selected timbers for the construction of permanent wooden buildings. Sample buildings have been constructed at the Forestry School, Sunyani, at the Department of Forestry Offices at Kumasi, and at the North Labone Estates in Accra. At the 1979 Trade Fair one such wooden house of *Piptadeniastrum africanum* (Danhoma) was displayed. Other timbers suitable for wooden houses include *Albizia ferruginea* (Awiemfosamina), *Aningeria altissima* (Samfena-nini), *Trilepisium madagascariense* (*Bosqueia angolensis*) (Okure), *Celtis wightii* (*brownii*) (Esafufuo), *Milicia excelsa* (*Chlorophora excelsa*) Iroko (Odum); *Nauclea diderrichii* (Kusia), and *Symphonia globulifera* (Ehureke). Wooden houses are erected on concrete foundations or on stilts. This practice checks the incidence of termite attack as does the treating of wood with chemical preservatives.

Other house-building materials

These are burnt bricks with cement mortar, cement blocks with mortar, concrete panels, or slates. Invariably, however, these materials are supplemented with wood. The roofing rafters, the face boards, the doors, windows, and their frames (and recently louvres) are normally made of wood. The correct choice of suitable timber for building is important. This selection depends on the part of the building for which the wood is required. Rafters and roofing frames are of thick, durable beams and account for a large proportion of the timber used in building. Suitable timbers for rafters include *Borassus aethiopum* Fan Palm, *Casuarina equisetifolia* Whistling Pine, *Milicia excelsa* (*Chlorophora excelsa*) Iroko (Odum); *Eriocoelum pungens* (Onibona), *E. racemosum* (Onibonakokoo), *Harungana madagascariensis* (Okosoa), *Lecaniodiscus cupanioides* (Dwindwera), *Mitragyna stipulosa* (Subaha-akoa), *Morus mesozygia* (Wonton), *Piptadeniastrum africanum* (Danhoma), *Sterculia rhinopetala* Sterculia Brown (Wawabimma); *Ziziphus mucronata* Buffalo Thorn, and *Z. spina-christi* var. *microphylla*. *Oxytenanthera abyssinica* Bamboo and the petioles of *Raphia hookeri* Wine Palm are used as rafters in mud huts.

Timbers used for doors, windows, and the frames include *Afzelia africana* (Papao), Iroko, *Daniellia oliveri* Africana Copaiba Balsam Tree (Sanya); *Distemonanthus benthamianus* African Satinwood (Bonsamdua); *Mitragyna stipulosa* (Subaha-akoa), *Hannoa klaineana* (Fotie), *Pellegriniodendron diphyllum* (Fetefele), *Pseudocedrela kotschyi* Dry-Zone Cedar, and *Trachylobium verrucosum* East African Copal. *Canarium schweinfurthii* Incense Tree is used for decorative doors. Unlike the conventional solid, panel doors, flush doors consist of a hollow, diagonally braced frame covered on both sides with plywood. Some flush doors are sound-proofed. Plywood is being used to replace glass louvres in view of the high cost of importation and the equally high incidence of glass louvre theft.

Ceilings may be of plywood or hardboard. *Afzelia africana* (Papao) is suitable for ceiling battens since it does not easily split when nailed. Sometimes the ceiling is of strips of wood such as *Terminalia ivorensis* (Emire).

Other parts of the building for which wood is used are the steps or stairways, rails, pillars, and floors. Timbers recommended for stairways and flooring because of their smooth wear and resistance to abrasion include *Afzelia africana* (Papao), *Canarium schweinfurthii* Incense Tree, *Carapa procera* Crabwood, *Celtis mildbraedii* (Esa), *C. adolfi-friderici* (Esakosua), *C. zenkeri* (Esakokoo), *Milicia excelsa* (*Chlorophora excelsa*) Iroko (Odum); *Erythrophleum suaveolens* (*guineense*) Ordeal Tree (Potrodom); *Holoptelea grandis* Orange-barked Terminalia (Nakwa); *Mitragyna stipulosa* (Subaha-akoa), and *Lovoa trichilioides* African Walnut. Others are *Gilbertiodendron limba* (Tetekon), *Nauclea diderrichii* (Kusia), *Nesogordonia papaverifera* (Danta), *Pericopsis elata* (*Afrormosia elata*) (Kokrodua), *Piptadeniastrum africanum* (Danhoma), *Klainedoxa gabonensis* var. *oblongifolia* (Kroma), *Distemonanthus benthamianus* African Satinwood (Bonsamdua); *Strombosia glaucescens* (Afena), *Symphonia globulifera* (Ehureke), and *Bussea occidentalis* (Kotoprepre). High-class timbers like *Entandrophragma candollei* Candollei, Unscented Mahogany, Omu; *E. cylindricum* West African Cedar, Sapele; and *E. utile* Utile are also used for flooring. *Cylicodiscus gabunensis* African Greenheart (Denya) and *Lophira alata* Red Ironwood (Kaku) are suitable for heavy duty flooring for warehouses and factories. (For panelling see Chapter 5.)

Building Accessories

Building platforms or scaffolding are erected with wooden poles and planks. *Triplochiton scleroxylon* (Wawa) and *Alstonia boonei* (Sinduro), being light and reasonably strong, are used as planks for scaffolding with *Oxytenanthera abyssinica* or *Bambusa vulgaris* Bamboo as poles. Other timbers useful as planks are *Antrocaryon micraster* (Aprokuma), *Aporrhiza urophylla*, *Hannoa klaineana* (Fotie), *Lannea acida* (Kuntunkuri), *L. velutina* (Sinsa), and *Sterculia oblonga* Yellow Sterculia (Ohaa). *Triplochiton* and *Alstonia* are used as moulding for gutters, lintels, pillars, and general concrete works.

Construction

VEHICLES

Horse-drawn carts and wheelbarrows

Simple, everyway utility vehicles like horse-drawn carts, wagons, and wheelbarrows are normally constructed with wood. Timbers used to construct carts and wheelbarrows include *Angoeissus leiocarpus* (Sakanee), *Azadirach'a*

4. BUILDING AND CONSTRUCTION

indica Neem Tree, *Berlinia occidentalis* (Kwatafo-mpaboa), *Thespesia populnea* (Frefi), and *Tamarindus indica* Indian Tamarind. *Thespesia* is also used for wheels. Other timbers for wheel-making are *Cassia sieberiana* African Laburnum, *Tetrapleura tetraptera* (Prekese), *Petersianthus macrocarpus* (*Combretodendron macrocarpum*) (Esia), *Ziziphus mucronata* Buffalo Thorn, and *Terminalia catappa* Indian Almond.

Mammy trucks and coachwork

In Ghana and Nigeria there are many wooden trucks of all sizes. The popular *Trotro*, mammy truck, is not only constructed with wood: apart from the engine, gearbox, and chassis, almost everything else is solid wood. Timbers for constructing lorry bodies include *Celtis mildraedii* (Esa), *C. adolfi-friderici* (Esakosua), *C. zenkeri* (Esakokoo), *Milicia excelsa* (*Chlorophora excelsa*) Iroko (Odum); *Distemonanthus benthamianus* African Satinwood (Bonsamdua); *Nesogordonia papaverifera* (Danta), *Terminalia catappa* Indian Almond, and *T. ivorensis* (Emire). *Triplochiton* serves as flooring for motor lorry bodies.

Luxury coach builders are today using hardwoods. A description of one such coach built by Plaxtons (Scarborough) Ltd includes the following: 'Niangon was widely used in the body framework, such as bottom frames, side frames, in the back, roof, and cab, with Iroko to a lesser extent in the side framing and the back. Another hardwood used in smaller quantities was Yang, and the interior polish-work was of mahogany.' Niangon is *Heritiera utilis* (*Tarrietia utilis*) (Nyankom); Iroko is *Milicia excelsa* (*Chlorophora excelsa*) (Odum), and mahogany is probably either *Khaya ivorensis* African Mahogany (Dubini) or *K. anthotheca* White Mahogany. All these are merchantable timbers in West Africa (see Chapter 1). Yang is *Dipterocarpus* species. The two luxury coach-building firms in Ghana — Willowbrook (Ghana) Ltd and Neoplan (Ghana) Ltd — both use hardwoods.

Aircraft propellers

The wood of *Terminalia superba* Afara (Ofram) has been used in South Africa for the manufacture of aircraft propellers. Generally, nine elements were specially cut to size, glued, and then machined. Groulez and Wood (1985) observe that the wood stood up very well to the very stringent demands made on it in this role because of its good impact resistance and uniformity, enabling pieces to be selected with little variation in density within them.

Railway coachwork

For the construction of railway coachwork the following timbers are used: *Cola nitida* Bitter Cola (Bese); *Coula edulis* Gaboon Nut (Bodwue); *Khaya ivorensis* African Mahogany (Dubini); *K. senegalensis* Dry-Zone Mahogany, *Guarea-cedrata* Scented Guarea (Kwabohoro); *Lovoa trichilioides* African Walnut, *Mansonia altissima* (Oprono), *Nesogordonia papaverifera* (Danta),

Sacoglottis gabonensis (Tiabutuo), and Iroko. The floors of the coaches are built with *Nauclea diderrichii* (Kusia) and *Piptadeniastrum africanum* (Danhoma).

Boat-building

Timber still plays an important part in boat-building. Recommended for surf boats are *Anacardium occidentale* Cashew Tree (Atea); *Acacia nilotica* var. *tomentosa*, *Berlinia occidentalis* (Kwatafo-mpaboa), *Coula edulis* Gaboon Nut (Bodwue); *Erythrophleum suaveolens* (*guineense*) Ordeal Tree (Potrodom); *Entandrophragma utile* Utile, *Guarea thompsonii* Black Guarea (Kwadwuma); *Heritiera utilis* (*Tarrietia utilis*) (Nyankom), *Prosopis africana* (Sanga), *Milicia excelsa* (*Chlorophora excelsa*) Iroko (Odum), and *Sterculia foetida*. *Nesogordonia papaverifera* (Danta) serves as ribs for surfboats. Others suitably used are *Lonchocarpus sericeus* Senegal Lilac (Keklengbe); *Uapaca heudelotii* (Kuntanakoa), *U. guineensis* Sugar Plum (Kuntan); and *Syzygium guineense* var. *guineense* (Sunya).

Ship-building

Hard and heavy timbers are required for the planking of ships' decks and the construction of rails. Suitable timbers include *Afzelia africana* (Papao), *Milicia excelsa* (*Chlorophora excelsa*) Iroko (Odum); *Irvingia gabonensis* Wild Mango, *Klainedoxa gabonensis* var. *oblongifolia* (Kroma), *Pericopsis elata* (*Afrormosia elata*) (Kokrodua), and *Tectona grandis* Teak. *Nesogordonia* serves for the bulkheads of ships.

BRIDGE-BUILDING

There are two basic methods for bridging with hardwoods. Whole logs can be felled to span the stream or river. Otherwise sawn timber serves as the deck of the bridge supported by concrete trestles. Low-classed but durable hardwoods like *Antiaris toxicaria* (*africana*) Bark Cloth Tree (Kyenkyen); *Burkea africana* (Pinimo), *Mammea africana* African Mammy Apple (Bompagya); *Pterygota macrocarpa* Pterygota (Kyereye); and *Strombosia glaucescens* (Afena) are felled across a stream as a means of crossing from one side to the other. In the savanna areas *Borassus aethiopum* Fan Palm is particularly suitable as a bridging log since it is termite-proof and resistant to salt water. Heavy and durable timbers are used in the construction of all-purpose bridges. They include *Breonadia salicina* (*Adina microcephala*), *Afzelia africana* (Papao), *Erythrophleum suaveolens* (*guineense*) Ordeal Tree (Potrodom); *Piptadeniastrum africanum* (Danhoma), and *Cylicodiscus gabunensis* African Greenheart (Denya). The rest are *Milicia excelsa* (*Chlorophora excelsa*) Iroko (Odum); *Cynometra ananta* (Ananta), *Nauclea diderrichii* (Kusia), and *Newbouldia laevis* (Sasanemasa). For light bridging — that is, bridges exclusively for the use of man and beast — timbers like *Aningeria altissima* (Sanfenanini), *Trilepisium madagascariense* (*Bosqueia angolensis*) (Okure), *Terminalia ivorensis* (Emire),

4. BUILDING AND CONSTRUCTION 87

and *T. superba* (Ofram) are suitable. Hammock suspension bridges are constructed with the climbing palms *Calamus deeratus* and *Eremospatha macrocarpa*, or with the strong vines of *Loeseneriella* species (*Hippocratea* species).

SLEEPERS

Timber plays an important part as sleepers in the construction of railway lines. Suitable timbers for sleepers are *Afzelia africana* (Papao), *Anopyxis klaineana* (Kokote), *Calpocalyx brevibracteatus* (Atrotre), *Coula edulis* Gaboon Nut (Bodwue); *Cylicodiscus gabunensis* African Greenheart (Denya); *Distemonanthus benthamianus* African Satinwood (Bonsamdua); *Erythrophleum ivorense* Sasswood Tree, *E. suaveolens* (*guineense*) Ordeal Tree (Potrodom); *Heritiera utilis* (*Tarrietia utilis*) (Nyankom), *Klainedoxa gabonensis* var. *oblongifolia* (Kroma), *Lophira alata* Red Ironwood (Kaku); *Manilkara obovata* (*multinervis*) (Berekankum), and *Nauclea diderrichii* (Kusia). The rest are *Ongokea gore* (Bodwe), *Ostryderris stuhlmannii* (Kaman Godui), *Maranthes glabra* (*Parinari glabra*) (Punini), *Petersianthus macrocarpus* (*Combretodendron macrocarpum*) Stinkwood Tree (Esia); *Piptadeniastrum africanum* (Danhoma), *Prosopis africana* (Sanga), *Rhizophora* species Red Mangrove, *Sacoglottis gabonensis* (Tiabutuo), and *Tieghemella heckelii* (Baku). *Aubrevillea kerstingii* (Danhomanua) and *A. platycarpa* are suitable for sleepers, but in view of their scarcity and limited distribution their exploitation for this purpose would exhaust the stock. Afara, Iroko, and *Nauclea* are used as cross-ties by the Ghana Railways. Wooden sleepers last a long time, especially when treated. They are comparably cheaper than metal ones and give a cushioned ride by absorbing and minimizing the impact of the wheels on the rails.

TELEGRAPH POLES

The Posts and Telecommunications Department and, to some extent, the Electricity Corporation of Ghana use poles of *Strombosia glaucescens* (Afena) impregnated with preservatives as telegraph and electricity poles respectively. Other timbers utilized as poles are *Borassus aethiopum* Fan Palm, *Diospyros kamerunensis* (Omenewa), *D. sanza-minika* (Sanzamulike), *Celtis mildbraedii* (Esa), and *Nesogordonia papaverifera* (Danta). Telegraph pole cross-arms are usually of *Nesogordonia* or *Nauclea diderrichii* (Kusia). Treated, sawn timber is also used as poles. The use of timber for electric poles is also practised in developed countries like the United States.

PILES

Timbers that are fairly durable under water or are resistant to water and marine borers are used for harbour construction and wharf piles. They include *Allanblackia parvifolia* (*floribunda*) Tallow Tree (Sonkyi); *Conocarpus erectus* Button Wood (Koka); *Coula edulis* Gaboon Nut (Bodwue); *Cylicodiscus gabunensis* African Greenheart (Denya); *Detarium senegalense* Tallow Tree (Takyikyiriwa); *Distemonanthus benthamianus* African Satinwood (Bonsamdua); *Erythrophleum ivorense* Sasswood Tree and *E. suaveolens* (*guineense*) Ordeal Tree (Potrodom). Others are the Fan Palm, *Klainedoxa gabonensis* var. *oblongifolia* (Kroma), *Lophira alata* Red Ironwood, (Kaku); and *Nauclea diderrichii* (Kusia). The rest are *Mammea africana* African Mammy Apple (Bompagya); *Piptadeniastrum africanum* (Danhoma), *Rhizophora* species Red Mangrove, and *Trichilia monadelpha* (*heudelotii*) (Tanduro). *Mitragyna inermis* (Kukyafie) and *M. stipulosa* (Subaha-akoa) are equally resistant to water.

PIT-PROPS

Tectona grandis Teak, a native of Burma, is widely grown in the tropics. In Ghana, plantations around Obuasi in Ashanti are specifically intended for the supply of pit-props and shaft guides for the underground mines. Other suitable timbers are *Diospyros sanz-aminika* (Sanzamulike), *Maesopsis eminii* (Onwam-dua), *Morinda lucida* Brimstone Tree (Kankroma), and *Lophira alata* Red Ironwood (Kaku). The rest are *Detarium senegalense* Tallow Tree, *Allanblackia parviflora* (*floribunda*) (Sonkyi), *Pentadesma butyraceum* Tallow Tree (Abotoasebie); *Nauclea diderrichii* (Kusia), *Piptadeniastrum africanum* (Danhoma), *Strombosia glaucescens* (Afena), and *Tieghemella heckelii* (Baku).

FENCES AND PALINGS

Roundwood or sawn timber is a common fencing material. Hard, heavy, and durable timbers are preferred; however, soft woods like *Myrianthus arboreus* (Anyankoma) and *Musanga cecropioides* Umbrella Tree (Dwumma) are occasionally used as palings on farms. Fences are often coated with preservative or paint to protect the wood from borers and the weather. Timbers used for fencing include *Afzelia africana* (Papao), *Terminalia ivorensis* (Emire), *T. superba* Afara (Ofram); *Pentaclethra macrophylla* Oil Bean Tree (Ataa); *Tamarindus indica* Indian Tamarind, *Uapaca heudelotii* (Kuntan-akoa), and *Zanha golungensis*. Others are *Celtis* species (Esa), *Milicia excelsa* (*Chlorophora excelsa*) Iroko (Odum); *Piptadeniastrum africanum* (Danhoma), *Erythrophleum ivorense* Sasswood Tree, *E. suaveolens* (*guineense*) Ordeal Tree (Potrodom); *Lecaniodiscus cupanioides* (Dwindwera), *Bridelia stenocarpa* (*micrantha*) (Opam), and *Malacantha alnifolia* (Fafaraha). Also suitable for fencing are *Coula edulis* Gaboon Nut (Bodwue); *Aulacocalyx jasminiflora* (Sanbrim), *Adenodolichos paniculatus*, *Pentadesma butyraceum* Tallow Tree (Abotoasebie); *Detarium senegalense* Tallow Tree (Takyikyiriwa), and *Xylopia quintasii* Elo (Asimba). *Oxytenanthera abyssinica* Bamboo is split and interwoven for fencing. The petioles of *Borassus aethiopum* Fan Palm, *Phoenix reclinata* Wild Date Palm, *Raphia hookeri* Wine Palm, *Cocos nucifera* Coconut Palm, *Elaeis guineensis* Oil Palm, and *Hyphaene thebaica* Dum Palm are used for

4. BUILDING AND CONSTRUCTION

fencing. Woven material may also be used for fencing. The stems of *Strophanthus gratus* (Omaatwa), *S. hispidus* Arrow Poison (Amamfohama); *S. sarmentosus* (Adwokuma), and the climbing palms *Calamus deeratus* (Demmere) and *Eremospatha macrocarpa* are woven into reed screens for fencing. The stems of *Marantochloa* species (Awora), *Sarcophrynium* species, *Thalia welwitschii*, *Thaumatococcus danielli* Katemfe (Awuram-asie), and *Typha domingensis* (*australis*) Bulrush are useful as fencing material. A number of grasses, especially those with well-developed stems, are useful as fencing material directly or as fence mats. These are *Andropogon gayanus* varieties *gayanus*, *tridentatus*, *squamulatus*, and *bisquamulatus*; *A. tectorum*, *A. tenuiberbis*, *A. macrophyllus*, and *A. pteropholis*. Others are *Pennisetum unisetum* (*Beckeropsis uniseta*), *Chasmopodium caudatum*, *Chloris robusta*, and *Cymbopogon giganteus* var. *giganteus*. The rest are *Echinochloa crus-pavonis*, *Hyparrhenia rufa*, *H. smithiana* var. *major*, *H. cyanescens*, *H. rudis*, *H. welwitschii*, *H. involucrata* var. *involucrata*, *Pennisetum purpureum* Elephant Grass, *P. americanum* Bulrush Millet, *Phragmites karka* Common Reed, and *Sorghum* species Guinea Corn.

BARRELS

Timbers employed for barrel or cask construction include *Cola lateritia* var. *maclaudi* (Watapuobere), *Mitragyna stipulosa* (Subaha-akoa), *Pachypodanthium staudtii* (Fale), *Pseudocedrela kotschyi* Dry-Zone Cedar, *Rhizophora* species Red Mangrove, *Terminalia catappa* Indian Almond, and *T. superba* Afara (Ofram).

SIGN AND DISPLAY BOARDS

Both the stands and supports of sign and display boards, commonly used for advertisements in the cities and on the outskirts of towns, are mostly constructed of wood.

FIVE
Furnishings

Wood is not only the most suitable and most widely used material for furniture, but it is also easily available and comparatively cheaper than metal. From the simple kitchen stool to the furnishings in average sitting-rooms, or the more sophisticated furniture for special occasions and state functions, wood is indisputably the material of first choice. The ease with which wood may be sawn and planed without sacrificing strength and durability, its ability to take nails, and the fact that it polishes and finishes well to add lustre, puts it in this unique category. It is also easily available from pit sawers, local timber markets, sawmills, and wood dealers. These factors contribute to the comparatively cheap price. Virtually any piece of timber can be utilized to make a suitable item of furniture, be it bits and pieces of packing case or plywood, off-cuts, or seasoned and selected timbers. Seasoning involves the drying out of the excess water content in the timber. This may be done naturally by stacking the boards or beams in an open airy place, or artificially with heaters or hot air.

Wood provides the furniture and furnishings for our homes, offices, factories; schools, colleges, and universities; hospitals, churches, and theatres. Similarly, the furniture for parks and gardens, camp chairs, tables, beds, etc., is made of suitable wood. The furnishings in yachts, boats, and ships may be designed with wood. Some buses, coaches, and motor cars are furnished with wood today (see Chapter 4). The suitability of wood for furniture, joinery, and internal decoration and fittings, and the frequency with which it is employed to this effect, makes it common — in fact, so common that it is taken for granted and sometimes overlooked. It is the material of woodworkers, sculptors, carvers, and amateur and professional carpenters. One of the objectives of the Ghana Interntational Furniture and Woodworkers Exhibition (GIFEX '85) in April 1985 was to display the variety of items that may be designed with local timbers. Despite the advances of modern science and technology, and the resulting development of man-made materials, wood is likely to remain the first choice for furniture for a very long time to come.

Furniture and Joinery

In carpentry and joinery, the quality of timber used and the degree of workmanship together determine the class of furniture produced. However,

5. FURNISHINGS

furniture firms are now producing high-class furniture with lower-grade or unmerchantable timber. Apparently the emphasis is more on the workmanship than the quality of the timber used. The majority of local carpenters still insist on merchantable timbers as raw material. In view of the importance of these timbers to the economy as a foreign exchange earner (see Chapter 1) it would be advisable to meet local needs from timbers with less export potential. Merchantable timbers preferred for high-class furniture, cabinet-making, and veneers include *Entandrophragms angolense* Gedu Nohor (Edinam); *E. cylindricum* West African Cedar, Sapele (Apenkwa); *E. utile* Utile, *Guarea cedrata* Scented Guarea (Kwabohoro); *G. thompsonii* Black Guarea (Kwadwuma); *Heritiera utilis* (*Tarrietia utilis*) (Nyankom); *Khaya anthotheca* White Mahogany (Krumben); *K. ivorensis* African Mahogany (Dubini); *Lovoa trichilioides* African Walnut (Dubinibiri); *Mansonia altissima* (Oprono); *Mitragyna ciliata* Abura, Poplar (Subaha); *Pericopsis elata* (*Afrormosia elata*) (Kokrodua), and *Tieghemella hechelii* (Baku). *Aningeria robusta* (samfena), previously little known, has recently shot up as a first-class timber for high-class furniture. The light brown discolouring in the heartwood of *Terminalia superba* Afara (Ofram) gives a decorative effect and is in demand for furniture. The white woods used for high-class furniture are *Turraeanthus africanus* Avodire and *Terminalia ivorensis* (Emire). The following timbers are equally suitable for quality furniture: *Canarium schweinfurthii* Incense Tree (also for veneers), *Carapa procera* Crabwood, *Copaifera salikounda* Bubinga (Entedua) (also for veneers); *Berlinia grandiflora* (Tetekono), *Trilepisium madagascariense* (*Bosqueia angolensis*) (Okure), *Cassia sieberiana* African Laburnum, *Coula edulis* Gaboon Nut (Bodwue), and *Daniellia oliveri* African Copaiba Balsam Tree (Sanya). Others are *Maesopsis eminii* (Onwam-dua), *Nauclea diderrichii* (Kusia) (also for veneers), *Nesogordonia papaverifera* (Danta), *Pseudocedrela kotschyi* Dry-Zone Cedar (Kurubeta), *Prosopis africana* (Sanga), *Pterocarpus erinaceus* Senegal Rose Wood Tree, African Kino; *Pycnanthus angolensis* African Nutmeg (Otie); and *Sterculia rhinopetala* Sterculia Brown (Wawabimma). The rest are *Strombosia glaucescens* var. *lucida* (Afena), *Tamarindus indica* Indian Tamarind, *Tectona grandis* Teak, *Zanthoxylum viride* (*Fagara viridis*) (Oyaanini), and *Ziziphus spina-christi* var. *microphylla*.

For cheap furniture, timbers used include *Acacia sieberiana* var. *villosa* (Kulgo), *Adenanthera pavonina* Bead Tree, Red Sandalwood; *Breonadia salicina* (*Adina microcephala*), *Afzelia bella* var. *glacior*, *Albizia coriaria* (Awiemfoseminaakoa), *A. lebbeck* Lebbek, *Alstonia boonei* (Sinduro) (also for veneers), *Anopyxis klaineana* (Kokote), *Antrocaryon micraster* (Aprokuma), *Blighia sapida* Akee Apple (Akye); *Burkea africana* (Pinimo), *Celtis mildbraedii* (Esa), and *Chrysophyllum delevoyi* (*albidum*) White Star Apple. Others are *Cola nitida* Bitter Cola (Bese); *Cordia millenii* Drum Tree, *Cylicodiscus gabunensis* African Greenheart (Denya); *Diospyros mespiliformis* West African Ebony, *Erythrophleum suaveolens* (*guineense*) Ordeal Tree (Potrodom), and *Erythroxylum mannii* Landa (Pepesia). Also used for simple furniture are *Ficus vallis-choudae* Aloma

Bli), *Khaya senegalensis* Dry-Zone Mahogany, *Mammea africana* African Mammy Apple (Bompagya); *Manilkara multinervis* subsp. *lacera* (*lacera*) African Pearwood, *Mitragyna inermis* (Kukyafie), *M. stipulosa* (Subaha-akoa), *Parinari excelsa* Guinea Plum (Ofam) and *Pycnanthus angolensis* African Nutmeg (Otie). The rest are *Sacoglottis gabonensis* (Tiabutuo), *Sterculia oblonga* Yellow Sterculia (Ohaa), *Symphonia globulifera* (Ehureke), *Treculia africana* African Bread Fruit (Brebretim); *Trichilia emetica* subsp. *suberosa* (*roka*) (Kisiga), *Triplochiton scleroxylon* (Wawa), *Uapaca guineensis* Sugar Plum (Kuntan); and *Zanha golungensis*. Chairs, tables, and other items of furnishings for both outdoor and indoor use are made with *Oxytenanthera abyssinica* Bamboo and the climbing palms *Laccosperma opacum* (*Ancistrophyllum opacum*) Rattan Palm, *L. secundiflorum* (*A. secundiflorum*) (Ayike), *Calamus deeratus* (Demmere) and *Eremospatha macrocarpa*. The midribs of *Raphia* species Wine Palm, Bamboo-palm are useful for lazy chairs.

The furnishings in church buildings consist mainly of the pulpit and the pews, both usually wooden. The lectern may be of wood as well. Durable timbers like *Milicia excelsa* (*Chlorophora excelsa*) Iroko (Odum); *M. regia* (*C. regia*) (Odum-nua), *Afzelia africana* (Papao), *Distemonanthus benthamianus* African Satinwood (Bonsamdua); *Daniellia thurifera* Niger Copal Tree (Kwanga) and *Morus mesozygia* (Wonton) are used for pews.

Organs, Pianos and Harmoniums

Wood plays an important part in the building of organs, pianos, and harmoniums. Manufacturers use the choicest timbers, as the price of an instrument is greatly influenced by the quality of wood used. *Entandrophragma cylindricum* Sapele (Apenkwa) and *Swartzia madagascariensis* Snake Bean are excellent for piano manufacture. The resonance box, keyboard, and hammers in a piano are all of wood. Electronic organs are now being made with chipboard.

Frames, Cabinets for Electronic Sets and Equipment

Electronic products like radios, radiograms, tape and video recorders, music centres, television sets, and loudspeakers are sometimes finished in wood. Hi-fi systems may use shelves or racks of wood. Taylor (1959) reports that *Terminalia ivorensis* (Emire) is used for radio cabinets in the United States. Veneers and curls are used to give a special finish to radiograms and cabinet works generally. Curls are veneers sliced from wood cut at the fork of a tree, and have a characteristic V- or Y-shaped interlocking design. Like figured wood, curls are highly priced.

Wood is sometimes used as a finishing for wall and table clocks and sewing machines. *Pericopsis elata* (*Afrormosia elata*) (Kokrodua) serves well as the base in the manufacture of sewing machines.

Veneers and Plywood

Besides the export of timber as primary products there is also the export of processed timber as self-assembly furniture, veneers, and plywood. Veneers are used in the manufacture of plywood. The middle part or core is of less economically important timbers like *Antiaris toxicaria (africana)* Bark Cloth Tree (Kyenkyen); *Albizia zygia* (Okoro), *Celtis mildbraedii* (Esa), *Daniellia ogea* Gum Copal Tree, and sometimes *Triplochiton scleroxylon* (Wawa), although it is one of the merchantable timbers. The core is covered on both sides by veneers from quality timbers and the layers glued together. There are three main veneer and plywood factories in Ghana. These are the African Timber and Plywood (Gh.) Ltd Samreboi, Gliksten West Africa Ltd Sefwi Wiawso, and Takoradi Veneer and Lumber Company (formerly F. Hills Gh. Ltd) at Takoradi. Veneers and plywood play an important part in the furniture industry.

Panelling

Wood is used to furnish or decorate interior walls as panels. Suitable timbers for panelling include *Copaifera salikounda* Bubinga (Entedua); *Entadrophragma angolense* Gedu Nohor (Edinam); *E. candollei* Omu, *E. cylindricum* West African Cedar, Sapele (Apenkwa); *Guarea cedrata* Scented Guarea (Kwabohoro); *Khaya ivorensis* African Mahogany (Dubini); *Lovoa trichilioides* African Walnut, *Mitragyna stipulosa* (Subaha-akoa), *Pericopsis elata* (*Afrormosia elata*) (Kokrodua), *Terminalia ivorensis* (Emire), and *Turraeanthus africanus* Avodire. As well as in houses, wood is used for decoration and pannelling in railway coaches, yachts, boats, and ships. *Tieghemella heckelii* (Baku) and *Turraeanthus africanus* Avodire have both been successfully used for the interior of Ghana Railways coaches and in the ocean liner *Queen Mary*. *Khaya ivorensis* African Mahogany (Dubini) and *Distemonanthus benthamianus* African Satinwood (Bonsamdua) are equally used as cabin fittings. (For flooring see Chapter 4.)

SIX
Fuel

The demand for wood as a source of fuel for industrial and especially for domestic purposes is likely to be even greater as fossil fuels become scarcer and more expensive. To many rural dwellers wood is not only the cheapest source of fuel, and the most convenient and accessible, but also the only one; except, of course, charcoal, which is also derived from trees. With no other means of preparing the evening meal, fuel-wood is an essential item in rural life. Traditionally women keep the house and cook the meals and, owing to the dominant role which fuel-wood plays in food preparation, women (and to some extent young girls) feature prominently in its collection. According to *CERES,* January–February 1983, fuel-wood accounts for more than a fifth of all energy used in developing nations. In West Africa it is, in fact, by far the largest source of domestic energy.

Normally farmers and hunters, when returning to their homes, will collect firewood and carry it with them. For this reason, a large quantity of half-charred branches is deliberately left on new farms when these are being prepared for crops. Clearing the forest for farming is a major source of fuel-wood supply. In addition fuel-wood is obtainable as dead wood on the forest floor, from the branches of felled timber trees, and also from felled small trees which are not big enough to be converted into timber or used for other purposes. Usually no fees are paid for these collections, fuel-wood in rural areas, unlike urban and city centres, being almost entirely free. In addition, fuel-wood in rural areas is obtained from sources within walking distance. Certain species of trees are preferred for fuel, for example *Celtis mildbraedii* (Esa) and *C. zenkeri* (Esa-kokoo), both of which give a hot flame and little smoke. *Corynanthe pachyceras* (Pamprama) and *Trichilia tessmannii* (*lanata*) (Tanuronini) are also popular fuel-woods, but in their absence collectors make do with any equally suitable and available middle-storey trees. The emergents or upper canopy trees, the tallest trees in the forest, are usually left for shade in farms. Below these are the middle-storey trees mostly cut during farming, which constitute the bulk of fuel-wood.

The proportion of urban dwellers using fuel-wood and charcoal is still considerably high despite the availability of alternative forms of cooking fuel such as electricity and gas. Though the running cost of these alternatives is comparatively cheaper, especially of electricity, a modern cooker is beyond the purchasing power of the average worker. The influx of young, able-bodied men and women to the city centres for a taste of urban life-style has contributed to the concentration of the majority of the population in these few centres. The main source of the fuel-wood needs of these urban dwellers is the rural areas — from which it must be transported long distances by road and rail. Consequently the price is steadily rising, and the average family in an urban centre (such as in the Sahara region) spends a fourth of its income on wood and charcoal required for fuel. Another reason for the high price is the demand. Fuel-wood plantations or energy cropping on the outskirts of some of the towns, cities, and urban centres, and the provision of wood-lots in new townships, provide only a meagre proportion of the needs of these dwellers. Energy cropping generally means the cultivation of trees, sugar-cane, oil-producing plants and so on, for energy purposes. In addition to the primary objective of supplementing fuel-wood needs, these plantations protect crops and animals from the effects of desiccating winds, stabilize the local climate, and prevent soil erosion. They provide hunting grounds for game and a source of wild plants of botanical interest. Their aesthetic value is appreciated, as well.

Azadirachta indica Neem Tree and *Cassia siamea*, both introduced, are two of the common plantation trees along the coast. In the Guinea savanna woodland *Tectona grandis* Teak, *Gmelina arborea*, trees from the India-Burma Thailand region, and *Anogeissus leiocarpus* (Sakanee) are planted in addition. For convenience, the plantation trees are normally cut before they overgrow into large stands, stacked into cords, and sold as such. Normally two men cut and stack a cord a day. A cord is a measure of cut wood, usually 128 cubic feet. The billets are 4 feet long and stacked 4 feet high and 8 feet wide. The coppice system of regeneration is employed in plantations. Fuel-wood is also obtained from the scrubs, thickets, and woodland in the savanna. *Zanthoxylum xanthoxyloides* (*Fagara zanthoxyloides*) Candlewood and *Dialium guineense* Velvet Tamarind are preferred trees. With the heavy demand for fuel-wood and the constant practice of clearing the land for agricultural purposes, it may now take a whole day's trek, in some cases, to fetch a head-load of fuel-wood in the savanna zone.

Fuel-wood from the forest and the plantations is often converted into charcoal — a process whereby the wood is burned under controlled conditions and in a limited air supply. A large proportion of felled wood from parts of the Subri Forest Reserve near Daboase in Ghana is used to burn for charcoal. The clearings produced at this reserve are to be planted with *Gmelina arborea*, the produce of which will supply a proposed paper factory. Charcoal is lighter and less bulky than fuel-wood, so it is easily transported and stored. It is also more efficient, producing a steady heat, and has the advantage of burning with little or no smoke. These characteristics make charcoal a popular and much preferred fuel. It is, however, comparatively expensive.

Trees suitable for fuel-wood are generally equally suitable for charcoal. However, *Erythrophleum suaveolens* (*guineense*) Ordeal Tree (Potrodom), which is a good fuel-wood, produces poor quality charcoal. Trees with higher specific gravity make better charcoal than those with lower specific gravity. *Azadirachta indica* Neem Tree, for example, makes a better charcoal than *Cassia siamea*.

Alcohol extraction from sugar-cane as a source of fuel for automobiles is practised. Brazil, for example, has proposed running its automobiles exclusively on such alcohol; and Australia, New Zealand, and South Africa are seriously considering the commercial use of sugar-cane alcohol as fuel. The possibility of using wood wastes as raw material for fuel to drive combustion engines is also under research. In Ghana gas produced from a maxi-bag of charcoal has been used to fuel a Land Rover vehicle from Takoradi to Accra and back — a distance of about 320 miles.

Suitable Trees for Fuel-wood

The type of fuel-wood available in an area is determined by the vegetation, that is, the presence or absence of plants, their distribution and their frequency. In West Africa there are four types of vegetation: the Sahel, Sudan and Guinea savanna woodland to the north; and the high forest to the south. The forest is characterized by a rich distribution of a wide range of trees suitable for fuel-wood. As such there is more to choose from compared to the savanna and Sahel areas where there is scanty and sparsely distributed vegetation with comparatively fewer trees. The limited availability of suitable fuel-wood and, as is often the case, the remote location have contributed to the use of less suitable materials as fuel. These are the fronds of *Elaeis guineensis* Oil Palm and that of *Cocos nucifera* Coconut Palm. The husks and shells of Coconut Palm, the dried fruits of *Delonix regia* Flamboyante, Flame Tree and the cobs and chaff of *Zea mays* Maize and of *Oryza sativa* Rice are used as fuel in emergency.

Plants used as brushwood in the savanna areas are *Grewia carpinifolia* (Ntanta), *Uvaria chamae* (Akotompotsen), *Griffonia simplicifolia* (Kagya), *Tiliacora dinklagei* (Susanfo), *T. funifera* (*warneckei*) (Katopa), and *Carissa edulis* (Botsu). Others are *Mallotus oppositifolius* (Satadua), *Deinbollia pinnata* (Woteegbogbo), *Vernonia amygdalina* Bitter Leaf, *Dichrostachys cinerea* (*glomerata*) Marabou Thorn, and the small branches of *Azadirachta indica* Neem Tree. The clippings of the decorative *Bougainvillea* are used, so are the branches of *Pithecelobium dulce* Madras Thorn after feeding the leaves to goats and sheep. The rest are *Securinega virosa* (Nkanaa), *Byrsocarpus coccineus* (Awennade), *Combretum* species, *Clausena anisata* Mosquito Plant, *Allophyllus africanus* (Hokple), *Chaetacme aristata* (Kodia), and *Leucaena leucocephala* (*glauca*) Leucaena. Brushwood is used by bakers to raise the temperature of the oven after the first set of loaves is baked, since it blazes rapidly. Wood shavings are used for the same purpose. Brushwood is also used in general household cooking in the absence of roundwood. As it is light and slender,

6. FUEL

large quantities are required for cooking.

SAVANNA AND WOODLAND FUEL-WOOD

The introduced fuel-woods are *Cassia siamea*, *Azadirachta indica* Neem Tree, *Eucalyptus* species, *Gmelina arborea*, *Tectona grandis* Teak, *Casuarina equisetifolia* Whistling Pine, and *Dalbergia sissoo*. These are the main stands established in fuel-wood plantations in addition to the indigenous *Anogeissus leiocarpus* (Sakanee). The following trees are better priced as fuel-wood in the savanna areas and the transitional zones: *Hymenostegia afzelii* (Takorowa), *Lecaniodiscus cupanioides* (Dwindwera), *Lonchocarpus sericeus* Senegal Lilac, *Albizia* species, *Pterocarpus santalinoides* (Hote), *Zanthoxylum xanthoxyloides* (*Fagara zanthoxyloides*) Candle Wood, and *Millettia* species. *Laguncularia racemosa* White Button Wood (Abin); *Avicennia africana* Mangrove, and *Rhizophora racemosa* Red Mangrove, all plants of the coastal swamps and mangroves, are useful sources of fuel-wood. These trees are preferred because they are efficient, burning with an intense heat of long duration.

As fuel-wood *Morinda lucida* Brimstone Tree (Kankroma) produces much ash. Fallen branches of both *Vitellaria paradoxa* (*Butyrospermum paradoxum* subsp. *parkii*) Shea Butter Tree and *Parkia bicolor* (Asoma) are useful as fuel-wood. Other savanna trees useful as fuel-wood are *Acacia albida* (Gozanga), *Albizia lebbeck* East Indian Walnut, *Afzelia africana* (Papao), *Cassipourea barteri*, *Combretum nigricans* var. *elliotii*, *Chrysobalanus icaco* (*ellipticus* and *orbicularis*) (Abeble), *Conocarpus erectus* Button Wood (Koka); *Cordia rothii*, *Crossopteryx febrifuga* African Bark (Pakyisie); *Dodonaea viscosa* Switch Sorrel, and *Drypetes floribunda* (Bedibesa). Others are *Manilkara multinervis* subsp. *lacera* (*lacera*) African Pearwood, *Ficus platyphylla* Gutta-Percha Tree, *Millettia thonningii* (Sante), *Quassia undulata* (*Hannoa undulata*) (Kunmuni), and *Talbotiella gentii* (Kpeteple) (perhaps the only true endemic species to Ghana). The rest are *Syzygium guineense* var. *guineense* (Sunya), *Terminalia glaucescens* (Ongo), *T. avicennioides* (Petni), *Uapaca togoensis* (Dzogbedzro), *Vitex doniana* (Afua), *Ziziphus mauritiana* Indian Jujube, and *Phoenix reclinata* Wild Date Palm.

HIGH FOREST FUEL-WOOD

The forest vegetation is rich in fuel-wood. Soft woods like *Musanga cecropioides* Umbrella Tree or *Sterculia tragacantha* African Tragacanth (Sofo) are unsuitable as fuel-wood, and very hard ones like *Bussea occidentalis* (Kotoprepre) are equally unfavourable. Branches that break off in a storm or otherwise, however, serve as first-class fuel-wood. Trees of medium hardness are generally useful as fuel-wood. The preferred ones include *Psydrax subcordata* (*Canthium subcordatum*) (Tetiadupon), *Uapaca heudelotii* (Kuntanakoa), *Sacoglottis gabonensis* (Tiabutuo), *Trichilia monadelpha* (*heudelotii*) (Tanduro), *Petersianthus macrocarpus* (*Combretodendron macrocarpum*)

Stinkwood Tree (Esia); and *Funtumia* species. Others are *Lannea welwitschii* (Kumanini), *Albizia zygia* (Okoro), *Lepisanthes senegalensis* (*Aphania senegalensis*) (Akisibaka), *Anopyxis klaineana* (Kokoti) (favourite fuel in the mines), *Blighia unijugata* (Akyebiri), *B. welwitschii* (Akyekobiri), *Carapa procera* Crabwood, *Cola gigantea* var. *glabrescens* (Watapuo), *Cremaspora triflora* (Otu), *Dacryodes klaineana* (Adwea), *Daniellia oliveri* African Copaiba Balsam Tree, *Distemonanthus benthamianus* African Satinwood (Bonsamdua); *Heisteria parvifolia* (Sikakyia), and *Macaranga hurifolia* (Kpazina). The rest are *Manilkara obovata* (*multinervis*) (Berekankum), *Morus mesozygia* (Wonton), *Napoleonaea leonensis* (*Napoleona leonensis*), *Newbouldia laevis* (Sasanemasa), *Ochna multiflora, Octoknema borealis* (Wisuboni), *Oxyanthus unilocularis* (Kwaetawa), *O. speciosus* (Korantema), *Protomegabaria stapfiana* (Agyahere), *Strombosia glaucescens* var. *lucida* (Afena), *Swartzia fistuloides* (Asomanini), *Uapaca guineensis* Sugar Plum (Kuntan); and *Zanthoxylum gilletii* (*Fagara macrophylla*) (Okuo). *Theobroma cacao* Cocoa and *Cola nitida* Cola Nut, Bitter Cola also serve as good sources of fuel-wood. Roundwood is by far the major source of fuel-wood supply. Investment in fuel-wood is a profitable enterprise for both fuel-wood dealers and crop farmers.

SAWN WOOD OR TIMBER

The off-cuts and sapwood from sawmills are useful as fuel-wood. Most of the timbers are hardwoods and provide useful material for fuel for both industrial and domestic purposes.

OTHER SOURCES OF FUEL

These are the by-products of sawn timber and carpentry — sawdust, wood shavings, and wood wool. A sawdust stove has been designed which uses this by-product efficiently. It is, however, not yet popular — much of the sawdust still goes to waste. The branches of *Cocos nucifera* Coconut Palm, the husks and shells, are used extensively as fuel, especially in areas like the south-western part of Ghana, where the crop is widely cultivated and fuel-wood is scarce. Similarly, branches, kernels, shells, and the pericarp of *Elaeis guineensis* Oil Palm are used as fuel. The shells of the kernels are especially preferred and almost exclusively used by local blacksmiths as fuel. Bagasse, the dried stems of crushed *Saccharum officinarum* Sugar-cane and a waste product in sugar manufacture, is used as fuel for smoking fish and other purposes. Millet and maize stalks and rice chaff are occasionally used as a fuel substitute, depriving the soil of useful sources of plant nutrients. In extreme scarcity (especially in the Sahel), when all the crop and livestock residues have been used or destroyed by sweeping fires, dried dung is burnt as fuel. This practice further deprives agricultural land of a cheap but nevertheless important source of plant nutrients.

6. FUEL

Sources of Fuel-wood

FORESTS

Timber and fuel-wood together make up the major produce of the forest — the volume of fuel-wood exceeding by far all the timber, both locally used and exported (see Chapter 1). Forest Reserves, unreserved forests, and protected timber lands are all sources of fuel-wood — the greater part being obtained from the latter two. In Ghana, the forest area is the main source of the country's fuel-wood supply from where roundwood cut into billets is transported to the savanna areas, particularly the southern savanna where there is a concentration of population.

SAVANNA WOODLAND

In Ghana, this covers about two-thirds of the country, and it is the next important source of fuel-wood after the forest. Although the area is about twice that of the forest zone, the supply is much less because of the sparsely distributed vegetation.

PLANTATIONS

A third source of fuel-wood, in order of importance and supply, is the plantation (see Table 6.1). Apart from the Kumasi Town Plantation, the fuel-wood plantations in Ghana are sited near population centres in the savanna zone to help meet the fuel needs of the urban dwellers. Though these fuel-wood plantations are small individually, they are nevertheless important locally. Exotic species are usually planted, and the coppice system of regeneration is employed. In this system the shoots spring up from the sides of the cut stools. Plantations are normally established many miles from the cities and towns, but eventually these grow and gradually expand in closer proximity to these centres. Urbanization, therefore, now threatens the existence of some of these plantations, like the Kumasi Town Plantation, just as farming activities threaten afforestation in the forest zone.

Charcoal

Charcoal is a popular source of fuel in urban centres despite the relatively high cost compared to fuel-wood. The long distances over which charcoal is transported to urban centres accounts for its high price; a bagful at the producing centres is less than half the price charged in the cities. Some homes in the urban centres, for this reason, manufacture their own charcoal when fuel-wood is available. Charcoal manufacture is an important rural industry. Traditionally, the time-consuming and laborious shallow pit method is used.

The Utilization Branch of the Department of Forestry in Ghana is

Table 6.1 Fuel-wood plantations in Ghana

Name	Area in hectares	Location
Achimota Fuel-wood Plantation F.R.	404	Achimota
Agali Fuel-wood Plantation	980 approx	Dzodze
Ankaful Fuel-wood Plantation F.R.	243	Ankaful
Bazura Bridge Fuel-wood Plantation	218	Bazua
Brimsu Plantation	1070	Brimsu
Bumbugu Extension Block 1 F.R.	41	Bumbugu/Bawku
Bumbugu Plantation F.R.	414	" "
Inchaban Waterworks Plantation	254	Inchaban
Karkaa Plantation	240 approx	Nandom/Lambusie
Komenda Fuel-wood F.R.	250	Komenda
Kpandu Plantation	180 approx	Kpandu
Kpong Plantation	40 approx	Kpong Agric. Station
Kumasi Town Fuel-wood Plantation	170 approx	Kumasi
Mumford Fuel-wood Plantation	267	Apam
Prampram Fuel-wood Plantation	150 approx	Prampram
Red Volta East F.R. (part only)	21811	Zuongo
Sekondi Waterworks Blocks I and II	750 approx	Angu
Senya Fuel-wood F.R.	104	Senya Bereku
Tamale Fuel-wood Block I and II	215	Tamale
Tamale Waterworks Plantation	115 approx	"
Upper Tamne Blocks I and II	1729	Pusiga
Winneba State Fuel-wood Plantation F.R.	161	Winneba
Yendi Education Plantation F.R.	259	Yendi
Yendi Town Plantation	75	"
Zawse Plantation	228	Zawse/Bawku

educating charcoal manufacturers on how to improve production with portable kilns at centres like Opon Mansi, Abotoase, Adjina, and Kumasi. However, the initial cost of a portable kiln is not within the purchasing power of most rural charcoal traders. The Charcoal Research team has made a promising start with clay kilns, which have the advantage of low initial capital cost. They are efficient, carbonization is complete, and there are high yields of quality charcoal. The Research Unit is exploring the possibility of a combination of clay and pit kilns. Charcoal is also made by the retort method, which has been tried in Ghana. The wood is placed inside two long cylinders made from 55-gallon oil drums and placed side by side, each with a door that closes tightly and some means to let tar and gases escape. Heating is from the outside, with no air entering the drums. When heated to the right temperature the wood begins to carbonize and turns to charcoal which can be collected after cooling. With this method the charcoal is ready the next day instead of after seven to ten days by the local method. The retort can be built to allow collection of valuable by-products.

The charcoal project at Daboase, in the Western Region of Ghana, is most probably the largest single one in the country. For carbonization it uses mixed tropical hardwoods logged from the degraded Subri Forest Reserve, and produces about 2000 33 kg bags in six months, 77 per cent of this production coming from private entrepreneurs.

In the absence of electric cookers or fossil fuel, charcoal is preferred to fuel-wood since it is lighter and more compact, therefore easier to store and cheaper to transport. It burns with a steady and concentrated heat and, being completely carbonized, without smoke or soot — the greatest advantage over fuel-wood. However, in preparing charcoal half the wood's energy is wastefully burned away. It also has the disadvantage of being more costly. Charcoal may be used in a comparatively small room, provided it is fairly well ventilated. It also serves as a source of heat for ironing, both for the box iron and the solid cast-iron type. Charcoal stores indefinitely.

FOREST TREES FOR CHARCOAL

All hardwoods used as fuel-wood are generally suitable for the manufacture of charcoal. *Albizia zygia* (Okoro), *Erythrophleum ivorense* Sasswood Tree, and *Corynanthe pachyceras* (Pamprama) are popular charcoal trees. The charcoal from *A. adianthifolia* (Pampena) is particularly preferred by goldsmiths, and that of *Uapaca corbisieri* (*esculenta*) (Kuntammiri) by blacksmiths. Charcoal prepared from *Piptadeniastrum africanum* (Danhoma) gives good heat. Other high forest trees suitable for making charcoal are *Afrosersalisia afzelii* (Bakunini), *Coula edulis* Gaboon Nut (Bodwue); *Diospyros* species, *Pentaclethra macrophylla* Oil Bean Tree (Ataa); *Parinari excelsa* Guinea Plum (Ofam); *Maranthes glabra* (*P. glabra*) (Punini), *M. polyandra* (*P. polyandra* var. *polyandra*) (Abrabesi), *Trichilia prieuriana* (Kakadikro), *Trema orientalis* (*guineensis*) (Sesea), and *Phyllocosmus africanus* (*Ochthocosmus africanus*). *Cylicodiscus gabunensis* African Greenheart (Denya), a high forest species, makes a good, heavy charcoal. However, due to its hard nature, the whole log must be burned gradually on the forest floor — the charcoal being chipped off and collected as it forms.

SAVANNA AND WOODLAND TREES FOR CHARCOAL

Savanna trees that give good charcoal include *Acacia nilotica* var. *tomentosa*, *A. polyacantha* subsp. *campylacantha* African Catechu, *Anacardium occidentale* Cashew Nut, *Blighia sapida* Akee Apple (Akye); *Burkea africana* (Pinimo), *Balanites aegyptiaca* Desert Date, *Calotropis procera* Sodom Apple, *Combretum micranthum* (Landaga), *Dialium guineense* Velvet Tamarind, *Dichrostachys cinerea* (*glomerata*) Marabou Thorn, *Faurea speciosa* (Se ngo se bari), *Ozoroa reticulata* (*Heeria insignis*) (Nasia), and *Parinari curatellifolia* (Atena). The rest are *Parkinsonia aculeata* Jerusalem Thorn, *Ximenia americana* Wild Olive, *Daniellia oliveri* African Copaiba Balsam Tree, *Dendrocalamus strictus* Male Bamboo, *Maytenus senegalensis* (Kumakuafo), *Prosopis africana* (Sanga), *Psidium guajava* Guava, *Swartzia madagascariensis* Snake Bean, and *Tamarindus indica* Indian Tamarind. *Oxytenanthera abyssinica* Bamboo and the shells of *Cocos nucifera* Coconut Palm make good charcoal.

Uses of Charcoal and Fuel-wood

Charcoal has a few industrial and market applications. As fuel it is used to dry fish, extract metals, and manufacture lime and cement. By-products in the production of charcoal by the retort method may be used for weather-proofing, preventing rust, or protecting houses from termites. The bulk of charcoal is used domestically for cooking and heating.

Fuel-wood is used as a source of heat energy by some commercial enterprises such as bakeries and confectioners, metal smelting, ceramic factories, commercial *gari* manufacturers, distilleries, canneries, hotels, restaurants; and for drying fish, tobacco, grain, copra, and other agricultural products. Fuel-wood may be used to heat boilers in steam locomotive engines. Wood gas has also been used to power automobiles. By far the largest commercial users of fuel-wood are the sawmills. Some sawmills, like African Timber and Plywood at Samreboi in the Western Region of Ghana, utilize the off-cuts, sapwood and, sawdust as fuel to heat boilers which generate electricity. These drive the band and circular saws and provide lighting for both factory and township. Timber for veneers is softened by boiling and then sliced into sheets; and timber off-cuts are used as fuel. The bulk of fuel-wood supply is however used for domestic purposes.

SEVEN
Tools and Crafts

The direct benefits of the forest include the products that may be obtained from it. Taylor (1962) considered these products under two categories:

1. major forest produce — the woody material or timber from the trees; and
2. minor forest produce — any produce other than the timber (see Chapter 1).

Many small-scale rural industries owe their existence to these minor forest products and trees which are too small to be converted into timber. They are a convenient choice as raw materials for innumerable crafts and household utility articles. Some of the reasons for this are the comparative ease with which they may be worked and the ability to obtain them locally at a relatively cheap price. Very often the manufacturing centre is sited near the source of material. This pre-empts time and cost wastage on transportation. Canoes, for example, are hollowed out of *Triplochiton scleroxylon* (Wawa), a huge forest tree, hundreds of miles from the sea. The finished product is conveyed by road or rail to the coast.

There are many such successful rural industries, large and small, engaged in carving stools, drums, utensils, and dolls; weaving baskets, chairs, mats, hats, traps, and doormats; manufacturing gun stocks, tool handles, and brushes; and making musical instruments, bamboo and gourd crafts, etc. These crafts are mostly for local use, but, some of them, like the famous Bolga baskets from Ghana, are already being exported. It would, of course, prove economically worth while to develop foreign markets for these and other products. Assistance schemes designed to encourage production would enable the ensuing demand to be met.

Tools

Many tools have wooden parts, while others are entirely made of wood.

GUN STOCKS

The price of a gun is largely determined by the choice of wood, the very special steps taken by the gunmakers in its treatment, and the level of craftmanship in its overall design and finish. Webley and Scott, the well-known gunmakers in Birmingham, use Walnut from the South of France for their double-barrel and single-barrel shotguns and Beech from Central Europe for air rifles and bolt action rifles. The wood from the centre of the tree — heart wood — has the correct degree of hardness essential in a gun stock. To prevent distortion and warping the wood is seasoned to an 8 per cent moisture content to stabilize it before machining. (For seasoned wood, see Chapter 5.) In West Africa timbers for gun stocks include *Lovoa trichilioides* African Walnut, *Nesogordonia papaverifera* (Danta), *Carapa procera* Crabwood, *Diospyros mespiliformis* West African Ebony (Okisibiri); *Pseudocedrela kotschyi* Dry-Zone Cedar, *Cola nitida* Cola Nut, Bitter Cola (Bese); *C. millenii* (Anasedodowa), *Hexalobus crispiflorus* (Etwa prada), and *Uvariastrum pierreanum* (Ankumakaba).

CARPENTERS' PLANES

Wood used for planes include *Citrus aurantium* Sour or Seville Orange, *Gardenia ternifolia* (Peteprebi), *Hunteria elliotii* (Bisi), *Lecaniodiscus cupanioides* (Dwindwera), *Picralima nitida* (Ekuama), *Pleiocarpa bicarpellata* (Kakali), *P. mutica*, and *P. pycnantha* var. *tubicina* (Kakapimbe) (the last three also for combs).

HOE HANDLES

Trees used are *Acacia sieberiana* var. *villosa* (Kulgo), *Baphia nitida* Camwood (Odwen) (also for axes); *Dialium guineense* Velvet Tamarind (Asenamba); *Dalbergia saxatilis* (Nuodolega) (also for axes), *Gardenia ternifolia* (Peteprebi) (also for other tools), *Lonchocarpus sericeus* Senegal Lilac (Keklengbe); *Microdesmis puberula* (Fema), *Monodora tenuifolia* (Motokuradua) (also for axes), *Ochna afzelii* (Okoli Awotso), *O. schweinfurthiana* (Bibie), *Parkia clappertoniana* West African Locust Bean, *Psidium guajava* Guava (also for other tools), *Rothmannia whitfieldii* (Sabobe), *Trichilia emetica* subsp. *suberosa* (*roka*) (Kisiga) (also for axes), and *Ziziphus mucronata* Buffalo Thorn (also for axes).

AXE HANDLES

Suitable trees used are *Burkea africana* (Pinimo), *Christiana africana* (Sesedua), *Diospyros kamerunensis* (Omenewa) (also for other tools), *Eriocoelum kerstingii* Monkey Akee, *Hibiscus tiliaceus* (Nwohwea), *Hymenostegia afzelii* (Takorowa), *Monodora myristica* (Ayerewamba), *Placodiscus pseudostipularis* (Ankyewobiri), *Prosopis africana* (Sanga) (also for other tools), *Swartzia madagascariensis* Snake Bean, and *Thevetia peruviana* 'Milk Bush', Exile Oil Plant. (See also Hoe handles above.)

TOOL HANDLES (GENERAL)

Trees used include *Acacia* species, *Pericopsis laxiflora* (*Afrormosia laxiflora*) Satinwood, *Antiaris toxicaria* (*africana* and *welwitschii*) Bark Cloth Tree

7. TOOLS AND CRAFTS

(Kyenkyen); *Balanites aegyptiaca* Desert Date (Gongu); *Trilepisium madagascariense* (*Bosqueia angolensis*) (Okure), *Caloncoba echinata* (Gorli) (also for combs), *Calpocalyx brevibracteatus* (Atrotre), *Carpolobia* species (also for combs), *Cassia sieberiana* African Laburnum, *Celtis africana* (Yisa), *Chrysophyllum delevoyi* (*albidum*) White Star Apple (Akasaa); *Petersianthus macrocarpus* (*Combretodendron macrocarpum*) Stinkwood Tree (Esia); *Combretum glutinosum* (Nkunga), *C. paniculatum* (Omeha), and *Crescentia cujete* Calabash Tree (Dweraba). Others are *Dalbergia hostilis* (Wota), *D. saxatilis* (Nuodolega), *Dichrostachys cinerea* (*glomerata*) Marabou Thorn, *Diospyros* species Ebony, *Dodonaea viscosa* Switch Sorrel (Fomitsi); *Ehretia trachyphylla* (Okyini), *Gardenia erubescens* (Dasuli), *Heinsia crinata* Bush Apple, *Heisteria parvifolia* (Sikakyia), *Hexalobus monopetalus* var. *monopetalus*, *Hunteria umbellata* (also for combs), *Khaya ivorensis* African Mahogany (Dubini); *Lecaniodiscus cupanioides* (Dwindwera), *Malacantha alnifolia* (Fafaraha), *Millettia thonningii* (Sante), *Napoleonaea vogelii* (*Napoleona vogelii*) (Obua) (also used for *tapori*, masher), and *Nesogordonia papaverifera* (Danta). The rest are *Omphalocarpum elatum* (Timatibre), *Oxytenanthera abyssinica* Bamboo, *Piliostigma thonningii* (Opitipata), *Rinorea ilicifolia*, *Schrebera arborea* (Kokofobene), *Spondias mombin* Hog Plum, (Ataaba); *Teclea verdoorniana* (Owebiribi), and *Xylopia quintasii* Elo, (Asimba).

WOODEN PLOUGHS

Trees used for ploughs include *Acacia farnesiana*, *A. nilotica* var. *tomentosa*, and *Maerua crassifolia* — all savanna plants since bullock ploughing is practised mainly in the savanna region. *Ziziphus mucronata* Buffalo Thorn, *Albizia adianthifolia* (Pampena), and *Trema orientalis* (*guineensis*) (Sesea) are used for yokes.

MALLETS AND CUDGELS

Heavy and durable timbers are used for mallets. Some of these are *Acacia nilotica* var. *tomentosa*, *Casuarina equisetifolia* Whistling Pine, *Dalbergia saxatilis* (Nuodolega), *Diospyros mespiliformis* West African Ebony (Okisibiri); *Hymenocardia acida* (Sabrakyie), *Prosopis africana* (Sanga), *Rinorea ilicifolia*, *Tamarindus indica* Indian Tamarind, *Terminalia avicennioides* (Petni), *T. glaucescens* (Ongo), and *T. macroptera* (Kwatiri).

SPEAR SHAFTS

Suitable poles for spear shafts include those from *Acacia sieberiana* var. *villosa* (Kulgo), *Cola laurifolia* Laurel-Leaved Cola, *Grewia venusta* (both also for bows), *C. mollis*, *C. villosa*, *Morelia senegalensis*, *Greenwayodendron oliveri* (*Polyalthia oliveri*) (Duabiri), *Teclea verdoorniana* (Owebiribi), *Xylopia quintasii* Elo (Asimba); and *Ziziphus spina-christi* var. *microphylla*. *Dendrocalamus strictus* Male Bamboo is equally suitable.

BOWS

These are made from flexible sticks. They include *Acacia ataxacantha*, *Cnestis ferruginea* (Akitase), *Cola caricifolia* Monkey Cola, *Dichrostachys cinerea* (*glomerata*) Marabou Thorn, *Glyphaea brevis* (Foto), *Grewia bicolor*, *Heisteria parvifolia* (Sikakyia), *Hymenodictyon floribundum* (Amandidua), *Lannea acida* (Kuntunkuri), *Mallotus oppositifolius* (Satadua), *Manilkara obovata* (*multinervis*) (Berekankum), and *Massularia acuminata* (Pobe). The rest are *Picralima nitida* (Ekuama), *Pterocarpus erinaceus* Senegal Rose Wood Tree, African Kino; *Scaphopetalum amoenum* (Nsoto), *Securidaca longepedunculata* (Kpaliga), *Strophanthus preussii* (Dietwa), *S. sarmentosus* (Adwokuma), *Syzygium guineense* var. *guineense* (Sunya), *Terminalia glaucescens* (Ongo) (roots), *T. macroptera* (Kwatiri), *Xylopia acutifolia* (Dwombobre), *X. aethiopica* Spice Tree (Hwenetia); *Ziziphus mucronta* Buffalo Thorn, and *Oxytenanthera abyssinica* Bamboo. The following are suitable for bow strings: the fibre of *Calotropis procera* Sodom Apple and *Triumfetta cordifolia* Burweed (Ekuba); the stems of *Oncinotis glabrata* and *O. nitida*; and the stem peeling of bamboo.

ARROWS

Sticks from the following trees are useful for arrows: *Albizia adianthifolia* (Pampena), *Cordia rothii*, *Jaundea pubescens*, *Picralima nitida* (Ekuama), *Heisteria parvifolia* (Sikakyia), and *Sesbania sesban* Egyptian Sesban (Tingkwanga). Suitable grasses can also be used as arrow shafts. These are *Pennisetum unisetum* (*Beckeropsis uniseta*), *Panicum subalbidum*, *Phragmites karka* Common Reed, *Olyra latifolia*, thin stems of *Oxytenanthera abyssinica* Bamboo and *Saccharum spontaneum* var. *aegyptiacum*. Others are *Chasmopodium caudatum* and *Loudetia phragmitoides*.

Crafts

Indigenous crafts in West Africa include carving, basketry, matting, bark cloth making, seed craft, broom making, sponge preparation, fibre craft, and the manufacture of musical instruments.

CARVING

Carved objects include stools, drums, boats, mortars, utensils, ladles, combs, and masks. Others are paddles, dolls, puppets and images, *oware* (a game of two usually played with Bonduc counters), and *tapori* (masher). Stools and drums have a historical significance and are pertinent to the cultural development of many ethnic groups in the region. Quarcoo (1968) observed that the Akan stool can exemplify the way of life as much as kinship patterns, religious beliefs, and history, not only in Akan land today, but also among other non-Akan groups in Ghana. He added that the styling, context, and function of the stool in contemporary Ghana reflect tradition and development trends in the social structure.

7. TOOLS AND CRAFTS

Stools

Timbers of even grain, medium weight, and easy to work and finish well are used to carve stools. *Holarrhena floribunda* False Rubber Tree (Sese) (also for utensils, *oware*, and combs) is considered the best white wood for carving native stools. Others are *Cordia platythyrsa* (Tweneboa), *Alstonia boonei* (Sinduro) (both for spoons, utensils, and *oware*), *Funtumia africana* False Rubber Tree (also for plates, bowls, and ladles), *F. elastica* West African Rubber Tree (Funtum) (also for *oware*), and *Christiana africana* (Sese-dua) (also for basins). Also useful in stool carving are *Vitellaria paradoxa* (*Butyrospermum paradoxum* subsp. *parkii*) Shea Butter Tree, *Ceiba pentandra* Silk Cotton Tree (Onyina); *Diospyros mespiliformis* West African Ebony (also for combs), *Quassia undulata* (*Hannoa undulata*) (Kunmuni) (also for bowls and troughs), *Lannea acida* (Kuntunkuri) (also for plates and utensils), *Pterocarpus erinaceus* African Kino (also for spindles), *Ricinodendron heudelotii* (Wamma) (also for spoons, ladles, bowls, and plates), *Sterculia tragacantha* African Tragacanth (Sofo) (also for basins); *Terminalia superba* (Ofram), *Triplochiton scleroxylon* (Wawa) (also for plates and platters), and *Vitex doniana* Black Plum (Afua) (also for spoons). Like the traditional kente cloth in Ghana, stools are popular gift items.

Drums

These are made from hollowed-out tree stumps or can also be made from barrel staves (see Chapter 4), more common in the Volta Region of Ghana. In hollowed-out drums, trees that give the required vibration and resonance are the best. The *fontomfrom* and *atumpan* drums are usually used by the Ashanti and southern tribes of Ghana and the *dondo* by the Hausa and other northern tribes of West Africa. *Cordia platythyrsa* (Tweneboa) is the wood used for the majority of Akan drums (also for spoons). Others are *C. millenii* Drum Tree (also for bowls and utensils), *Alstonia boonei* (Sinduro) (also for spoons, bowls, basins, and plates), *Borassus aethiopum* Fan Palm, *Cleistopholis patens* Salt and Oil Tree (Ngo ne nkyene), *Nauclea diderrichii* (Kusia) (popularly used for mortars), *Diospyros mespiliformis* West African Ebony (Okisibiri), and *Omphalocarpum elatum* (Timatibre) (also for bowls). Others are *Vitex grandifolia* (Supuwa), *V. micrantha* (Nyamele-buruma) (popularly used for Koranic writing boards), *Crateva adansonii* (*religiosa*) (Chelum Punga), and *Quassia undulata* (*Hannoa undulata*) (Kunmuni) (also for bowls and troughs). The rest are *Milicia excelsa* (*Chlorophora excelsa*) Iroko (Odum); *Daniellia oliveri* African Copaiba Balsam Tree (also for bowls and basins), *Distemonanthus benthamianus* African Satinwood (Bonsamdua); *Zanthoxylum leprieuri* (*Fagara leprieuri*) (Oyaa), *Z. gilletii* (*Fagara macrophylla*) (Okuo), *Kigelia africana* Sausage Tree (also for spoons), *Maesopsis eminii* (Onwam-dua), *Pseudocedrela kotschyi* Dry-Zone Cedar (also for bowls), and *Phoenix reclinata* Wild Date Palm. *Cedrela mexicana* Mexican Cedar, *Bombax buonopozense* Red-Flowered Silk

Cotton Tree (also for utensils), *Ceiba pentandra* Silk Cotton Tree (also for plates), *Musanga cecropioides* Umbrella Tree, and *Ricinodendron heudelotii* (Wamma) (also for spoons, ladles, bowls, and plates) also meet acoustic requirements and are durable. *Cussonia kirkii* var. *kirkii* (*bancoensis*) (Kwaebrofre) is used for side-drums. Disodium tetraborate (borax) is an effective preservative used to treat wood before carving.

Baphia nitida Camwood, (Odwen); the climbing palms *Laccosperma opacum* (*Ancistrophyllum opacum*) Rattan Palm, *L. secundiflorum* (*A. secundiflorum*) (Ayike), and *Microdesmis puberula* (Fema) serve as drumsticks. Sticks from *Mallotus oppositifolius* (satadua) are also utilized. Climbing palms are useful as drum-strings. *Napoleonaea vogelii* (*Napoleona vogelii*) (Obua) is used for drum pegs. Some large gourds like those of *Cucurbita maxima* Squash Gourd and *Crescentia cujete* Calabash Tree are used for single-membrane drums. Drums have long been used as a means of communication. The Ghana Broadcasting Corporation and Television use drums to herald the news — and as their motif.

State umbrellas for chiefs are closely associated with stools and drums. Trees used as ribs for umbrellas include *Harrisonia abyssinica* (Penku) (also for supports of chief's palanquins or litters), *Cuviera nigrescens*, *Crescentia cujete* Calabash Tree, *Baphia nitida* Camwood (Odwen); *B. pubescens* (Odwenkobiri), and *Cussonia kirkii* var. *kirkii* (*bancoensis*) (Kwaebrofre).

Mortars

Mortars are carved from *Terminalia ivorensis* (Emire), *Morus mesozygia* (Wonton), *Milicia excelsa* (*Chlorophora excelsa*) Iroko (Odum); *Khaya* species Mahogany, *Nesogordonia papaverifera* (Danta), *Entandrophragma* species West African Cedars, and *Afzelia africana* (Papao). By far the most popular wood used for mortars in Ghana is *Nauclea diderrichii* (Kusia). *N. pobeguinii* (Sukusia) is equally suitable for mortars but comparatively rare. In the absence of these trees others are used. These include *Acacia albida* (Gozanga), *A. nilotica* var. *tomentosa* (also for bowls), *Anacardium occidentale* Cashew Nut, *Balanites aegyptiaca* Desert Date (Gongu) (also for bowls); *Baphia nitida* Camwood (Odwen); *Vitellaria paradoxa* (*Butyrospermum paradoxum* subsp. *parkii*) Shea Butter Tree (also for bowls), *Canarium schweinfurthii* Incense Tree, *Cassia sieberiana* African Laburnum, *Petersianthus macrocarpus* (*Combretodendron macrocarpum*) Stinkwood Tree (Esia); *Corynanthe pachyceras* (Pamprama), *Crossopteryx febrifuga* African Bark (also for utensils and spoons), *Dacryodes klaineana* (Adwea), *Daniellia oliveri* African Copaiba Balsam Tree, and *Diospyros abyssinica* (Gblitso). Others are *Drypetes ivorensis*, *Ficus capensis* Fig (Nwadua); *F. umbellata* (Gyedua), *Lecaniodiscus cupanioides* (Dwindwera), *Lophira alata* Red Ironwood (Kaku); *Manilkara multinervis* subsp. *lacera* (*lacera*) African Pearwood (also for bowls), and *Mimusops elengi*, an Asiatic plant. The rest are *Mitragyna stipulosa* (Subaha-akoa), *Necepsia afzelii* (also for

7. TOOLS AND CRAFTS

spoons), *Omphalocarpum elatum* (Timatibre) (also for bowls), *Ophiobotrys zenkeri* (Abuana), *Pachystela brevipes* (Aframsua), *Parinari curatellifolia* (Atena), *Parkia clappertoniana* West African Locust Bean (also for bowls), *Blighia welwitschii* (Akyekobiri), *Prosopis africana* (Sanga), *Pterocarpus erinaceus* African Kino, *Sapium ellipticum* (Tomi), *Sclerocarya birrea* (Nanogba), *Stereospermum kunthianum* (Sonontokwakofo), *Swartzia fistuloides* (Asomanini), *Tabernaemontana crassa* (Pepae), *Tamarindus indica* Indian Tamarind, and *Terminalia glaucescens* (Ongo). The two main types of mortar are the flat, shallow ones (*wodur*) for pounding *fufu* and the slender, deep ones (*abewodur*) for pounding corn, rice, boiled palm fruits, *kokonte* (dried cassava chips), etc. There are large and small models of both types. Mortars are indispensable household items in the kitchen.

Canoes and paddles

Canoes provide the native means of sea and inland fishing and water transport. Canoes are probably the largest single wooden carving (except, perhaps, the totem poles of the American Indians). Some are 40–60 feet long and 8–10 feet wide at the broadest part. The suitable wood used for carving canoes is *Triplochiton scleroxylon* (Wawa) (also for platters), but *Terminalia ivorensis* (Emire) and *Heritiera utilis* (*Tarrietia utilis*) (Nyankom) are sometimes used — especially in the south-western rain-forest area in Ghana where *Triplochiton* does not grow. Other trees used in the construction of canoes are *Borassus aethiopum* Fan Palm, *Cordia millenii* Drum Tree (also for utensils), *C. platythyrsa* (Tweneboa) (popularly used for drums), *Entandrophragma cylindricum* Sapele (Apenkwa); *E. utile* Utile, *Ficus vogeliana* (Opanto), *Khaya grandifoliola* Broad-Leaved Mahogany, *Musanga cecropioides* Umbrella Tree, *Piptadeniastrum africanum* (Danhoma), *Antiaris toxicaria* (*africana* and *welwitschii*) Bark Cloth Tree (Kyenkyen); *Pseudocedrela kotschyi* Dry-Zone Cedar (also for bowls), *Omphalocarpum elatum* (Timatibre) (also for bowls), *Sacoglottis gabonensis* (Tiabutuo), and *Syzygium guineense* var. *littorale* (Avunle).

Suitable woods for carving paddles include *Carapa procera* Crabwood, *Hexalobus crispiflorus* (Etwa prada), *Markhamia lutea* (Sisimasa), *Picralima nitida* (Ekuama), *Pterocarpus erinaceus* Senegal Rose Wood Tree, African Kino; *Thespesia populnea* (Frefi), *Xylopia aethiopica* Spice Tree (Hwenetia), *X. parviflora* (Gyambobre), *X. quintasii* Elo (Asimba); the roots of *Musanga cecropioides* Umbrella Tree, and the buttresses of *Triplochiton scleroxylon* (Wawa).

Masks and Dolls

Wood for such carvings includes *Diospyros mespiliformis* West African Ebony (Okisibiri); *Discoglypremna caloneura* (Fetefre), *Ricinodendron heudelotii* (Wamma), *Nauclea vanderguchtii*, *N. xanthoxylon*, *Picralima nitida* (Ekuama) and *Swartzia madagascariensis* Snake Bean.

Bowls, Plates, Utensils, Cutlery, etc.

Woods used in carving domestic utensils include *Acacia sieberiana* var. *villosa* (Kulgo), *Afzelia africana* (Papao), *Carpolobia* species, *Cola gigantea* var. *glabrescens* (Watapuo), *C. lateritia* var. *maclaudi* (Watapuobere), *Dichapetalum madagascariense* (*guineense*) (Antro), *Discoglypremna caloneura* (Fetefre), *Drypetes floribunda* (Bedibesa), *Gardenia erubescens* (Dasuli), *Hildegardia barteri* (Akyere), *Holoptelea grandis* Orange-Barked Terminalia (Nakwa), and *Hunteria eburnea* (Kanwene-akoa) (also for shuttles and combs). Others are *Irvingia gabonensis* Wild Mango (Abesebuo); *Kigelia africana* Sausage Tree (Nufuten); *Malacantha alnifolia* (Fafaraha) (also for shuttles), *Mitragyna inermis* (Kukyafie), *Morelia senegalensis*, and *Myrianthus serratus* (Bangama). The rest are *Pentaclethra macrophylla* Oil Bean Tree, *Picralima nitida* (Ekuama) (also for shuttles), *Rinorea oblongifolia* (Mpawu) (also for combs), *Tabernaemontana crassa* (Pepae), *Thespesia populnea* (Frefi), and *Ziziphus mauritiana* Indian Jujube.

Hair Combs

Wooden combs often depict traditional symbolic motifs. Selected woods for comb carving include *Cussonia kirkii* var. *kirkii* (*bancoensis*) (Kwaebrofre), *Dialium dinklagei* (Awendade), *Diospyros gabunensis* Flint Bark Tree, *D. sanzaminika* (Sanzamulike), *Hunteria eburnea* (Kanwenakoa), *Pentadesma butyraceum* Tallow Tree (Abotoasebie); *Pleiocarpa pycnantha* var. *tubicina* (Kakapimbe), and *Rinorea dentata*.

Gourd Crafts

A number of domestic utensils and other utility articles are carved from *Lagenaria siceraria* Bottle Gourd, *Cucurbita pepo* Pumpkin, and *C. maxima* Squash Gourd, *Crescentia cujete* Calabash Tree, and *Afraegle paniculata* (Obuobi). The Bottle Gourd is used to make spoons, ladles, sieves, drinking utensils, or calabash and syringe (*bentoa*). It also serves as a container (*danka*) for water, palm wine, and other fluids. The Pumpkin is fashioned as a container (*apakyi*) for storing clothing, selling *aboloo* (steamed cassava dough), and other items. In addition to these articles, the gourd is used to make various decorations. It serves as a float for anglers. The shell of the Coconut Palm can also be fashioned into a vase or cup.

Bamboo Crafts

A variety of articles are produced from *Bambusa vulgaris* and *Oxytenanthera abyssinica* Bamboo, the craft being an important industry in countries like China and Japan. In West Africa, bamboo craft encompasses such items as drinking cups, jugs, and mugs; serving trays and trolleys; flower pots, vases,

7. TOOLS AND CRAFTS   111

and holders. See also Tool Handles (above), Basketry, Matting, Musical Instruments (below) and Chapter 5.

BASKETRY AND RELATED CRAFTS

Basket weaving is an important rural industry. Ngyiresi, a small town on the Accra–Kumasi road in Ghana, is famous for the manufacture of assorted baskets. Two main types are recognized: heavy-duty, utility baskets for conveying foodstuffs, vegetables, fish, and so on from the producing centres to the markets, and light, fancy ones for shopping or holding fruits. The split or whole stems or twigs of the following plants are suitable for heavy-duty baskets: *Abrus precatorius* Prayer Beads, *Adansonia digitata* Baobab Tree, *Azadirachta indica* Neem Tree, *Keetia hispida* (*Canthium hispidum*) (Homa-ben), *Psydrax horizontalis* (*C. horizontale*), *Cissampelos mucronata*, *Combretum micranthum* (Landaga), *Grewia carpinifolia* (Ntanta), *Griffonia simplicifolia* (Kagya), *Manniophyton fulvum* (Hunhun), *Pteleopsis hylodendron*, *Quisqualis indica* Rangoon Creeper (also for fishing traps), *Uvaria chamae* (Akotompotsen), *U. ovata* (Akotompo), *Tiliacora dinklagei* (Susanfo), *Triclisia dictyophylla* (*gilletii*), and *Sechium edule* Chayote. The roots of *Musanga cecropioides* Umbrella Tree, and *Dichrostachys cinerea* (*glomerata*) Marabou Thorn are equally suitable for heavy-duty basketry. Climbing palms are used to make both heavy-duty baskets and light fancy ones. The stems are usually split, and for fancy types only the outer bark is used. This is sometimes dyed or varished. Climbing palms include *Laccosperma opacum* (*Ancistrophyllum opacum*) Rattan Palm, *L. secundiflorum* (*A. secundiflorum*) (Ayike), *Calamus deeratus* (Demmere), and *Eremospatha macrocarpa*. The palms are also used in weaving chairs, tables, babies' cots, wig stands, hats, and fishing traps. The midribs of palm trees are used for lazy chairs, the peelings serving as material for the manufacture of heavy-duty baskets. Palm trees include *Borassus aethiopum* Fan Palm (also for fishing traps and hats — especially the large market mammy type), *Phoenix reclinata* Wild Date Palm (also for hats — either plaited and sewn, or woven; and for fans), *Hyphaene thebaica* Dum Palm (also for large market mammy hats), *Cocos nucifera* Coconut Palm, *Raphia hookeri* Wine Palm (also for hats), and *Elaeis guineensis* Oil Palm (also for fishing traps). The core obtained after the peeling of the midrib of palm trees is utilized for toy vehicles and birdcages. The peeled stems of *Oxytenanthera abyssinica* Bamboo and those of *Dendrocalamus strictus* Male Bamboo are useful for basketry. The famous Bolga baskets of commerce from northern Ghana are manufactured with twines plaited with split stems of *Phragmites karka* Common Reed, the twines being first dyed with bright colours. The high standard of workmanship contributes to the fame of the Bolga basket industry. In addition to baskets, hats, fans, bags of various sizes, and purses are made with the Common Reed.

The straw of *Digitaria gayana*, *Elionurus elegans*, *E. hirtifolius*, *Vetiveria fulvibarbis*, and *V. nigritana* when dyed are plaited to weave earrings, armlets, and hats; and *Loudetia togoensis* is plaited into armlets worn by women. Other grasses for decorative wear include *Rhytachne rottboellioides* for toy hoops and hats and *R. triaristata* for plaiting fine worked ornamental coverings (along with leather) for bottles. The stems of *Panicum subalbidum* are used for fans. The sedge *Cyperus articulatus* is beaten flat and woven into baskets called *birefi*. The leaves of *Pandanus abbiwii* Screw Pine (Nton); twigs of *Triumfetta rhomboidea* Burweed, and stems of *Marantochloa purpurea, M. leucantha, M. ramosissimum, Sarcophrynium brachystachys, Thalia welwitschii*, and *Thaumatococcus daniellii* Katemfe are employed in weaving fishing traps and in basketry; they are all in the family Marantaceae. At Lake Bosomtwe in Ashanti in Ghana, only fishing traps of Marantaceous plants are traditionally allowed to be used in the lake. Other plants used to weave fishing traps are the twigs of *Leucaena leucocephala* leucaena, *Morelia senegalensis*, and the grass *Saccharum spontaneum* var. *spontaneum*.

MATTING

Mats are mainly used as a bedding — but they also serve as screens, blinds, wall hangings, or as material on which grains and other items of food are dried in the sun. Mats are mainly woven from the stems or fibres of plants. The fibre of *Acacia kamerunensis* (*pennata*), *Dichrostachys cinerea* (*glomerata*) Marabou Thorn, and *Cleistopholis patens* Salt and Oil Tree (Ngo ne nkyene) are used for matting to dry cocoa beans. Other fibres woven into mats are *Celtis integrifolia* Nettle Tree, *Entada africana* (Kaboya), *E. pursaetha* Sea Bean, *Hildegardia barteri* (Akyere), *Sterculia setigera, Strophanthus* species, and *Cocos nucifera* Coconut Palm. The stems of climbing palms *Laccosperma opacum* (*Ancistrophyllum opacum*) Rattan Palm and *L. secundiflorum* (*A. secundiflorum*) (Ayike) and the split stems of *Calamus deeratus* (Demmere) and *Eremospatha macrocarpa* are used for matting. The split stems of *Thalia welwitschii, Dendrocalamus strictus* Male Bamboo, and that of *Oxytenanthera abyssinica* Bamboo, stems of *Typha domingensis* (*australis*) Bulrush, and leaves of *Pandanus abbiwii* Screw Pine (Nton) and *P. veitchi* are also useful material for matting. The petioles of plants in the family Marantaceae when split longitudinally, scraped of pith, and dyed are used for matting. They include *Marantochloa purpurea, M. leucantha, M. ramosissimum, Sarcophrynium brachystachys, S. prionogonium*, and *Thaumatococcus daniellii* Katemfe. Sedges used for matting include *Cyperus articulatus* (also tied up in bundles and used for mattresses), *Fimbristylis* species, *Hypolytrum purpurascens, Scleria boivinii, S. naumanniana*, and *S. depressa*. A number of grasses are used for coarse matting. These include *Andropogon gayanus* varieties *gayanus, tridentatus, squamulatus*, and *bisquamulatus*; and *A. tectorum*. Others are *Schizachyrium exile, Vetiveria fulvibarbis*, and *V. nigritana*. Some grasses, however, make high-quality mats. These are *Aristida adscensionis, Cymbopogon giganteus* var.

7. TOOLS AND CRAFTS

giganteus, Eragrostis cilianensis Stink Grass, *Hyperthelia dissoluta* (after the sharp spikelets have fallen), *Imperata cylindrica* Lalang, *Panicum anabaptistum, P. phragmitoides, P. turgidum, Pennisetum pedicellatum, P. americanum* Bulrush Millet, *Phragmites karka* Common Reed, *Rottboellia exaltata,* and *Sorghum* species Guinea Corn (leaf-sheaths woven to make mats). The straw of *Eragrostis ciliaris, E. gangetica, E. tremula,* and *Heteropogon contortus* Spear Grass are dyed and woven into small mats for covering food. The peelings of the midrib of palm trees are useful as material for matting, after drying. Palm trees include *Raphia hookeri* Wine Palm, *Hyphaene thebaica* Dum Palm, *Phoenix reclinata* Wild Date Palm, *Cocos nucifera* Coconut Palm, *Elaeis guineensis* Oil Palm, and *Borassus aethiopum* Fan Palm. The fibre or coir of the coconut fruit is woven into mats, carpets, and doormats; or used as cordage.

BARK CLOTH

The inner bark of *Antiaris toxicaria* (*welwitschii* and *africana*) Bark Cloth Tree (Kyenkyen) is beaten, washed, and dried for native cloth. Bark cloth was particularly used in Ashanti (in the former Gold Coast, now Ghana), and formed a considerable trade in forest villages of the Ivory Coast. Other trees which have been used for bark cloth include *Adansonia digitata* Baobab, *Artocarpus communis* Bread Fruit (Dziiball); *Aningeria robusta* (Samfena), *Omphalocarpum procerum* (Gyatofo-Akongua), and *Tabernaemontana brachyantha*. In addition the bark of several species of *Ficus* was useful as cloth. Examples are *F. glumosa* (Galinziela) in Sudan, *F. natalensis* Bark Cloth Tree (Akabofuni); *F. ovata, F. platyphylla* Gutta-Percha Tree, *F. polita* (Blohunyi) in Congo, *F. dekdekena* (*thonningii*) (Gamperoga) especially in the Ivory Coast and Togo, *F. vallis-choudae* (Aloma-Bli), and *F. vogeliana* (Opanto) in the Ivory Coast.

SEED CRAFT

The seeds of some plants are decorative and useful as beads, necklaces, bracelets, and rosaries; or as counters for children or for games. Seeds for rosaries include *Ensete gilletii* Wild Banana (also for necklaces and wristlets), *Canna indica* Indian Shot (also for necklaces), the false fruits of *Coix lacryma-jobi* Job's Tears (also worn as necklaces when mourning children or on religious or ritual occasions), *Melia azedarach* Persian Lilac, and *Ziziphus spina-christi* var. *microphylla*. The black and red seeds of *Abrus precatorius* Prayer Beads, Crab's Eyes (Anyen-enyiwa) are beautiful — they are used in rosaries, as necklaces or in beadwork — but they are also poisonous (see Chapter 9). Girls pricked by the needle while stringing the seeds are ill for a long time without knowing the cause. *A. precatorius* contains abrine, a poisonous phytotoxin.

Beads and necklaces are made from the seeds of *Trachyphrynium braunianum* and the sedge *Scleria boivinii*. The seeds of *S. naumanniana* and *S.*

depressa are also useful for necklaces. Other decorative seeds strung as necklaces are the black seeds with white aril of *Cardiospermum grandiflorum* and *C. halicacabum* Heart Seed, Balloon Vine; the large, white seeds of *Operculina macrocarpa* (Abia); the red seeds of *Adenanthera pavonina* Bead Tree, and the white, shining seeds of the grass *Olyra latifolia*. The seeds of *Pseudospondias microcarpa* var. *microcarpa* (Katawani), *Rauvolfia vomitoria* (Kakapenpen) (also for bracelets), and *Thevetia peruviana* Exile Oil Tree, 'Milk Bush' are also suitable for necklaces.

Seeds for beadwork include those of *Cleistopholis patens* Salt and Oil Tree (Ngo ne nkyene); *Commiphora africana* African Bdellium (Narga); *Detarium senegalense* Tallow Tree, *Erythrina senegalensis* Coral Flower, *E. fusca* (an Asiatic species), *Entada pursaetha* Sea Bean, and *Afzelia bella* var. *glacior* (Papaonua). The rest are seeds of *Zanthoxylum xanthoxyloides* (*Fagara zanthoxyloides*) Candle Wood (Kanto); *Grewia carpinifolia* (Ntanta), *Dialium guineense* Velvet Tamarind (Asenamba); *Majidea fosteri* (Ankyewa), *Monodora myristica* (Ayerew-amba), and the seed-shells of *Canarium schweinfurthii* Incense Tree. The hard nuts of *Hyphaene thebaica* Dum Palm and those of *Borassus aethiopum* Fan Palm as vegetable ivory are used for beads. The seeds of *Omphalocarpum ahia* (Duapompo), *O. elatum* (Timatibre), and *O. procerum* (Gyatofo akongua) are strung as ornaments and rattlers at dances. The fruits of *Raphia hookeri* Wine Palm are hollowed out and used by Boy Scouts and Girl Guides as scarf rings (woggles) or used with other decorative beads as necklaces.

Seeds of *Leucaena leucocephala* (*glauca*) Leucaena, *Pentaclethra macrophylla* Oil Bean Tree (Ataa); *Pithecellobium dulce* Madras Thorns, *Samanea saman* Rain Tree, *Chrysophyllum delevoyi* (*albidum*) White Star Apple, *C. beguei* (Atabene), *C. africanum* Star Apple, *Donella welwitschii* (*C. welwitschii*), and *Tieghemella heckelii* (Baku) are useful as counters. In the *oware* and *dawa* games seeds of *Balanites aegyptiaca* Desert Date, *Paramacrolobium coeruleum*, *Nelumbo nucifera*, an introduced water plant, and *Caesalpinia bonduc* Bonduc (Oware-amba) (the most popular) are used as counters. The fruit of *Oncoba spinosa* Snuff-box Tree (Asratoa) is used as a container for snuff, as suggested by the name. The seeds of *Entada pursaetha* Sea Bean are also hollowed out and made into snuff-pots when *Oncoba* fruits are not available. Others used as snuff-boxes are the fruit of *Afraegle paniculata* (Obuobi), *O. brachyanthera*, and the seeds of *Hyphaene thebaica* Dum Palm. *Dioclea reflexa* Marble Vine (Nte) seeds are used in a spinning game called *nte*, according to folklore a favourite game of dwarfs. Some varieties of *Canavalia ensiformis* Sword Bean are also used in *nte* games.

BROOMS

Sweeping brooms are everyday utility articles in the house. They are usually made from the midribs of the leaflets of palm trees such as those of *Elaeis guineensis* Oil Palm, *Cocos nucifera* Coconut Palm, *Phoenix reclinata* Wild Date

7. TOOLS AND CRAFTS

Palm, *Hyphaene thebaica* Dum Palm, and *Raphia hookeri* Wine Palm. The stems of *Asclepias curassavica* Blood Flower, *Hibiscus micranthus, Scoparia dulcis* Sweetbroom Weed, *Indigofera simplicifolia* (Nyagahe); the twigs of *Drypetes chevalieri* (Katrika), *Eugenia leonensis, Olax subscorpioidea* (Ahoohenedua), *Tephrosia linearis*, and the bark of *Scaphopetalum amoenum* (Nsoto) also serve as material for making brooms. The twigs of *Wissadula amplissima*, the roots of *Kohautia senegalense* (Chenchen-dibiga), and the petioles of *Tacca leontopetaloides* South Sea Arrowroot are useful as brooms. The whole plant of the sedge *Bulbostylis pilosa* and *B. metralis* are used as brooms. The main flowering stalks of some grasses are collected and bundled together as brooms. *Eragrostis* species are particularly useful: they include *E. aspera, E. atrovirens, E. ciliaris, E. cylindriflora, E. egregia, E. domingensis, E. namaquensis* var. *diplachnoides, E. namaquensis* var. *namaquensis, E. scotelliana, E. squamata*, and *E. turgida*. Other grasses useful as brooms are *Heteropogon contortus* Spear Grass, *Panicum anabaptistum, P. congoense, P. dinklagei, P. fluviicola, P. maximum* Guinea Grass, *P. pansum, P. paucinode, P. phragmitoides*, and *P. porphyrrhizos*. The stems of *Oryza sativa* Rice serve as brooms. Grasses with very slender stalks are used as whisks to drive away flies and sun flies.

WOOL, SPONGE AND FIBRE CRAFT

Wool for stuffing cushions, pillows, mattresses, and general padding and upholstery work is mainly obtained from *Ceiba pentandra* Silk Cotton Tree. The floss, commercially known as kapok, may also be substituted for cotton wool for surgical purposes. The Silk Cotton Tree is among the tallest trees of the forest and only a meagre proportion of the annual kapok produced is actually collected and utilized — a large proportion rots away and is wasted. Other floss-producing plants are *Asclepias curassavica* Blood Flower, *Rhodognaphalon brevicuspe* (*Bombax brevicuspe*) (Onyina-koben), *B. buonopozense* Red-Flowered Silk Cotton Tree, *Calotropis procera* Sodom Apple, *Cochlospermum planchoni, C. tinctorium* (Kokrosabia), *Imperata cylindrica* Lalang, *Funtumia africana* False Rubber Tree, *F. elastica* West African Rubber Tree, *Pleioceras barteri* (Bakapembe), and *Typha domingensis* (*australis*) Bulrush. The floss of West African Rubber Tree is whiter and preferred to that of *Ceiba, Bombax*, and *Rhodognaphalon*.

The stems of the forest climber *Momordica angustisepala* (Ahensaw) are beaten, washed, and dried to produce *ahensaw* or *sapowpa*, a much preferred bathing and washing sponge. For this purpose, the plant is often cultivated in farms and villages. Other plants that yield sponge are *Grewia mollis* (Yualega), *Lasianthera africana, Millettia irvinei* (Ahaemete), *Telfairea occidentalis* Fluted Pumpkin, *Triaspis odorata* (Kwaemu Sabrakyee), and the roots of *Parkia clappertoniana* West African Locust Bean. (For chewing sponges see Chapter 1.) Fibre for sweeping, ceiling, and general utility brushes are obtained from the coir of *Cocos nucifera* Coconut Palm, the leaves of *Agave sisalana* Sisal

Hemp, the bast of *Hildegardia barteri* (Akyere) and that of *Hibiscus rostellatus*, and the roots of *Pandanus abbiwii* Screw Pine (Nton). Scrubbing brushes are made from the spine-like excrescences on the margins of the petioles of *Borassus aethiopum* Fan Palm and the 'bass' or piassava fibre from the vascular bundles of the leaf-sheaths of *Raphia hookeri* Wine Palm. The outer portion of the rachis of *Elaeis guineensis* Oil Palm yields a strong, flexible fibre which is also suitable for brushes. Other materials for scrubbing brushes are the bark of *Millettia barteri*, the fibrous roots of *Vetiveria fulvibarbis*, *V. nigritana*, and those of the sedge *Mariscus alternifolius*. *Mitragyna ciliata* Poplar (Subaha) is suitable for brush handles.

MUSICAL INSTRUMENTS

Wood, wood products and plant material have been used for musical instruments from ancient times, and are still used today.

Woodwind instruments

Flutes (*Atenteben*) are made with the climbing palm *Calamus deeratus* and the giant grass *Oxytenanthera abyssinica* Bamboo. The bigger types of *Atenteben* with lower notes are called *Odurogya*. Flutes may be in the key of B^b, D, F, or C. Other grasses with hollow stems like *Pennisetum unisetum* (*Beckeropsis uniseta*), *Panicum subalbidum*, *Phragmites karka* Common Reed, and *Olyra latifolia* are used for flutes, pipes, and whistles. Stems of *Hypselodelphis poggeana* and *H. violacea* have hollow stems and are suitable for whistles. *Vitex doniana* (Afua) is used for trumpets. A long, narrow form of *Lagenaria siceraria* Bottle Gourd is used for flutes. The Bottle Gourd is also used for horns. The hollowed-out seeds of *Entada pursaetha* Sea Bean make a small wind instrument. Other woodwind instruments are the piccolo, clarinet, oboe, recorder, bassoon, and contrabassoon. The bamboo is used for organ pipes in the Philippines, and *Dalbergia* species African Blackwood for clarinets in Tanzania.

Stringed instruments

Stringed instruments like the violin, viola, double-bass, and box guitar are often designed and made with plywood. *Ceiba pentandra* Silk Cotton Tree is suitable for violins. The fingerboard, tailpiece, pegs, and bow frog of the violin are made from *Diospyros* species Ebony. The fibre of *Adansonia digitata* Baobab and *Parkia clappertoniana* West African Locust Bean are used as strings for musical instruments. The fibre of *Raphia hookeri* Wine Palm or raffia fibre or bast is also twisted into strings for musical instruments and the aerial roots of *Vanilla crenulata* for guitar strings. The piano, harp, and harpsichord are also made of wood.

7. TOOLS AND CRAFTS 117

The Xylophone

Woods used for the keyboard of the xylophone include *Cordia platythyrsa* (Tweneboa), *Enantia polycarpa* African Yellow Wood (Duasika), and *Pterocarpus santalinoides* (Hote). The dry fruits of *Crescentia cujete* Calabash Tree, *Lagenaria siceraria* Bottle Gourd, or *Cucurbita maxima* Squash Gourd are used for the resonance box or resonator.

Percussion and Other Musical Instruments

The dried fruits of *Oncoba spinosa* Snuff-Box Tree (Asratoa) joined by string serve as castanets, a popular musical instrument among young Hausa girls. The fruits are also used as leg rattlers by dancers. The cabaca is made with dried fruits of *Lagenaria siceraria* Bottle Gourd with the false fruits of *Coix lacryma-jobi* Job's Tears as rattling beads. Other rattling seeds are *Cassia podocarpa* (Nsuduru), *Ricinodendron heudelotii* (Wamma), *Swartzia madagascariensis* Snake Bean, and the fruits of *Crotalaria pallida* (*falcata* and *mucronata*) (Peagoro). The *ashewa* is a resonance box with metal or cane tongues, similar to the sansas or 'thumb piano' — so called because it is usually played with the thumbs. *Ashewa* music is popular with rural communities in Ashanti and southern Ghana.

Other musical instruments made of wood include the maracas, maraca sticks, claves, and the cog-rattle. The shell of the coconut is useful as a musical instrument. This is played by tapping two pieces of the shell together or rubbing them on paper. Coconut drum is made from a sound half-shell with a stretched damp cloth over it. The full diversity of musical instruments that can be made from plants and their products is not discussed here.

EIGHT
Potions and Medicines

The use of plants and their extracts for healing by fetish priests, native doctors, and other 'specialists' was the main method of treating various illnesses before the advent of Western medicine. The practice continues still, especially among rural communities (an estimated 82 per cent of the population in developing countries) who, in any case, may not have access to a hospital or health post. (In the remote rural areas of Ghana, for example, there may be only one medical doctor to 70,000 people, while in the urban centres like Accra it is 1:4000. Generally, health services reach only 30 per cent of the population.) The practice of herbal healing is equally practised in the urban areas as a result of the shortage of imported drugs. A WHO

Table 8.1 Plant-derived drugs employed in Western medicine

Source: Krogasgaard-Larsen/Christensen/Kofod (editors), 1984

Acetyldigoxin	Galanthamine	Quercetin
Adoniside	Glycyrrhizin	Quinidine
Aescin	Hemsleyadin	Quinine
Ajmalicine	Hesperidin	Rescinnamine
Allantoin	Hesperidin methyl chalcone	Reserpine
Anabasine	Hydrastine	Rorifone
Andrographolide	Hyoscyamine	Rotundine
Anisodamine	Kawain	Rutin
Arecoline	Khellin	Saligenin
Atropine	Lanatoside C	α-Santonin
Berberine	Leurocristine	Scillarens A and B
Bromelain	α-Lobeline	Scopolamine
Brucine	Morphine	Sennosides A and B
Caffeine	Narcotine	Silybin
Cephaeline	Necendrographolide	Sparteine
Cissampeline	Nicotine	Strychnine
Cocaine	Ouabain	Tetradine
Codeine	Pachycarpine	Tetrahydrocannabinol
Colchicine	Palmatine	Theobromine
Curcumin	Papain	Theophylline
Danthron	Papaverine	Tubocurarine
Deserpidine	Peonol	Vincaleukoblastine
Digitoxin	Physostigmine	Vincamine
Digotoxin	Picrotoxin	Xanthotoxin
L-Dopa	Pilocarpine	Yohimbine
Emetine	Protoveratrines A and B	
Ephedrine	Pseudoephedrine	

survey completed in 1983 confirmed that developing states are more interested than ever in making use of traditional, indigenous resources in implementing their primary health care (PHC) programmes.

The skill of healing with herbs is acquired informally and improved upon with practice. The ingredients or constituents of a particular prescription, and its preparation, are usually the herbalists' copyright, which is secretly and jealously guarded. Illiterate herbalists die, regrettably, with this wealth of secret knowledge. The efficacy or otherwise of herbal medicine depends on the active part or parts in it and their pharmacological effect. Aikman (1974) noted that a surprising number of present-day drugs still come from natural products. It is estimated (Krogsgaard-Larsen and others, 1984) that fewer than 100 plant-derived drugs of defined structure are in common use today throughout the world (Table 8.1). It is interesting that even though laboratory synthesis has been described for most of them, fewer than ten of these well-established drugs are produced commercially by synthesis. Ever since 1806, when the first active alkaloid of a natural drug called morphine was separated from *Papaver somniferum* Poppy by Friedrich Sertürner, analyses have shown that many plants contain active medicinal elements or alkaloids (see Table 8.2).

Table 8.2 Plant sources of medicinal elements

Species	Active principles	Nature	Remarks
Adansonia digitata	adansonin	alkaloid	antidote to *Strophanthus* arrow poison
Afraegle paniculata	dictamnine	alkaloid	—
Aframomum melegueta	paradol	resinous body	—
Alstonia boonei	echitamine	alkaloid	malaria remedy
Anacardium occidentale	cardol anacardic acid	alkaloid	—
Andira inermis	berberine angeline andirine	alkaloid	—
Annona squamosa	anonaine	alkaloid	—
Anthocleista nobilis	brucine loganine	alkaloid	—
Baphia nitida	isosantalene	terpenoid	—
Breonadia salicina	mitraphylline	alkaloid	—
Caesalpinia bonduc	bonducin guilandinin	alkaloids	

Table 8.2 Cont'd.

Species	Active principles	Nature	Remarks
Calotropis procera	asclepin mudarin		emeto-cathartic
Canarium schweinfurthii	phellandrene	terpene	—
Carapa procera	tulukunin carapin	alkaloids	—
Carica papaya	papa-yotin	proteolytic	digestive of albumiferment noids
Cassia absus	isochaksine chaksine absin	alkaloids toxalbumen	— —
C. alata	chrysophanic acid	anthraquinone	—
C. occidentalis	chrysarobin	oil (anthraquinone)	—
C. obtusifolia	amodin	anthraquinone	—
Catharanthus roseus	leurocristine	alkaloid	antileukemia
Cedrela mexicana	cadinene	terpene	—
Clausena anisata	atanisatine clausanitine mupamine	alkaloids	—
Coffea arabica C. canephora	caffeine	alkaloid	heart stimulant and diuretic
Cola acuminata C. nitida	caffeine theobromine kolatine	alkaloids	— " —
Combretum micranthum	catechin combretannin	alkaloid gallic tannin	diuretic principle
Copaifera salikounda	coumarin	phenolic	—
Corynanthe pachyceras	corynanthine corynantheine corynanthidine corynantheidine yohimbine	alkaloids	sympatholitic action and local anaesthetic action Enlarges blood vessels. Used in Veterinary medicine as aphrodisiac.
Crossopteryx febrifuga	crossopterine crossoptine B-quinovine	alkaloids glycoside	—

8. POTIONS AND MEDICINES

Table 8.2 Cont'd.

Species	Active principles	Nature	Remarks
Crotalaria retusa	cystisine?	alkaloid	
Diospyros canaliculata	plumbagin	naphthoquinone	—
D. mespiliformis	plumbagin	naphthoquinone	—
Enantia polycarpa	berberine	alkaloid	preparing yellow dye
Erythrophleum species	erythrophleine cassaine cassaidine norcassaidine homophleine pinitol luteolin	alkaloids acid alcohol flavonoid	poisonous, adapted in medicine for treating heart disease and spasmodic asthma
Erythroxylum coca	cocaine hygrine	alkaloids	paralyses peripheral nerves, dilates pupil of eye and powerful anaesthetic stimulant
Holarrhena floribunda	conessine	alkaloid	against dysentery amoebae
Jatropha curcas	curcin	toxalbumen	—
Lantana camara	lantanine	alkaloid	quinine-like, antispasmodic
Mallotus oppositifolius	rottlerin	phenolic	
Maytenus buchananii	maytansine	alkaloid	anti-cancer chemical
M. senegalensis	dulcite flavanol holoside	flavonoid	
Milicia excelsa (*Chlorophora excelsa*)	chloropherine	phenolic	
Mitragyna inermis	mitrinermine mitrincomine	alkaloids	
M. stipulosa	mitraphylline	alkaloid	
Morinda morindoides	oxymethoxymethyl anthraquinone morindanol hentriacontane	anthraquinone anthraquinone hydrocarbon	—
Moringa oleifera	pterygospermine		antibiotic
Mundulea sericea	rotenone	rotenoid	—

Table 8.2 Cont'd.

Species	Active principles	Nature	Remarks
Nicotiana tabacum	nicotine	alkaloid	stimulates peripheral nerves
Parkia biglobosa	soumara	fat	—
Parquetina nigrescens	periplocin strophanthidin	steroidal glycosides	—
Paullinia pinnata	timboin	—	—
Pausinystalia yohimbe	yohimbinine yohimbine	alkaloids	enlarges blood vessels, also tonic and aphrodisiac
Pergularia daemia	daemine	alkaloid	—
Physostigma venenosum	physostigmine eserine eseranine eseridine calabarine physovenine	alkaloids	
Picralima nitida	akuammine	alkaloid	sympathicosthenic and local anaesthetic like cocaine
Piliostigma thonningii	flavonic heteroside quercitoside	flavonoids	effective purgative
Piper guineense	chavicine piperine		—
P. nigrum	piperidine piperine		responsible for peppery taste
Plumbago zeylanica	plumbagin plumbagol	napthaquinone	antibiotic
Premna hispida	premnine ganarine	alkaloids	sympathomemetic
Psydrax subcordata	calmatambin	glycoside	—
Punica granatum	pelletierine	alkaloid	expulsion of tape worms
Quassia amara	quassin	quassinoid	—
Solanum erianthum (*S. verbascifolium*)	solanine	alkaloid	—
Sophora occidentalis	cystisine	alkaloid	—
Teclea verdoorniana	tecleanone	alkaloid	—
Tetracera alnifolia	syringine	phenolic glucoside	—

8. POTIONS AND MEDICINES

Table 8.2 Cont'd.

Species	Active principles	Nature	Remarks
Theobroma cacao	theobromine alkaloids caffeine	—	heart stimulant diuretic
Triclisia dictyophylla	triclisine tricliseine	alkaloids	
Vernonia amygdalina	vernonine	glucoside	heart action like digitaline
Ximenia americana	sambunigrin	—	—
Xylopia aethiopica	avoceine	fat and resin	rich in protein
Zanthoxylum viride	decarine	alkaloids	—
Z. xanthoxyloides	fagaridine fagarine skimmianin	alkaloids	

The folklore of Africa, like that of other continents, abounds in plant remedies. In Ghana, Dr Irvine reported more than 800 woody plants alone, not to mention herbaceous ones. Systematic screening of plants is a means of finding the active principle to verify their efficacy. Many of these plants may contain only traces of medicinal elements, explaining probably why large doses have to be administered, but a few may yield promising results. The success rate in discovering new drugs is about 1 in 5–10,000. In 1884 a young Viennese doctor discovered that cocaine is an effective local anaesthetic in cataract and other eye operations. A pharmaceutical firm dispatched Dr Henry Hurd Rusby to South America to collect coca and other drug plants. It is reported that he brought back not only coca supplies but also some 45,000 other plant specimens for medical research, and that many provided new sources of worthwhile drugs. Dr Jonathan Hartwell, an organic chemist and head of the Natural Products Section of the cancer institute's Drug Development Branch in Maryland, compiled a list of folklore remedies and assembled data on more than 3000 plant species. The extracts of several of these give anti-cancer promise in tests on laboratory animals and cell cultures. Dr Robert Perdue, chief of the medicinal plant resources laboratory of the US Agriculture Research Services, led several trips to Ethiopia and Kenya between 1960 and 1976. Of the 20,000 plants screened, *Maytenus buchananii*, which grows in Kenya's Shimba Hills, yielded 1.5 milligrams per kilo of maytansine — a unique chemical compound that shows promise in the fight against cancer, particularly against pancreatic cancer. The College of Pharmacy of the University of Illinois at Chicago, a WHO collaboratory centre for traditional medicine, has gathered information from world literature on the chemistry, pharmacology, and ethno-pharmacology of some 70,000 plants and the key data has been computerised. When the computer is fed with the botanical or the local name of the plant NAPRALERT (natural products alert) supplies the medicinal properties.

Plant screening is not without its problems. The USA has spent about $615 million on drug research and Britain about $70 million. Large-scale screening in developing countries under present economic conditions is practically impossible. However, with co-operation among the present institutions already engaged in various plant analyses, some new ground may be broken. In Ghana the institutions are the University of Ghana, Legon; the University of Science and Technology, Kumasi; the University of Cape Coast; the Ghana Medical Schools; the Centre for Scientific and Industrial Research (CSIR); and the Centre for Scientific Research into Plant Medicine (CSRPM) at Mampong, Akwapim.

The CSRPM was established on 2 November 1973 and given legal backing on 10 July 1975 by NRC Decree 344 with Dr Oku Ampofo as Director. Among others the Centre was charged with the selection by screening of plant materials reputed to have medicinal properties and their preparation as tinctures, extracts, decoctions for use in treating patients, for the purpose of verifying the claims made. There are similar institutions in Nigeria, Ivory Coast, and other West African states. In this way a large body of information on medicinal plants and their efficacy is being built up in a scientific way.

There is already the export of plant material, either for direct screening or for pharmacological or allied purposes, mainly to Britain, Europe, and America. The International Trade Centre (ITC) in Geneva is interested in finding out what suitable medicinal plants are available in West Africa and have carried out a survey into the ones with commercial export potential. This is the first co-ordinated and internationally sponsored attempt to look at this commercial export potential of medicinal plants in Africa. It is hoped, however, that the export of these raw materials would be an interim measure only, and that eventually the possibility of extracting and refining the active parts locally, both to feed our pharmaceuticals industry and for export, would be the ultimate goal.

For a few group of plants like *Griffonia simplicifolia* (Kagya), *Voacanga africana* (Ofuruma), *Heliotropium indicum* Indian Heliotrope, *Catharanthus roseus* Madagascar Periwinkle, *Rauvolfia vomitoria* (Kakapenpen), *Physostigma venenosum* Calabar Bean, and *Corynanthe pachyceras* (Pamprama) there has been a breakthrough and tons of the raw materials for the manufacture stage are now being contracted. *Griffonia* is the largest single plant material exported from Ghana in recent times for medicinal purposes — with an estimated 75–80 tonnes of seeds per annum mainly to Germany. Judd and others (1977) isolated a BSII lectin, a second haemogglutinin from the seeds. The lectin has affinity for type III polyagglutinable red cells and is useful in blood grouping tests. Other plants exported include *Gymnema sylvestre, Ancistrocladus abbreviatus, Dennettia tripetala, Hunteria eburnea,* (Kanwen-akoa), *Tabernaemontana psorocarpa, Strophanthus gratus* (Omaatwa), *Duparquetia orchidacea* (Pikeabo), *Adhatoda robusta, Centella asiatica* Indian Water Navelwort, *Gloriosa superba* Climbing Lily, and *Calotropis procera* Sodom Apple. However, the vast majority of plants are at the moment, used locally for various ailments.

Indigenous Medicinal Practices

Plant medicine is prepared and administered as tinctures, infusions, concoctions, decoctions, and extracts; or as enemas and poultices — with or without other ingredients. Usually the added ingredients include *Capsicum annuum* Pepper, *Zingiber officinale* Ginger, natron (local carbonate of soda), *Monodora myristica* (Ayerew-amba), *Xylopia aethiopica* Ethiopian Pepper, Spice Tree (Hwenetia), and *Aframomum melegueta* Melegueta, Guinea-grains. Unless otherwise stated infusions, concoctions, and decoctions are taken internally. *Woody Plants of Ghana* by F.R. Irvine and *The Useful Plants of West Tropical Africa* by J.M. Dalziel (A–D revised by H.M. Burkill) are generally the sources of information on the woody and herbaceous plants respectively unless stated otherwise. It is advisable to consult practising herbalists or qualified elders in all cases of herbal treatment. Self-medication is dangerous in view of the sketchy information on dosages and concentrations and the incidence of collecting the wrong plant.

ABORTION

Plants used to procure abortion are root of *Nauclea latifolia* African Peach, (Sukisia); *Aframomum sulcatum, Plumbago indica* Red Plumbago, *P. zeylanica* Ceylon Leadwort, (Opapohwea) given internally; *Diospyros mespiliformis* West African Ebony as ingredient, *Gossypium arboreum* Cotton, and root of *Momordica charantia* African Cucumber as ingredient with the pounded seeds in water inserted locally. Root infusion of *Phytolacca dodecandra* (Ahoro) with bark of *Piptadeniastrum africanum* (Danhoma) as enema and whole plant of *Alternanthera pungens* (*repens*) as ingredient in prescription. Infusion of dried leaves of *Carica papaya* Papaw as purge; leaf decoction of *Mareya micrantha* (Odubrafo); leaf of *Necepsia afzelii* boiled in water as purge; leaf of *Entada africana* (Kaboya) with native soap applied locally; dried leaves of *Ficus asperifolia* Sandpaper Tree with seeds of *Xylopiastrum villosum* (*Xylopia villosa*) Elo Pubescent, (Oba) ground in oil and salt; whole plant of *Heliotropium indicum* Indian Heliotrope mixed with clay and applied locally or *Trianthema portulacastrum* (Jain, 1968); leaf infusion of *F. vogeliana* (Opanto) and leaf of *Balanites aegyptiaca* Desert Date (Gongu) with unripe fruit, root, and bark decoction. Excess of edible seeds of *Sterculia foetida* (a native of tropical Asia) taken internally as a purge; seeds of *Sesamum indicum* Sesame, Beniseed;

crushed seeds of *Pentaclethra macrophylla* Oil Bean Tree with little brown ants taken internally; seed oil of *Turraeanthus africanus* Avodire; latex of *Calotropis procera* Sodom Apple both externally and internally; juice of freshly crushed *Parquetina nigrescens* (Aba-kamo) or latex (could be fatal) and decoction of *Tephrosia vogelii* Fish Poison Plant with the green pods. Bark decoction of *Maesobotrya barteri* var *sparsiflora* (Apotrewa) and *Trichilia monadelpha* (*heudelotii*) (Tanduro) with pulp; bark of *Pterocarpus erinaceus* Senegal Rose Wood Tree as ingredient in prescription; the violently emeto-purgative action of the bark decoction or pulp of *Khaya senegalensis* Dry-Zone Mahogany sweetened with honey and bark (and especially seeds) of *Pleioceras barteri* (Bakapembe). *Waltheria indica* (Sawai) and *Microglossa pyrifolia* (Asommerewa) and ground *Calliandra portoricensis* (Nkabe) are also used to cause abortion.

The following plants prevent miscarriage or abortion. Leaf of *Ethulia conyzoides* in food; leaf decoction of *Cissampelos mucronata* and *C. owariensis* (Akuraso), *Paullinia pinnata* (Toa-ntini) or a decoction of ground-up roots and leaves with local spices and Guinea-grains; root decoction of *Cnestis ferruginea* (Akitase) as drink or enemas; decoction of root and leaf of *Morinda lucida* Brimstone Tree as drink by pregnant women; root, bark, and leaf of *Pterocarpus santalinoides* (Hote); bark of *Bridelia stenocarpa* (*micrantha*) (Opam); resin from bark of *Trichoscypha arborea* (Anaku); bark infusion of *Ricinodendron heudelotii* (Wamma); powdered root and root decoction of *Hunteria umbellata;* leaf decoction of *Antidesma laciniatum* var. *laciniatum* and *A. membranaceum* as bath, and a lotion of root pith of *Alchornea cordifoia* Christmas Bush, (Gyamma) with young leaves, white clay, and peppers as enema. However, eating the leaves of *Alchornea* is said to cause abortion.

ABSCESSES (see Boils)

AMENORRHOEA (see Menstrual Troubles)

ANALGESIC (RELIEVING PAIN)

Plants used to relieve pain are the pounded bark of *Campylospermum reticulatum (Ouratea reticulata)* applied, or that of *Allanblackia parviflora (floribunda)* Tallow Tree, (Sonkyi); the root-pulp of *Securinega virosa* (Nkanaa) with palm oil and shea butter applied; summits of female flowering *Cannabis sativa* Marijuana (Boulos, 1983); pulp of leafy stems of *Premna hispida* (Aunni) with lemon applied; leaf of *Maerua angolensis* (Pugodigo); watery decoction of root, bark, and leafy stems of *Rothmannia longiflora* (Saman-kube) as drink, lotion, and bath; and unspecified parts of *Bridelia scleroneura* (Ba-Udiga), *Mitragyna inermis* (Kukyafie), *Physalis angulata* (Ayensu, 1978), and *Zanthoxylum xanthoxyloides (Fagara zanthoxyloides)* Candle Wood; whole plant of *Cyathula prostrata* (Ayensu, 1978) and *Palisota hirsuta* (Nzahuara) (Ayensu, 1978).

ANAEMIA (LACK OF BLOOD)

Prescriptions are the bark decoction of *Nauclea latifolia* African Peach, (Sukisia); *Coula edulis* Gaboon Nut, (Bodwue); and of *Entada pursaetha* Sea Bean as sitz-baths or enemas; a cold infusion of bark, root, and wood of *Detarium senegalense* Tallow Tree; root decoction of *Boerhavia diffusa* Hogweed (Ayensu, 1978), *Phyllanthus muellerianus* (Potopoleboblo) and a decoction of the root and leaf of *Mallotus oppositifolius* (Satadua); root decoction or blood-red exudate from trunk of *Amphimas pterocarpoides* (Yaya) as drink; pounded leaves of *Adansonia digitata* Baobab as tonic and leaf-pulp of *Triclisia patens*. The peeled stems of *Hibiscus lunariifolius* form an ingredient in prescriptions for anaemia.

ANAESTHETIC — LOCAL

Plant parts used are the bark of *Corynanthe pachyceras* (Pamprama) due to the alkaloid corynanthine, *Erythrophleum suaveolens* (*guineense*) Ordeal Tree, (Potrodom) due to the alkaloid erythrophleguine; seeds of *Picralima nitida* (Ekuama) due to the alkaloid akuammine; summits of female flowering *Cannabis sativa* Marijuana (Boulos, 1983); an unspecified part of *Mareya micrantha* (Odubrafo) and of *Mitragyna stipulosa* (Subaha-akoa) (Ayensu, 1978).

ANTHELMINTIC (EXPULSION OF WORMS)

Specific for hookworms are the root of *Antidesma venosum* (Mpepea) and the boiled roots of *Abrus precatorius* Prayer Beads. The following plants are specific for tapeworms; whole plant infusion of *Heliotropium indicum* Indian Heliotrope, *Celosia leptostachya* (*laxa* and *trigyna*), *Chenopodium ambrosioides* American or Indian Wormseed, Sweet Pigweed; seed kernels of *Cucurbita pepo* Pumpkin, Vegetable Marrow and *C. maxima* Squash Gourd and half-ripe fruits of *Ananas comosus* Pineapple; also the fruit of *Capparis brassii* (*thonningii*) (Kpitipkiti) or *Allophylus africanus* (Hokple); bark extract of *Harungana madagascariensis* (Okosoa) or *Khaya senegalensis* Dry-Zone Mahogany; shoots of *Sterculia tragacantha* African Tragacanth, (Sofo); leaves of *Erythrococca anomala* or crushed leaf of *Ozoroa reticulata* (*Heeria insignis* and *pulcherrima*) (Nasia) boiled in milk or cold infusion of *Cassia sieberiana* African Laburnum (though dangerous); root decoction of *Ziziphus mauritiana* Indian Jujube, that of *Kigelia africana* Sausage Tree, (Nufuten); and those of *Aframomum melegueta* Melegueta, (Famu-wisa) and *A. sulcatum*.

The following plants are for roundworms, threadworms, and other intestinal parasites: bark decoction of *Neostenanthera hamata* (Silikawkawle), *Microdesmis puberula* (Fema), *Alstonia boonei* (Sinduro), *Jatropha curcas* Physic Nut and powdered bark, and *Lonchocarpus laxiflorus* (Nalenga) with natron; macerated *Mollugo nudicaulis* with lime; whole plant decoction of *Celosia argentea*, *Kohautia senegalensis* (Chenchen-dibiga), *Stachytarpheta cayennensis*

Brazilian Tea as purge; *Eleusine indica* (Amico, 1977), and *Crinum zeylanicum* (*ornatum*) (Ngalenge); bark infusion of *Greenwayodendron oliveri* (*Polyalthia oliveri*) (Duabiri) and *Holoptelea grandis* Orange-barked Terminalia, (Nakwa); powdered bark of *Albizia adianthifolia* (Pampena) with honey, *Andira inermis* Dog Almond, *Lannea kerstingii* (Kobewu) with others, *Craterispermum laurinum*, *Crossopteryx febrifuga* African Bark (Pakyisie); *Psychotria peduncularis* (Kwesidua), chewed bark of *Picralima nitida* (Akuama); an enema of bark pulp of *Anthocleista nobilis* Cabbage Palm, (Hohoroho) and pulverized bark of *Vernonia conferta* (Flakwa) with *Sesamum indicum* Beniseed.

Others are root decoction of *Combretum micranthum* (Landaga), *Pentadesma butyraceum* (*butyracea*) Tallow Tree, (Abotoasebie); *Dichrostachys cinerea* (*glomerata*) Marabou Thorn, *Acacia sieberiana* var. *villosa* (Kulgo) with leaf decoction, *Amphimas pterocarpoides* (Yaya) or red stem juice, *Treculia africana* var. *africana* African Breadfruit, *Opilia celtidifolia* (Benkasa), *Millettia thonningii* (Sante) with bark decoction, *Zanthoxylum xanthoxyloides* (*Fagara zanthoxyloides*) Candle Wood, *Newbouldia laevis* (Sasanemasa), whole plant decoction of *Euphorbia hirta* Australian Astham Herb for cases in children (Jain, 1968); a decoction of the bitter root of *Vernonia cinerea* and the boiled root and leaf of *Spigelia anthelmia* Worm Weed especially when fresh, followed by an active purge like Epsom Salts — also used in Europe. The list includes the root infusion of *Cassia nigricans*, *Ximenia americana* Wild Olive, and *Pycnanthus angolensis* African Nutmeg, (Otie) with that of *Cassia occidentalis* Negro Coffee and Guinea-grains as drink. Boiled root of *Stephania abyssinica* var. *abyssinica*, *S. dinklagei*, *Cassia obtusifolia* (*tora*) Foetid Cassia or whole plant, *Eriosema griseum* (Trindobaga) with other plants, *Raphiostylis beninensis* (Akwakora-gyahene) with stems and leaves, *Dalbergiella welwitschii*, *Citrus aurantiifolia* Lime chewed and eaten, and *Bersama abyssinica* subsp. *paullinioides* (Duantu) as purge. Decoction of leaf and fruit of *Morinda morindoides*; leaf of *Cleistopholis patens* Salt and Oil Tree, *Combretum racemosum* (Wota), *Balanites aegyptiaca* Desert Date, (Gongu) with unripe fruit and root, *Cassia alata* Ringworm Shrub with line, *Desmondium adscendens* var. *adscendens* (Akwanfanu), *Celtis integrifolia* Nettle Tree, (Samparanga); *Ficus exasperata* (Nyankyeren) as ingredient, *Ziziphus mucronata* Buffalo Thorn chewed and swallowed, *Clausena anisata* Mosquito Plant, *Calotropis procera* Sodom Apple as ingredient, *Plumbago zeylanica* Ceylon Leadwort, (Opapohwea) in soup, *Cordia millenii* Drum Tree; boiled and powdered leaf of *Maytenus senegalensis* (Kumakuafo) in milk; young shoots of the climbing palm *Laccosperma secundiflorum* (*Ancistrophyllum secundiflorum*) (Ayike); leaf decoction of *Cissus populnea* (Agyako); seeds of *Xylopia aethiopica* Spice Tree, (Hwentia); *Azadirachta indica* Neem Tree and *Cleome gynandra* (*Gynandropsis gynandra*); fruit of *Garcinia kola* Bitter Cola, (Tweapea); *Hedranthera barteri* and *Pavetta crassipes* (Nyenyanke); 'sulphur oil' from *Allium sativum* Garlic; seed of *Portulaca oleracea* Purslane, Pigweed; spicular hairs of *Mucuna pruriens* var. *pruriens* Cow Itch (Apea) in syrup; unspecified parts of *Trema orientalis* (*guineensis*) (Sesea), *Curcuma domestica* Turmeric, *Carissa edulis* (Botsu) (also for cattle), *Microglossa pyrifolia* (*Asommerewa*), and *Melia azedarach* Persian Lilac, Bead Tree; oil of *Carapa procera* Crabwood; latex of *Landolphia owariensis* White Rubber wine as drink or enema and decoction, or cold infusion of leaf of *Gongronema latifolium* (Nsurogya) with lime.

ANTISEPTIC

Powdered root of *Hoslundia opposita* (Asifuaka) with Guinea-grains and potassium nitrate; root-pulp infusion of *Strychnos spinosa* Kaffir Orange (Akankoa); scented roots of *Euadenia trifoliolata* with ingredients; the powdered bark or decoction of *Hymenocardia acida* (Sabrakyie); bark extract of *Enantia polycarpa* African Yellow Wood; rind of *Citrus aurantiifolia* Lime (Ayensu, 1978); whole plant of *Cyathula prostrata* (Ayensu, 1978) or *Palisota hirsuta* (Ayensu, 1978) and unspecified part of *Griffonia simplicifolia* (Kagya) and *Zanthoxylum xanthoxyloides* (*Fagara zanthoxyloides*) Candle Wood.

APHRODISIAC

Root or chewed root of *Penianthus zenkeri, Sphenocentrum jollyanum* (Krakoo), *Euadenia trifoliolata* with other ingredients, *Grewia carpinifolia* (Ntanta) with *Cyperus esculentus* Tiger Nuts, *Margaritaria discoidea* (*Phyllanthus discoideus*) (Pepea), *Securinega virosa* (Nkanaaa) with the leaves, *Uapaca guineensis* Sugar Plum, (Kuntan); *Cissus populnea* (Agyako), *Harrisonia abyssinica* (Penku) with tiger nuts or palm kernels, *Carissa edulis* (Botsu), *Gardenia erubescens* (Dasuli) as ingredient, *Hedyotis corymbosa* (*Oldenlandia corymbosa*), *Cenchrus biflorus* as ingredient in preparation, chewed sweet root of *Abrus precatorius* Prayer Beads, and tuberous root of *Vernonia guineensis* chewed raw; internal part of rhizome of *Cymbopogon schoenanthus* subsp. *schoenanthus* and *C. schoenanthus* subsp. *proximus* eaten; root and rootbark of *Carpolobia alba* (Afiafia); root, bark and fruit of *Garcinia afzelii* (nsokodua) taken internally; root decoction of *Glyphaea brevis* (Foto) with salt and red pepper (also flower and flower-buds); chewed young tap root of *Cola gigantea* var. *glabrescens* (Watapuo); steeped root of *Mansonia altissima* (Oprono) as enema; dried, pulverized root-bark of *Alchornea floribunda* with food; ground roots of *Eriosema* species with milk; root decoction of *Hymenocardia acida* (Sabrakyie), *Paullinia pinnata* (Toantini) in palm wine or a suck of leaf-tips or root, *Cnestis ferruginea* (Akitase) as drink, and decoction of root and leaf of *Newbouldia laevis* (Sasanemasa); pulped root of *Massularia acuminata* (Pobe) as enema; cut up and boiled swollen roots of *Dracaena surculosa* var. *surculosa* (Mobia) with tiger nuts; pounded root and stem of *Caesalpinia bonduc* Bonduc, Nicker Nut, (Owareamba) in palm wine; root-bark infusion of *Cassia sieberiana* African Laburnum; root of *Acacia kamerunensis* (*pennata*) (Oguaben) boiled with flour; root-juice of *Uraria picta* (Heowe) as ingredient; root-bark of *Pseudocedrela kotschyi* Dry-Zone Cedar; crushed root of *Rauvolfia vomitoria* (Kakapenpen) in warm water; root of *Strophanthus hispidus* Arrow Poison taken in spirit and woody rootstock of *Fadogia agrestis* (Buruntirikwa) as ingredient; tubers of *Tacca leontopetaloides* South Sea Arrowroot, and the false bulbs of some species of *Eulophia*, *Habenaria*, *Angraecum*, *Ansellia*, *Bulbophyllum*, *Listrostachys*, *Polystachya*, etc., all in the Orchidaceae family, as ingredients in aphrodisiac preparations.

Other prescriptions are powdered bark of *Isolona campanulata*; bark and leaf decoction of *Heritiera utilis* (*Tarrietia utilis*) (Nyankom) as drink; powdered bark of *Antidesma laciniatum* var. *laciniatum* (Kpoploti) in water or palm wine; bark decoction of *Bridelia atroviridis* (Asaraba) and *Trichilia monadelphia* (*heudelotii*) (Tanduro); bark of *Erythrina mildbraedii* (Nfona), *Calotropis procera* Sodom Apple, and *Corynanthe pachyceras* (Pamprama); pounded bark of *Crossopteryx febrifuga* African Bark, (Pakyisie) in porridge and tender shoots of *Trachyphrynium braunianum* chewed with palm kernels.

Others are leafy twigs and root of *Rhigiocarya racemifera*; leafy twigs of *Rinorea ilicifolia* with salt and *Capsicum*; crushed new shoot or infusion of root, bark, and stem of *Microdesmis puberula* (Fema); stem pulp of *Adenia rumicifolia* var. *miegei* (*lobata*) (Peteha) as enema; decoction of leafy stems and leaf of *Griffonia simplicifolia* (Kagya); leaf decoction of *Albizia zygia* (Okoro), *Urera obovata* (Osurosuso) as enema, *Zanthoxylum leprieuri* (*Fagara leprieuri*) (Oyaa) as drink in the morning, and *Machaerium lunatum* (*Drepanocarpus lunatus*) (Nkako); dried, powdered leafy tops of *Gouania longipetala* (Homabiri) in palm wine; summits of female flowering *Cannabis sativa* Marijuana (Boulos, 1983); young leaves and shoots of *Klainedoxa gabonensis* var. *oblongifolia* (Kroma); powdered leaf of *Deinbollia pinnata* (woteegbogbo) with salt in palm wine; sun-dried leaf or root-bark of *Eriocoelum racemosum* (Onibonakokoo) with that of *Paullinia pinnata* (Toa-ntini) and vegetable salt; leaf and leafy tips of clove-like scented *Strychnos afzelii* (Duapepere) and leaf of *Vernonia conferta* (Flakwa) steeped in palm wine.

The rest are the edible fruit of *Capparis corymbosa*; fruit pulp of *Euadenia eminens* (Dinsinkoro); fruit of *Craterispermum laurinum*; seeds of *Cola acuminata* Commercial Cola Nut Tree; gum of *Acacia hockii* Shittim Wood and chewed wood of *Erythrina senegalensis* Coral Flowers; unspecified part of *Ficus capensis* (Nwadua) and *Tetracera alnifolia* (Akotopa); whole plant of *Sesamum indicum* Sesame, Beniseed; *Stylosanthes fruticosa* (*mucronata*) and *Mallotus oppositifolius* (Satadua) are ingredients.

APOPLEXY (see Stroke)

APPETIZER (including Stomachic)

Tonic prepared from bark of *Symphonia globulifera* (Ehureke), *Alchornea cordifolia* Christmas Bush, (Gyamma); *Albizia zygia* (Okoro) and *Spathodea campanulata* African Tulip Tree, (Kokoanisua); sweet-tasting bark of *Maesobotrya barteri* var. *sparsiflora* (Apotrewa) chewed; aromatic bark and leaf

8. POTIONS AND MEDICINES

of *Zanthoxylum viride* (*Fagara viridis*) (Oyaanini), boiled bark of *Lannea kerstingii* (Kobewu) as drink; powdered bark or bark decoction of *Ziziphus mauritiana* Indian Jujube; bark decoction of *Hunteria eburnea* (Kanwen-koa), and decoction or infusion of macerated inner bark of *Dichrostachys cinerea* (*glomerata*) Marabou Thorn; boiled succulent stems of *Cissus quadrangularis* Edible-stemmed Vine, (Kotokoli) sweetened with sugar; decoction of *Centaurea perrottetii* Star Thistle, *Celosia leptostachya* (*laxa* and *trigyna*), rhizomes of *Zingiber officinale* Ginger, whole plant of *Momordica charantia* African Cucumber, *Phyllanthus fraternus* subsp. *togoensis* (*niruri*) (Ombatoatshi) as bitter stomachic, root of *Cardiospermum grandiflorum* and *C. halicacabum* Heart Seed, Balloon Vine; leaves of *Crassocephalum crepidioides* and ashes of *Pistia stratiotes* Water Lettuce taken with food; juice and leaves of *Solenostemon monostachys* (Sisiworodo) taken; *Allium sativum* Garlic taken with millet as drink; cold infusion of bitter root of *Aristolochia albida*; root decoction of *Glyphaea brevis* (Foto); root and fruit pulp of *Cassia sieberiana* African Laburnum; boiled root of *Lonchocarpus cyanescens* West African Indigo, Yoruba Indigo with ingredients and root (probably root-resin) of *Pachycarpus lineolatus*. Leaf of *Ocimum gratissimum* Tea Bush, (Nunum); fresh young shoots and buds of *Harungana madagascariensis* (Okosoa); chopped leaf of *Zanthoxylum leprieuri* (*Fagara leprieuri*) (Oyaa) added to food, and pounded leaf of *Commiphora africana* African Bdellium, (Narga) with bulrush millet and milk. Seed of *Piper guineense* West African Pepper, (Soro-wisa); fruit-rind of *Citrus aurantiifolia* Lime; gum resin of *Boswellia dalzielii* Frankincense Tree, (Kabona), and fruit decoction of *Nauclea latifolia* African Peach, (Sukisia) for indigestion and vomiting. Chewed roots of *Sphenocentrum jollyanum* (Krakoo), *Mussaenda erythrophylla* Ashanti Blood (Damaramma) and *Vernonia amygdalina* Bitter Leaf, (Bonwen); powdered bark decoction of *Coula edulis* Gaboon Nut, (Bodwue); pulverized leaves of *Cassia nigricans*; leaf of *Combretum micranthum* (Landaga), leaf infusion of *Guiera senegalensis* with Tamarind pulp; leaves of *Trema orientalis* (*guineensis*) (Sesea) cooked with groundnuts; leaf decoction of *Bersama abyssinica* subsp. *paullinioides* (Duantu); the pulp of *Adansonia digitata* Baobab as a seasoner, and the berries of *Solanum indicum* subsp. *distichum* Children's Tomato to the sick. The fruits of *Solanum anomalum* (Nsusua) and *S. nigrum* (Nsusuabiri) are also given to the sick to restore the appetite or as a digestive tonic. Pounded bark of *Pycnanthus angolensis* African Nutmeg; dried, pounded bark of *Beilschmiedia mannii* Spicy Cedar with rice; a cold infusion of the bark of *Khaya senegalensis* Dry-Zone Mahogany with natron and salt as tonic for horses; the bark of *Newbouldia laevis* (Sasanemasa) for horses and that of *Pseuderanthemum tunicatum* as medicine for cattle.

ARTHRITIS (ARTICULAR RHEUMATISM) (see Rheumatism)

ASCITES (see Dropsy)

ASTHENIA (WEAKNESS OR LOSS OF STRENGTH)

Decoction of whole plant of *Paullinia pinnata* (Toa-ntini) as drink or vapour bath; decoction of *Sida cordifolia* (Moro, 1984–5); pulped leafy stems of *Machaerium lunatum* (*Drepanocarpus lunatus*) (Nkako) as liniment; cold root decoction of *Gardenia ternifolia* (Peteprebi) for restoring failing strength; cold infusion of bark, root, and wood of *Detarium senegalense* Tallow Tree as a restorative for weakness; root decoction of *Cassia occidentalis* Negro Coffee boiled with butter and an unspecified part of *Morus mesozygia* (Wonton).

ASTHMA

Leaf decoction or dried leaf of *Cordia millenii* Drum Tree smoked as tobacco; dried root of *Deinbollia pinnata* (Woteegbogbo) in palm soup or soda water (Ayensu, 1978); decoction of young roots of *Borassus aethiopum* Fan Palm; root pulp of *Turraea heterophylla* (Ahunanyankwa) as enema; bark decoction of *Harungana madagascariensis* (Okosoa); roots and leaves of *Boerhavia diffusa* and *B. repens* Hogweed in moderate doses (Jain, 1968); flowers of *Cassia alata* Ringworm Shrub (Ayensu, 1978); ground mature flowers of the grass *Rottboellia exaltata* with natron inhalled on attack (Dokosi, 1969); chewed raw inner bark and roasted roots of *Napoleonaea leonensis* (*Napoleona leonensis*); bark of *Gardenia ternifolia* (*Peteprebi*); seed oil of *Ricinus communis* Castor Oil Plant (Ayensu, 1978); root of *Desmodium gangeticum* var. *gangeticum* (Ayensu, 1978); macerated or boiled leaf of *Kalanchoe integra* var. *crenata* (Aporo); infusion or decoction of *Euphorbia hirta* Australian Asthma Herb — special reputation; dry leaf of *Datura stramonium* Apple of Peru smoked as cigarette or pulverized and burnt as inhalant to relieve; leaf of *Evolvulus alsinoides* as cigarette; smoke from pithy stems and leaf of *Calotropis procera* Sodom Apple; dried powdered leaf of *Desmodium adscendens* var. *adscendens* (Akwamfanu) in warm water (Ayensu, 1978); powdered leaves of *Adansonia digitata* Baobab taken orally to prevent crisis; chewed leaves of *Abrus precatorius* Prayer Beads (Moro, 1984–5); unspecified part of *Trema orientalis* (*guineensis*) (Sesea) and decoction of bark of *Alstonia boonei* (Sinduro) mixed with the roots and bark of cola and fruits of *Xylopia parviflora* (Gyambobre) with hard potash as drink; dried, ground root of *Thonningia sanguinea* with honey (Avumatsodo, 1984–5); bulb of *Allium sativum* Garlic (Evans and others, 1983) a leaf infusion of *Dialium guineense* Velvet Tamarind as drink for chest complaints with shortage of breath, and unspecified parts of *Musa paradisiaca* var. *sapientum* Banana and *Physalis angulata* (Totototo) (Ayensu, 1978).

BACKACHE

A leaf decoction of *Markhamia lutea* (Sisimasa), *M. tomentosa* (Tomboro), and *Cordia senegalensis* (Kyeneboa); an enema of the bark infusion of *Spathodea campanulata* African Tulip, (Kokoanisua); *Buchholzia coriacea* (Esono-bise), or enema of the leaf pulp of *Rothmannia longiflora* (Saman-kube), or the whole

8. POTIONS AND MEDICINES

plant of *Parquetina nigrescens* (Aba-kamo); the bark decoction of *Lophira lanceolata* and root decoction of *Fadogia agrestis* (Buruntirikwa); a massage of the leaf pulp of *Erythrococca anomala* or the bark lotion of *Lonchocarpus sericeus* Senegal Lilac, and the fruit of *Detarium senegalense* Tallow Tree.

BILIOUSNESS (see Dizziness)

BLADDER TROUBLE/KIDNEY DISEASE

Bark decoction of *Spathodea campanulata* African Tulip, (Kokoanisua); fruit pulp of *Bixa orellana* Anatto; root decoction of *Cassia occidentalis* Negro Coffee; cold decoction of *Scoparia dulcis* Sweetbroom Weed (also for gravel); root of *Oncoba spinosa* Snuff-Box Tree, (Asratoa); root bark of *Bridelia ferruginea* (Opam-fufuo); decoction of rhizomes of *Cynodon dactylon* Bahama or Bermuda Grass (Boulos, 1983); leaf infusion of *Acacia sieberiana* var. *villosa* (Kulgo); leaf-juice of *Griffonia simplicifolia* (Kagya) as enema and an unspecified part of *Abrus precatorius* Prayer Beads. Half-ripe fruits of *Ananas comosus* Pineapple; fruits of *Tribulus terrestris* Devel's Thorn or whole plant as diuretic; the bark and leaves of *Kigelia africana* Sausage Tree, (Nufuten) alone or with *Olax subscorpioidea* (Ahoohenedua), *Clerodendrum capitatum* (Tromen), *Carapa procera* Crabwood, and the seeds of *Xylopia aethiopica* Spice Tree, (Hwenetia); the dried leaf powder of *Adansonia digitata* Baobab is used as a preventive measure.

BLEEDING, STOP (see Styptic)

BLINDNESS, TO CAUSE (OR ALMOST)

A drop or two, in the eye, of latex of *Anthostema aubryanum* (Kyirikesa), *Elaeophorbia drupifera* (Akane), *Tabernaemontana crassa* (Pepae), *Calotropis procera* Sodom Apple, and *Hura crepitans* Sandbox Tree; the powdered fruit shells of *Balanites aegyptiaca* Desert Date, (Gongu); the bark of *Securidaca longepedunculata* (Kpaliga) and an unspecified part of *Ostryderris stuhlmannii* (Kaman-Godui). For latex in the eye the antidote is application of mucilaginous plants or washing with plenty of water.

BLOOD, PURIFYING

Leaves of *Morinda lucida* Brimstone Tree, (Kankroma); *Demodium velutinum* (Koheni-koko), *Ziziphus mucronata* Buffalo Thorn, *Abrus precatorius* Prayer Beads, and fruit of *Momordica charantia* African Cucumber (Ayensu, 1978); decoction of rhizomes of *Cynodon dactylon* Bahama or Bermuda Grass (Boulos, 1983); decoction of root, leaf, and fruit of *Tephrosia purpurea*; leaf decoction of *Gossypium arboreum* Cotton; leaf juice of *Crotalaria retusa* Devil Bean, and pulverized roots and bark of *Millettia thonningii* (Sante) boiled.

BOILS, ABSCESSES, AND ERUPTIONS

Lotion of *Xylopia aethiopica* Spice Tree, (Hwenetia) and *Caloncoba echinata* (Gorli); pounded seeds of *Xylopiastrum villosum (Xylopia villosa)* Elo Pubescent, (Oba) as poultice; pounded leaf of *Syzygium guineense* var. *macrocarpum* (Kultia); leaf of *Hyptis suaveolens* Bush Tea-Bush, *Amaranthus viridis* Wild Amaranth as poultice, *Sphaeranthus senegalensis, Pupalia lappacea* (Akukuaba) with palm oil, *Bryophyllum pinnatum* Resurrection Plant, *Alternanthera nodiflora* and *A. sessilis* as poultice; macerated leaf of *Sterculia tragacantha* African Tragacanth, (Sofo) as poultice; whole plant of *Portulaca oleracea* Purslane, Pigweeed crushed with natron and oil to bring to a head, and crushed plant of *Celosia leptostachya* and *Datura metel* Metel applied. Leaf infusion of *Phyllanthus muellerianus* (Potopoleboblo) as wash; leaf and bark of *Rhodognaphalon brevicuspe (Bombax brevicuspe)* (Onyinakoben) as poultice; leaf of *Dalbergiella welwitschii* for 'tumbo fly' boils in cattle; pulverized leaf of *Eriosema glomeratum* as dressing; leaf decoction of *Maytenus senegalensis* (Kumakuafo) for dental abscesses; charred leaf, root, and root-bark of *Calotropis procera* Sodom Apple with ointment; leaf and wood ashes of *Psydrax venosa (Canthium venosum)* with oil as poultice; decoction or infusion of leaf of *Mitragyna inermis* (Kukyafie) as dressing; young leaf of *Buchholzia coriacea* (Esono-bise) as poultice; pounded leaf of *Ritchiea reflexa* (Aayerebi) and *Dichrostachys cinerea (glomerata)* Marabou Thorn to abscesses; leaf decoction of *Erythrococca anomala* to wash abscesses; leaf paste of *Ziziphus mucronata* Buffalo Thorn applied externally; pounded leaf-less, jointed stems of *Cissus aralioides* (Asirimu) as poultice to bring to a head and leaf, root, and stem of *Paullinia pinnata* (Toa-ntini) with Ginger and Guinea-grains. Boiled roots of *Boerhavia diffusa* and *B. repens* Hogwed as poultice to bring to a head; scraped pulp of *Cucurbita pepo* Pumpkin and *C. maxima* Squash Gourd applied as poultice and boiled rice used to rub skin eruptions.

Bark of *Vernonia conferta* (Flakwa) to reduce abscesses; inner bark of *Didymosalpinx abbeokutae* as poultice; bark of *Securinega virosa* (Nkanaa) for simple abscesses; preparation of bark and fruit of *Drypetes ivorensis* applied externally; bark of *Penianthus zenkeri* as dressing; bark of *Stereospermum kunthianum* (Sonontokwakofo) for skin eruptions, *Pericopsis laxiflora (Afrormosia laxiflora)* Satinwood for boils, and bark decoction of *Margaritaria discoidea (Phyllanthus discoideus)* (Pepea) as drink for boils. Decoction of root of *Ampelocissus grantii* (Enotipsi) with bark of *Khaya senegalensis* Dry-Zone Mahogany and natron taken with food; pounded inner bark of *Cola lateritia* var. *maclaudii* (Watapuobere) with clay for cases in the nostril; fruit-juice of *Solanum incanum* Egg plant with pounded leaf applied and ground seeds of *Marantochloa purpurea* (Sugugwa), *M. leucantha* (Sibere), and *M. filipes* as paste applied. Paste of seeds of *Abrus precatorius* Prayer Beads with salt to boils; seeds of *Myrianthus arboreus* (Anyankoma) for boils; fruit of *Coula edulis* Gaboon Nut, (Bodwue); pulverized fruit of *Combretum micranthum* (Landaga)

8. POTIONS AND MEDICINES 135

with oil as ointment for abscesses; fruit of *Microdesmis puberula* (Fema) as preventive; the copious latex of *Secamone afzelii* (Kotohume) locally applied to boils and the salt made from the young roots of *Rhizophora* species Red Mangrove with palm oil used as ointment for sores produced by boils. An unspecified part of *Cordia myxa* Sapistan Plum, Assyrian Plum as emollient plaster to bring abscesses to a head; ground seeds of *Tamarindus indica* Indian Tamarind with water to carbuncular boils; unspecified parts of *Cassia occidentalis* Negro Coffee, *Heritiera utilis* (*Tarrietia utilis*) (Nyankom), and *Daniellia oliveri* African Copaiba Balsam Tree for abscesses.

BRONCHIAL TROUBLE

Bark decoction of *Xylopia aethiopica* Spice Tree, (Hwenetia); *Petersianthus macrocarpus* (*Combretodendron macrocarpum*) Stinkwood Tree, (Esia) as expectorant; *Phyllanthus muellerianus* (Potopoleboblo) as beverage and enema and *Antiaris toxicaria* (*africana* and *welwitschii*) Bark Cloth Tree, (Kyenkyen) with root of *Erythrina senegalensis* Coral Flowers and *Carapa procera* Crabwood as gargle or drink. Root decoction of *Isolona campanulata*, *Cochlospermum tinctorium* (Kokrosabia) as beverage, bath, and liniment; *Manniophyton fulvum* (Hunhun) with young shoots, *E. senegalensis* with bark as drink or bath, often with lemon-juice, spices, and Guinea-grains, and decoction of young roots of *Borassus aethiopum* Fan Palm. Root of *Piper guineense* West African Pepper (Soro-wisa); root pulp of *Securinega virosa* (Nkanaa) with palm oil or shea butter as ointment; infusion of crushed roots of *Entada abyssinica* (Sankasaa); bulb of *Allium sativum* Garlic for chronic cases (Evans, 1983), infusion of bark and leaves of *Hymenocardia acida* (Sabrakyie) and seed of *Garcinia kola* Bitter Cola (Tweapea). Pressed fruit of *Beilschmiedia mannii* Spicy Cedar as counter-irritant, *Piliostigma thonningii* (Opitipata) as ingredient, and boiled palm cabbage of *Elaeis guineensis* Oil Palm eaten with pepper and salt.

Heated leaves of *Nicotiana tabacum* Tobacco ground with oil as chest rub; leaf of *Sesamum indicum* Sesame, Beniseed as remedy; infusion of *Eleusine indica* for treating the spitting of blood from the bronchi, larynx, lungs, or trachea (haemoptysis); whole plant of *Euphorbia hirta* Australian Asthma Herb for inflammation of the respiratory tract; leaf decoction of *Adenia rumicifolia* var. *miegei* (*lobata*) (Peteha), *Discoglypremna caloneura* (Fetefre), *Alchornea cordifolia* Christmas Bush, (Gyamma) as beverage or bath for the soothing, anti-spasmodic effect; *Dalbergia saxatilis* (Nuodolega) or *Dalbergiella welwitschii* as beverage and vapour bath, and decoction of leafy twigs of *Microglossa afzelii* (Pofiri) and *Combretum molle* (Gburega) as beverage or bath. Infusion of young shoots of *Lophira lanceolata* as lotion and drink; powdered leaf of *Cassia occidentalis* Negro Coffee with lemon tree roots and Guinea-grains in palm wine; leaf-juice of *Tamarindus indica* Indian Tamarind with ginger; *Clausena anisata* Mosquito Plant as antiseptic, *Deinbollia pinnata*

(Woteegbogbo) as drink and pulped young leaves as chest rub, and infusion of *Lantana camara* Wild Sage. Bark pulp of *Millettia zechiana* (Selena) with sea salt and Guinea-grains diluted in warm water; bark of *Entada africana* (Kaboya); bark and leaf decoction of *Trema orientalis* (*guineensis*) (Sesea) as gargle, inhalation, drink, lotion, bath, or vapour bath; leaves of *Evolvulus alsinoides* smoked as cigarette and unspecified parts of *Ficus capensis* (Nwadua) and *F. vallis-choudae* (Aloma-bli).

BRONCHITIS (see Bronchial trouble)

BRUISES (see Sprains)

BURNS (see Sprains)

CANCER

Maytansine from *Maytenus buchananii* and other species of *Maytenus* for cases of the pancreas particularly and the roots of *Catharanthus roseus* Madagascar Periwinkle for cases of the blood (leukaemia); the seed extract of *Rauvolfia vomitoria* (Kakapenpen) and *Voacanga africana* (Ofuruma) for genera cases. Powdered roots of *Xylopia aethiopica* Spice Tree, Ethiopian Pepper, (Hwenetia) (Ayensu, 1978); root of *Adenia rumicifolia* var. *miegei* (*lobata*) (Peteha) with pepper and Guinea-grains and leaf of *Musa paradisiaca* Plantain both for cases of the nose (Ayensu, 1978); powdered root of *Boerhavia diffusa* Hogweed (Ayensu, 1978); leaf of *Ficus asperifolia* Sandpaper Tree, *Momordica charantia* African Cucumber (Ayensu, 1978) and unspecified part of *Hilleria latifolia* both for cases of the breast. External application of ground bean pods of *Abrus precatorius* Prayer Beads in water for cases of the skin (epithelioma) (Ayensu, 1978); bark of *Spondias mombin* Hog Plum, Ashanti Plum in water for cases of the uterus (Ayensu, 1978); the stem twigs of *Ocimum basilicum* (Eme)(Ayensu, 1978); unspecified parts of *Zanthoxylum xanthoxyloides* (*Fagara zanthoxyloides*) Candle Wood or *Z. gilletii* (*F. macrophylla*) (Okuo) (Lewis and Elvis-Lewis, 1977); infusion of *Plumbago zeylanica* Ceylon Leadwort, (Opapohwea) in oil applied and unspecified part of *Ricinus communis* Castor Oil Plant for cases of the stomach (Ayensu, 1978).

CARMINATIVE (FLATULENCE)

Seeds of *Piper guineense* West African Pepper (Soro-wisa), *Aframomum melegueta* Melegueta (Famu-wisa), and *Xylopia aethiopica* Spice Tree, (Hwenetia); leaf decoction of *Pterygota macrocarpa* Pterygota (Kyereye) or *Quisqualis indica* Rangoon Creeper; inner bark of *Scytopetalum tieghemii* (Aprim); bark of young shoots of *Cinnamomum zeylanicum* Cinnamon; fruit and root of *Solanum aethiopicum*; fruit-rind of *Citrus aurantiifolia* Lime; rhizomes of *Zingiber officinale* Ginger and seeds of *Myristica fragrans* Nutmeg (Evans, 1983); leaf of

Baphia nitida Camwood, (Odwen) taken internally; ashes of *Pupalia lappacea* (Akukuaba) with water as drink; fruits of *Syzygium cumini* Java Plum, Jambolan (Evans, 1983); powdered bark of *Sterculia rhinopetala* Sterculia Brown, (Wawabimma) taken internally and root of *Crotalaria retusa* Devil Bean with wine.

CATARACT (see Eye Diseases)

CATARRH (CORYZA)

Powdered leaf of *Drypetes chevalieri* (Katrika) as snuff; leaf-juice of *Dialium guineense* Velvet Tamarind (Asenamba) with salt and pepper; bark of young shoots of *Cinnamomum zeylanicum* Cinnamon; bulb of *Allium sativum* Garlic (Evans, 1983); root-bark of *Uvaria chamae* (Akotompotsen); root-tips of *Securidaca longepedunculata* (Kpaliga); powdered leaf and juice of *Rhigocarya racemifera* as drops; ground and boiled roots of *Conocarpus erectus* Button Wood; macerated gummy bark of *Sterculia setigera* (Pumpungo); leaf-juice or infusion of *Ocimum gratissimum* Tea Bush, Fever Plant, (Nunum) and *Lantana trifolia*; infusion of whole plant of *L. camara* Wild Sage; decoction of *Leonotis nepetifolia* var. *nepetifolia* or *L. nepetifolia* var. *africana* to steam the head as relief; chewed rhizomes of *Zingiber officinale* Ginger; crushed bark of *Newbouldia laevis* (Sasanemasa) with natron; juice of warmed leaves of *Calotropis procera* Sodom Apple as snuff; whole plant of *Desmodium gangeticum* var. *gangeticum*, *Euphorbia hirta* Australian Asthma Herb, and *Phytolacca dodecandra* (Ahoro) (Evans, 1983) and powdered bark of *Cola gigantea* var. *glabrescens* (Watapuo). The dry fruit powder of *Solanum incanum* Egg Plant insufflated through a tube into the nose of horses and *Hyptis pectinata* (Peaba) mixed with *Guiera senegalensis* and indigo plant and boiled for horses to inhale vapour for diseases accompanied by mucous catarrh.

CHANCRE/SYPHILITIC SORES (see Venereal Diseases)

CHEST COMPLAINTS AND PAINS

Bark decoction of *Anogeissus leiocarpus* (Sakanee) with pepper, *Piliostigma thonningii* (Opitipata), *Musanga cecropioides* Umbrella Tree, (Dwumma); *Canarium schweinfurthii* Incense Tree, *Carapa procera* Crabwood, *Crossopteryx febrifuga* African Bark (Pakyisie) in palm wine; *Ficus sycomorus* (*gnaphalocarpa*) (Kankanga) and latex, *Morinda lucida* Brimstone Tree with root, and decoction of outer bark of *Newbouldia laevis* (Sasanemasa) with pepper. Bark of *Myrianthus arboreus* (Anyankoma); chewed bark of *Craterispermum laurinum*; boiled bark of *Mitragyna ciliata* Abura (Subaha) with pepper and *Piper guineense* West African Pepper (Soro-wisa) and chewed bark of *Pachypodanthium staudtii* (Fale) with kola nuts. Young leafy twigs and root of *Annona senegalensis* var. *senegalensis* Wild Custard Apple as tisane; an infusion of *Ipomoea coptica* or *Schwenckia americana* (Agyingyinsu) for children, *Desmodium adscendens* var. *adscendens* (Akwamfanu) as bath, and *Sesamum indicum* Beniseed with salt as soothing remedy; tea-like infusion of leaf of *Lippia multiflora* Gambian Tea Bush, (Saa-nunum); leaf of *Harungana madagascariensis* (Okosoa) and that of *Microglossa afzelii* (Pofiri) with Guinea-grains in rum; leaf infusion of *Dialium guineense* Velvet Tamarind (Asenamba); *Abrus precatorius* Prayer Beads and *Guiera senegalensis* with *Combretum micranthum* (Landaga) as febrifuge. Leaf decoction of *Phyllanthus muellerianus* (Potopoleboblo), root decoction of *Cassia occidentalis* Negro Coffee as stimulant, *Anthocleista vogelii* (Aworabontodee) and boiled root of *Combretum mucronatum (smeathmanii)* with pepper and ashes as drink. Root of *Glyphaea brevis* (Foto), *Tamarindus indica* Indian Tamarind, and *Pseuderanthemum tunicatum*. Fruit of *Cordia myxa* Sapistan Plum, Assyrian Plum; *Detarium senegalense* Tallow Tree and inhaled vapour of whole plant infusion of *Acacia pentagona*; chewed seeds of *Garcinia kola* Bitter Cola, (Tweapea); *Cola acuminata* Commercial Cola Nut Tree, and *Picralima nitida* (Ekuama), and unspecified parts of *Secamone afzelli* (Kotohume) with shea butter, *Trema orientalis (guineensis)* (Sesea) and *Desmodium gangeticum* var. *gangeticum*; also the boiled gum of *Acacia nilotica* var. *tomentosa* in water as drink.

CHICKEN POX

Small quantities of bark of *Erythrophleum suaveolens (guineense)* Ordeal Tree (Potrodom) applied; the fresh leaves of *Manihot esculenta* Cassava; pulverized leaves of *Euphorbia convolvuloides* with palm oil applied, said to dry up eruptions quickly, and an unspecified part of *Crossopteryx febrifuga* African Bark (Pakyisie) applied.

CHOLAGOGUE (Stimulating liver and increasing bile)

An infusion of *Hoslundia opposita* (Asifuaka) as tisane and plant decoction of *Flacourtia flavescens* Niger Plum (Amugui). An infusion of *Cola acuminata* Commercial Cola Nut Tree is used in Central African Republic to get rid of bile.

CIRCUMCISION WOUNDS (see Wounds)

COLDS

Leaf infusion of *Acacia sieberiana* var. *villosa* (Kulgo), *Lippia multiflora* Gambian Tea Bush (Saa-nunum); *Leucas martinicensis*, *Vitex doniana* Black Plum, (Afua); *Clerodendrum buchholzii* (Taasendua) with pepper and *Guiera senegalensis* with that of *Combretum micranthum* (Landaga) as nasal douche. Dried leaf of *Dissotis rotundifolia* (Adowa-alie); dried, pulverized, and burnt leaf of *Monechma ciliatum* (Bunglale) inhaled; crushed *Hyptis spicigera* applied

8. POTIONS AND MEDICINES

to the head for relief; aromatic leaf of *Ocimum gratissimum* Tea Bush, Fever Plant, (Nunum); powdered leaf of *Dialium dinklagei* (Awendade) chewed with kola; leaf-juice of *Ficus exasperata* (Nyankyeren) with lemon-juice and crushed leaf of *Allophyllus africanus* (Hokple) applied to nostrils. Leaf decoction of *Alchornea cordifolia* Christmas Bush (Gyamma) as antispasmodic, *Cnestis ferruginea* (Akitase), and *Cordia millenii* Drum Tree or dried leaf smoked as tobacco; decoction of leafy tips of *Microglossa afzelii* (Pofiri) and of *Ximenia americana* Wild Olive.

Root decoction of *Hoslundia opposita* (Asifuaka) as drink or gargle; root of *Clematis hirsuta;* bulbs of *Allium sativum* Garlic (Evans, 1983); infusion of root scrapings of *Mimosa pigra* Sensitive Plant, (Kwedi); decoction of root, fruit, and leaf of *Tephrosia purpurea* and powdered root of *Microglossa pyrifolia* (Asommerewa) as snuff. Bark of *Uncaria africana, Zanthoxylum gilletii (Fagara macrophylla)* (Okuo), *Entada africana* (Kaboya), *Piliostigma thonningii* (Opitipata) and bark of young shoots of *Cinnamomum zeylanicum* Cinnamon; spicy bark of *Buchholzia coriacea* (Esono-bise) with sawdust as snuff; powdered bark of *Hymenocardia acida* (Sabrakyie); bark of *Chaetacme aristata* (Kodia) sucked and decoction gargled; bark decoction of *Crossopteryx febrifuga* African Bark (Pakyisie) in palm wine; infusion or decoction of bark of *Ficus dekdekena (thonningii)* (Gamperoga) and of *F. lutea (vogelii)* (Fonto); seed of *Piper guineense* West African Pepper (Soro-wisa); infusion of *Cadaba farinosa* and of *Stylosanthes fruticosa (mucronata)* and *Hexalobus monopetalus* var. *monopetalus*.

COLIC (GRIPING, BELLY PAIN)

Leaf infusion of *Combretum micranthum* (Landaga), *Ceiba pentandra* Silk Cotton Tree (Onyina); *Abrus precatorius* Prayer Beads and tea-like infusion of *Lippia multiflora* Gambian Tea Bush (Saa-nunum) as laxative; also of young shoots of *Lophira lanceolata* as relief. Leaf of *Leea guineensis* (Okatakyi), *Lawsonia inermis* Henna in milk with dates and gum arabic as drink; leaf of *Phyllanthus fraternus* subsp. *togoensis (niruri)* (Ombatoatshi) for allaying the griping of dysentery and painful spasmodic affections of the intestines; leaf juice of *Clerodendrum umbellatum* (Niabiri) and leaf of *Oldenlandia affinis* subsp. *fugax*; leaf-juice or decoction of *Ficus exasperata* (Nyankyeren) as enema; leaf decoction of *Allophylus africanus* (Hokple), *Palisota hirsuta* (Nzahuara) taken internally, and chewed leaf, fruit, and seed of *Paullinia pinnata* (Toa-ntini); juice of pressed leaves of *Hyptis suaveolens* Bush Tea-Bush along with lime-juice as drink and juice and leaves of *Solenostemon monostachys* (Sisiworodo) taken; decoction of leaf and fruit of *Morinda morindoides* and a preparation of leaf and root of *M. lucida* Brimstone Tree (Kankroma) as a bitter astringent tonic.

Root of *Ectadiopsis oblongifolia*; root decoction of *Acridocarpus smeathmannii* (Alasaayo) as purgative, *Vernonia nigritiana* (Gyakuruwa), *Nauclea latifolia* African Peach (Sukisia) as stomachic, and *Maytenus senegalensis* (Kumakuafo) with rice and Baobab pulp and butter; boiled root of *Mitragyna stipulosa* (Subaha-akoa) eaten; rhizomes of *Zingiber officinale* Ginger (Evans, 1983), ground root and leaf of *Cassia mimosoides* (Langrinduo) with Guinea-grains as a relief and powdered root of *Lantana camara* Wild Sage in milk. Bark of *Ziziphus mauritiana* Indian Jujube, and enema of fresh bark of *Canarium schweinfurthii* Incense Tree, and bark chips of *Cola acuminata* Commercial Cola Nut Tree internally; bark infusion of *Grewia mollis* (Yualega), *Hymenocardia acida* (Sabrakyie), *Hannoa klaineana* (Fotie), and *Anthocleista nobilis* Cabbage Palm in enemas; hot decoction of bark and leaf of *Vernonia conferta* (Flakwa) and *Daniellia oliveri* African Copaiba Balsam Tree with leaf taken internally. Pounded, dry bark and young twigs of *Newbouldia laevis* (Sasanemasa) with *Xylopia* and other species given as a decoction for uterine colic; fruit and root of *Solanum aethiopicum* as remedy; fruits of *Syzygium cumini* Java Plum, Jambolan (Evans, 1983); decoction or cold infusion of pounded leafy stems of *Gongronema latifolium* (Nsurogya) with lime-juice; fresh seed of *Saba florida* (Akontoma) or *Piper guineense* West African Pepper (Soro-wisa); seed of *Garcinia kola* Bitter Cola (Tweapea) to prevent or relieve; sap of *Tetracera alnifolia* (Akotopa) alone or with water taken internally, crushed root of *Crotalaria retusa* Devil Bean with spices; infusion of rind or bark of *Citrus limon* Lemon; watery decoction of an unspecified part of *Zanthoxylum xanthoxyloides (Fagara zanthoxyloides)* Candle Wood with pepper as enema; dried, pulverized root-bark of *Calotropis procera* Sodom Apple added to soup as stomachic and an unspecified part of *Clematis hirsuta*.

CONJUNCTIVITIS (see Eye Diseases)

CONSUMPTION (see Tuberculosis)

CONVULSIONS (FITS, EPILEPSY)
Smoke of burnt leaf of *Pentaclethra macrophylla* Oil Bean Tree; leaf juice or lotion or decoction of *Steganotaenia araliacea* (Pienwogo), *Cussonia arborea (barteri)* (Saa-borofere), *Raphiostylis beninensis* (Akwakora gyahene), *Jatropha curcas* Physic Nut, and *Croton zambesicus* (Dodwatu); an enema of the leaf-juice of *Leea guineensis* (Okatakyi); infusion of *Boerhavia diffusa* and *B. repens* Hogweed; the leaf decoction of *Anthocleista nobilis* Cabbage Palm with lemon and leaf infusion of *Abrus precatorius* Prayer Beads with palm oil. A decoction of leaf and bark of *Rinorea ilicifolia* with *Palisota hirsuta* (Nzahuara); the bark decoction of *Newbouldia laevis* (Sasanemasa) and *Lonchocarpus sericeus* Senegal Lilac; root decoction of *Sansevieria liberica* and of *Cochlospermum tinctorium* (Kokrosabia) as liniment; root-bark of *Rauvolfia vomitoria* (Kakapenpen) and root extract of *Maytenus senegalensis* (Kumakuafo) as drink; leaf infusion of *Desmodium adscendens* var. *adscendens* (Akwamfanu) as bath; the bark of *Ekebergia senegalensis*, *Blighia sapida* Akee Apple, (Akye); *Pseudocedrela kotschyi* Dry-Zone Cedar and bark of *Ficus capensis* (Nwadua) as

ingredients in the preparations; *Ehretia cymosa* (Okosua) and *Hoslundia opposita* (Asifuaka) as ingredients in Agbo infusions; decoction of *Terminalia macroptera* (Kwatiri) given to bulls and epileptic men and an unspecified part of *Annona senegalensis* subsp. *onlotricha* (*arenaria*) (Aboboma).

CORYZA (see Catarrh)

COUNTER-IRRITANT (REVULSIVE)

The crushed leaves of *Mikania chevaliera* (*cordata*) Climbing Hemp Weed, *Gynandropsis speciosa* for local pain, *Momordica foetida* (Sopropo) as rub to relieve, and *Premna quadrifolia* (Atantaba); the seeds of *Aframomum melegueta* Melegueta, (Famu-wisa) crushed and rubbed on body; *Piper guineense* West African Pepper (Soro-wisa) and fruit of *Capsicum annuum* Pepper; leaves and stem of *Xylopia aethiopica* Spice Tree (Hwenetia) and *Caesalpinia bonduc* Bonduc (Oware-amba); root of *Uvaria chamae* (Akotompotsen) with pepper; root and bark of *Moringa oleifera* Horse-Radish Tree; root of *Plumbago zeylanica* Ceylon Leadwort (Opapohwea); bark and leaves of *Pisonia aculeata* (Mpintinko); root of *Reissantia indica* (*Hippocratea indica*) with hot oil and bark infusion of *Anthonotha macrophylla* (Totoro) with palm nuts. Leaf or root bark of *Rauvolfia vomitoria* (Kakapenpen) with spices and *Allium sativum* Garlic as rub.

COUGH AND WHOOPING COUGH

Leaf decoction of *Ocimum gratissimum* Tea Bush, (Nunum); *Tricalysia reticulata* (Kwakenya), *Vernonia amygdalina* Bitter Leaf (Bonwen); *V. colorata*, *Tetracera potatoria* (Twihama), *Mollugo nudicaulis* (cold), *Crossopteryx febrifuga* African Bark (Pakyisie); *Ficus sagittifolia*, *Cordia millenii* Drum Tree or dried leaves smoked like tobacco, *Ximenia americana* Wild Olive, *Psidium guajava* Guava boiled with lemon grass, *F. sycomorus* (*gnaphalocarpa*) (Kankanga), and *Triplotaxis stellulifera* (Kokoo) for cases in children; also *Acanthus montanus* False Thistle for cases involving women and children; bark decoction of *Psydrax subcordata* (*Canthium subcordatum*) (Teteadupon), *Canarium schweinfurthii* Incense Tree, *Nauclea latifolia* African Peach (Sukisia) sometimes with spices; *Spondias mombin* Hog Plum, Ashanti Plum as emetic for severe cases; *Bridelia stenocarpa* (*micrantha*) (Opam) or powdered bark in palm wine, *Macaranga heterophylla* (Opamkokoo) as beverage and for bathing, and chips of bark of *Cola acuminata* Commercial Cola Nut Tree with salt and seeds of *Xylopia aethiopica* Spice Tree, (Hwenetia). Bitter bark of *Khaya ivorensis* African Mahogany (Dubini); bark infusion of *Milicia excelsa* (*Chlorophora excelsa*) Iroko (Odum); *Acacia albida* (Gozanga) and *Anogeissus leiocarpus* (Sakanee) or decoction. Bark or powdered bark of *Uvaria afzelii* and root, *Diospyros mespiliformis* West African Ebony as fumigant, *Corynanthe pachyceras* (Pamprama) chewed with water as drink, *Zanthoxylum gilletii*

(*Fagara macrophylla*) (Okuo), *Teclea verdoorniana* (Owebiribi), chewed *Carapa procera* Crabwood, *Khaya anthotheca* White Mahogany (Krumben), *Treculia africana* var. *africana* African Breadfruit, *Heisteria parvifolia* (Sikakyia) eaten, *Necepsia afzelii* applied to chest, and raw inner bark and roasted roots of *Napoleonaea leonensis* (*Napoleona leonensis*) chewed.

Leaf infusion of *Dinophora spenneroides*, *Lippia multiflora* Gambian Tea Bush (Saa-nunum) drunk like a strong tea, *Aspilia africana* Haemorrhage Plant for children, *Scoparia dulcis* Sweetbroom Weed, *Elytraria marginata* both as remedy for children, and steeped leaf of *Ritchiea reflexa* (Aayerebi) inhaled. Dried leaf of *Dissotis rotundifolia* (Adowa-alie); leaf juice of *Pisonia aculeata* (Mpintinko) with spices, *Ficus exasperata* (Nyankyeren) with lemon-juice, and plant juice of *Paullinia pinnata* (Toa-ntini). Young leaf of *Donella welwitschii* (*Chrysophyllum welwitschii*) and bark; powdered leaf of *Cadaba farinosa* with millet flour chewed or crushed leaves of *Abrus precatorius* Prayer Beads mixed with lime juice; leaf of *Cassia obtusifolia* (*tora*) Foetid Cassia with ingredients, *Hugonia planchonii* chewed or infusion, *Pupalia lappacea* (Akukuaba) in soup, *Schwenckia americana* (Agyingyinsu) for children; powdered leaf of *Dialium dinklagei* (Awendade) chewed with kola; pulverized leaves of *Desmodium adscendens* var. *adscendens* (Akwamfanu) with Guinea-grains applied to chest after scratching the skin; leaf of *Elaeophorbia drupifera* (Akane), *Platostoma africanum* (Siresireke) with seed as remedy for children, and fermented leaf of *Mareya micrantha* (Odubrafo) with rum and coconût as drink. Dried inflorescence of *Cymbopogon giganteus* var. *giganteus* as ingredient, fragrant rhizomes of *Cyperus articulatus* (Kokyi) chewed, or *C. rotundus* Nut Grass as cure for children's cases; bulb of *Allium sativum* Garlic (Evans, 1983); seeds of *Buchholzia coriacea* (Esono-bise), *Abutilon guineense* (Odonno-bea), *A. mauritianum* (Nwaha) and mucilagenous seed covering of *Sherbournea bignoniiflora* (Kyerebeteni) steamed with lime, *S. calycina*, *Cordia myxa* Sapistan Plum, unripe fruit of *Uapaca guineensis* Sugar Plum, (Kuntan), and pods of *Stereospermum kunthianum* (Sonontokwakofo) chewed with salt; fruit decoction of *Usteria guineensis* (Kwaemeko), *Solanum torvum* (Saman-ntoroba), and decoction or chewed succulent stems or pounded fruits of *Costus afer* Ginger Lily.

Roots of *Bryophyllum pinnatum* Resurrection Plant, *Ficus asperifolia* Sandpaper Tree chewed, *Acacia gourmaensis* (Gowuraga), *Albizia zygia* (Okoro) mixed with food, *Mussaenda erythrophylla* Ashanti Blood with Guinea-grains, and *Penianthus zenkeri* chewed with twigs. Root decoction of *Hymenostegia afzelii* (Takorowa) and *Tephrosia purpurea* with leaf and fruit. Burned pithy stem and leaf of *Calotropis procera* Sodom Apple inhaled; powdered gum of *Daniellia thurifera* Niger Copal Tree (Kwanga) in water as drink; boiled gum of *Acacia nilotica* var. *tomentosa* in water taken internally, and pulverized whole plant of *Waltheria indica* (Sawai) with hot water as drink. Unspecified parts of *Annona senegalensis* subsp. *onlotricha* (*arenaria*) (Aboboma), *Microglossa pyrifolia* (Asommerewa), *Trichilia emetica* subsp. *suberosa* (*roka*) (Kisig?),

Margaritaria discoidea (*Phyllanthus discoideus*) (Pepea), *Cassia kirkii* var. *guineensis* rubbed on chest, and decoction of whole plant of *Mikania chevalieri* (*cordata*) Climbing Hemp Weed and root decoction of *Ataenidia conferta* (Konkon sibere).

CRAWCRAW

Oil or camwood of *Elaeis guineensis* Oil Palm; oil extract of *Ricinus communis* Castor Oil Plant as remedy, *Jatropha curcas* Physic Nut applied, *Balanites aegyptiaca* Desert Date; resinous latex of *Milicia excelsa* (*Chlorophora excelsa*) Iroko (Odum); pulverized gum resin of *Bombax buonopozense* Red-Flowered Silk Cotton Tree with oil, and pounded yellowish-red resin of *Vismia guineensis* (Kosowa-nini) as ointment. Leaf of *Vitex micrantha* (Nyamele-buruma), *Rothmannia longiflora* (Saman-kube), *Parquetina nigrescens* (Aba-kamo), *Morinda morindoides* as lotion, and *Tacazzea apiculata* as rub; scented leaf of *Heinsia crinita* Bush Apple pounded, dried, and applied as poultice; ashes of *Cyathula prostrata* (Mpupua) applied; boiled leaves and twigs of *Psorospermum corymbiferum* var. *kerstingii* (E-kariga); the red resin with the pounded, dried leaves and roots of *P. corymbiferum* var. *corymbiferum*; latex of *Mangenotia eburnea* and pounded leaves of *Asystasia calycina* with lime-juice applied. Bark infusion and latex of *Pycnanthus angolensis* African Nutmeg, (Otie); bark of *Trema orientalis* (*guineensis*) (Sesea), *Rhizophora* species Red Mangrove, *Symphonia globulifera* (Ehureke) extract, and *Harungana madagascariensis* (Okosoa) extract; pounded bark of *Mareya micrantha* (Odubrafo) with white clay; pulverized bark of *Ximenia americana* Wild Olive and bark and root decoction of *Daniellia oliveri* African Copaiba Balsam Tree taken internally. Root of *Mammea africana* African Mammy Apple, (Bompagya); root decoction of *Acridocarpus smeathmannii* (Alasaayo); decoction of *Ageratum conyzoides* Billy-goat Weed (Efumomoe) and of *Hygrophila auriculata* (Eyitro) as remedy; fragrant rhizomes of *Cyperus articulatus* (Kokyi) ground with clay; rhizome of *Mariscus sumatrensis* (*alternifolius*) ground and applied; tubers of *Solenostemon monostachys* (Sisiworodo) as medium and an unspecified parts of *Strychnos spinosa* Kaffir Orange (Akankoa).

CUTS (see Wounds)

DENTITION (TEETHING)

Leaf infusion of *Pericopsis laxiflora* (*Afrormosia laxiflora*) Satinwood; leaf of *Bertiera racemosa* (Mantannua) used to prepare remedy; latex of *Jatropha curcas* Physic Nut, (Adadze) applied to gums; leaf juice of *Sansevieria liberica* African Bowstring Hemp with fluid from edible snails and native soap applied by Yorubas to relieve pains, and an unspecified part of *Waltheria indica* (Sawai). *Cardiospermum grandiflorum* and *C. halicacabum* Heart Seed, Balloon Vine are used by pregnant women in northern Nigeria to rub their teeth so that the child will have easy teething.

DIABETES

Decoction of *Catharanthus roseus* Madagascar Periwinkle or *Nelsonia canescens* (Moro, 1984-5); a tea-like infusion of *Bridelia ferruginea* (Opam-fufuo) (Ayensu, 1978); leaf of *Gymnema sylvestre*; unripe fruit of *Musa paradisiaca* var. *sapientum* Banana with other drugs (Ayensu, 1978); bark of *Myrianthus arboreus* (Anyankoma) (Ayensu, 1978) and unspecified parts of *Momordica charantia* African Cucumber and *Costus schlechteri* (Ayensu, 1978); also the edible tuber of *Solenostemon rotundifolius* Hausa, Frafra, or Salaga Potato as food.

DIAPHORETIC/SUDORIFIC (PROMOTING PERSPIRATION)

Hot infusion of *Ocimum gratissimum* Tea Bush (Nunum); *O. basilicum* (Eme) (Ayensu, 1978); leaf infusion of *Lippia multiflora* Gambian Tea Bush (Saanunum); *Cassia occidentalis* Negro Coffee and aromatic infusion of leaves of *Lantana camara* Wild Sage sometimes with *Ocimum*; rhizome decoction of *Cynodon dactylon* Bermuda or Bahama Grass (Boulos, 1983); root decoction of *Leea guineensis* (Okatakyi); rhizomes of *Zingiber officinale* Ginger (Evans, 1983); leaf decoction of *Indigofera suffruticosa* West Indian Indigo; tincture of rootbark of *Plumbago zeylanica* Ceylon Leadwort (Ayensu, 1978); bark decoction or dried leaf powder of *Adansonia digitata* Baobab; bark of *Zanthoxylum xanthoxyloides* (*Fagara zanthoxyloides*) Candle Wood in lotion and fumigation; bark decoction of *Mitragyna inermis* (Kukyafie) and unspecified parts of *Ricinus communis* Castor Oil Plant (Ayensu, 1978), *Ageratum conyzoides* Billygoat Weed, (Efumomoe) (Ayensu, 1978), and *Amaranthus spinosus* Prickly Amaranth (Ayensu, 1978); also hot decoction of *Brillantaisia lamium* and *B. nitens* in sitz-baths.

DIARRHOEA/DYSENTERY

Bark decoction of *Xylopia aethiopica* Ethiopian Pepper (Hwenetia); *Justicia flava* (Ntumenum) for children, *Combretum molle* (Gburega), *Allanblackia parviflora* (*floribunda*) Tallow Tree, (Sonkyi); *Pentadesma butyraceum* (Abotoasebie), *Monodora tenuifolia* (Motokuradua) or root, *Harungana madagascariensis* (Okosoa) or root, *Albizia ferruginea* (Awiemfosemina), *Tamarindus indica* Indian Tamarind, *Acacia hockii* Shittim Wood or pounded bark sweetened with honey, *Pterocarpus erinaceus* Senegal Rose Wood Tree with resin, *Trilepisium madagascariense* (*Bosqueia angolensis*) (Okure) with leaves, *Coula edulis* Gaboon Nut, (Bodwue) taken as stomachic or in enemas; *Manilkara obovata* (*multinervis*) (Berekankum), *Saba senegalensis* (Sono-nantin) with leaf, *Ziziphus mauritiana* Indian Jujube as stomachic, *Canarium schweinfurthii* Incense Tree as ingredient, *Schumanniophyton magnificum* as enema, and *Holarrhena floribunda* False Rubber Tree, (Sese) or macerated

8. POTIONS AND MEDICINES 145

bark in palm wine. Bark infusion of *Pandanus abbiwii* Screw Pine (Nton); *P. veitchi*, *Urena lobata* Congo Jute (Asonsom) with root, *Piliostigma thonningii* (Opitipata), *Ficus lutea* (*vogelii*) (Fonto), *Sclerocarya birrea* (Nanogba) with natron, and *Crossopteryx febrifuga* African Bark (Pakyisie). Bark or bark extract of *Smeathmannii pubescens* (Turunnua), *Cinnamomum zeylanicum* Cinnamon as intestinal astringent, *Annona senegalensis* var. *senegalensis* Wild Custard Apple with roots, *Monotes kerstingii* boiled, *Syzygium guineense* var. *guineense* (Sunya) with root and leaf, *Strephonema pseudocola* (Awuruku), *Rhizophora* species Red Mangrove especially in children, *Bridelia ferruginea* (Opam-fufuo) or root decoction with cassava flour, *Acacia farnesiana* with leaf taken internally, *A. nilotica* var. *tomentosa* or with infusion of pods, *Pentaclethra macrophylla* Oil Bean Tree as enema, *Erythrina senegalensis* Coral Flowers with root, *Irvingia gabonensis* Wild Mango (Abesebuo) or shavings in roasted banana; *Breonadia salicina* (*Adina microcephala*), *Mitragyna ciliata* Abura, Poplar (Subaha) and leaf; *Vernonia conferta* (Flakwa), *Spathodea campanulata* African Tulip Tree (Kokoanisua); *Stereospermum kunthianum* (Sonotokwakofo), *Vitex doniana* Black Plum (Afua) with leaf taken internally; *Kigelia africana* Sausage Tree (Nufuten) with leaf alone or with ingredients; bark juice of *Entada abyssinica* (Sankasaa) in milk and pounded, macerated bark of *Dichrostachys cinerea* (*glomerata*) Marabou Thorn in cold water with spices and lime-juice taken internally.

Extract of pounded bark of *Maesobotrya barteri* var. *sparsiflora* (Apotrewa); boiled *Trichodesma africanum* with cereal, shea butter, and natron as paste taken internally; decoction of whole plant of *Eleusine indica* (Amico, 1977); decoction of *Blumea aurita* var. *aurita* as enema and decoction of *Alternanthera pungens* (*repens*) (Mpatowansoe) or *Celosia argentea* (Nkyewodue) as enema; decoction of bark resin of *Amphimas pterocarpoides* (Yaya), decoction of *Desmodium gangeticum* var. *gangeticum* as beverage, *D. incanum* (*canum*) or pulped plant; small doses of bark of *Sesbania grandiflora* and decoction of *Stachytarpheta cayennensis* Brazilian Tea with natron for both humans and horses.

Leaf or leaf extract of *Ocimum gratissimum* Tea Bush (Nunum) with *Paullinia pinnata* (Toa-ntini), and that of *Triumfetta cordifolia* Burweed, (Ekuba); *Dichapetalum pallidum* (Folie), *Desmodium adscendens* var. *adscendens* (Akwamfanu) mixed with roasted corn and salt, *Millettia thonningii* (Sante), *Myrianthus arboreus* (Anyankoma) with that of *Alchornea cordifolia* Christmas Bush as drink, *Zanthoxylum xanthoxyloides* (*Fagara zanthoxyloides*) Candle Wood, *Nauclea latifolia* African Peach (Sukisia) eaten with Guinea-grains and followed by a drink of warm water; young leaf of *Anacardium occidentale* Cashew Nut; leaf pulp of *Rothmannia longiflora* (Samankube) in enema and leaves of *Cassià occidentalis* Negro Coffee chewed or cooked with groundnuts or *Irvingia*. Leaf decoction of *Annona muricata* Sour Sop, *Anogeissus leiocarpus* (Sakanee), *Croton zambesicus* (Dodwatu), *Securinega virosa* (Nkanaa) or whole plant with *Terminalia*; *Bauhinia rufescens* (Jinkiliza) with fruit and millet,

Detarium senegalense Tallow Tree as enema, *Newbouldia laevis* (Sasanemasa) with root, *Urera mannii* (Ahyehyew-nsa), *U. obovata* (Osurosuso), *Nauclea diderrichii* (Kusia), and *Cassia alata* Ringworm Shrub as a drink or in enemas. Dried and pulverized leaf of *Crossandra guineensis* boiled with cereal pap as remedy; decoction of *Schwenckia americana* (Agyingyinsu) with natron taken by both mother and child if breast milk is purging the child; leaf infusion of *Sterculia tragacantha* African Tragacanth (Sofo); *Mimosa pudica* Sensitive Plant, *Abutilon guineense* (Odonno-bea) as fomentation to relieve pain, *Gossypium arboreum* Cotton with lime-juice as relief, *Microdesmis puberula* (Fema) with leaves of *Triumfetta*, and *Diospyros mespiliformis* West African Ebony with fruit.

Infusion of young shoots of *Lophira lanceolata*; lotion of leafy stems and bark of *Hymenocardia acida* (Sabrakyie) as beverage; decoction of leafy stems of *Ficus capensis* (Nwadua) and of *Griffonia simplicifolia* (Kagya) with leaves.

Root or root extract of *Uvaria chamae* (Akotompotsen), *Oncoba spinosa* Snuff-box Tree (Asratoa); *Combretum zenkeri* (Tadatso) with *Aframomum* and peppers, *Quisqualis indica* Rangoon Creeper with seed, *Margaritaria discoidea* (*Phyllanthus discoideus*) (Pepea), *P. muellerianus* (Potopoleboblo) cooked with maize meal, *Chrysobalanus icaco* (*orbicularis*) (Abele) with leaf and seed oil, *Mezoneuron benthamianus* (Akoobowerew), and *Vernonia nigritiana* (Gyakuruwa). Root-bark of *Garcinia afzelii* (Nsokodua) dried and pulverized with spices and that of *Pseudocedrela kotschyi* Dry-Zone Cedar; bitter root-bark and stems of *Vernonia amygdalina* Bitter Leaf and *V. colorata*; bitter, acrid root-bark of *Calotropis procera* Sodom Apple and root-bark or stem decoction of *Morinda lucida* Brimstone Tree with spices. Root or root-bark decoction of *Terminalia avicennioides* (Petni) as beverage and probably enema, *Ricinodendron heudelotii* (Wamma), *Glyphaea brevis* (Foto), *Pericopsis laxiflora* (*Afrormosia laxiflora*) Satinwood, *Ehretia cymosa* (Okusua), *Maytenus senegalensis* (Kumakuafo) with natron, *Ximenia americana* Wild Olive with leaf, *Cnestis ferruginea* (Akitase) as enema, *Fadogia agrestis* (Buruntirikwa), and *Vernonia guineensis* or leaf decoction or a decoction of the whole plant. Powdered or pulverized root of *Rhodognaphalon brevicuspe* (*Bombax brevicuspe*) (Onyinakobin) taken internally, *Ceiba pentandra* Silk Cotton Tree with gum, *Guiera senegalensis* boiled, and *Paullinia pinnata* (Toa-ntini) with leaf and spices. Pounded or pulped root of *Psidium guajava* Guava with water taken internally, *Jatropha curcas* Physic Nut as suppository with fruit of *Xylopia, Harrisonia abyssinica* (Penku), or leafy tops in hot water as enemas, *Khaya ivorensis* African Mahogany (Dubini) with black peppercorns, *Piper*, and juice as enemas, or *Rauvolfia vomitoria* (Kakapenpen) and *Massularia acuminata* (Pobe) as enemas.

Infusion of pulped root of *Desmodium velutinum* (Koheni-koko) or leaf decoction; root-bark infusion of *Allophylus africanus* (Hokple); hot infusion of root-bark of *A. spicatus* (Kotamenyati) and infusion of root-bark of *Mangifera indica* Mango. An infusion of the bark of the Mango (Moro, 1984–5) and that

8. POTIONS AND MEDICINES

of *Annona senegalensis* var. *senegalensis* Wild Custard Apple (Burkill, 1985) are specific for cholera. Seeds of *Celosia argentea* (Nkyewodue), *Aframomum melegueta* Melegueta (Famu-wisa); *Cola laurifolia* Laurel-leaved Cola; powdered nuts of *C. acuminata* Commercial Cola Nut Tree; galled fruits of *Terminalia macroptera* (Kwatiri); fruit pulp of *Bixa orellana* Anatto, *Dialium guineense* Velvet Tamarind with other ingredients; fruit-rind of *Punica granatum* Pomegranate; fruit of *Piper umbellatum* (Amuaha); fruit infusion of *Parinari excelsa* Guinea Plum (Ofam); the burs of *Cyathula prostrata* (Mpupua), fruit-juice of *Citrus aurantiifolia* Lime or with hot spices and powdered, moistened fruit of *Ziziphus spina-christi* var. *spina-christi* with kernels. Fruit of *Cyrtosperma senegalense* as ingredient in remedies; reddish gum of *Moringa oleifera* Horse-Radish Tree, Oil of Ben Tree; leaf latex of *Parquetina nigrescens* (Aba-kamo); juice of pulped leafy stems of *Drypetes chevalieri* (Katrika) in water sieved as drink; chewed stems of *Flacourtia flavescens* Niger Plum (Amugui); decoction of roasted seeds of *Adansonia digitata* Baobab; boiled *Chrozophora senegalensis* with cereal foods; infusion of wood chips of *Trema orientalis* (*guineensis*) (Sesea); sitz-bath of the juice of *Citrus limon* Lemon in water and sap of *Musa paradisiaca* Plantain alone or with food or pounded roots in enemas. Unspecified parts of *Annona senegalensis* subsp. *onlotricha* (*arenaria*) (Aboboma) as purgative, *Hexalobus monpetalus* var. *monopetalus*, *Triclisia distyophylla* (*gilletii*), *Macaranga heudelotii* (Apazina) and the pith from the young stems of *Alchornea cordifolia* Christmas Bush (Gyamma) chewed with salt and natron. Whole plant of *Evolvulus alsinoides* and of *Mallotus oppositifolius* (Satadua) especially the fruits. The root of *Hedyotis corymbosa* (*Oldenlandia corymbosa*) causes diarrhoea in goats and sheep.

The whole plant of *Zea mays* Maize and corms of *Gladiolus daleni* (*psittacinus*), *G. klattianus*, *G. gregarius*, and *G. unguiculatus* Sword Lily, Cornflag are used for mucous diarrhoea in horses by rectal injection.

DISLOCATION (see Fractures)

DIURETIC (PROMOTING URINATION)

The leaf or leaf-juice or leaf infusion of *Piper umbellatum* (Amuaha), *Phyllanthus reticulatus* var. *reticulatus*, *P. reticulatus* var. *glabra* (Awobe), *Piliostigma thonningii* (Opitipata), *Moringa oleifera* Horse-Radish Tree, Oil of Ben Tree; *Guiera senegalensis*, *Mangifera indica* Mango, *Dialium dinklagei* (Awendade), *Melia azedarach* Persian Lilac, Bead Tree; *Alchornea cordifolia* Christmas Bush (Gyamma); *Steganotaenia araliacea* (Pienwogo), *Argemone mexicana* Mexican or Prickly Poppy, and infusion of stems and roots of *Nymphaea lotus* Water Lily. Hot infusion of *Scoparia dulcis* Sweetbroom Weed; macerated leaf of *Portulaca oleracea* Purslane, Pigweed in water; juice of *Bryophyllum pinnatum* Resurrection Plant; tisane from *Zea mays* Maize (Evans, 1983); decoction of *Pennisetum purpureum* Elephant Grass (Amico, 1977),

Eleusine indica (Amico, 1977), *Cynodon dactylon* Bahama Grass (Amico, 1977), roots of *Stenotaphrum secundatum* Buffalo Grass (Amico, 1977), and roots and inflorescence of *Imperata cylindrica* Lalang (Amico, 1977). The leaf decoction of *Psorospermum corymbiferum* var. *corymbiferum, Solanum nigrum* (Nsusuabiri) (also depurative), *Combretum micranthum* (Landaga), *Vernonia amygdalina* Bitter Leaf (Bonwen); *V. colorata, Jatropha curcas* Physic Nut, *Luffa acutangula*, and *Spilanthes filicaulis* Para or Brazil Cress.

Stem pulp of *Adenia rumicifolia* var. *miegei* (*lobata*) (Peteha) in enemas; fruit juice of *Spondias mombin* Hog Plum and *Anacardium occidentale* Cashew Nut; root or leaf decoction or infusion of *Ziziphus mucronata* Buffalo Thorn; root extract of *Maytenus senegalensis* (Kumakuafo); root, leaf, or fruit decoction of *Tephrosia purpurea*; root preparation of *Abrus precatorius* Prayer Beads; root of *Dichrostachys cinerea* (*glomerata*) Marabou Thorn in milk; root of *Croton membranaceum* in spirit (special reputation); boiled roots of *Vernonia nigritiana* (Gyakuruwa) and *Securidaca longepedunculata* (Kpaliga); root of *Smilax kraussiana* (Kokora), *Asparagus africanus* (Adedende), *A. flagellaris* and root-bark infusion of *Cassia sieberiana* African Laburnum. Root decoction of *Premna quadrifolia* (Atantaba) with pepper as drink; root decoction of *Nauclea latifolia* African Peach, (Sukisia) in enemas or drink and of *Bridelia ferruginea* (Opam-fufuo). Half-ripe fruits of *Ananas comosus* Pineapple; fruit pulp of *Parkia biglobosa* (Duaga); fresh plant of *Amaranthus caudatus*; macerated *Parquetina nigrescens* (Aba-kamo) in palm-wine with *Secamone*; pounded roots of *Funtumia africana* False Rubber Tree in palm-wine and theobromine in seed of *Theobroma cacao* Cocoa.

Bark of *Abutilon guineense* (Odonno-bea); bark decoction of *Bridelia atroviridis* (Asaraba) and *Mitragyna inermis* (Kukyafie); powdered bark of *M. stipulosa* (Subaha-akoa) with warm water; bark and leaf decoction of *Terminalia macroptera* (Kwatiri); pulped bark of *Pseudospondias microcarpa* var. *microcarpa* (Katawani) with water or palm-wine; dried, powdered stem of *Calotropis procera* Sodom Apple with Tamarind water and bark-pulp of *Afzelia africana* (Papao) with that of Tamarind and Satinwood. Unspecified parts of *Spilanthes filicaulis* Para or Brazil Cress, *Desmodium gangeticum* var. *gangeticum*, *Hoslundia opposita* (Asifuaka), *Opilia celtidifolia* (Benkasa), *Microglossa afzelii* (Pofiri), *Maesopsis eminii* (Onwamdua), *Cassia occidentalis* Negro Coffee, *Ficus capensis* (Nwadua), and *Paullinia pinnata* (Toa-ntini).

DIZZINESS/GIDDINESS/BILIOUSNESS

Leaf decoction of *Hedranthera barteri, Waltheria indica* (Sawai), *Hymenocardia acida* (Sabrakyie) with honey; bark decoction of *Xylopia aethiopica* Ethiopian Pepper, (Hwenetia); a cold infusion of *Ipomoea coptica*, and root with fruit-pulp of *Cassia sieberiana* African Laburnum. The leaves of *Sorindeia grandifolia* are also used.

PLATE 1   Chapter 1   FOREST AND CONSERVATION
Entandrophragma cylindricum Sapele (Apenkwa)
Milicia regia Iroko (Odum)
Tieghemella heckelei (Baku)
Triplochiton scleroxylon (Wawa)
Synsepalum dulcificum Miraculous Berry (Asaa)
Thaumatococcus daniellii (Katemfe)

Photos all F.N. Hepper

PLATE 2   Chapter 2   FOOD AND FODDER

Cajanus cajan Pigeonpea
Citrus sinensis Orange

Colocynthis lanatus Water Melon

Capsicum annuum Red pepper
Colocasia esculenta Eddoes, Dasheen, Taro
  (Kookoo, Ntwibo)
Manihot esculenta Cassava (Bankye)

Photos all F.N. Hepper

PLATE 3   Chapter 2   FOOD AND FODDER

Pennisetum americanum Bulrush Millet (Ewio)
Sechium edule Chayote
Solanum macrocarpon (Atropo)
Sorghum bicolor Guinea Corn (Atoko)
Vigna subterranea Bambara Groundnut
Xanthosoma mafaffa Cocoyam (Kontomire)

Photos D. Abbiw & F.N. Hepper

PLATE 4   Chapter 2   FOOD AND FODDER

Roasting plantains (*Musa paradisiaca*)
Palm oil (*Elaeis guineensis*)
Roasting *gari* from ground fermented cassava dough (*Manihot esculenta*)

Plantains (*Musa paradisiaca*) for sale
Raphia sp. being tapped for palm wine
Pounding *fufu* (*Manihot esculenta*)

Photos D. Abbiw & F.N. Hepper

PLATE 5 Chapter 3 INDUSTRIAL OR CASH CROPS

Cocos nucifera Coconut Palm

Cola nitida Bitter Cola (Bese)

Coffea canephora Rio Nunez or Robusta Coffee

Elaeis guineensis Oil Palm

Theobroma cacao Cocoa (pod cut open)

Photos all F.N. Hepper

PLATE 6   Chapter 4   BUILDING AND CONSTRUCTION

Wooden railway sleepers (see p. 87)

Wooden bus shelter

Telegraph poles of *Strombosia glaucescens* var *lucida* (Afena) (see p. 87)

Mud and timber house with thatched roof at Agbogba (see p. 79)

Modern wooden house (see p. 83)

Wooden coachwork for vehicles (see p. 85)

Photos all D. Abbiw

PLATE 7   Chapter 5   FURNISHINGS
Wooden console for organ (see p. 92)   Laminated stool of many species of timber (see p. 90)

Photos D. Abbiw

Chapter 6   FUEL

Mangroves with *Avicennia* and *Rhizophora*

Blighia welwitschii (Akyekobiri)

Azadirachta indica Neem Tree

Traditional open cooking fire using timber (see p. 8, 94)

Photos F.N. Hepper & D. Abbiw

PLATE 8   Chapter 7   TOOLS AND CRAFTS
Alstonia boonei (Sinduro)
Borassus aethiopum Fan Palm
Lagenaria siceraria Calabash, Bottle Gourd, White Pumpkin

Andropogon gayanus
Cochlospermum planchonii
Pandanus abbiwii Screw Pine (Nton)

Photos F.N. Hepper & D. Abbiw

PLATE 9 Chapter 7 TOOLS AND CRAFTS
Baskets using *Cyperus articulatus* (Kokyi) Bolga baskets (see p. 111)
Gourd containers (see p. 110) Wooden drums and carvings (see p. 107)
Wooden xylophone (see p. 117) Dugout canoes (see p. 109)

Photos D. Abbiw & F.N. Hepper

PLATE 10   Chapter 8   POTIONS AND MEDICINES

Acacia pentagona

Catharanthus roseus Madagasacar Periwinkle

Duparquetia orchidacea (Pikeato)

Cardiospermum halicacabum Heart Seed, Balloon Vine

Dialium guineense Velvet Tamarind (Asenamba)

Flacourtia flavescens Niger Plum (Amugui)

Photos all F.N. Hepper

PLATE 11 Chapter 8 POTIONS AND MEDICINES
Griffonia simplicifolia (Kagya)　　*Jatropha curcas* Physic Nut
Khaya senegalensis Dry-zone Mahogany　　*Lonchocarpus sericeus* Senegal Lilac (Keklengbe)
Mallotus oppositifolius (Satadua)　　*Morinda lucida* Brimstone Tree (Kankroma)

Photos all F.N. Hepper

PLATE 12   Chapter 8   POTIONS AND MEDICINES
Opilia celtidifolia (Benkasa)
Pycnocoma macrophylla (Akofie-kofi)
Thonningia sanguinea
Pterocarpus erinaceus Senegal Rose Wood Tree
Voacanga africana (Ofuruma)
Zanthoxylum xanthoxyloides Candle Wood

Photos all F.N. Hepper

PLATE 13   Chapter 8   POTIONS AND MEDICINES
Adansonia digitata Baobab
Aloe buettneri (Sere-berebe)
Crossopteryx febrifuga African Bark (Pakyisie)
Cyrtosperma senegalense
Euphorbia balsamifera Balsam Spurge (Aguwa)
Ricinus communis Castor Oil Plant

Photos all F.N. Hepper

PLATE 14 Chapter 9 POISONS, TANNINS AND DYES

Bixa orellana Anatto

Erythophloeum suaveolens Ordeal Tree (Potrodom)

Lonchocarpus cyanescens West African Indigo, Yoruba Indigo (Akase)

Paullinia pinnata (Toa-ntini)

Strophanthus sarmentosus (Adwokuma)

Tetracera alnifolia (Akotopa)

Photos all F.N. Hepper

PLATE 15 Chapter 10 AMENITY LANDSCAPING AND GARDENING

Combretum racemosum (Wota)
Dissotis rotundifolia (Adowa-alie)
Mussaenda erythrophylla Ashanti Blood (Damaramma)

Cnestis ferruginea (Akitase)
Hildegardia barteri (Akyere)
Scadoxus multiflorus Blood Flower, Fire-ball Lily (Ngobo)

Photos all F.N. Hepper

PLATE 16 Chapter 11 WEEDS
Chromolaena odorata Siam Weed
Mikania scandens Climbing Hemp Weed
Mechanical weed control (see p. 250)
Tapinanthus bangwensis Mistletoe
Striga hermonthica (Wumlim)
Stuffing a mattress with dried weeds (see p. 251)

Photos F.N. Hepper & D. Abbiw

DROPSY (ASCITES)

Leaf decoction of *Microglossa pyrifolia* (Asommerewa), *Mussaenda elegans* (Damaram), and *Strephonema pseudocola* (Awuruku) as beverage; a drink or enema of the leaf decoction of *Pycnanthus angolensis* African Nutmeg, (Otie); leaf-juice or decoction of *Mareya micrantha* (Odubrafo); leaf decoction of *Cassia alata* Ringworm Shrub, (Nsempii) as purgative; *Lantana camara* Wild Sage in baths and hot fomentation; oral application of *Sapium grahamii* (Pampiga) preparation; crushed seeds of *Jatropha curcas* Physic Nut boiled with cereal pap as purge; whole plant of *Boerhavia diffusa* Hogweed and *Trianthema portulacastrum* as diuretic (Jain, 1968); root bark of *Maesopsis eminii* (Onwamdua) as diuretic and a drink of *Cassia occidentalis* Negro Coffee. Pulverized fruits of *Ananas comosus* Pineapple; decoction of *Mirabilis jalapa* Marvel or Peru and *Ipomoea pes-caprae* subsp. *brasiliensis* Goat's Foot Convolvulus.

DYSMENORRHOEA (see Menstrual Troubles)

DYSPEPSIA (see Stomach Ache)

EARACHE (INCLUDING OTITIS — INFLAMATION)

Leaf-juice of *Clitoria ternatea* Blue Pea, *Erythrococca anomala*, *Desmodium incanum (canum)* for the analgesic properties or *Diospyros mespiliformis* West African Ebony; bark decoction of *Strychnos spinosa* Kaffir Orange (Akankoa) and root ashes of *Raphia hookeri* Wine Palm; root decoction of *Euadenia eminens* (Disinkoro) and *E. trifoliolata* as drops or pulp or leafy stem-tips of *Premna hispida* (Aunni) with lemon applied locally. Leaf-juice of *Bryophyllum pinnatum* Resurrection Plant, *Kalanchoe integra* var. *crenata* (Aporo), *Portulaca oleracea* Purslane, Pigweed; *Cleome viscosa* Wild Mustard, *C. gynandra* (*Gynandropsis gynandra*) Cat's Whiskers, *Sansevieria liberica* African Bowstring Hemp, *Grangea maderaspatana* Marcella, *Maesa lanceolata*, *Microglossa afzelii* (Pofiri), *Margaritaria discoidea* (*Phyllanthus discoideus*) (Pepea), *Brillantaisia lamium* or *B. nitens* (Guare-ansra) squeezed into ear to relieve symptoms due to water in the ear after bath, *Stachytarpheta cayennensis* Brazilian Tea for sores in the ear, juice of roasted leaves of *Palisota hirsuta* (Nzahuara) as drops and remedy for deafness, and juice of fresh warmed leaves of *Bidens pilosa* Bur Marigold. Leaf of *Scoparia dulcis* Sweetbroom Weed applied warm locally; crushed leaf of *Lycopersicon esculentum* (*lycopersicum*) Tomato; decoction of *Phyllanthus pentandrus* and watery extract of ashes instilled in ear; leaf decoction of *Adansonia digitata* Baobab and application of the hot leaf decoction of *Lippia multiflora* Gambian Tea Bush (Saa-nunum). Bark of *Cocos nucifera* Coconut Palm; soft inner bark of *Newbouldia laevis* (Sasanemasa); bark juice of *Berlinia confusa* and *Buchholzia coriacea* (Esonobise) and sap of inflorescence of *Musa paradisiaca* Plantain as drops. Root juice of *Byrsocarpus coccineus* (Awennade) and *Ritchiea reflexa* (Aayerebi); the berries of *Solanum indicum* subsp. *distichum* Children's Tomato and crushed seeds of *Aframomum melegueta* Guinea-grains applied as paste to the head.

ELEPHANTIASIS (ENLARGEMENT OF LIMBS)

Decoction of pounded leaves of *Mussaenda elegans* (Damaram); pounded bark of *Duparquetia orchidacea* (Pikeabo) with ingredients applied; decoction or cold infusion of macerated inner bark of *Dichrostachys cinerea* (*glomerata*) Marabou Thorn taken internally, pounded and warmed bark of *Ricinodendron heudelotii* (Wamma) applied, and ground root of *Annona senegalensis* var. *senegalensis* Wild Custard Apple applied as paste and leaf decoction taken (Moro, 1984–5).

EMETIC (INDUCING VOMITING)

Root of *Clitoria ternatea* Blue Pea, *Ziziphus mauritiana* Indian Jujube, *Thevetia peruviana* Milk Bush, Exile Oil Plant; *Boerhavia repens* and *B. diffusa* Hogweed both with leaves; root decoction of *Mimosa pudica* Sensitive Plant; root, leaf, and leafy tops of *Opilia celtidifolia* (Benkasa); bitter root-bark of *Rauvolfia vomitoria* (Kakapenpen) (as the name implies), and pounded, boiled roots of *Vernonia nigritiana* (Gyakuruwa). Whole plant infusion of *Luffa acutangula* in large doses, *Momordica charantia* African Cucumber, and *Heliotropium ovalifolium*; leaf decoction of *Xylopia aethiopica* Ethiopian Pepper (Hwenetia); *Alchornea cordifolia* Christmas Bush, (Gyamma); *Ficus sagittifolia*, *Vernonia amygdalina* Bitter Leaf, and *V. colorata*. Leaf infusion of *Tephrosia vogelii* Fish Poison Plant, *Trema orientalis* (*guineensis*) (Sesea), and crushed leaves of *Ageratum conyzoides* Billy-goat Weed, (Efumomoe) in water. Leaf of *Moringa oleifera* Horse-Radish Tree, Oil of Ben Tree; leaf-juice of *Pergularia daemia* (Jain, 1968); seed of *Argemone mexicana* Mexican or Prickly Poppy, *Jatropha gossypiifolia* Physic Nut and *Carapa procera* Crabwood; crushed seed of *Strychnos aculeata* in water and an overdose of crushed seeds of *Dicranolepis persei* (Prahoma); seed-cake of *Trichilia emetica* subsp. *suberosa* (*roka*) (Kisiga) as antidote to poison, decoction or infusion of pounded seeds of *Blighia unijugata* (Akyebiri); small doses of lime-juice of *Citrus aurantiifolia* Lime; fruit-rind of *Strychnos spinosa* Kaffir Orange (Akankoa) and fruit pulp of *Rothmannia longiflora* (Saman-kube).

Bark decoction of *Pycnanthus angolensis* African Nutmeg (Otie); *Harungana madagascariensis* (Okosoa), *Tetrapleura tetraptera* (Prekese), *Ficus vogeliana* (Opanto), and *Khaya senegalensis* Dry-Zone Mahogany or pulp sweetened with honey. Bark of *Ceiba pentandra* Silk Cotton Tree, *Sacoglottis gabonensis* (Tiabutuo), *Uapaca guineensis* Sugar Plum (Kuntan), and *Sesbania grandiflora* in large doses; fresh bark of *Boswellia dalzielii* Frankincense Tree (Kabona) eaten; cold bark infusion of *Nauclea latifolia* African Peach (Sukisia); *Erythrophleum suaveolens* (*guineense*) Ordeal Tree (Potrodom) and ground

8. POTIONS AND MEDICINES

bark of *Pycnocoma cornuta* with lime-juice and water. The juice of young leaves of *Solanum aethiopicum* as a sedative for intractable cases of uterine origin. Unspecified parts of *Maesopsis eminii* (Onwamdua), *Byrsocarpus coccineus* (Awennade), and *Asclepias curassavica* Blood Flower.

The following plants are anti-emetic and arrest vomiting; roots of *Euphorbia hirta* Australian Asthma Herb (Jain, 1968); the seeds of *Myristica fragrans* Nutmeg (Evans, 1983); bark decoction of *Acacia albida* (Gozanga); leaf infusion of *Combretum micranthum* (Landaga); bark pulp of *Afzelia africana* (Papao) with that of Tamarind and Satinwood, and the root of *Ximenia americana* Wild Olive. The action from the nuts of *Cola acuminata* Commercial Cola Nut Tree is that of caffeine used to prevent vomiting in fevers.

ENTERITIS (INFLAMMATION OF THE SMALL AND LARGE INTESTINES)

Leaf infusion of *Baphia nitida* Camwood (Odwen) as drink; leaf decoction of *Abrus precatorius* Prayer Beads; root and leaf decoction of *Lippia multiflora* Gambian Tea Bush (Saa-nunum); bark infusion of *Pandanus veitchi* Screw Pine; gummy sap of *Harungana madagascariensis* (Okosoa) as enema, and an unspecified part of *Smeathmannia pubescens* (Turunnua) and of *Guiera senegalensis* for acute attacks.

EPILEPSY (see Convulsions)

EPISTAXIS (NOSE BLEEDING) (see Styptic)

ERUPTIONS (see Boils)

EYE DISEASES

Cataract (partial or total opacity on the eye lens)

Sap of crushed leaves of *Aspilia africana* Haemorrhage Plant with salt and lime-juice as drops; juice of *Stachytarpheta cayennensis* Brazilian Tea as drops; root of *Lonchocarpus cyanescens* West African or Yoruba Indigo; root juice of *Microglossa pyrifolia* (Asommerewa) as eye drops, the residue from crushed roots with shea butter melted and smeared on eyelids at bedtime; roasted, powdered seed and seed decoction of *Cassia occidentalis* Negro Coffee and *C. absus* Four-Leaved Henna with leaf lotion; powdered seeds of *Caesalpinia bonduc* Bonduc (Oware-amba) applied to the eye to prevent spots forming over cornea; seeds of *Entada abyssinica* (Sankasaa) applied (also for diseases of the back of the eye) and warmed leaf-juice of *Ruspolia hypocrateriformis* (Wonane) as drops (Moro, 1984–5).

Sore Eyes, Conjunctivitis, Iritis, Ophthalmia, Trachoma, etc.

Leaf-juice of the following plants as drops or wash: *Emilia coccinea*, *Bryophyllum pinnatum* Resurrection Plant, *Crassocephalum rubens* (Banfa-

banfa), *Heliotropium indicum* Indian Heliotrope, Cock's Comb; *Pentodon pentandrus, Cola caricifolia* Fig-Leaved Cola, *Bidens pilosa* Bur Marigold, *Hoslundia opposita* (Asifuaka), *Ageratum conyzoides* Billy-goat Weed, (Efumomoe); *Mikania chevalieri (cordata)* Climbing Hemp Weed, *Physalis angulata* (Totototo), *Commelina diffusa, Ocimum gratissimum* Tea Bush (Nunum); *Mussaenda afzelii, Pistia stratiotes* Water Lettuce, *Blumea aurita* var. *aurita, Motandra guineensis* (Amafohae), *Sansevieria liberica* African Bowstring Hemp, *Lippia multiflora* Gambian Tea Bush (Saa-nunum); *Malacantha alnifolia* (Fafaraha), *Strychnos spinosa* Kaffir Orange (Akankoa); *Securidaca welwitschii*; *Cnestis ferruginea* (Akitase), *Steganotaenia araliacea* (Pienwogo), *Manotes longiflora* (Awohanya), *Blighia sapida* Akee Apple, (Akye); *Teclea verdoorniana* (Owebiribi), *Clausena anisata* Mosquito Plant; *Cissus polyantha, Gouania longipetala* (Homabiri), *Eriosema psoraleoides, Acridocarpus smeathmannii* (Alasaayo), *Samanea dinklagei*, and *Stachytarpheta cayennensis* Brazilian Tea. Leaf infusion of *Uvaria chamae* (Akotompotsen), *Lantana trifolia*, and *Dissotis rotundifolia* (Adowa-alie) as lotion; leaf of *Scoparia dulcis* Sweetbroom Weed applied warm locally and leaves of *Solenostemon rotundifolius* Hausa, Frafra, or Salaga Potato as drops or wash; leaf decoction of *Newbouldia laevis* (Sasanemasa), *Premna hispida* (Aunni) as eye lotion, *Lantana camara* Wild Sage applied externally, *Strophanthus sarmentosus* (Adwokuma), *Vitellaria paradoxa* (*Butyrospermum paradoxum* subsp. *parkii*) Shea Nut Tree, *Alchornea cordifolia* Christmas Bush (Gyamma); *Ximenia americana* Wild Olive and *Margaritaria discoidea* (*Phyllanthus discoideus*) (Pepea).

Root decoction of the following plants as lotions and eye bath or drops: *Euadenia trifoliolata, Hymenocardia acida* (Sabrakyie), *Cochlospermum tinctorium* (Kokrosabia), *Bauhinia rufescens* (Jinkiliza), *Waltheria indica* (Sawai), and *Argemone mexicana* Mexican or Prickly Poppy. Root decoction of *Smilax kraussiana* (Kokora) steamed over eyes; boiled root of *Solanum incanum* Egg Plant (in small amounts) as wash (corrosive in large doses); peeled root of *Ceratotheca sesamoides* (Bungu) applied for stye; crushed root of *Capparis polymorpha* (Sansangwa) with that of *Waltheria* applied to eyelids and root infusion of *Albizia adianthifolia* (Pampena) applied.

Boiled leaf of *Celtis africana* (Yisa) as fomentation; leaf-juice or latex of *Ficus exasperata* (Nyankyeren) as drops; crushed leaf of *Ricinus communis* Castor Oil Plant with water as lotion; warm lotion of leaf of *Pericopsis laxiflora* (*Afrormosia laxiflora*) Satinwood; pounded leaves and sometimes seeds of *Abrus precatorius* Prayer Beads or leaf-juice of *A. pulchellus*; leaf of *Nymphaea lotus* Water Lily as lotion; macerated leaf and bark of *Parkia bicolor* (Asoma) as lotion; crushed leaves and flowers of *Cassia singueana* as lotion; juice from cut stems of *Paullinia pinnata* (Toa-ntini); juice of stem of *Cardiospermum grandiflorum* and *C. halicacabum* Heart Seed, Balloon Vine; juice from petioles of *Cola gigantea* var. *glabrescens* (Watapuo); juice of *Rinorea subintegrifolia* (Atobegyaso); juice of *Phyllanthus reticulatus* var. *reticulatus*; sap or leaf infusion of *P. muellerianus* (Potopoleboblo) and to remove foreign body; stem juice of *Anchomanes*

8. POTIONS AND MEDICINES

difformis (Atoe) as drops; latex of *Euphorbia hirta* Australian Asthma Herb as drops, juice extract of rhizomes of *Zingiber officinale* Ginger as drops; stem-juice of *Coix lacryma-jobi* Job's Tears used for irritations due to injury; and latex from stem of *Manihot esculenta* Cassava.

Juice of bark of *Zanthoxylum xanthoxyloides* (*Fagara zanthoxyloides*) Candle Wood; decoction of crushed bark of *Carapa procera* Crabwood; juice of pulped bark or leaf-juice of *Turraea heterophylla* (Ahunanyankwa) as drops; liquor from mashed bark of *Deinbollia grandifolia* (Potoke); resinous bark of *Pseudospondias microcarpa* var. *microcarpa* (Katawani); decoction of pounded bark of *Spondias mombin* Hog Plum, Ashanti Plum; bark of *Vernonia conferta* (Flakwa) and bark decoction of *Tamarindus indica* Indian Tamarind as lotion. Infusion or decoction of pod and bark or bark alone of *Acacia nilotica* var. *tomentosa*; fruit of *Mimosa pigra* Sensitive Plant (Kwedi); flowers of *Canna indica* Indian Shot; decoction of flower-heads of *Echinops longifolius*; decoction of roasted and ground seeds of *Lepidagathis heudelotiana* as wash; seed decoction of *Datura metel* Hairy Thorn Apple, Metel; pulverized seeds of *Eriosema glomeratum* as dressing; seeds of *Dyschoriste perrottetii* used to remove foreign bodies; crushed seeds of *Aneilema lanceolatum* placed in the eye to induce tears to remove foreign body; vapour of boiled gum of *Commiphora africana* African Bdellium (Narga); leaf of *Calotropis procera* Sodom Apple; leaf of *Parquetina nigrescens* (Aba-kamo) drained of latex as eye-salve; juice from cut fruits of *Massularia acuminata* (Pobe) as drops, boiled fruits of *Citrus aurantiifolia* Lime with orange juice as lotion or juice alone as drops; and *Euadenia eminens* (Disinkoro) or *Desmodium incanum* (*canum*) as drops.

Seeds of *Physostigma venenosum* Calabar Bean are used in ophthalmic practice to correct dilation caused by atropine, homatropine, or cocaine; and also in glaucoma to decrease intra-ocular pressure. The fruit-juice of *Turraea vogelii* is used for filaria (i.e. *Loa loa*, transmitted by a fly, *Chrysops*) in the eye. The juice expressed from the leaves of *Cissus populnea* (Agyako) is a well-known remedy for the blinding serum spat at the eyes by the Black-necked Cobra (*Naja nigricollis*).

FAINTING, REVIVE

Decoction of rhizome of *Zingiber officinale* Ginger; pulverized corms of *Gladiolus daleni* (*psittacinus*), *G. klattianus*, *G. gregarius*, and *G. unguiculatus* Sword Lily, Corn-flag in nostrils; sneezing as sign of recovery; root of *Smilax kraussiana* (Kokora) applied to head or root decoction sometimes with *Leptadenia hastata* (Benaduru); leaf-juice of *Physalis ungulata* (Totototo) dropped into eyes and nostrils to revive children; piece of grated root of *Albizia zygia* (Okoro) inserted into nostrils; a sniff of ground bark of *Erythrophleum suaveolens* (*guineense*) Ordeal Tree (Potrodom); decoction of root, leaf, and fruit or leaf alone of *Cassia occidentalis* Negro Coffee as lotion for fumigation and patient washed and rubbed with fluid; root extract of *Maytenus senegalensis* (Kumakuafo) given for fainting fits; the irritating root-sap of *Securidaca longepedunculata* (Kpaliga) inserted in the nostrils to restore consciousness; gum of a tree called Ope-igbo in Yoruba (probably *Dracaena mannii*) applied to the head and pulverized bark of *Dichapetalum toxicarium* West African Ratbane (Ekum-nkura) as sneezing powder to restore consciousness.

FEBRIFUGE (FEVER REMEDY)

Leaf infusion of *Hyptis pectinata* (Peaba), *H. suaveolens* Bush Tea-bush, *Ocimum gratissimum* Tea Bush (Nunum); *Aeolanthus pubescens, Justicia flava* (Ntumenum), *Cleistopholis patens* Salt and Oil Tree, *Canna indica* Indian Shot, *Campylospermum reticulatum* (*Ouratea reticulata*) as wash, *Combretum micranthum* (Landaga) and *Phyllanthus muellerianus* (Potoboleboblo) both with the root, *Croton penduliflorus* (Nyamerem) applied externally, *Cassia singueana* or decoction, *Entada abyssinica* (Sankasaa) with root or leaf decoction, *Morinda lucida* Brimstone Tree with salt, *Leucas martinicensis* as wash or steam bath, *Cymbopogon citratus* Lemon Grass, *C. giganteus* var. *giganteus* with lime-juice, *Platostoma africanum* (Siresireke), *Ipomoea involucrata* as stimulant or preventive or hot drink taken to ward off cold, *Pupalia lappacea* (Akukuaba), *Boerhavia repens* and *B. diffusa* Hogweed, *Heliotropium indicum* Indian Heliotrope, Cock's Comb; *Momordica charantia* African Cucumber as wash and decoction taken internally, *M. cissoides* as lotion, and leaf infusion of *Ampelocissus grantii* (Enotipsi) with onions as drink or wash.

Leaf decoction of *Annona muricata* Sour Sop, *Adenia rumicifolia* subsp. *miegei* (*lobata*) (Peteha), *Pulicaria crispa, Combretum platypterum* (Ohwirem), *Conocarpus erectus* Button Wood (Koka); *Terminalia macroptera* (Kwatiri) as a hot wash or fumigation, *Scaphopetalum amoenum* (Nsoto) in enemas, *Croton zambesicus* (Dodwatu) both internally and as a wash, *Ricinodendron heudelotii* (Wamma) as beverage, bath, or vapour bath; *Albizia zygia* (Okoro), *Parinari curatellifolia* (Atena), *Maranthes polyandra* (*P. polyandra*) (Abrabesi) both as beverage and bath, *Cassia laevigata, Acacia kamerunensis* (*pennata*) (Oguaben), *Securinega virosa* (Nkanaa), *Crotalaria retusa* Devil Bean or fresh juice taken internally, *Milicia excelsa* (*Chlorophora excelsa*) Iroko, (Odum) as wash; *Gouania longipetala* (Homabiri), *Azadirachta indica* Neem Tree with root, *Melia azedarach* Persion Lilac, Bead Tree; *Allophylus africanus* (Hokple) as lotion for children, *Diospyros mespiliformis* West African Ebony taken internally and as wash, *Craterispermum laurinum* or bark, *Microglossa pyrifolia* (Asommerewa), *Premna hispida* (Aunni) with root, *Vernonia biafrae* (Hu) as bath, *Markhamia lutea* (Sisimasa) and bark, *Lantana camara* Wild Sage, *Hygrophila auriculata* (Eyitro) as bath, *Ipomoea asarifolia* taken and as a wash for chills, or face steamed over hot decoction of the plant boiled with husks of Bulrush Millet; *Brillantaisia lamium* or *B. nitens* (Guare-ansra) steamed over patient to induce perspiration or leaves used as wash, and tea-like leaf infusion of *Lippia multiflora* Gambian Tea Bush (Saa-nunum). Decoction of *Leonotis nepetifolia* var. *nepetifolia* or *L. nepetifolia* var. *africana* (Nyeddo) to steam the head as relief and taken internally, *Pistia stratiotes* Water Lettuce as fumigant, whole plant of *Eleusine indica* (Amico, 1977); juice of *Nelsonia canescens* squeezed into the eyes; a copious cool drink of whole plant of *Scoparia dulcis* Sweetbroom Weed macerated in warm water for cases with headache or bruised leaves rubbed on the skin; leafy twigs decoction of *Alchornea cordifolia* Christmas Bush (Gyamma); *Pterocarpus erinaceus* Senegal Rose Wood Tree, *Psorospermum corymbiferum* var. *corymbiferum*, *P. febrifugum* var. *ferrugineum*, *Turraea heterophylla* (Ahunanyankwa) as drink or bath, *Mimosa pigra* Sensitive Plant, (Kwedi) or infusion as wash; *Cnestis ferruginea* (Akitase), *Clerodendrum capitatum* (Tromen), and *Daniellia oliveri* African Copaiba Balsan Tree as beverage, bath, or vapour bath.

Leaf of *Uvaria afzelii*, *Gilbertiodendron limba* (Tetekon), *Grewia mollis* (Yualega) and bark, *Piliostigma thonningii* (Opitipata), *Ehretia cymosa* (Okosua) in Yoruba Agbo infusions, *Ximenia americana* Wild Olive and root, *Strophanthus gratus* (Omaatwa) as lotion and rub, *Microglossa afzelii* (Pofiri) as lotion, *Plumbago zeylanica* Ceylon Leadwort in soup, *Newbouldia laevis* (Sasanemasa), *Cassia nigricans* and root as anti-periodic and quinine substitute, *Nauclea diderrichii* (Kusia) with those of *N. latifolia* African Peach, (Sukisia); bitter leaf of *Oxyanthus speciosus* (Korantema) as beverage and leaf of *Scaevola plumieri* (Wogbo) in water for washing patients. Leaf of *Dissotis rotundifolia* (Adowa-alie); pounded leaf of *Manihot esculenta* Cassava applied to head; lotion of *Alternanthera pungens* (*repens*) Khaki Weed, (Mpatowansoe) applied to head; fresh juice of *Solenostemon monostachys* (Sisiworodo) squeezed into the nostrils for cases in children; juice of *Allium ascalonicum* Shallot rubbed on the body especially in children's cases, or a mixture of onions, palm oil, and pepper heated in the sun as drink; leaf-juice of *Psychotria peduncularis* (*Cephaelis peduncularis*) (Kwesidua) as drink, *Uncaria talbotii* (Akoo-ano) and *Tarenna thomasii* as drink with leaf-pulp as rub.

Pounded leafy twigs of *Tetracera alnifolia* (Akotopa) in palm-wine; leaf or bark infusion of *Adansonia digitata* Baobab (reported taken by Adanson); paste of dried, pulverized leaves of *Heteropteris leona* mixed into paste with water as poultice applied to the top of the head; ground, boiled shoots of *Dialium guineense* Velvet Tamarind taken with food; boiled shoots of *Detarium senegalense* Tallow Tree given with food; leaf, root, bark, and seed of *Pterocarpus santalinoides* (Hote) as lotion or by inhaling leaf and root of *Cocculus pendulus*; stewed leaf and buds of *Mangifera indica* Mango taken internally; mashed leaves of *Macrosphyra longistyla* (Zetitsui) with freshly-made local soap and pounded leaf of *Vitex doniana* Black Plum (Afua) applied to body.

Root decoction of *Clematis hirsuta*, *Cochlospermum tinctorium* (Kokrosabia), *Acridocarpus smeathmannii* (Alasaayo), *Hymenocardia acida* (Sabrakyie), *Caesalpinia bonduc* Bonduc (Oware-amba); *Treculia africana* var. *africana* African Breadfruit, *Ziziphus mauritiana* Indian Jujube, and *Imperata cylindrica* Lalang (Amico, 1977). Root of *Uvaria chamae* (Akotompotsen), *Cassia occidentalis* Negro Coffee with bark as quinine substitute, *Crateva adansonii* (*religiosa*) (Chelum Punga) with natron and Guinea-corn pap, *Napoleonaea vogelii* (*Napoleona vogelii*) (Obua), *Bauhinia rufescens* (Zinkiliza), *Olax subscorpioidea* (Ahoohenedua), *Zanthoxylum xanthoxyloides* (*Fagara zanthoxyloides*) Candle Wood boiled with cereal foods, *Strychnos spinosa* Kaffir Orange (Akankoa) with those of *Pericopsis laxiflora* (*Afrormosia laxiflora*) Satinwood, *Pleiocarpa mutica* (Kanwen), *Picralima nitida* (Ekuama) with bark, *Gardenia ternifolia* (Peteprebi) boiled with millet flour, *Vernonia nigritiana* (Gyakuruwa); root, bark, and leafy tips of *Rothmannia longiflora* (Saman-kube) as drinks, lotions, and baths, and root and bark decoction of *Morinda morindoides* or in boluses prepared from powdered roots with palm oil.

Root infusion of *Abutilon guineense* (Odonno-bea), tuberous root of *Smilax kraussiana* (Kokora) as vapour bath; infusion or decoction of *Cadaba farinosa;* juice of *Dracaena fragrans* with palm oil as liniment rubbed on body; warm lotions and infusions of bark of *Anogeissus leiocarpus* (Sakanee); dried and powdered bark of *Margaritaria discoidea* (*Phyllanthus discoideus*) (Pepea) as rub or a decoction as drink; infusion of *Uraria picta* (Heowe) as wash; root-bark of *Citrus aurantiifolia* Lime, *Pseudocedrela kotschyi* Dry-Zone Cedar; infusion of inflorescence of *Cymbopogon schoenanthus* subsp. *schoenanthus, C. schoenanthus* subsp. *proximus,* and seeds of *Panicum maximum* Guinea Grass; decoction of *Ageratum conyzoides* Billy-goat Weed, (Efumomoe); powdered seed or seed oil of *Securidaca longepedunculata* (Kpaliga), *Oncoba spinosa* Snuff-box Tree (Asratoa) taken internally; *Entada pursaetha* Sea Bean, powdered seeds of *Ficus asperifolia* Sandpaper Tree in glass of water; seeds of *Helictonema velutina* (*Hippocratea velutina*); seed decoction of *Strychnos afzelii* (Duapepere) with *Ocimum* and ingredients; fruit-juice or root of *Spondias mombin* Hog Plum, Ashanti Plum sometimes with leaves of *Ximenia, Premna, Ficus,* and *Alchornea*; fruit-juice of *Aframomum latifolium* or *A. melegueta* Guinea-grains sucked or decoction of plant and root with leaves of *Morinda lucida* Brimstone Tree as bath to restore strength after attack; fruit-pulp of *Bixa orellana* Anatto; decoction of pods of *Acacia nilotica* var. *tomentosa;* fruit infusion of *Tetrapleura tetraptera* (Prekese) with Shea butter and ingredients as wash; *Zornia latifolia* mixed with Shea butter as rub for chills and a mixture of the seed oil of *Ricinus communis* Castor Oil Plant with kerosene to fight malaria fever.

Bark of *Lophira alata* Red Ironwood (Kaku) and *L. lanceolata* or infusion of young shoots, *Sacoglottis gabonensis* (Tiabutuo) as preventive, *Tamarindus indica* Indian Tamarind as tonic, *Entada africana* (Kaboya) alone or with various essence oils as *Hymenocardia acida* (Sabrakyie), *Piliostigma thonningii* (Opitipata), and *Combretum molle* (Gburega) drunk as a decoction or used in bath; bark of *Quassia undulata* (*Hannoa undulata*) (Kunmuni), *H. klaineana* (Fotie), *Carapa procera* Crabwood, *Mitragyna ciliata* Abura, (Subaha); *M. stipulosa* (Subaha-akoa), *Lecaniodiscus cupanioides* (Dwindwera), and *Crossopteryx febrifuga* African Bark (Pakyisie) — as the name implies, a well-known remedy. Bitter bark of root and stem of *Vernonia amygdalina* Bitter

8. POTIONS AND MEDICINES

Leaf (Bonwen) and *V. colorata*; bitter bark of *Thevetia peruviana* 'Milk Bush', Exile Oil Plant; bark pulp of *Afzelia africana* (Papao) with that of Tamarind and Satinwood; pounded bark of *Celtis africana* (Yisa); ground bark of *Octoknema borealis* (Wisuboni) as rub and bark pulp of *Blighia unijugata* (Akyibiri) steeped in water as drink. Bark infusion of *Ceiba pentandra* Silk Cotton Tree (Onyina); *Acacia albida* (Gozanga), *Rauvolfia vomitoria* (Kakapenpen), and *Corynanthe pachyceras* (Pramprama). Bark decoction of *Tetrorchidium didymostemon* (Aboagyedua), *Parkia biglobosa* (Duaga), *Boswellia dalzielii* Frankincense Tree (Kabona) as wash; *Khaya anthotheca* White Mahogany (Krumben); *K. senegalensis* Dry-Zone Mahogany or cold infusion, *Holarrhena floribunda* False Rubber Tree (Sese) or macerated in palm-wine, *Hunteria eburnea* (Kanwen-akoa) and *Mitragyna inermis* (Kukyafie) with leafy tips.

Decoction of *Dalbergia hostilis* (Wota), *Vernonia guineensis* or with lime-juice, and *Pseudarthria hookeri* (Kwaeni) or *V. nigritiana* (Gyakuruwa); decoction of upper leaves of *Carpolobia lutea* (Otwewa) for washing children; a drink prepared with *Setaria longiseta* in children's cases and decoction of *Blumea mollis*, *B. aurita* var. *foliolosa*, and *Emilia sonchifolia*. Unspecified parts of *Urena lobata* Congo Jute, *Desmodium gangeticum* var. *gangeticum*, *D. velutinum* (Koheni-koko), *Bridelia scleroneura* (Ba-Udiga), *Reissantia indica* (*Hippocratea indica*), *Dodonaea viscosa* Switch Sorrel (Fomitsi), and *Alafia barteri* (Momorehemo).

FEVER (see Febrifuge)

FITS (see Convulsions)

FLATULENCE (see Carminative)

FRACTURES/DISLOCATION

Crushed or pulped leaves of *Chasmanthera dependens*, *Mareya micrantha* (Odubrafo), *Parinari curatellifolia* (Atena), *Maranthes polyandra* (*P. polyandra*) (Abrabesi), and *Rauvolfia vomitoria* (Kakapenpen) as dressing; *Eleusine indica* as strap (Boulos, 1983); bark decoction of *Alstonia boonei* (Sinduro); roots, leaves, and seeds of *Paullinia pinnata* (Toa-ntini) with ginger applied before dressing; powdered bark and leaves of *Allophylus africanus* (Hokple) and pounded roots of *Rhodognaphalon brevicuspe* (*Bombax brevicuspe*) (Onyina-koben).

FRAMBOESIA (see Yaws)

GALACTOGOGUE (see Lactogenic)

GASTRO-INTESTINAL PAINS OR DISORDERS

Bark decoction of *Acacia nilotica* var. *tomentosa*, *Boswellia dalzielii* Frankincense Tree, *Lannea kerstingii* (Kobewu), and *L. velutina* (Sinsa). Bark of *Pachypodanthium staudtii* (Fale) chewed with Cola nuts; powdered bark of *Khaya senegalensis* Dry-Zone Mahogany in food, bark decoction of *Piper guineense* West African Pepper (Soro-wisa) as enema, and *Trichilia monadelpha* (*heudelotii*) (Tanduro) with pulp and bark of *Lophira lanceolata*. Seeds of *Aframomum melegueta* Guinea-grains in enemas; berries and young shoots or buds of *Harungana madagascariensis* (Okosoa); infusion of bitter bark of *Pseudocedrela kotschyi* Dry-Zone Cedar; pulp of leafy stems of *Abrus precatorius* Prayer Beads taken internally, decoction of *Zanthoxylum xanthoxyloides* (*Fagara zanthoxyloides*) Candle Wood with pulp as enema, and *Grewia carpinifolia* (Ntanta) as beverage. Decoction of *Leonotis nepetifolia* var. *nepetifolia* or *L. nepetifolia* var. *africana* (Nyeddo); leafy stems of *Macaranga hurifolia* (Kpazina) with those of *Baphia nitida* Camwood (Odwen) steeped in water and taken internally as purgative; decoction of stem tops of *Ficus vallis-choudae* (Aloma-bli) as drink and infusion of *Cissus quadrangularis* Edible-Stemmed Vine (Kotokoli) and of *Leucas martinicensis*. Hot or cold root infusion of *Ozoroa reticulata* (*Heeria insignis* and *H. pulcherrima*) (Nasia), root-bark of *Bridelia ferruginea* (Opam-fufuo); root decoction of *Paullinia pinnata* (Toa-ntini) as purgative; root and leaf decoction of *Lippia multiflora* Gambian Tea Bush (Saa-nunum) as drink; root of *Vernonia amygdalina* Bitter Leaf (Bonwen) with red natron; boiled root of *Nauclea latifolia* African Peach (Sukisia) with lemon-juice or *Bauhinia* leaves sweetened with sugar or honey; decoction of root of *Pachycarpus lineolatus* with natron and bitter root-bark of *Rauvolfia vomitoria* (Kakapenpen) as enema. Decoction of root and leaf of *Premna hispida* (Aunni) and *Crossopteryx febrifuga* African Bark (Pakyisie); root decoction of *Maytenus senegalensis* (Kumakuafo) with natron; leaf decoction of *Mitragyna inermis* (Kukyafie) sometimes with *Cassia obtusifolia* (*tora*) Foetid Cassia used both externally and internally; leafy decoction of *Cissus populnea* (Agyako); leaf-juice or decoction of *Ficus exasperata* (Nyankyeren) as enema; leaf infusion of *Baphia nitida* Camwood (Odwen) as drink; leaf of *Berlinia confusa* and an unspecified parts of *Acacia hockii* Shittim Wood with *A. sieberiana* var. *villosa* (Kulgo); unspecified parts of *Anacardium occidentale* Cashew Nut, *Combretum micranthum* (Landaga), and *Guiera senegalensis*. The ash of *Pistia stratiotes* Water Lettuce is taken with food for cases associated with worms and fully ripened bananas are often prescribed as diet to patients since they are completely digestible.

GENITO-URINARY TROUBLES (see Venereal Diseases)

GIDDINESS (see Dizziness)

GONORRHOEA (see Venereal Diseases)

8. POTIONS AND MEDICINES

GRIPING (BELLY PAIN) (see Colic)

GUINEA WORM SORES

Pounded, powdered, or mashed leaves of *Tapinanthus bangwensis* Mistletoe, *Mimosa pudica* Sensitive Plant, *Strophanthus gratus* (Omaatwa), *Coffea ebracteolata* (Nhwesono), *Ipomoea asarifolia, Cassia podocarpa* (Nsuduru), *Sapium grahamii* (Pampiga), *Adansonia digitata* Baobab, *Ritchiea reflexa* (Aayerebi), *Aloe schweinfurthii* and *A. buettneri* (Sere-berebe) warmed and applied; leaf of *Ricinus communis* Castor Oil Plant rendered soft by heating and applied to sores to facilitate extraction; cladoles of *Asparagus africanus* (Adedende) and *A. flagellaris* crushed with milk and applied and mashed leaves of *Hibiscus lunariifolius* applied. Pulverized, dried leaf and root of *Annona senegalensis* var. *senegalensis* Wild Custard Apple; leaf poultice of *Guiera senegalensis;* pounded leaf of *Elaeophorbia drupifera* (Akane) with salt and onions to assist extraction; lotion of crushed leaves of *Jatropha curcas* Physic Nut in hot water and plaster of pounded leaves of *Mareya micrantha* (Odubrafo) with pepper and lemon. Fresh fruits of *Crinum zeylanicum* (*ornatum*) pounded and applied; roasted and ground seeds of *Monodora myristica* (Ayerew-amba); seeds of *Cassia occidentalis* Negro Coffee with onions and salt; a drop of latex of *Calotropis procera* Sodom Apple applied during extraction; latex of *Ficus leprieurii* (Amangyedua) and sap of *Strychnos aculeata* as rub.

Pulverized root of *Argemone mexicana* Mexican or Prickly Poppy with onion; boiled roots of *Boerhavia diffusa* and *B. repens* Hogweed for extraction; powdered root of *Aristolochia albida* with seeds of *Lepidium sativum* The Common Cress with garlic and natron applied, *Pericopsis laxiflora* (*Afrormosia laxiflora*) Satinwood and *Olax subscorpioidea* (Ahoohenedua). Pulverized stems of *Strophanthus hispidus* Arrow Poison applied; preparation of stem infusion of *Euphorbia lateriflora* (Kamfo-barima) in which affected foot is soaked; juice expressed from crushed bark of *Spathodea campanulata* African Tulip, (Kokoanisua) with cola fruit; decoction of unspecified part of *Desmodium gangeticum* var. *gangeticum* as vapour bath; decoction of *Cathormion altissimum* (Abobonkakyere) and an unspecified part of *Garcinia* species. The root decoction of *Combretum mucronatum* (*smeathmannii*) has been found to be an effective cure by the Centre for Scientific Research into Plant Medicine (CSRPM) at Mampong in Ghana.

Global 2,000, of which former US President, Jimmy Carter, is president, is working with the World Health Organization (WHO), the UN Development Programme (UNDP), the US Agency for International Development (USAID), and the Bank of Credit and Commerce International (BCCI) to eradicate guinea worm in Africa.

HAEMATEMENSIS (VOMITING OF BLOOD) (see Styptic)

HAEMATURIA (BLOOD PASSED IN URINE)

Root of *Desmodium velutinum* (Koheni-koko) with pepper as enema, scraped roots of *Capparis erythrocarpos* (Apana) with shea butter applied locally (Moro, 1984-5), *Hyphaene thebaica* Dum Palm and root-bark of *Uvaria chamae* (Akotompotsen), whole plant of *Portulaca oleracea* Purslane, Pigweed with aromatic Labiate, and whole plant decoction of *Asparagus africanus* (Adendende) and *A. flagellaris* as remedy; leaf of *Cassia occidentalis* Negro Coffee; leaf extract of *Pavetta crassipes* (Nyenyanke), watery decoction of bark and leaf of *Morinda lucida* Brimstone Tree (Kankroma) and ground bark of *Dichrostachys cinerea* (*glomerata*) Marabou Thorn applied locally, or decoction of *Parkia clappertoniana* West African Locust Bean followed by application of ground bark (Moro, 1984-5).

HAEMOPTYSIS (SPITTING OF BLOOD) (see Styptic)

HAEMORROIDS (see Piles)

HAEMOSTATIC (ARRESTING BLEEDING) (see Styptic)

HEADACHE (MIGRAINE)

The application of the leaves of the following plants as poultice, paste, or compress to head or temple: *Xylopia aethiopica* Ethiopian Pepper (Hwenetia); *Cassia occidentalis* Negro Coffee, *Moringa oleifera* Horse-Radish Tree, *Dichrostachys cinerea* (*glomerata*) Marabou Thorn, *Antidesma venosum* (Mpepea), *Pseudarthria hookeri* (Kwaheni), *Bryophyllum pinnatum* Resurrection Plant, *Croton lobatus* (Akonansa), *Pulicaria crispa*, *Struchium sparganophora*, *Manihot esculenta* Cassava, *Leonotis nepetifolia* var. *nepetifolia* or *L. nepetifolia* var. *africana*, *Jacquemontia tamnifolia* (Boeboe) with scented spices, *Hyptis suaveolens* Bush Tea-Bush, *Strychnos spinosa* Kaffir Orange (Akankoa); *Calotropis procera* Sodom Apple, *Haumaniastrum lilacinum* and *Helictonema velutina* (*Hippocratea velutina*). Crushed leaves of *Gynandropsis pentaphylla* as counter-irritant; whole plant decoction of *Panicum maximum* Guinea Grass as relief (Amico, 1977); scraped pulp of *Cucurbita pepo* Pumpkin and *C. maxima* Squash Gourd applied as poultice and smoke from burned *Cymbopogon giganteus* var. *giganteus* inhaled as relief. Leaf of *Crateva adansonii* (*religiosa*) (Chelum Punga) as counter-irritant; inhaled vapour from leaf of *Lophira lanceolata* steeped in boiling water; dried leaves of *Guiera senegalensis* with that of *Melanthera scandens* as snuff; lotion of leaf and fruit of *Ceiba pentandra* Silk Cotton Tree as wash; paste of dried, pulverized leaf of *Heteropteris leona* applied as poultice; leaf decoction of *Alchornea cordifolia* Christmas Bush, (Gyamma) as anti-spasmodic and powdered leaf of *Erythrococca anomala* with pepper as snuff

Fresh juice of *Solenostemon monostachys* (Sisiworodo) squeezed into the

8. POTIONS AND MEDICINES

nostrils for cases in children; juice of *Platostoma africanum* (Siresireke) squeezed into the eyes, *Ethulia conyzoides* squeezed into the eyes and *Basilicum polystachyon* squeezed into the nostrils of children. A copious cold drink of leaves of *Scoparia dulcis* Sweetbroom Weed macerated in water or bruised leaves rubbed on the skin; fumes from burned leaves of *Sansevieria liberica* African Bowstring Hemp inhaled to relieve feverish cases and leaf infusion of *Ampelocissus grantii* (Enotipsi) with onions as drink or wash. Leaf decoction of *Aspilia africana* Haemorrhage Plant as lotion, *Crassocephalum crepidioides* as lotion and *Striga hermonthica* (Wumlim) with salt applied to head. Vapour from leaf decoction of *Microdesmis puberula* (Fema) inhaled or as a vapour bath; leaf infusion of *Phyllanthus maderaspatensis*; dried, pulverized leaf of *Calliandra portoricensis* (Nkabe) as snuff; steeped leaf of *Cylicodiscus gabunensis* African Greenheart, (Denya); leaf lotion of *Celtis integrifolia* Nettle Tree, (Samparanga); warmed leaf of *Ficus asperifolia* Sandpaper Tree rubbed on head and freshly crushed leaf and root of *Ximenia americana* Wild Olive applied locally. Leaf juice of *Gouania longipetala* (Homabiri) and *Clausena anisata* Mosquito Plant each applied as nasal drops; pulped leaf twigs of *Pseudocedrela kotschyi* Dry-Zone Cedar; vapour from boiled leaf of *Mallotus oppositifolius* (Satadua); crushed leaves of *Hyptis spicigera* applied to the head as relief; crushed leaves of *Allophylus africanus* (Hokple) inhaled to cause sneezing; pulp of leafy twigs of *Blighia sapida* Akee Apple (Akye) applied to forehead; powdered leaf of *Zanthoxylum chevalieri* (*Fagara pubescens*) as snuff; decoction of young leaves of *Vitellaria paradoxa* (*Butyrospermum paradoxum* subsp. *parkii*) Shea Nut Tree as vapour bath; scented leaf of *Heinsia crinita* Bush Apple rubbed on forehead or pounded and dried applied as poultice; crushed leaf of *Vernonia biafrae* (Hu) as snuff and pounded leaf of *Ehretia cymosa* (Okosua) applied with that of *Newbouldia laevis* (Sasanemasa) and Guinea-grains.

Roots of *Securidaca longepedunculata* (Kpaliga) with shea butter as massage or the powdered root and seed as snuff; root decoction or pulverized root of *Elaeis guineensis* Oil Palm; ground root-bark of *Capparis erythrocarpos* (Apana) applied; pounded root of male *Carica papaya* Pawpaw with food; fragrant rhizomes of *Cyperus articulatus* (Kokyi) with Guinea-grains as rub on the forehead; root decoction of *Pericopsis laxiflora* (*Afrormosia laxiflora*) Satinwood; vapour bath of boiled root of *Myrianthus arboreus* (Anyankoma) with Guinea-grains; charred, powdered root of *Ekebergia senegalensis* as snuff; an enema of root pulp of *Turraea heterophylla* (Ahunanyankwa) in water; pounded root of *Paullinia pinnata* (Toa-ntini) in water sniffed up the nostrils; root-juice or pounded root of *Cnestis ferruginea* (Akitase) applied to nostrils and pounded root and juice of *Microglossa pyrifolia* (Asommerewa) as snuff or tea-like infusion of the leaves taken. Bark of *Newbouldia laevis* (Sasanemasa); sundried, powdered fruit of *Nauclea latifolia* African Peach (Sukisia) applied; few drops of latex of *Tabernaemontana crassa* (Pepae) in nostrils; juice of the fruit of *Citrus aurantiifolia* Lime applied to the temple; pounded bark of *Celtis africana* (Yisa); flowers and fruits of *Eriosema psoraleoides* as calming inhalant; lotion of *Desmodium incanum* (*canum*) and fruit juice of *Abrus precatorius* Prayer Beads with ginger.

Fresh fruits of *Cissus producta* crushed and tied around forehead or a rub of the dry berries; fruiting heads and flowers of *Spilanthes filicaulis* Para or Brazil Cress rubbed on forehead; fruit-juice of some varieties of *Capsicum annuum* Pepper squeezed into the eyes as cure and crushed seeds of *Aframomum melegueta* Melegueta applied as paste to the head; fruit of *Piliostigma thonningii* (Opitipata) as ingredient; ground, powdered bark of *Erythrophleum suaveolens* (*guineense*) Ordeal Tree (Potrodom) as snuff; infusion of powdered bark and buds of *Daniellia oliveri* African Copaiba Balsam Tree; the yellowish-brown substance in pith of old stems of *Jatropha curcas* Physic Nut as snuff; charred twigs of *Hymenocardia acida* (Sabrakyie) as rub; chewed seeds of *Monodora myristica* (Ayerew-amba) rubbed on forehead; decoction of bark and leaf of *Beilschmiedia mannii* Spicy Cedar as lotion; spicy bark of *Buchholzia coriacea* (Esono-bise) as snuff; pounded bark of *Carpolobia lutea* (Otwewa) as snuff; powdered seeds of *Garcinia kola* Bitter Cola (Tweapea) with other drugs; bark of *Cola gigantea* var. *glabrescens* (Watapuo); paste of seeds of *Gossypium arboreum* Cotton applied and bark or leafy stem infusion of *Bridelia ferruginea* (Opam-fufuo) as beverage, bath, and vapour bath. Unspecified parts of *Ficus capensis* (Nwadua) and *Trichilia emetica* subsp. *suberosa* (*roka*) (Kisiga) applied.

HEART DISEASE, HEART TROUBLE, HEARTBURN, PALPITATIONS

A mixture of *Tamarindus indica* Indian Tamarind, *Afzelia africana* (Papao), and iron filings as drink; leaf-juice of *Abrus precatorius* Prayer Beads in palm-wine; leaf decoction of *Trichilia monadelpha* (*heudelotii*) (Tanduro); mashed leaves of *Chrysanthellum indicum* var. *afroamericanum* taken in shea butter or any other oil (Burkill, 1985); leaf ashes of *Newbouldia laevis* (Sasanemasa) with salt; young crushed shoots of *Zanthoxylum gilletii* (*Fagara macrophylla*) (Okuo) steeped in lemon-juice; crushed bark of *Milicia excelsa* (*Chlorophora excelsa*) Iroko (Odum) in water or palm-wine; bark and seeds of *Bussea occidentalis* (Kotoprepre); decoction of leaves of *Machaerium lunatum* (*Drepanocarpus lunatus*) (Nkako); mucilagenous seed-covering of *Sherbournea bignoniiflora* (Kyere-beteni) and an unspecified part of *Uraria picta* (Heowe). Leaves of *Portulaca oleracea* Purslane, Pigweed macerated in water as tonic; leaves of *Celosia leptostachya* eaten raw; leaves of *Stachytarpheta cayennensis* Bastard Vervain, Brazilian Tea; root of *Boerhavia diffusa* Hogweed (Ayensu, 1978); root of *Clausena anisata* Mosquito Plant (Ayensu, 1978); whole plant of *Euphorbia hirta* Australian Asthma Plant to depress (Jain, 1968); juice of unspecified part of *Pisonia aculeata* Cockspur with pepper and other plants (Ayensu, 1978) and unspecified part of *Cyathula prostrata* (Mpupua) with *Synedrella nodiflora*, clay, and Guinea-grains (Ayensu, 1978). *Vernonia nigritiana* (Gyakuruwa) contains a glycocide, vernonine, having a heart action like that of digitaline but weaker.

8. POTIONS AND MEDICINES

HEPATITIS (INFLAMMATION OF THE LIVER AND LIVER PROBLEMS)

Bark infusion of *Cleistopholis patens* Salt and Oil Tree as drink or vapour-baths; bark decoction of *Harungana madagascariensis* (Okosoa) with *Cajanus cajan* Pigeonpea as a warm drink or enema; root or bark decoction of *Newbouldia laevis* (Sasanemasa) or bark applied internally and bruised leaves applied as poultice with fruits of *Xylopia*. An unspecified part of *Citrus aurantiifolia* Lime for infectious cases (Ayensu, 1978) and *Oldenlandia affinis* subsp. *fugax* for inflammation of the spleen. Pounded leaves or roots of *Cassia occidentalis* Negro Coffee; root of *Boerhavia diffusa* Hogweed (Ayensu, 1978); fruit of *Momordica charantia* African Cucumber (Ayensu, 1978), also recommended for the spleen; a poultice of *Paullinia pinnata* (Toa-ntini) with boiling water applied to the side and the fruit of *Balanites aegyptiaca* Desert Date (Ayensu, 1978).

HERNIA (PROTRUSION OF AN ORGAN — USUALLY ASSOCIATED WITH ABDOMINAL CAVITY; ORCHITIS)

Bark decoction of *Milicia excelsa* (*Chlorophora excelsa*) Iroko (Odum) or *Zanthoxylum leprieuri* (*Fagara leprieuri*) (Oyaa); root decoction of *Cochlospermum tinctorium* (Kokrosabia) or *Clerodendrum capitatum* (Tromen); leaf decoction of *Acacia sieberiana* var. *villosa* (Kulgo) or *Carica papaya* Pawpaw; leaf vapour bath of *Vernonia conferta* (Flakwa) or *Trema orientalis* (*guineensis*) (Sesea); hot application of root and leaf decoction of *Newbouldia laevis* (Sasanemasa); powdered roots of *Afzelia africana* (Papao) in millet beer; an enema of leaf juice of *Acacia pentagona*; a tonic of the bark infusion of *Parkia clappertoniana* West African Locust Bean; bark-pulp of *Blighia sapida* Akee Apple (Akye) eaten with ginger; young shoots of *Gouania longipetala* (Homabiri) with *Zanthoxylum gilletii* (*Fagara macrophylla*) (Okuo), pepper, and salt; roots of *Cassia sieberiana* African Laburnum with other plants and fresh leaves of *Pseuderanthemum tunicatum* or decoction of the whole plant. The roots of African Laburnum are used for strangulated hernia or ground root of *Annona senegalensis* var. *senegalensis* Wild Custard Apple applied as paste (Moro, 1984–5); the leaves of *Heinsia crinita* Bush Apple cooked with rice is given to children for umbilical hernia.

HICCUPS

Chewed bark of *Tamarindus indica* Indian Tamarind (Moro, 1984–5); roasted and powdered fruits of *Guiera senegalensis* with salt to disguise the bitter taste taken in small doses; leaf-juice of *Strombosia glaucescens* var. *lucida* (Afena) as drink and the chewed leaf of *Phyllanthus fraternus* subsp. *togoensis* (*niruri*) (Ombatoatshi).

HYPERTENSION (HIGH BLOOD PRESSURE)

Decoction of the dried leaves of *Byrsocarpus coccineus* (Awennade), *Carica papaya* Pawpaw, *Gomphrena celosioides*, *Cymbopogon citratus* Lemon Grass, *Persea americana* Avocado Pear, *Musa paradisiaca* Plantain, *Tamarindus indica* Indian Tamarind, and *Clausena anisata* Mosquito Plant or chewed fruit-pulp of *Hyphaene thebaica* Dum Palm (Moro, 1984–5). The roasted, dried seeds of *Cassia occidentalis* Negro Coffee brewed as a drink is also recommended as a remedy. The dried stems and roots of *Aristolochia indica* (an introduced plant) has activity on blood pressure (Jain, 1968).

IMPOTENCE

Bark decoction of *Zanthoxylum xanthoxyloides* (*Fagara zanthoxyloides*) Candle Wood (Moro, 1984–5); leaf of *Turraea heterophylla* (Ahunanyankwa) taken internally as bitters; tuber of *Icacina olivaeformis* (*senegalensis*) False Yam and roots of *Byrsocarpus coccineus* (Awennade) with other roots; chewed roots of *Acridocarpus smeathmannii* (Alasaayo) or tea-like infusion of leaves of *Annona senegalensis* var. *senegalensis* Wild Custard Apple at bedtime (Moro, 1984–5); bark of *Newbouldia laevis* (Sasanemasa) with clay and red pepper (Ayensu, 1978); latex of *Funtumia elastica* West African Rubber Tree (Ayensu, 1978); leaf of *Microdesmis puberula* (Fema) (Ayensu, 1978) and unspecified part of *Microglossa pyrifolia* (Asommerewa) (Ayensu, 1978). The bark decoction of *Commiphora africana* African Bdellium (Narga) is used for sterility in men.

INCONTINENCE OF URINE

Root or leaf decoction or infusion of *Ziziphus mucronata* Buffalo Thorn; root decoction of *Jatropha curcas* Physic Nut or the ends of the stems chewed; pounded roots of *Funtumia africana* False Rubber Tree in palm-wine or water and infusion of *Vernonia cinerea*.

INDIGESTION (see Stomach Ache)

INSECTICIDES

Pulverized or powdered seeds of *Piper guineense* West African Pepper (Sorowisa); *Datura stramonium* Apple or Peru, *Annona glauca* var. *glauca* (Mampihege), *A. senegalensis* var. *senegalensis* Wild Custard Apple, *A. muricata* Sour Sop, *A. squamosa* Sweet Sop and *Monodora myristica* (Ayerew-amba); pulverized seeds of *Caloncoba echinata* (Gorli) and *Azadiracta indica* Neem Tree for lice; bark decoction of *Pachypodanthium staudtii* (Fale) and leaf of *Calotropis procera* Sodom Apple for lice; leaf and flowers of *Psydrax parviflora* (*Canthium vulgare*) (Kwae-susua) and scented leaf of *Heinsia crinita* Bush Apple pounded and dried as poultice against lice. Bruised leaves of *Clausena anisata* Mosquito Plant and that of *Chromolaena odorata* (*Eupatorium odoratum*) Siam Weed to repel mosquitoes; leaves of *Tabernaemontana crassa* (Pepae) as thatch to drive away cockroaches; leaves of *Tephrosia vogelii* Fish Poison Plant sometimes with prepared indigo to keep away ticks, fleas, lice, etc.; leaves of

8. POTIONS AND MEDICINES

Griffonia simplicifolia (Kagya) in hen-coops to drive away lice; leaves of *Erythrophleum suaveolens* (*guineense*) Ordeal Tree (Potrodom) put in stored corn to keep away insects; leaves of *Paropsia adenostegia* (*Androsiphonia adenostegia*) (Nkatie) with palm oil for lice and juice of *Citrus aurantiifolia* Lime for crablouse.

Pulverized bark of *Avicennia africana* Mangrove, (Asoporo) with palm oil as ointment for lice; juice of young, crushed, roasted shoots of *Strophanthus hispidus* Arrow Poison to kill head lice; crushed root and bark of *Rauvolfia vomitoria* (Kakapenpen) for lice and vermin; yellow gum of *Anacardium occidentale* Cashew Nut or the shell-nut oil as a mosquito larvicide and against insects in woodwork and books; fat of *Melia azedarach* Bead Tree, Persian Lilac (the repellent substance being meliatin); burned gum resin of *Boswellia dalzielii* Frankincense Tree (Kabona) and *Canarium schweinfurthii* Incense Tree to drive away flies and mosquitoes; fruit and crushed kernels of *Quassia undulata* (*Hannoa undulata*) (Kunmuni) for lice and an unspecified part of *Lonchocarpus sericeus* Senegal Lilac. Seed fat of *Pentadesma butyraceum* (*butyracea*) Tallow Tree (Abotoasebie) for lice and jiggers; young shoots of *Grewia carpinifolia* (Ntanta) to remove and prevent lice; pounded fruit, seed, and leaf of *Heteropteris leona* for lice; latex of *Euphorbia lateriflora* (Kamfobarima) for lice; burned seeds of *Detarium senegalense* Tallow Tree to drive away mosquitoes; seeds of *Dioclea reflexa* Marble Vine (Ntehama) in killing head lice; leaves of *Eriosema psoraleoides* rubbed on dogs to prevent lice and wood smoke of *Acacia hockii* Shittim Wood as fumitory for insects, lice, etc.

Leaves of *Leucas martinicensis*, *Blumea aurita* var. *aurita*, and whole plant of *Evolvulus alsinoides* burned in rooms to get rid of mosquitoes; *Hyptis spicigera* put in layers below bundles of millet to keep away termites and burned to get rid of mosquitoes; dried and pulverized leaves of *Triplotaxis stellulifera* (Kokoo) with palm oil as remedy for head lice; whole plant of *Leonotis nepetifolia* var. *nepetifolia* or *L. nepetifolia* var. *africana* (Nyeddo) put among stored corn to keep away vermin; leaves of *Ipomoea batatas* Sweet Potato, (Santom) as hair-wash for lice or mixed with charcoal and black earth to plaster floor and walls; crushed leaf of *Gloriosa superba* and *G. simplex* Climbing Lily to kill head lice and leaf of *Solanum nigrum* (Nsusuabiri) as poultice to destroy head lice. Pulverized fruits of *Capsicum annuum* Pepper with *Xylopia aethiopica* Ethiopian Pepper (Hwenetia) for cola weevils; oil from seeds of *Argemone mexicana* Mexican or Prickly Poppy used to drive away white ants; crushed leaves or broken stems of *Manihot esculenta* Cassava spread in the path of driver ants to disperse them and bruised leaves of *Calotropis procera* Sodom Apple for keeping off ants. Root tubers of *Cyperus rotundus* Nut Grass yield an essential oil for repelling insects; mucilaginous juice of *Sesamum indicum* Sesame, Beniseed to destroy head lice; *Ricinus communis* Castor Oil Plant planted round houses to expel mosquitoes; and pulped leaf of *Datura metel* Hairy Thorn Apple, Metel with clay for flooring to destroy jiggers. Nicotine, an extract from *Nicotiana tabacum* Tobacco, is recognized in Europe and elsewhere as a powerful insecticide for horticulture and sheep-dips (see also Jiggers below).

INTERCOSTAL PAINS

Ointment of pulped bark of *Pericopsis laxiflora* (*Afrormosia laxiflora*) Satinwood with shea butter, lemon, and pepper; the pulped young leaf of *Deinbollia pinnata* (Woteegbogbo) as rub, followed by a drink of the leaf-juice; pulped leaf of *Desmodium velutinum* (Koheni-koko) as rub; pulped leaf stem of *Machaerium lunatum* (*Drepanocarpus lunatus*) (Nkako) as liniment; powdered leaf of *Dialium dinklagei* (Awendade) chewed with cola; leaves and wood ashes of *Psydrax venosa* (*Canthium venosum*) applied; root decoction of *Cochlospermum tinctorium* (Kokrosabia) as liniment; pulped root of *Zanthoxylum viride* (*Fagara viridis*) (Oyaanini) as massage; root of *Clausena anisata* Mosquito Plant pounded with Guinea-grains and lime as liniment; bark of *Erythroxylum mannii* Landa, (Pepesia) with lemon and peppercorn as liniment; bark pulp of *Blighia sapida* Akee Apple (Akye) as liniment; a poultice of the ground and heated bark of *Trema orientalis* (*guineensis*) (Sesea); gum and pulp of *Daniellia oliveri* African Copaiba Balsam Tree as rub and unspecified parts of *Grewia bicolor*, *Caesalpinia bonduc* Bonduc, and *Adenia rumicifolia* var. *miegei* (*lobata*) (Peteha).

INTOXICANT

The ground bark of *Erythrophleum suaveolens* (*guineense*) Ordeal Tree, (Potrodom) in palm-wine; bark shavings of *Musanga cecropioides* Umbrella Tree (Dwumma); chewed seeds of *Lagenaria breviflora* (*Adenopus breviflorus*) (Anuwatre) while smoking tobacco; fruit-pulp of *Tetrapleura tetraptera* (Prekese) in palm-wine; leaves of *Morinda morindoides* in strengthening palm-wine; boiled root of *Pericopsis laxiflora* (*Afrormosia laxiflora*) Satinwood alone or added to palm-wine; a cold infusion of *Ipomoea coptica* and an unspecified part of *Pterocarpus erinaceus* Senegal Rose Wood Tree, African Kino. The root and fruit-pulp of *Cassia sieberiana* African Laburnum are remedies for intoxication. (See also Chapter 2.)

ITCH (MANGE)

Leaf decoction of *Phytolacca dodecandra* (Ahoro), *Anogeissus leiocarpus* (Sakanee), and *Piliostigma thonningii* (Opitipata); leaf of *Cardiospermum grandiflorum*, *C. halicacabum* Heart Seed, Balloon Vine, and *Gardenia ternifolia* (Peteprebi) as liniment, leaf pulp of *Premna luscens* and *Hoslundia opposita* (Asifuaka) as rub; fresh leaf of *Ocimum gratissimum* Tea Bush (Nunum) with shea-butter; leaf of *Grewia carpinifolia* (Ntanta) with Guinea-grains as rub; decoction of leafy stems of *Antidesma venosum* (Mpepea); leaf decoction or latex of *Rauvolfia vomitoria* (Kakapenpen); pulverized bark of *Avicennia africana* Mangrove (Asoporo) with palm-oil as ointment; pulverized seed of

8. POTIONS AND MEDICINES

Cordia millenii Drum Tree with palm-oil and seeds of *Caloncoba echinata* (Gorli).

Crushed bark of *Buchholzia coriacea* (Esono-bise) as rub; bark extract of *Harungana madagascariensis* (Okosoa); boiled bark and root of *Mammea africana* African Mammy Apple (Bompagya) and *Symphonia globulifera* (Ehureke) as wash; oil of *Jatropha curcas* Physic Nut applied; bark of *Microdesmis puberula* (Fema); root, fresh leaf, or pulverized seed of *Cassia obtusifolia* (*tora*) Foetid Cassia with lime and senna plant; resin of *Daniellia oliveri* African Copaiba Balsam Tree; boiled bark of *Detarium senegalense* Tallow Tree as lotion; crushed bark of *Albizia adianthifolia* (Pampena) with kaolin as liniment; bark of *Pentaclethra macrophylla* Oil Bean Tree (Ataa) as liniment; pounded bark of *Trichilia emetica* subsp. *suberosa* (*roka*) (Kisiga); root-bark decoction of *Diospyros mespiliformis* West African Ebony with other ingredients; ashes of fruits of *Morinda morindoides*; pulverized bark of *Pausinystalia lane-poolei* and unspecified parts of *Markhamia lutea* (Sisimasa) and *M. tomentosa* (Tomboro).

IRITIS (see Eye Diseases)

JAUNDICE

Sap from pulped leaf of *Dichapetalum madagascariensis* (*guineense*) (Antro) with the leaves of *Cassia occidentalis* Negro Coffee and *Hoslundia opposita* (Asifuaka) as nasal drops; decoction of leafy stems of Negro Coffee taken internally or leaf-juice used as eye drops; scraped roots of *Byrsocarpus coccineus* (Awennade) macerated in water or with Guinea-grains in palm-wine and dried, powdered root of *Diospyros mespiliformis* West African Ebony with salt and palm oil given as pellets. Bark decoction of *Lawsonia inermis* Henna; bark and root of *Picralima nitida* (Ekuama), *Lonchocarpus laxiflorus* (Nalenga), and *Morinda morindoides* as tonic; decoction of bark and root of *M. lucida* Brimstone Tree as drink; bark or leaf decoction of *Craterispermum laurinum* taken internally; fresh bark of *Canarium schweinfurthii* Incense Tree; bark infusion of *Swartzia madagascariensis* Snake Bean and *Parquetina nigrescens* (Aba-kamo) as ingredient along with *Uncaria talbotii* (Akoo-ano), *Alstonia boonei* (Sinduro), *Holarrhena floribunda* False Rubber Tree, (Sese); *Nauclea diderrichii* (Kusia) and *Harungana madagascariensis* (Okosoa). The leaf-juice of *P. nigrescens* is useful as eye drops or drink or bath. Leaf concoction of *Uvaria doeringii* (Agbana); the leaves and young shoots of *Olax subscorpioidea* (Ahoohenedua) soaked in water with lime-juice and plantain stem and *Cymbopogon citratus* Lemon Grass added as drink and poured over the patient; a bath of the dried inflorescence of *C. giganteus* var. *giganteus* as preventive; leaf of *Lantana camara* Wild Sage with banana leaf and that of *Morinda lucida* Brimstone Tree and other herbs boiled and patient steamed over; *Vernonia nigritiana* (Gyakuruwa) with *Lippia multiflora* Gambian Tea Bush (Saa-nunum) and *Cassia occidentalis* Negro Coffee as decoction; a decoction of *V. guineensis* alone or with lime-juice, with *Pseudarthria hookeri* (Kwaheni), or with *V. nigritiana*; leaf-decoction of *Microglossa pyrifolia* (Asommerewa), *Maesobotrya barteri* var. *sparsiflora* (Apotrewa), and *Dialium dinklagei* (Awendade) as beverage.

Boiled leaf of *Jatropha curcas* Physic Nut with lime as drink or wash with *Lippia multiflora* Gambian Tea Bush, *Blighia sapida* Akee Apple (Akye); *Cissus populnea* (Agyako), *Ficus vallis-choudae* (Aloma-bli), and *Terminalia glaucescens* (Ongo) as ingredients. Root or root-bark of *Uvaria chamae* (Akotompotsen) boiled with spices as drink; root-bark infusion of *Moringa oleifera* Horse-Radish Tree, Oil of Ben Tree; pounded root of *Cochlospermum tinctorium* (Kokrosabia) in water or millet beer; root decoction of *Boerhavia diffusa* Hogweed as diuretic (Jain, 1968); root-bark of *Lophira lanceolata*; root of *Harungana madagascariensis* (Okosoa), *Alchornea cordifolia* Christmas Bush, (Gyamma) with other ingredients, and bitter roots of *Pleiocarpa mutica* (Kanwen); bitter root-bark of *Rauvolfia vomitoria* (Kakapenpen) with spices as enema and decoction of roots and leaves of *Ximenia americana* Wild Olive as drink.

Decoction of *Paullinia pinnata* (Toa-ntini) with other ingredients; resinous bark of *Deinbollia grandifolia* (Potoke) and *Pseudospondias microcarpa* var. *microcarpa* (Katawani); decorticated seed of *Citrus aurantiifolia* Lime; leaf decoction of *Maesopsis eminii* (Onwandua) prescribed as diuretic and purgative; bark of *Erythrina senegalensis* Coral Flowers with food or decoction of root and bark as drink or bath with lemon-juice, spices, and Guinea-grains; leaf-juice of *Baphia nitida* Camwood (Odwen) as eye drops; decoction of leafy stems or bark and root of *Pericopsis laxiflora* (*Afrormosia laxiflora*) Satinwood as drink or bath or vapour bath and decoction of leaves and probably bark of *Anogeissus leiocarpus* (Sakanee) together with *Trema* and *Nauclea* as drink.

JIGGERS

Affected foot soaked in a preparation of stem infusion of *Euphorbia lateriflora* (Kamfobarima); juice from the fruit of *Ficus vogeliana* (Opanto) applied; dark yellow oil of *Carapa procera* Crabwood applied; stem core of *Musa paradisiaca* Plantain warmed and wrapped round feet to soften for the extraction and scraped pulp of *Manihot esculenta* Cassava applied as poultice. (See also Insecticides, above.)

KIDNEY DISEASE (see Bladder Trouble)

LABOUR, DIFFICULT

The following plants facilitate childbirth and hasten delayed labour. Leaf poultice of *Calotropis procera* Sodom Apple applied to the abdomen; leaf decoction of *Glyphaea brevis* (Foto), *Hyptis pectinata* (Peaba), *Cassia alata*

Ringworm Shrub, and *Microglossa pyrifolia* (Asommerewa); leaf of *Celtis integrifolia* Nettle Tree (Samparanga) with that of *Anthonotha macrophylla* (Totoro) as ingredient; leaf-juice of *Dialium guineense* Velvet Tamarind, (Asenamba); leaf of *Piliostigma thonningii* (Opitipata) and *Newbouldia laevis* (Sasanemasa) in palm soup and young shoots of *Smilax kraussiana* (Kokora) cooked with seeds of *Cucumis edulis* Cucumber as food. Infusion of *Celosia leptostachya, Aspilia africana* Haemorrhage Plant, *Centaurea praecox* Star Thistle (Balinyiri) to relieve pains and *Physalis angulata* (Totototo) taken in childbirth; decoction or infusion of *Solenostemon monostachys* (Sisiworodo) with peppers in later part of pregnancy to relieve pains; decoction of *Borreria stachydea* (Barungwini); decoction of *Asystasia gangetica* taken internally or with peppers as enemas during later stages of pregnancy to lighten pains at childbirth and leaf and roots of *Sansevieria liberica* African Bowstring Hemp given during labour as abortifacient.

Bark decoction of *Erythrina senegalensis* Coral Flowers, *Harungana madagascariensis* (Okosoa) with that of *Antiaris toxicaria* (*africana*) Bark Cloth Tree, (Kyenkyen) as enema; *Detarium senegalense* Tallow Tree, *Vitellaria paradoxa* (*Butyrospermum paradoxum* var. *parkii*) Shea Nut Tree as drink or sitz-bath and bark infusion of *Acacia albida* (Gozanga). Root infusion of *Triumfetta rhomboidea* Burweed, *Cochlospermum tinctorium* (Kokrosabia), and *Aeglopsis paniculata* (Obuobi); root bark extract of *Zanthoxylum xanthoxyloides* (*Fagara zanthoxyloides*) Candle Wood in rum; root decoction of *Maytenus senegalensis* (Kumakuafo), *Amorphophalus abyssinicus* and crushed seeds of *Dicranolepis persei* (Prahoma) in rum, lime-juice, palm-wine, or water. Infusion of *Pollia condensata* as lotion in pregnancy for a quick delivery; *Argemone mexicana* Mexican or Prickly Poppy as drink at the commencement of labour pains; leaves of *Lippia multiflora* Gambian Tea Bush (Saa-nunum) boiled with palm-nuts to assist delivery; leaf-pulp of *Clerodendrum umbellatum* (Niabiri) in enemas to hasten the expulsion of the placenta after childbirth and roots of *Palisota hirsuta* (Nzahuara) in palm soup taken during pregnancy. An unspecified part of *Combretum molle* (Gburega) and the crushed bark of *Antiaris toxicaria* (*africana*) Bark Cloth Tree (Kyenkyen) with that of *Harungana madagascariensis* (Okosoa) and eggshells as enema in confinements to hasten expulsion of the placenta. Root ashes of *Elaeis guineensis* Oil Palm with salt and fresh palm oil as drink for the expulsion of the placenta or the crushed leaves of *Sida acuta* in water as drink (Moro, 1984–5). Young shoots of *Grewia carpinifolia* (Ntanta) fed to lambs to induce fertility and help easy delivery of the young.

LACTOGENIC (INDUCING BREAST MILK)

Decoction of *Nelsonia canescens* (Moro. 1984–5); root scrapings of *Capparis erythrocarpos* (Apana) applied to incisions on the breast; root of *Allophylus africanus* (Hokple), tubers of *Cyperus esculentus* Tiger Nut (Atadwe) (Boulos,

1983); steeped, raw, young roots of *Manihot esculenta* Cassava; decoction or infusion of root bark of *Dodonaea viscosa* Switch Sorrel (Fomitsi); dried, pulverized root bark of *Calotropis procera* Sodom Apple in soup; pounded bark of *Bombax buonopozense* Red-Flowered Silk Cotton Tree and *Kigelia africana* Sausage Tree (Nufuten). Bark decoction of *Pentaclethra macrophylla* Oil Bean Tree (Ataa); *Alstonia boonei* (Sinduro) and *Vernonia conferta* (Flakwa); leaf decoction of *V. amygdalina* Bitter leaf (Bonwen) and *V. colorata*; leaf of *Newbouldia laevis* (Sasanemasa) in palm soup; young leaf of *Urera oblongifolia* cooked with groundnut; young leaf of *Ficus capensis* (Nwadua) in palm soup; latex of *Landolphia dulcis*; latex and bark infusion of *Milicia excelsa* (*Chlorophora excelsa*) Iroko (Odum); trunk sap of *Musanga cecropioides* Umbrella Tree (Dwumma) boiled with maize and palm-wine from *Raphia hookeri* Wine Palm as drink. The bark and leaf infusion of *Bridelia grandis* is given to women in their confinement to purify the milk. Crushed leaf of *Ricinus communis* Castor Oil Plant applied to breast of nursing mothers or decoction taken internally; whole plant decoction of *Euphorbia hirta* Australian Asthma Plant (Jain, 1968); *E. convolvuloides* applied to breast; *Ampelocissus grantii* (Enotipsi) applied to breasts; the tuberous roots of *Ipomoea mauritiana* (Disinkoro) and the leaves of *Alternanthera pungens* (*repens*) Khaki Weed, (Mpatowa-nsoe). Leaves of *Guiera senegalensis*, *Phytolacca dodecandra* (Ahoro), *Cissus populnea* (Agyako) with *Pergularia daemia*, *Scoparia dulcis* Sweetbroom Weed, and *Launaea taraxacifolia* (*Lactuca taraxacifolia*) Wild Lettuce increase lactation in cattle.

LAXATIVE (see Purgative)

LEPROSY

Bark decoction of *Pycnanthus angolensis* African Nutmeg (Otie) as purgative; *Crateva adansonii* (*religiosa*) (Chelum Punga), *Zanthoxylum viride* (*Fagara viridis*) (Oyaanini) as drink and pulped leaf applied, *Anthocleista nobilis* Cabbage Palm as drink or vapour bath, *Corynanthe pachyceras* (Pamprama), *Acacia hockii* Shittim Wood, *Antiaris toxicaria* (*africana*) Bark Cloth Tree, (Kyenkyen) with that of *Calotropis procera* Sodom Apple as drink or bath, *Milicia excelsa* (*Chlorophora excelsa*) Iroko (Odum) as drink or sitz-bath with that of *Alchornea cordifolia* Christmas Bush (Gyamma); *Annona senegalensis* var. *senegalensis* Wild Custard Apple and *Microglossa pyrifolia* (Asommerewa), and decoction of *Diospyros monbuttensis* Yoruba Ebony (Akyirinian) with leafy tips and those of *Cassia occidentalis* Negro Coffee and *Lippia multiflora* Gambian Tea Bush (Saanunum) as drink or bath. Bark of *Rhizophora* species Red Mangrove, *Terminalia avicennioides* (Petni) with that of *T. glaucescens* (Ongo); *Cola gigantea* var. *glabrescens* (Watapuo) as ingredient, *Necepsia afzelii* pounded and applied, *Parinari congensis* (Tulingi) as purgative, *Ficus capensis* (Nwadua) as ingredient in a complex preparation, and *Daniellia oliveri* African Copaiba

8. POTIONS AND MEDICINES

Balsam Tree with that of *Sterculia setigera* (Pumpungo) with millet as sun-dried balls eaten by lepers. Bark of *Pentaclethra macrophyla* Oil Bean Tree pounded and applied, *Treculia africana* var. *africana* African Breadfruit (Brebretim); *Lannea kerstingii* (Kobewu) with that of *L. velutina* (Sinsa) used externally, *Spondias mombin* Hog Plum, Ashanti Plum applied; *Diospyros canaliculata* Flint Bark freshly pulped and applied, *Vitex doniana* Black Plum (Afua); bark extract of *Enantia polycarpa* African Yellow Wood (Duasika); bark, root, or leaf of *Piliostigma thonningii* (Opitipata) in treatment and infusion of *Ficus vogeliana* (Opanto) as wash.

Leaf decoction of *Securidaca longepedunculata* (Kpaliga); leaf decoction of *Centella asiatica* Indian Water Navelwort due to presence of asiaticoside (Jain, 1968) or whole plant (Evans, 1983); decoction of leafy stems of *Scytopetalum tieghemii* (Aprim) as bath; fresh leaves of *Pupalia lappacea* (Akukuaba) applied to sores after bleeding them; fresh leaves of *Lawsonia inermis* Henna with lime-juice applied; leaves of *Acanthospermum hispidum* Star Bur (Peteku-nsoe) applied; boiled leaf of *Mareya micrantha* (Odubrafo) applied; and powdered, dried *Striga hermonthica* (Wumlim) with bark of *Acacia nilotica* applied to the ulcers or as an ointment for the unbroken skin or large doses of root decoction sometimes with *Acacia seyal* after general scarification and cupping.

Root of *Ceiba pentandra* Silk Cotton Tree (Onyina) applied; *Alchornea cordifolia* Christmas Bush (Gyamma); *Cassia sieberiana* African Laburnum and *Tamarindus indica* Indian Tamarind both with ingredients, *Lonchocarpus cyanescens* West African or Yoruba Indigo, *Plumbago indica* Red Plumbago as external remedy, *P. zeylanica* Ceylon Leadwort with that of *Eleusine indica* taken internally, *Ziziphus mauritiana* Indian Jujube as emeto-purgative, *Calotropis procera* Sodom Apple alone or with *Antiaris toxicaria* (*africana*) Bark Cloth Tree and *Cochlospermum tinctorium* (Kokrosabia), and powdered root of *Gardenia ternifolia* (Peteprebi) rubbed in small incisions in affected parts. Root decoction of *Bauhinia rufescens* (Jinkiliza) with bark, *Feretia apodanthera* (Bitinamusa), *Dichrostachys cinerea* (*glomerata*) Marabou Thorn, *Erythrina senegalensis* Coral Flowers as drink, bath, and rub and *Rauvolfia vomitoria* (Kakapenpen) or root infusion, pulp, or juice taken internally. Powdered root of *Zanthoxylum xanthoxyloides* (*Fagara zanthoxyloides*) Candle Wood; root-bark of *Psorospermum febrifugum* var. *ferrugineum*, *Hunteria eburnea* (Kanwen-akoa) as paste, and pounded, sun-dried root of *Zanthoxylum leprieuri* (*Fagara leprieuri*) (Oyaa) with lemon-juice applied as paste.

Seed oil of *Oncoba spinosa* Snuff-box Tree (Asratoa) and *Caloncoba echinata* (Gorli); decoction of pods of *Acacia nilotica* var. *tomentosa*; gum of *A. dudgeoni* (Gosei); gummy sap of *Harungana madagascariensis* (Okosoa) and latex of *Tabernaemontana crassa* (Pepae) for leprous wounds. Unspecified parts of *Clematis hirsuta*, *Lophira lanceolata* with ingredients, *Combretum aculeatum*, *Machaerium lunatum* (*Drepanocarpus lunatus*) (Nkako), *Acacia albida* (Gozanga), *Cassia occidentalis* Negro Coffe as drink, *C. alata* Ringworm Shrub, *Mitragyna inermis* (Kakyafie), *Ficus iteophylla* (Kwonkwia), *Desmodium velutinum* (Koheni-koko) as ingredient, *Parkia biglobosa* (Duaga), *Okoubaka aubrevillei* (Odee), and decoction of unspecified part of *Anogeissus leiocarpus* (Sakanee) with that of *Raphiostylis beninensis* (Akwakora-gyahene) and *Diospyros viridicans* (*kekemi*) Gaboon Ebony. Preparations of *Sapium grahamii* (Pampiga) applied orally and *Microglossa pyrifolia* (Asommerewa) in enemas as purgative. Cold decoction of *Guiera senegalensis* and the young flower-buds of *Parkia clappertoniana* West African Locust Bean, (Daudawa) used as preventative. The following plants serve as ingredients in preparations; *Saba senegalensis* (Sono-nantin), *Paullinia pinnata* (Toa-ntini), and *Pseudocedrela kotschyi* Dry-Zone Cedar.

LEUCORRHOEA (see Venereal Diseases)

LIVER PROBLEMS (see Hepatitis)

LUMBAGO (LOWER BACK PAINS)
A massage of leaf infusion, decoction, or pulp of *Pericopsis laxiflora* (*Afrormosia laxiflora*) Satinwood, *Clerodendrum volubile* (Kumamuno), *Zanthoxylum viride* (*Fagara viridis*) (Oyaanini), and *Desmodium velutinum* (Koheni-koko); pungent root of *Calliandra portoricensis* (Nkabe) or *Carissa edulis* (Botsu) with fresh ginger and water as enema; boiled fruits of *Kigelia africana* Sausage Tree, (Nufuten) with roots of *Anthocleista* species as drink; enema of roots of *Tylophora conspicua* with *Piper guineense* West African Pepper (Soro-wisa); pulped leaf of *Massularia acuminata* (Pobe) as liniment; hot decoction of root or leaf of *Combretum micranthum* (Landaga) as vapour bath or wash; an enema of pounded roots of *Mallotus oppositifolius* (Satadua); powdered root-bark of *Gardenia ternifolia* (peteprebi) in palm-wine as drink; hot bark of *Petersianthus macrocarpus* (*Combretodendron macrocarpum*) Stinkwood Tree (Esia); bark of *Cola gigantea* var. *glabrescens* (Watapuo); whole plant of *Ocimum gratissimum* Tea Bush (Nunum) as poultice; leaf decoction of *Bersama abyssinica* subsp. *paullinioides* (Duantu) as a drink; crushed fresh leaf or pounded boiled leaf of *Paullinia pinnata* (Toa-ntini) applied externally; an enema of root-pulp of *Turraea heterophylla* (Ahunanyankwa); root decoction of *Ziziphus mucronata* Buffalo Thorn; bark decoction of *Trichilia monadelpha* (*heudelotii*) (Tanduro) with pulp; unspecified part of *T. emetica* subsp. *suberosa* (*roka*) (Kisiga), *Oxyanthus tubiflorus*, and *Ficus capensis* (Nwadua).

Bark decoction of *Khaya ivorensis* African Mahogany (Dubini) as drink or bath; bark of *Musanga cecropioides* Umbrella Tree (Dwumma) warmed on hot cinders; a drink of crushed bark of *Milicia excelsa* (*Chlorophora excelsa*) Iroko (Odum) in water or palm-wine; bark ashes of *Afzelia africana* (Papao) with soap and shea butter as rub and unspecified parts of *Morus mesozygia* (Wonton) and *Xylopia aethiopica* Ethiopian Pepper (Hwenetia).

MADNESS (see Mental Troubles)

8. POTIONS AND MEDICINES

MALARIA (see Fever)

MANGE (see Itch)

MEASLES

Decoction of leaves of *Nauclea latifolia* African Peach (Sukisia) with those of *Rauvolfia vomitoria* (Kakapenpen) and *N. diderrichii* (Kusia) as wash (Ayensu, 1978); leaf or root decoction of *Byrsocarpus coccineus* (Awennade) with that of *Cajanus cajan* Pigeonpea followed by a rub of *Cajanus* leaves and in enemas (Moro, 1984–5); fresh leaf of *Clausena anisata* Mosquito Plant pounded with clay as rub; leaf decoction of *Celtis integrifolia* Nettle Tree (Samparanga) as beverage and macerated leaves of *Grewia carpinifolia* (Ntanta) with Guinea-grains as rub or leaf decoction of *Euphorbia hirta* Australian Asthma Herb (Moro, 1984–5). The powdered bark of *R. vomitoria* applied and boiled rice as rub for the rashes. An estimated 50,000 children die of measles in Ghana yearly. The figure for West Africa could be several million.

MENORRHAGIA (see Menstrual Troubles)

MENSTRUAL TROUBLES

Amenorrhoea (absence of menstruation)

The following plants are used as an emmenagogue, that is, for promoting menstruation: the large stipular sheath of *Musanga cecropioides* Umbrella Tree (Dwumma) and inflorescence boiled in soup; bark concoction of *Bombax buonopozense* Red-Flowered Silk Cotton Tree; bark of *Nauclea latifolia* African Peach (Sukisia); bark and seed of *Khaya senegalensis* Dry-Zone Mahogany; dried roots and stem decoction of *Aristolochia indica* (Jain, 1968); root of *Cochlospermum tinctorium* (Kokrosabia), *Gossypium arboreum* Cotton, *Ziziphus mauritiana* Indian Jujube as purgative, and root decoction of *Morinda lucida* Brimstone Tree (Kankroma). Pulped leaf of *Uncaria talbotii* (Akoo-ano); leaf-juice of *Melia azedarach* Persian Lilac, Bead Tree; leaf juice of *Pergularia daemia* as constituent of purgative (Jain, 1968); leaf of *Cardiospermum grandiflorum* and *C. halicacabum* Heart Seed, Balloon Wine; leaves, roots, and fruits of *Lawsonia inermis* Henna and ground, dried leaves of *Ficus asperifolia* Sandpaper Tree with seeds of *Xylopiastrum villosum* (*Xylopia villosa*) Elo Pubescent (Oba) in oil and salt. Decoction of *Ricinus communis* Castor Oil Plant; tubercles of *Cyperus rotundus* Nut Grass and decoction of roots of *Cynodon dactylon* Bahama or Bermuda Grass (Boulos, 1983); decoction of leaves and roots of *Passiflora glabra* (*foetida*) Stinking Passion Flower; decoction of fruit-fibres of *Adansonia digitata* Baobab; decoction of fruit of *Xylopia aethiopica* Ethiopian Pepper (Hwenetia) with that of *Newbouldia laevis* (Sasanemasa); flowers of *Ananas comosus* Pineapple or of *Musa paradisiaca* Plantain; seeds of *Sesamum indicum* Sesame, Beniseed; roots of *Ensete gilletii*

Wild Banana; decoction of *Boerhavia diffusa* and *B. repens* Hogweed; decoction of *Borreria stachydea* (Barungwini) and an unspecified part of *Vernonia nigritiana* (Gyakuruwa).

Dysmenorrhoea (pains and disorders of the cycle)

Bark decoction of *Anthocleista nobilis* Cabbage Palm as drink, bath, and vapour bath; bark decoction of *Harungana madagascariensis* (Okosoa) and *Khaya senegalensis* Dry-Zone Mahogany or pulp with honey as a violent emeto-purgative. Leaf decoction of *Alchornea cordifolia* Christmas Bush (Gyamma); *Paullinia pinnata* (Toa-ntini) with other herbs and leaf-juice of *Parquetina nigrescens* (Aba-kamo) with pepper as enema. Bark and stem of *Pleioceras barteri* (Bakapembe); bark of *Daniellia oliveri* African Copaiba Balsam Tree as a drink and enema; pounded bark of *Maesopsis eminii* (Onwan-dua) with salt from the salt-bush tree, dissolved in water as a drink; tubercles of *Cyperus rotundus* Nut Grass as analgesic (Boulos, 1985); decoction of root and stem of *Cussonia arborea* (*barteri*) (Saa-borofere); chewed twigs of *Combretum zenkeri* (Tadatso); a decoction or infusion of dried bark and twigs of *Newbouldia laevis* (Sasanemasa) pounded with *Xylopia* and other ingredients and an unspecified part of *Gossypium herbaceum* Cotton and other species (Evans, 1983).

Menorrhagia (excessive blood flow during menstruation)

Cabbage of *Elaeis guineensis* Oil Palm taken with food; root of *Uvaria chamae* (Akotompotsen); root and stem of *Stephania dinklagei* and powdered root and root decoction of *Hunteria umbellata*. The leaf-juice of *Pergularia daemia* (Jain, 1968) and infusion of *Eleusine indica* (Boulos, 1983). The bark decoction of *Zanthoxylum xanthoxyloides* (*Fagara zanthoxyloides*) Candle Wood is used for profuse bleeding after childbirth (Moro, 1984–5).

MENTAL TROUBLES

Seeds of *Datura metel* Hairy Thorn-apple, Metel with other ingredients for symptoms; root of *Uvaria chamae* (Akotompotsen) with Guinea-grains applied to fontanelles; powdered root of *Cnestis ferruginea* (Akitase) taken in ripe pawpaw; root and leaf decoction of *Psydrax subcordata* (*Canthium subcordatum*) (Tetia-dupon) as vapour bath with pulp as rub; root decoction of *Rauvolfia vomitoria* (Kakapenpen) as sedative to induce several hour's sleep; bark infusion of *Pericopsis laxiflora* (*Afrormosia laxiflora*) Satinwood; leaf of *Albizia zygia* (Okoro); crushed, red, enlarged calyx-lobes of *Mussaenda erythrophylla* Ashanti Blood in water as drink and an unspecified part of *Chrozophora senegalensis*.

MIGRAINE (see Headache)

8. POTIONS AND MEDICINES 175

NEURALGIA (PAIN ALONG THE NERVES)

Bark of *Erythroxylum mannii* Landa (Pepesia) with lemon and peppercorn; boiled and crushed leaves of *Cassia sieberiana* African Laburnum; root pulp of *Securinega virosa* (Nkanaa); crushed roots of *Euadenia trifoliolata* with lemon; decoction of leaves and stem of *Psorospermum corymbiferum* var. *corymbiferum*; crushed seeds of *Xylopia aethiopica* Ethiopian Pepper (Hwenetia) and pulped bark of *Buchholzia coriacea* (Esono-bise). Juice from heated leaves of *Alternanthera pungens* (*repens*) Khaki Weed (Mpatowa-nsoe) applied and sniffed up the nostrils; scraped pulp of *Cucurbita pepo* Vegetable Marrow, Pumpkin and *C. maxima* Squash Gourd applied as poultice; scrapings of the petioles of *Musa paradisiaca* Plantain; ground rhizomes of *Zingiber officinale* Ginger applied; poultice of roots of *Gloriosa superba* and *G. simplex* Climbing Lily applied as relief and whole plant decoction of *Asclepias* and *Passiflora* species (Evans 1983), *Caesalpinia bonduc* Bonduc (Oware-amba); and *Grewia bicolor*.

OEDEMAS, GENERAL (EXCESSIVE ACCUMULATION OF FLUID IN TISSUES)

Bark pulp of *Zanthoxylum xanthoxyloides* (*Fagara zanthoxyloides*) Candle Wood and *Spathodea campanulata* African Tulip (Kokoanisua) as rub or liniment; pulped bark of *Acacia polyacantha* subsp. *campylacantha* African Catechu; bark decoction of *Trichilia monadelpha* (*heudelotii*) (Tanduro) with pulp; pulped bark of *Ficus capensis* (Nwadua) seasoned with *Xylopia* peppers; doses of crushed beans of *Physostigma venenosum* Calabar Bean with water; leaf decoction of *Afzelia africana* (Papao) with those of *Syzygium guineense* var. *guineense* (Sunya) and *Xylopia* pepper as drink; leaf decoction of *Bridelia ferruginea* (Opam-fufuo) as drink or vapour bath; powdered leaf of *Dialium guineense* Velvet Tamarind (Asenamba) with food; pulped young roots and leaves of *Cissus populnea* (Agyako); leaf of *Clerodendrum volubile* (Kumamuno) as rub; enema of leaf pulp of *C. capitatum* (Tromen) with peppers and a decoction of bark and leaf of *Maytenus senegalensis* (Kumakuafo) as vapour bath or drink. Unspecified part of *Cassia occidentalis* Negro Coffee as drink and *Cola gigantea* var. *glabrescens* (Watapuo) with wood ashes as liniment. *Microglossa afzelii* (Pofiri), being a diuretic, is used for oedemas in pregnant women and a decoction of the root-bark of *Triplochiton scleroxylon* (Wawa) is also used for oedemas in pregnant women followed by a rub with the pulped bark.

OPHTHALMIA (see Eye Diseases)

ORCHITIS (see Hernia)

OTITIS (see Earache)

PALPITATIONS

Fresh bark of *Boswellia dalzielii* Frankincense Tree eaten as emetic to relieve symptoms; leaf concoctions of *Uvaria doeringii* (Agbana) and leaf of *Dialium guineense* Velvet Tamarind (Asenamba) with orange or pineapple juice. Macerated or boiled leaf of *Kalanchoe integra* var. *crenata* (Aporo); leaves of *Portulaca oleracea* Purslane, Pigweed macerated in water; *Solenostemon monostachys* (Sisiworodo) mixed with white clay taken as relief and bark decoction of *Spondias mombin* Hog Plum (Moro, 1984–5). (See also Heart disease.)

PARALYSIS

Root bark of *Glyphaea brevis* (Foto) with Guinea-grains as beverage and bath; root decoction of *Zanthoxylum xanthoxyloides* (*Fagara zanthoxyloides*) Candle Wood with other drugs; pulp of leafy stems of *Phyllanthus muellerianus* (Potopoleboblo) as rub; seeds of *Jatropha curcas* Physic Nut (Ayensu, 1978) and unspecified part of *Trema orientalis* (*guineensis*) (Sesea). (See also Stroke.)

PILES

Leaf infusion or concoction of *Uvaria doeringii* (Agbana), *Adenia rumicifolia* var. *miegei* (*lobata*) (Peteha), *Abutilon guineense* (Odonno-bea), *Anacardium occidentale* Cashew Nut, *Barleria opaca* (Mu), and *Lippia multiflora* Gambian Tea Bush (Saa-nunum); ground leaves of *Amaranthus spinosus* Spiny or Prickly Amaranth as ingredient in enema; powdered leaf and bark of *Alchornea cordifolia* Christmas Bush, (Gyamma); boiled leaf of *Ocimum gratissimum* Tea Bush, (Nunum) with *Spondias mombin* Hog Plum and *Ficus elastica* Rubber Plant; the root of *Allophylus africanus* (Hokple) and *Cassia sieberiana* African Laburnum; root decoction of *Terminalia macroptera* (Kwatiri) and *Pericopsis laxiflora* (*Afrormosia laxiflora*) Satinwood; the chewed roots of *Hymenostegia afzelii* (Takorowa); root-bark of *Uvaria chamae* (Akotompotsen); root decoction of *Cochlospermum tinctorium* (Kokrosabia) as sitz-bath and *Vernonia nigritiana* (Gyakuruwa) as relief; root-bark and leaf of *Pseudocedrela kotschyi* Dry-Zone Cedar as sitz-bath; root of *Carica papaya* Pawpaw ground with lime; root-pulp of *Clausena anisata* Mosquito Plant with pepper as enema; steeped roots of *Anthocleista nobilis* Cabbage Palm (Hohoroho) with pepper and Guinea-grains; an enema of root of *Plumbago zeylanica* Ceylon Leadwort; root or stem bark of *Morinda lucida* Brimstone Tree with spices as a drink; an enema of bark of *Newbouldia laevis* (Sasanemasa); an enema or drink of the boiled root and stem bark of *Kigelia africana* Sausage Tree (Nufuten); pounded bark of *Funtumia elastica* West African Rubber Tree in spirit as drink; bark decoction of *Ficus elegans* and *Cola gigantea* var. *glabrescens* (Watapuo) taken internally; bark of *C. caricifolia* Fig-Leaved Cola, Monkey Cola as enema with peppers or as decoction for sitz-bath; fresh bark of

8. POTIONS AND MEDICINES 177

Canarium schweinfurthii Incense Tree and fruit of *Nauclea latifolia* African Peach (Sukisia).

PLEURISY (see Neuralgia)

PNEUMONIA (INFLAMMATION OF THE LUNG)

Leaf extract of *Ageratum conyzoides* Billy-goat Weed (Efumomoe) rubbed on chest; steeped bark of *Acacia albida* (Gozanga) as bath and liniment; a decoction of roots and bark of *Erythrina senegalensis* Coral Flowers as drink or bath often with lemon-juice, spices, and Guinea-grains; chewed leaves of *Eriosema glomeratum* with pepper spat on scratches on patient's chest; crushed seeds of *Picralima nitida* (Ekuama) taken internally; decoction of *Vernonia colorata* or *V. amygdalina* Bitter Leaf (Bonwen) with *Argemone mexicana* Mexican or Prickly Poppy as drink; pounded seeds of *Cassia sieberiana* African Laburnum with the gum of *Erythrophleum africanum* African Black Wood (Bupunga); root decoction of *Cochlospermum tinctorium* (Kokrosabia) as liniment and an unspecified part of *Securinega virosa* (Nkanaa).

POISONING, GENERAL AND ARROW (see Chapter 9)

PURGATIVE (including Laxative)

Leaf decoction of *Ricinus communis* Castor Oil Plant, *Ocimum gratissimum* Tea Bush (Nunum); *Machaerium lunatum* (*Drepanocarpus lunatus*) (Nkako), *Garcinia gnetoides* (Tweapeakoa). *Eriosema glomeratum* with root and bark, *Phyllanthus muellerianus* (Potopoleboblo) and young twigs, *Griffonia simplicifolia* (Kagya) and leafy stems, *Desmodium adscendens* var. *adscendens* (Akwamfanu), *Clausena anisata* Mosquito Plant with roots, *Cussonia arborea* (*barteri*) (Saa-borofere), *Coffea liberica* Liberian or Monrovia Coffee with salt, *Morinda morindoides* with fruit, *Vernonia amygdalina* Bitter Leaf (Bonwen) and *V. colorata* sweetened with honey or sugar, *Cordia vignei* (Tweneboa-akoa), *Lantana camara* Wild Sage, *L. trifolia*, and decoction of leafy stems of *Scytopetalum tieghemii* (Aprim). Decoction of whole plant of *Blumea aurita* var. *aurita* as enema, *Alternanthera pungens* (*repens*) Khaki Weed (Mpatowa-nsoe) in enemas; *Tribulus terrestris* Devil's Thorn or *Centaurea perrottetii* Star Thistle (Balinyiri); *Momordica charantia* African Cucumber with natron, *M. cissoides* (Ntonto) as laxative for children, *Euphorbia hirta* Australian Asthma Plant with lime as laxative or pounded with water as enema for constipation and leaf decoction of *Aneilema beniniense*.

Leaf extract of *Eclipta alba* (*prostrata*) (Ntum) taken internally; leaf infusion of *Bridelia atroviridis* (Asaraba), *Tephrosia vogelii* Fish Poison Plant, tea-like leaf infusion of *Lippia multiflora* Gambian Tea Bush (Saa-nunum), and infusion of young leaves of *Vismia guineensis* (Kosowanini). Pounded leaf of *Ageratum conyzoides* Billy-goat Weed (Efumomoe) with *Ocimum* and pepper as enema; ground whole plant of *Cuscuta australis* as enema for young children and leaf infusion of *Synedrella nodiflora* taken as laxative. Leaf of *Uvaria afzelii, Pycnocoma cornuta* or *P. macrophylla* (Akofie-kofi) with bark, stem, and root; leaf of *Baphia nitida* Camwood (Odwen) with bark in enema; leaf of *Funtumia africana* False Rubber Tree with bark in enema and leaf of *Adenia rumicifolia* var. *miegei* (*lobata*) (Peteha) as enema. Leaf of *Ochna schweinfurthiana* (Bibie) as laxative, *Combretum collinum* subsp. *hypopilinum* (*hypopilinum*) (Chinchapula), *Berlinia confusa* as enema, *Cassia nodosa* Pink Cassia, *Tamarindus indica* Indian Tamarind and pods as laxative, *Abrus precatorius* Prayer Beads, *Opilia celtidifolia* (Benkasa) with leafy tops and roots, and *Sabicea calycina* (Anansentoromahama).

Leaf-juice of *Ehretia cymosa* as mild laxative for children, *Mareya micrantha* (Odubrafo) and *Cola caricifolia* Fig-leaved Cola macerated in water; pounded leaf and pericarp oil of *Piper umbellatum* (Amuaha) eaten by pregnant women as a laxative and leaf and fruit infusion of *Cassia podocarpa* (Nsuduru) with ashes, filtered and taken internally.

Root or root-bark or root extract of *Acacia macrothyrsa, Uvaria chamae* (Akotompotsen), *Sphenocentrum jollyanum* (Krakoo) chewed, *Securidaca longepedunculata* (Kpaliga) in small doses, *Lagenaria breviflora* (*Adenopus breviflorus*) (Anuwatre), *Combretum aculeatum, Cardiospermum grandiflorum* and *C. halicacabum* Heart Seed or Baloon Vine as laxative, *Glyphaea brevis* (Foto), *Balanites aegyptiaca* Desert Date, *Ricinodendron heudelotii* (Wamma) with pepper and salt, *Cassia alata* Ringworm Shrub, *Lepidagathis heudelotiana* with stems (also as depurative), *C. sophera* with fruit pulp, *C. obtusifolia* (*tora*) Foetid Cassia or the whole plant, *Albizia ferruginea* (Awiemfosemina) powdered with salt and taken internally, *Calliandra portoricensis* (Nkabe) with fresh ginger and water as enemas, *Dalbergiella welwitschii, Ziziphus mauritiana* Indian Jujube, *Z. mucronata* Buffalo Thorn, *Leea guineensis* (Okatakyi) with seed as drink, *Trichilia emetica* subsp. *suberosa* (*roka*) (Kisiga) soaked for seven days as enema, *Bersama abyssinica* subsp. *paullinioides* (Duantu) as purging vermifuge, *Cnestis ferruginea* (Akitase), *Pleiocarpa mutica* (Kanwen) ground with Guinea-grains in palm-wine as laxative, *Rauvolfia vomitoria* (Kakapenpen), *Calotropis procera* Sodom Apple in small doses as purgative in leprosy treatment, *Mitragyna inermis* (Kukyafie), *Nauclea latifolia* African Peach, (Sukisia), or bark and pounded, boiled roots of *Vernonia nigritiana* (Gyakuruwa).

Corms of *Gladiolus daleni* (*psittacinus*), *G. klattianus, G. gregarius,* and *G. unguiculatus* Sword Lily, Corn-flag in enemas with ginger for constipation or powder mixed with castor oil and lime-juice or fluid extract alone; bulb of *Crinum zeylanicum* (*ornatum*) (Ngalenge); tuberous roots of *Ipomoea mauritiana* (Disinkoro); roots of *Luffa cylindrica* Loofah Gourd, Vegetable Sponge; *Ethulia conyzoides* with pepper as enema for constipation and root stock of *Curculigo pilosa* (Bebenga). Root infusion of *Cassia nigricans* and *Chaetacme aristata* (Kodia); root-sap of *Securinega virosa* (Nkanaa) and root and seed of

8. POTIONS AND MEDICINES

Clitoria ternatea Blue Pea.

Bark decoction of *Pycnanthus angolensis* African Nutmeg (Otie); *Tetrorchidium didymostemon* (Aboagyedua), *Afzelia africana* (Papao), *Khaya senegalensis* Dry-Zone Mahogany or pulp sweetened with honey as emeto-purgative, *Omphalocarpum ahia* (Duapompo) and *O. elatum* (Timatibre) with fruit. Bark infusion of *Monodora brevipes* (Abotokuradua), *Berlinia grandiflora* (Tetekono), *Erythrophleum suaveolens* (*guineense*) Ordeal Tree (Potrodom); *Milletia thonningii* (Sante), *Milicia excelsa* (*Chlorophora excelsa*) Iroko (Odum); *Ongokea gore* (Bodwe) as drink or enema or pinch of pulverized bark licked with salt, *Lecaniodiscus cupanioides* (Dwindwera), and infusion of outer bark of *Vernonia conferta* (Flakwa). Bark or bark extract of *Coelocaryon oxycarpum* (Abruma), *Harungana madagascariensis* (Okosoa) as children's purgative, *Garcinia kola* Bitter Cola (Tweapea); *Margaritaria discoidea* (*Phyllanthus discoideus*) (Pepea); *Parinari congensis* (Tulingi), *Dialium dinklagei* (Awendade), *Tetrapleura tetraptera* (Prekese), and *Newbouldia laevis* (Sasanemasa) as enemas; *Lonchocarpus sericeus* Senegal Lilac and *Treculia africana* var. *africana* African Bread fruit (Brebretim) as laxative; *Reissantia indica* (*Hippocratea indica*) and *Zanthoxylum xanthoxyloides* (*Fagara zanthoxyloides*) Candle Wood as laxative; powdered bark of *Crossopteryx febrifuga* African Bark (Pakyisie) in palm-wine or water; pulped bark of *Blighia unijugata* (Akyebiri) steeped in water, *Pseudospondias microcarpa* var. *microcarpa* (Katawani) in palm-wine or water, and inner bark of *Samanea dinklagei*.

Infusion of roasted seeds of *Ipomoea turbinata* (*muricata*) as laxative; seeds of *I. aitonii* with those of *Hibiscus sabdariffa* Roselle; mucilaginous seeds of *Sesamum indicum* Sesame, Beniseed taken with salt; seeds of *I. nil* Morning Glory as a result of the presence of an acrid resin; crushed seeds of *Ricinus communis* Castor Oil Plant given with cooked cereal or in doses with groundnuts; seed or seed oil of *Argemone mexicana* Mexican or Prickly Poppy, *Hura crepitans* Sandbox Tree, *Jatropha curcas* Physic Nut, *Monodora myristica* (Ayerew-amba), *J. gossypiifolia*, *J. multifida* Coral Plant, *Holoptelea grandis* Orange-Barked Terminalia (Nakwa) as laxative; *Lannea acida* (Kuntunkuri), *L. nigritana* var. *nigritana* (Sinsabgbetiliga), and the bitter, nauseous seed oil of *Carapa procera* Crabwood which resembles castor oil in action. Fruit of *Solanum anomalum* (Nsusoa) as laxative and digestive, *Luffa acutangula* (Akatong) as a result of the principle luffein, *Vitex doniana* Black Plum (Afua); *Kigelia africana* Sausage Tree (Nufuten) and boiled fruit of *S. aculeatissimum* as enema. Kernels of *Thevetia peruviana* Milk Bush, Exile Oil Plant chewed; sweetish, dirty black pulp of *Cassia fistula* Golden Shower; ripe fruit of *Psidium guajava* Guava as laxative and boiled fruit of *Gongronema latifolium* (Nsurogya) in soup. Latex of *Elaeophorbia drupifera* (Akane) taken internally, *Euphorbia lateriflora* (Kamfo-barima), *Alstonia boonei* (Sinduro) with that of *Anthostema aubryanum* (Kyirikesa), *Plumeria rubra* var. *acutifolia* Red Frangipani, *Parquetina nigrescens* (Aba-kamo) or leaf-juice in enemas; bitter sap of stem and leaf of *Secamone afzelii* (Kotohume); sap of *Dracaena fragrans* applied to the breast as baby's purge; gum of *Daniellia ogea* Gum Copal Tree, *D. oliveri* African Copaiba Balsam Tree chewed and swallowed, and all parts of *Melia azedarach* Persian Lilac, Bead Tree and *Cassia occidentalis* Negro Coffee. Decoction of unspecified parts of *Hoslundia opposita* (Asifuaka) and *Flacourtia flavescens* Niger Plum; stem decoction of *Albizia zygia* (Okoro) and *A. adianthifolia* (Pampena) and infusion of the outer bark of *Vernonia conferta* (Flakwa). Unspecified parts of *Aptandra zenkeri* (Ayemtudua), *Maesopsis eminii* (Onwamdua), *Dodonaea viscosa* Switch Sorrel, (Fomitsi); *Asclepias curassavica* Blood Flower, *Indigofera simplicifolia* (Nyagahe) in enemas, *Sapium ellipticum* (Tomi) in small doses, *Phyllanthus fraternus* subsp. *togoensis* (*niruri*) (Ombatoatshi) as a bitter stomachic for constipation, *Annona senegalensis* subsp. *onlotricha* (*arenaria*) (Aboboma) and *Claoxylon hexandrum* (Sansammuro). Charred and pulverized whole plant of *Gardenia ternifolia* (Peteprebi) in palm-wine with Guinea-grains and shea-butter roots is an effective remedy. The roots of *Ximenia americana* Wild Olive cure purging, but the poisonous seeds purge and the leafy twigs are a laxative.

RESPIRATORY TROUBLES (see Bronchial Trouble)

REVULSIVE (see Counter-irritant)

RHEUMATISM

Infusion of bark and leafy stems of *Bridelia ferruginea* (Opam-fufuo) as beverage, bath, and vapour bath; bark decoction of *Mammea africana* African Mammy Apple (Bompagya); *Omphalocarpum ahia* (Duapompo), *Dracaena fragrans* and *Cylicodiscus gabunensis* African Greenheart (Denya) both as lotion; *Boswellia dalzielii* Frankincense Tree (Kabona) as wash; *Carpolobia alba* (Afiafia) taken internally and applied externally and bitter bark decoction of *Khaya ivorensis* African Mahogany (Dubini) as lotion. Crushed bark of *Erythrophleum suaveolens* (*guineense*) Ordeal Tree (Potrodom) with white clay applied; bark of *Zanthoxylum gilletii* (*Fagara macrophylla*) (Okuo), *Chytranthus carneus* (*villiger*) (Bobiri), *Nauclea latifolia* African Peach (Sukisia); *Kigelia africana* Sausage Tree (Nufuten); *Afraegle paniculata* (Obuobi) as Agbo infusion taken internally and as a wash; powdered bark of *Dialium dinklagei* (Awendade) mixed into paste with palm oil as rub; pounded bark of *Syzygium rowlandii* (Asibenyanya) with white clay and *Xylopia* or other spices as rub; pulverized bark of *Zanthoxylum leprieuri* (*Fagara leprieuri*) (Oyaa) as counter-irritant and bitter bark of *Pseudocedrela kotschyi* Dry-Zone Cedar.

Root decoction of *Isolona campanulata*, *Platostoma africanum* (Siresireke) with that of *Tephrosia linearis* used both internally and externally, and crushed root and leaf of *Solanum incanum* Egg Plant as rub; root of *Piper umbellatum* (Amuaha) in alcoholic drinks; root of *Securidaca longepedunculata* (Kpaliga) with ingredients and shea butter as rub; root of *Abrus precatorius* Prayer Beads or *Strophanthus hispidus* Arrow Poison; pounded roots of *Pericopsis laxiflora*

8. POTIONS AND MEDICINES   181

(*Afrormosia laxiflora*) Satinwood applied; root-pulp of *Erythrina senegalensis* Coral Flowers as rub; root-bark and leaf of *Celtis integrifolia* Nettle Tree, (Samparanga) or *Alstonia boonei* (Sinduro) applied; root and bark of *Holoptelea grandis* Orange-Barked Terminalia (Nakwa) beaten and applied; pounded root of *Clausena anisata* Mosquito Plant with lime and Guinea-grains applied; pulped root of *Zanthoxylum viride* (*Fagara viridis*) (Oyaanini) as massage; pulverized root-bark or bark of *Z. xanthoxyloides* (*F. zanthoxyloides*) Candle Wood applied as poultice and powdered roots of *Gardenia ternifolia* (Peteprebi) rubbed in small incisions in affected parts.

Leaf decoction of *Xylopia aethiopica* Ethiopian Pepper (Hwenetia); *Maerua angolensis* (Pugodigo), *Albizia zygia* (Okoro) as beverage and bath, *Pistia stratiotes* Water Lettuce as fumigant for pains, *Ipomoea asarifolia* taken and as a wash, *Schwenckia americana* (Agyingyinsu) taken internally and rubbed locally or a rub of pounded leaves with natron and shea butter; decoction of leafy twigs of *Alchornea cordifolia* Christmas Bush (Gyamma) as wash; leaf pulp of *Erythrococca anomala* or *Leea guineensis* (Okatakyi) as massage; powdered leaf of *Hymenocardia acida* (Sabrakyie) applied locally; steeped, pulped leaf of *Mareya micrantha* (Odubrafo) as rub; leaf lotion of *Piliostigma thonningii* (Opitipata) taken internally and applied externally; boiled leaves of *Costus afer* Ginger Lily as soothing fomentation for pains; warm fomentation of leaves and fruits of *Capsicum annuum* Pepper; crushed leaves of *Datura metel* Hairy Thorn Apple, Metel as poultice applied; warm poultice of *Celosia leptostachya* (*laxa* and *trigyna*); heated leaf of *Pseudarthria confertiflora* applied and leaves of *Wahlenbergia perrottetii* (*Cephalostigma perrottetii*) rubbed on limbs to relieve pain. Whole plant of *Polygonum senegalense* pounded with natron as rub, *Centella asiatica* Indian Water Navelwort (Evans, 1983); *Dissotis rotundifolia* (Adowa-alie) and *Ipomoea pes-caprae* subsp. *brasiliensis* Goat's Foot Convolvulus applied; chopped up leaves of *Rauvolfia vomitoria* (Kakapenpen) stewed in oil applied; crushed leaf of *Psydrax venosa* (*Canthium venosum*) with shea butter applied; leaf of *Pentodon pentandrus* (Buburanya) applied; infusion of crushed leaf of *Heinsia crinita* Bush Apple applied; leaf infusion of *Guiera senegalensis* with that of *Combretum micranthum* (Landaga) as febrifuge; decoction or infusion of leaf extract of *Mitragyna inermis* (Kukyafie) as lotion; leaf concoction of *Sabicea calycina* (Anansetoromahama) as fomentation; leaf decoction of *Microglossa afzelii* (Pofiri) as lotion; pounded leaf of *Cordia vignei* (Tweneboa-akoa) tied to affected parts and hot application of pounded leaves and roots of *Newbouldia laevis* (Sasanemasa). Decoction of stem-tips of *Strophanthus sarmentosus* (Adwokuma) applied. Leaves of *Plumbago zeylandica* Ceylon Leadwort and crushed, warmed leaves of *Clerodendrum buchholzii* (Taasendua) applied.

Fruit and seeds of *Paullinia pinnata* (Toa-ntini) with ginger applied; fruit oil of *Tribulus terrestris* Devil's Thorn as rub; 'sulphur oil' from *Allium sativum* Garlic as rub; crushed seeds of *Piper guineense* West African Pepper (Sorowisa) or *Physostigma venosum* Calabar Bean applied; oil of *Carapa procera*

Crabwood or *Ceiba pentandra* Silk Cotton Tree; oil of *Jatropha curcas* Physic Nut as ingredient in extract used as rubefacient; ointment prepared from the fruit of *Tetrapleura tetraptera* (Prekese); pressed fruit of *Beilschmiedia mannii* Spicy Cedar as counter-irritant; bitter fruit of *Citrus aurantium* Sour Orange taken internally and red heartwood of *Baphia nitida* Camwood, (Odwen) pounded and mixed with shea butter applied.

Latex of *Calotropis procera* Sodom Apple as rubefacient; leaf ashes of *Terminalia avicennioides* (Petni) with *Crinum* bulbs and butter as ointment and whole plant of *Ocimum gratissimum* Tea Bush (Nunum) as poultice. Unspecified parts of *Cassia occidentalis* Negro Coffee, *Adenia rumicifolia* subsp. *miegei* (*lobata*) (Peteha), *Tetracera alnifolia* (Akotopa), *Oxyanthus tubiflorus*, *Alafia scandens* (Momonimo), *Morus mesozygia* (Wonton), and a decoction of *Chrozophora senegalensis*. The tuber of *Smilax kraussiana* (Kokora) is useful for chronic cases and rheumatoid arthritis (Evans, 1985). Cortisone, obtained from *Strophanthus*, is used in treating articular rheumatism, otherwise called arthritis. *Cyathula prostrata* (Mpupua) is equally used for articular rheumatism.

RICKETS

Leaf decoction of *Carapa procera* Crabwood, *Heisteria parvifolia* (Sikakyia), *Mucuna flagellipes* (Tatwea), *Paullinia pinnata* (Toa-ntini) for children, *Piliostigma thonningii* (Opitipata), *Crossopteryx febrifuga* African Bark (Pakyisie) as enema for strengthening children; infusion of pounded and macerated fresh leaves of *Parquetina nigrescens* (Aba-kamo) as daily wash for children (Ayensu, 1978) and leaf decoction of *Tabernaemontana crassa* (Pepae) as liniment to strengthen; decoction of *Laportea aestuans* (*Fleurya aestuans*) (Hunhon) as lotion for children (Ayensu, 1978); cold infusion of *Crinum zeylanicum* (*ornatum*) (Ngalenge) as bath and remedy for general debility in children and the tisane of leaf tips of *Fadogia agrestis* (Buruntirikwa). Root decoction of *Vitex doniana* Black Plum (Afua) with bark and that of *Lannea acida* (Kuntunkuri); root-juice of *Tetracera alnifolia* (Akotopa) as enema for children; powdered bark of *Trichilia monadelpha* (*heudelotii*) (Tanduro); bark decoction of *Erythrina senegalensis* Coral Flowers and *Parkia biglobosa* (Duaga); bark of *Ficus capensis* (Nwadua) as ingredient and an unspecified part of *Annona senegalensis* subsp. *onlotricha* (*arenaria*) (Aboboma).

RINGWORM, SKIN DISEASES, AND AFFECTIONS (PARASITIC)

Leaf of *Terminalia macroptera* (Kwatiri), *Plumbago zeylanica* Ceylon Leadwort with lemon-juice; leaf-juice and fruit of *Alchornea cordifolia* Christmas Bush, (Gyamma); whole plant of *Centella asiatica* Indian Water Navelwort (Evans, 1985); latex of *Elaeophorbia drupifera* (Akane), *Tabernaemontana crassa* (Pepae), *Calotropis procera* Sodom Apple, and *Euphorbia lateriflora* (Kamfobarima) for ringworm of the scalp; seed of *Cassia absus* Four-Leaved Henna, Black Grain; bark of *Pterocarpus erinaceus* Senegal Rose Wood Tree, African Kino for

8. POTIONS AND MEDICINES

ringworm of the scalp; pulverized seed of *Cordia millenii* Drum Tree with palm oil; leaf of *Vernonia amygdalina* Bitter Leaf (Bonwen) and *V. colorata* as rub; pulped bark of *Trichilia emetica* subsp. *suberosa* (*roka*) (Kisiga); pulped fresh leaf-juice of *Cassia alata* Ringworm Shrub and root, and fresh leaf and pulverized seeds of *C. obtusifolia* (*tora*) Foetid Cassia with lime-juice and senna.

Bark decoction or infusion of *Spathodea campanulata* African Tulip (Kokoanisua); *Vitex simplicifolia* (Abisa) as lotion, *Holarrhena floribunda* False Rubber Tree (Sese), and *Psorospermum febrifugum* var. *ferrugineum*; pulverized bark of *Avicennia africana* Mangrove, (Asoporo) in warm water as paste; pulped bark of *Lannea nigritana* var. *nigritana* (Sinsabgbetiliga); bark of *Lonchocarpus sericeus* Senegal Lilac as lotion, hot or cold bark infusion of *Albizia adianthifolia* (Pampena); bark and leaf of *Acacia farnesiana* as lotion; charred bark of *Erythrophleum suaveolens* (*guineense*) Ordeal Tree (Potrodom); powdered bark of *Distemonanthus benthamianus* African Satinwood (Bonsamdua) mixed with redwood powder (padouk) in water and that of *Afzelia bella* var. *glacior* (Papaonua); bark of *Uapaca guineensis* Sugar Plum (Kuntan) as emetic, enema, or lotion and boiled bark of *Crossopteryx febrifuga* African Bark (Pakyisie).

Pounded leaves and roots of *Clematis hirsuta*; fresh leaves of *Lawsonia inermis* Henna with lime; leaf and bark of *Anopyxis klaineana* (Kokote); decoction of leaf-stem of *Scytopetalum tieghemii* (Aprim) as bath; crushed leaves of *Manihot esculenta* Cassava; leaf of *Cassia occidentalis* Negro Coffee used externally and internally; leaf-juice of *Acacia kamerunensis* (*pennata*) (Oguaben) as liniment; leaf extract of *Eriosema psoraleoides*; leaf of *Lonchocarpus cyanescens* West African or Yoruba Indigo (Akase) as dressing; leaf infusion of *Dracaena bicolor* (Gblieku); leaf and flowers of *Hoslundia opposita* (Asifuaka) and leaf decoction or latex of *Rauvolfia vomitoria* (Kakapenpen). Gummy sap of *Harungana madagascariensis* (Okosoa); sap of *Garcinia kola* Bitter Cola (Tweapea) and *Manniophyton fulvum* (Hunhun) for local application; oil of *Azadirachta indica* Neem Tree and *Jatropha curcas* Physic Nut; crushed seed of *Physostigma venenosum* Calabar Bean for local application; gum copal of *Daniellia thurifera* Niger Copal Tree; flower ashes of *Leonotis nepetifolia* var. *africana* (Nyeddo); paste of the spines of *Elaeis guineensis* Oil Palm with lemon for local application and an extract of *Ricinus communis* Castor Oil Plant and *Balanites aegyptiaca* Desert Date, Soap-berry Tree (Gongu).

Leaf of *Portulaca oleracea* Purslane, Pigweed, chewed with tiger nuts; whole plant of *Celosia leptostachya* (*laxa* and *trigyna*) applied; fruit of *Solanum dasyphyllum*; seed oil of *Luffa cylindrica* Loofah Gourd, Vegetable Sponge, applied; strong infusion of fruits of *Capsicum annuum* Pepper as lotion for ringworm of the scalp; whole plant of *Ampelocissus grantii* (Enotipsi) boiled with cereal pap; ground rhizomes of *Mariscus alternifolius* applied; latex of *Parquetina nigrescens* (Aba-kamo); juice of *Citrus aurantiifolia* Lime and *Allium sativum* Garlic applied. Pulped root of *Psorospermum corymbiferum* var. *corymbiferum* and *Mammea africana* African Mammy Apple (Bompagya); root decoction of *Acridocarpus smeathmannii* (Alasaayo) as ointment for pemphisus (a grave skin disease); root bark of *Bridelia ferruginea* (Opam-fufuo) used externally; powdered heartwood of *Baphianitida* Camwood (Odwen) with water and oil as ointment and root bark decoction of *Lannea acida* (Kuntunkuri) in baths and lotions. A decoction of the whole plant of *Sapium grahamii* (Pampiga) as bath; that of *Tephrosia vogelii* Fish Poison and unspecified parts of *Markhamia lutea* (Sisimasa) and *M. tomentosa* (Tomboro).

SCABIES

Leaf extract of *Eriosema psoraleoides* applied; leaf of *Cardiospermum grandiflorum* Heart Seed, Balloon Vine (Ayensu, 1978); bark extract of *Bridelia stenocarpa* (*micrantha*) (Opam) applied; milky juice decoction of *Plumbago zeylanica* Ceylon Leadwort (Ayensu, 1978); ashes of whole plant of *Cyathula prostrata* (Mpupua) applied; juice of *Jatropha curcas* Physic Nut (Ayensu, 1978); leaf pulp of *Ocimum gratissimum* Tea Bush, (Nunum) applied (Ayensu, 1978) and unspecified parts of *Crossopteryx febrifuga* African Bark (Pakyisie); *Albizia zygia* (Okoro), *Erythrophleum suaveolens* (*guineense*) Ordeal Tree (Potrodom), and *Psorospermum febrifugum* var. *ferrugineum* applied.

SCHISTOSOMIASIS (see Chapter 9)

SCORPION STING, ANTIDOTE TO (see Chapter 9)

SCURVY

Leaf paste of *Cassia alata* Ringworm Shrub with palm oil and whole plant of *Piper umbellatum* (Amuaha); the leaves of *Vernonia amygdalina* Bitter Leaf, (Bonwen) and *V. colorata* or *Portulaca oleracea* Purslane, Pigweed in soup; the leaves and seeds of *Hibiscus subdariffa* Red Sorrel, Roselle and the whole plant of *Blumea aurita* var. *foliolosa* and *Spilanthes filicaulis* Para or Brazil Cress; the pulp of *Tamarindus indica* Indian Tamarind; the whole plant of *Celosia argentea* (Nkyewodue); the chewed bark, leaves, and pods of *Acacia nilotica* var. *tomentosa*; the fruit-juice of *Anacardium occidentale* Cashew Nut and *Annona muricata* Sour Sop; the roots and bark of *Moringa oleifera* Horse-Radish Tree, Oil of Ben Tree; bark of *Harungana madagascariensis* (Okosoa) and a diet of bananas, as the antiscorbutic vitamin is plentifully present.

SINUSITIS

Powdered bark of *Millettia barteri* or *Newbouldia laevis* (Sasanemasa) as snuff; charred, powdered roots of *Ekebergia senegalensis* as snuff; leaf-juice of *Clausena anisata* Mosquito Plant as nasal drops; root-juice or powdered root of *Cnestis ferruginea* (Akitase) applied to the nostrils; powdered leaf of *Drypetes chevalieri* (Katrika) as snuff and whole plant of *Dissotis rotundifolia* (Adowa-alie) (Ayensu, 1978).

SKIN DISEASES, PARASITIC (see Ringworm)

SLEEPING SICKNESS (TRYPANOSOMIASIS)

The disease is transmitted by the tsetsefly and has rendered large areas of West and Central Africa virtually uninhabitable. Herbal treatments include the root and bark of *Annona senegalensis* var. *senegalensis* Wild Custard Apple; leaf decoction of *Cleistopholis patens* Salt and Oil Tree as drink and bath; root decoction of *Securidaca longepedunculata* (Kpaliga) and *Afzelia africana* (Papao); *Tamarindus indica* Indian Tamarind also with *Afzelia* and *Ficus;* root bark of *Glyphaea brevis* (Foto); bark decoction of *Haematostaphis barteri* Blood Plum; macerated leaves of *Cola caricifolia* Fig-Leaves Cola or *Cordia myxa* Sapistan Plum; root and leaf of *Opilia celtidifolia* (Benkasa); roots of *Ximenia americana* Wild Olive with those of Wild Custard Apple; root decoction of *Costus afer* Ginger Lily with the false bulbs of the orchids *Eulophia* species, *Habenaria* species, *Angraecum* species, *Bulbophyllum* species, *Ansellia* species, *Listrostachys* species, and *Polystachya* species and decoction of *Ageratum conyzoides* Billy-goat Weed (Efumomoe); *Lophira lanceolata, Lippia multiflora* Gambian Tea Bush (Saa-nunum), and *Vitex fosteri* (Otwentorowa) are used with other plants.

SMALLPOX (included as a historical record since this disease is now eradicated)

Root of *Grewia mollis* (Yualega) as an ingredient; bark, root, and leaf of *Piliostigma thonningii* (Opitipata); root of *Boerhavia diffusa* and *B. repens* Hogweed with seed of *Blighia sapida* Akee Apple (Akye) applied (Ayensu, 1978); thick emulsion of the fruit-pulp of *Adansonia digitata* Baobab put on the patient's eyes several times a day; powdered bark of *Zanthoxylum gilletii* (*Fagara macrophylla*) (Okuo) with *Carapa* oil both to cure and to prevent; a decoction of leafy tips or bark of *Khaya senegalensis* Dry-Zone Mahogany sometimes with powdered bark and leaves of *Dalbergia saxatilis* (Nuodolega) to promote the eruption; pulverized *Euphorbia convolvuloides* with palm oil applied which is said to dry up quickly and leaf pulp of *Sansevieria liberica* African Bowstring Hemp with other herbs or the expressed juice alone as ointment for sores. The seeds of *Sesamum alatum* and *S. indicum* Sesame, Beniseed, are forbidden to smallpox patients as the seeds resemble the postules.

SNAKE-BITE (see Chapter 9)

SNUFF/SNEEZING

Dried, cured, and powdered leaves of *Nicotiana tabacum* Tobacco and *N.*

rustica; powdered leaves of *Zanthoxylum chevalieri* (*Fagara pubescens*) (Oyaabere) to cure headache; powdered bark of *Milletia barteri* or *Newbouldia laevis* (Sasanemasa); dried, pulverized bark of *Calliandra portoricensis* (Nkabe), *Securidaca longepedunculata* (Kpaliga) with root and seed, *Erythrophleum suaveolens* (*guineense*) Ordeal Tree (Potrodom), and the yellowish-brown substance in the pith of old stems of *Jatropha gossypiifolia* Physic Nut; powdered leaf of *Drypetes chevalieri* (Katrika), *Cola gigantea* var. *glabrescens* (Watapuo), or *C. caricifolia* Fig-Leaves Cola mixed with tobacco and *Erythrococca anomala* with pepper; fruit ashes of *Musa paradisiaca* Plantain mixed with snuff; fresh leaf of *Vernonia biafrae* (Hu), *V. conferta* (Flakwa), and *Premna hispida* (Aunni); powdered roots of *Phyllanthus muellerianus* (Potoboleboblo), *Ekebergia senegalensis* and *Microglossa pyrifolia* (Asommerewa), and powdered or roasted, pulverized seeds of *Entada abyssinica* (Sankasaa).

SORE EYES (see Eye Diseases)

SORE GUMS/SORE MOUTH

Powdered leaf of *Spilanthes filicaulis* Para or Brazil Cress rubbed on lips and gums; leaf and bark infusion of *Mangifera indica* Mango or *Anacardium occidentale* Cashew Nut as lotion and mouthwash; leaf decoction of *Sorindeia grandifolia* as gargle for children, *Byrsocarpus coccineus* (Awennade), and latex of *Calotropis procera* Sodom Apple. The leaves of *Tristemma littorale* in water as mouthwash is used for stomatitis in children.

SORE THROAT

Roots of *Hoslundia opposita* (Asifuaka) and *Abrus precatorius* Prayer Beads; roots decoction of *Cassia occidentalis* Negro Coffee with seeds; boiled roots of *Prosopis africana* (Sanga) as poultice; bark decoction of *Rhizophora* species Red Mangrove, *Ficus dekdekena* (*thonningii*) (Gamperoga), and *Mangifera indica* Mango; bark of *F. capensis* (Nwadua) chewed with cola nuts, *Myrianthus arboreus* (Anyankoma) cooked with palm oil, and bark decoction of *Dialium guineense* Velvet Tamarind (Asenamba) as gargle. Leaf infusion of *C. nigricans*, *Abutilon guineense* (Odonno-bea), or *Microglossa afzelii* (Pofiri), *Solanum incanum* Egg Plant and young leaves of *Vismia guineensis* (Kosowanini) with Guinea-grains; leaf decoction of *Caesalpinia bonduc* Bonduc, (Oware-amba) as gargle and lime-juice from *Citrus aurantiifolia* Lime as gargle. Leaf of *Emilia sonchifolia* with Guinea-grains and lime-juice; *Platostoma africanum* (Siresireke) chewed with salt and *Scoparia dulcis* Sweetbroom Weed ground with Guinea-grains or clay and water as drink or suck of cut lime on which the powdered leaves have been sprinkled.

SORES (see Wounds)

8. POTIONS AND MEDICINES

SPRAINS, BURNS AND BRUISES

Pulverized or crushed leaves or leaf-juice of *Portulaca oleracea* Purslane, Pigweed; *Chasmanthera dependens, Terminalia avicennioides* (Petni), *Mareya micrantha* (Odubrafo), *Lawsonia inermis* Henna, *Pulicaria crispa, Cassia occidentalis* Negro Coffee, *Tamarindus indica* Indian Tamarind, *Gouania longipetala* (Homabiri), and *Paullinia pinnata* (Toa-ntini); dried, powdered leaves of *Clerodendrum splendens* (Ekenyieya); leaf ashes of *Morinda morindoides*; dried leaf of *Ipomoea aitonii* and *Merremia aegyptiaca* applied; leaf of *Bryophyllum pinnatum* Resurrection Plant applied and pounded leaves and stem of *Cissus quadrangularis* Edible-Stemmed Vine (Kotokoli) applied; pounded and warmed leaves of *Zingiber officinale* Ginger applied; *Momordica charantia* African Cucumber as liniment or dressing; slightly scorched leaves of *Laportea aestuans* (*Fleurya aestuans*) (Hunhon) applied, and unopened, terminal leaves of *Musa paradisiaca* Plantain moistened with palm oil as dressing.

Scraped pulp of *Cucurbita pepo* Pumpkin, Vegetable Marrow and *C. maxima* Squash Groud, *Dioscorea* species and *Manihot esculenta* Cassava as poultice applied; a lotion of *Dodonaea viscosa* Switch Sorrel (Fomitsi); seed oil of *Carapa procera* Crabwood; powdered kernels of *Irvingia gabonensis* Wild Mango (Abesebuo); roasted and crushed fruits of *Parkia biglobosa* (Duaga); bark of *Indigofera macrophylla* (Ogyafam); bark and leaf of *Microdesmis puberula* (Fema) pounded with spices; leaves of *Adenodolichos paniculatus* or *Lecaniodiscus cupanioides* (Dwindwera) with butter or steeped in oil and powdered heartwood of *Baphia nitida* Camwood (Odwen) with shea butter; root of *Cochlospermum tinctorium* (Kokrosabia) in shea butter; dry, powdered root bark of *Saba senegalensis* (Sono-nantin) for children's cases and latex of *Milicia excelsa* (*Chlorophora excelsa*) Iroko (Odum) applied.

STERILITY

Root of *Gardenia nitida* boiled with eggs, *Waltheria indica* (Sawai) cooked with rice, pounded root and leaves of *Paullinia pinnata* (Toa-ntini), and pulverized roots and stems of *Caesalpinia bonduc* Bonduc (Oware-amba) in palm-wine. Bark decoction of *Albizia adianthifolia* (Pampena) as enema, *Crossopteryx febrifuga* African Bark (Pakyisie) taken internally; bark of *Vitex doniana* Black Plum (Afua); pounded bark and leaf of *Erythrina senegalensis* Coral Flowers in soup; pulverized fresh leaves of *Calotropis procera* Sodom Apple and pepper inserted locally; leaves of *Trema orientalis* (*guineensis*) (Sesea) with those of *Sherbournea bignoniiflora* (Kyerebeteni) in soup, *Ageratum conyzoides* Billy-goat Weed (Efumomoe) inserted locally for uterus troubles; *Tribulus terrestris* Devil's Thorn used to induce conception and *Brillantaisia patula* used to ensure conception. Bark infusion of *Mitragyna ciliata* Abura, Poplar (Subaha) with the bark of *Coula edulis* Gaboon Nut (Bodwue), *Isolona* and *Barleria*; leaves of *Alchornea cordifolia* Christmas Bush (Gyamma) and other ingredients; fruit of *Physalis angulata* (Totototo) crushed with milk as remedy and an unspecified part of *Ficus sur (capensis)* (Nwadua) and *Albizia zygia* (Okoro) as purgative.

STOMACH ACHE, PAINS, DISORDERS AND INDIGESTION

Root of *Dracaena fragrans*, *Sphenocentrum jollyanum* (Krakoo), *Cochlospermum tinctorium* (Kokrosabia), *Acridocarpus smeathmannii* (Alasaayo), *Afraegle paniculata* (Obuobi), *Rothmannia longiflora* (Saman-kube), *Pseuderanthemum tunicatum*, *Clerodendrum capitatum* (Tromen) taken hot, *Ruspolia hypocrateriformis* (Wonane) as paste in beer, *Stereospermum kunthianum* (Sonontokwakofo) with the bark, *Nauclea latifolia* African Peach (Sukisia) and leaf; *Securidaca longepedunculata* (Kpaliga) with ingredients for drastic purge, *Eriosema psoraleoides* as extract, *Heisteria parvifolia* (Sikakyia) ground as enema, *Ziziphus mauritiana* Indian jujube as emeto-purgative, *Clausena anisata* Mosquito Plant boiled, *Uvaria chamae* (Akotompotsen), *Maytenus senegalensis* (Kumakuafo), and *Turraea heterophylla* (Ahunanyankwa) as extract. Powdered root of *Lantana camara* Wild Sage in milk; pounded root of *Zanthoxylum xanthoxyloides* (*Fagara zanthoxyloides*) Candle Wood with Guinea-grains and pepper as infusion; root of *Palisota hirsuta* (Nzahuara) with *Piper guineense* West African Pepper (Soro-wisa) as enema; root decoction of *Dichrostachys cinerea* (*glomerata*) Marabou Thorn or stem, *Acacia sieberiana* var. *villosa* (Kulgo), *Cassia sieberiana* African Laburnum, *Afzelia africana* (Papao) with pepper, *Manniophyton fulvum* (Hunhun) with leafy twigs, *Vitex doniana* Black Plum (Afua); *Harungana madagascariensis* (Okosoa) with bark and root infusion of *Fadogia agrestis* (Buruntirikwa).

Leaf infusion of *Psidium guajava* Guava, *Maerua crassifolia*, *Ocimum gratissimum* Tea Bush, (Nunum); *Carpolobia alba* (Afiafia), *C. lutea* (Otwewa) with twigs, *Lippia multiflora* Gambian Tea Bush (Saa-nunum); *Lasianthera africana*, *Aspilia africana* Haemorrhage Plant with white clay and *Entada africana* (Kaboya) as tonic. Leaf of *Antidesma venosum* (Mpepea), *Sesbania sesban* Egyptian Sesban, and *Lantana trifolia*; leaf decoction of *Hoslundia opposita* (Asifuaka) as purgative, *Alchornea cordifolia* Christmas Bush (Gyamma); *Pterygota macrocarpa* Pterygota (Kyereye); *Cathormion altissimum* (Abobonkakyere), *Machaerium lunatum* (*Drepanocarpus lunatus*) (Nkako), *Ficus asperifolia* Sandpaper Tree, soaked leaf of *Piper umbellatum* (Amuaha) as enema; pounded leaf and fruit of *Guiera senegalensis* boiled with natron taken internally; infusion of young leaves of *Vismia guineensis* (Kosowanini) as purgative and *Greenwayodendron oliveri* (*Polyalthia oliveri*) (Duabiri); concoction of crushed, leafy stems of *Glyphaea brevis* (Foto); hot concoction of leaf and root of *Cassia occidentalis* Negro Coffee; decoction of *Hygrophila auriculata* as remedy; decoction or infusion of leaf of *Solanum incanum* Egg Plant; young leaves of *Momordica foetida* (Sopropo) taken; juice of pressed leaves of *Hyptis suaveolens* Bush Tea-Bush along with lime-juice as drink; leaf of *Culcasia scandens* (Otwa-tekyirema); leaf of *Abrus precatorius* Prayer Beads as

8. POTIONS AND MEDICINES

sweetening agent; decoction of large, stipular bud sheaths of *Musanga cecropioides* Umbrella Tree (Dwumma) as drink or enema; pulped leafy twigs of *Pseudocedrela kotschyi* Dry-Zone Cedar; pressed leaf juice of *Spondias mombin* Hog Plum, Ashanti Plum warmed and taken internally and macerated *Parquetina nigrescens* (Aba-kamo) in palm-wine.

Bark decoction of *Nauclea diderrichii* (Kusia), *Picralima nitida* (Ekuama), *Anthocleista nobilis* Cabbage Palm (Hohoroho) in enemas and sitz-baths; *Trichilia monadelpha* (*heudelotii*) (Tanduro), *Crateva adansonii* (*religiosa*) (Chelum Punga), *Lonchocarpus laxiflorus* (Nalenga) with natron, *Bridelia stenocarpa* (*micrantha*) (Opam), *Dialium guineense* Velvet Tamarind (Asenamba) with spices, and *Cylicodiscus gabunensis* African Greenheart (Denya) as enemas. Inner bark of *Calpocalyx brevibracteatus* (Atrotre); liquor from steeped bark ash of *Acacia polyacantha* subsp. *campylacantha* African Catechu; bark of *A. kamerunensis* (*pennata*) (Oguaben) as enemas; bark extract of *Piliostigma thonningii* (Opitipata); charred bark of *Annona senegalensis* var. *senegalensis* Wild Custard Apple; bark of *Sesbania grandiflora*; bark infusion of *Ficus lutea* (*vogelii*) (Fonto) and *Lannea acida* (Kuntunkuri); weak bark infusion of *Bersama abyssinica* var. *paullinioides* (Duantu) with Sesame; dried, pounded bark of *Omphalocarpum ahia* (Duapompo) with white clay in water as drink and stem of *Gongronema latifolium* (Nsurogya).

Flowers of *Crossopteryx febrifuga* African Bark (Pakyisie); *Khaya senegalensis* Dry-Zone Mahogany, *Margaritaria discoidea* (*Phyllanthus discoideus*) (Pepea), and chewed flowers of *Marantochloa purpurea* (Sugugwa), *M. leucantha* (Sibere), and *M. ramosissimum* to relieve; latex of *Parquetina nigrescens* (Aba-kamo) as enemas; resinous latex of *Milicia excelsa* (*Chlorophora excelsa*) Iroko (Odum); ground seed of *Strychnos afzelii* (Duapepere) in soup; unpleasant-smelling pericarp of *Antrocaryon micraster* (Aprokuma) taken internally; the bark and leaves of *Kigelia africana* Sausage Tree (Nufuten) alone or with *Olax subscorpioidea* (Ahoohenedua), *Clerodendrum capitatum* (Tromen), *Carapa procera* Crabwood, and seeds of *Xylopia aethiopica* Ethiopian Pepper, Spice Tree (Hwenetia); leaves of *Alternanthera nodiflora* and *A. sessilis* as enema and unspecified parts of *Combretum molle* (Gburega), *Carica papaya* Pawpaw, *Adenia rumicifolia* var. *miegei* (*lobata*) (Peteha), *Tetracera alnifolia* (Akotopa), and *Uraria picta* (Heowe). *Solenostemon monostachys* (Sisiworodo) is rubbed on the of pregnant women to prevent child from having stomach ache.

The following plants are used for indigestion or dyspepsia (disturbed digestion): bitter infusion of tender stalks and young leaves of *Lippia multiflora* Gambian Tea Bush (Saa-nunum); leaf decoction of *Spilanthes filicaulis* Para or Brazil Cress as digestive tonic; leaves of *Glyphaea brevis* (Foto) cooked with palm oil; leaf-juice of *Leea guineensis* (Okatakyi); bark extract of *Piliostigma thonningii* (Opitipata); ashes of *Pistia stratiotes* Water Lettuce taken with food as stomachic; decoction of *Rhinacanthus virens* var. *virens* (Kwaduko), root decoction of *Cochlospermum tinctorium* (Kokrosabia), and dried roots and stem decoction of *Aristolochia indica* in small doses (Jain, 1968).

STOMACHIC (see Appetizer)

STOMATITIS (see Sore Gums/Sore Mouth)

STROKE (APOPLEXY OR PARALYSIS)

A wash with the dried leaves of *Clausena anisata* Mosquito Plant in water followed by a drink of the roots of *Paullinia pinnata* (Toa-ntini), *Caesalpinia bonduc* Bonduc (Oware-amba), and the stem bark of *Zanthoxylum xanthoxyloides* (*Fagara zanthoxyloides*) Candle Wood in alcohol (Moro, 1984–5). (See also Paralysis.)

STYPTIC (ARREST BLEEDING)

Sap of *Pycnanthus angolensis* African Nutmeg (Otie), *Harungana madagascariensis* (Okosoa), *Jatropha curcas* Physic Nut, and *Ongokea gore* (Bodwe); latex of *Conocarpus erectus* Button Wood; leaf-juice of *Carica papaya* Pawpaw, *Platostoma africanum* (Siresireke), *Mallotus oppositifolius* (Satadua), *Desmodium incanum* (*canum*), *Diospyros mespiliformis* West African Ebony, *Ficus exasperata* (Nyankyeren), *F. sagittifolia*, *Bertiera racemosa* (Mantannua), *Psychotria peduncularis* (*Cephaelis peduncularis*) (Kwesidua), *Holarrhena floribunda* False Rubber Tree (Sese); *Blumea aurita* var. *aurita* and the sap of the grass *Pennisetum subangustum* or ground leaf of *Pupalia lappacea* (Akukuaba) (Moro, 1984–5); fresh leaf of *Diospyros soubreana* (Otweto); fruit, bruised leaves, and flowers of the fresh plant of *Aspilia africana* Haemorrhage Plant; powdered bark of *Stereospermum acuminatissimum* (Tokwa-kufuo) and *Detarium senegalense* Tallow Tree; leafy stems of *Abrus precatorius* Prayer Beads and *Baphia nitida* Camwood (Odwen); juice of pods and bark of *Acacia nilotica* var. *tomentosa* and roots of *Paullinia pinnata* (Toa-ntini) with *Piper guineense* West African Pepper (Soro-wisa).

Root and inflorescence of *Imperata cylindrica* Lalang (Amico, 1977). Boulos (1983) recommends the flowers of Lalang for haemoptysis and epistaxis; root of *Uvaria chamae* (Akotompotsen) is for epistaxis (nose-bleeding), haematemensis (the vomiting of blood), haemoptysis (the spitting of blood from the bronchi, larynx, lungs, trachia), and similar ailments. Moro (1984–5) says that haematemensis is also cured by chewing the bark of *Waltheria indica* (Sawai). The bark of *Allophylus africanus* (Hokple) is for nose-bleeding. Others for general cases are the juice of fresh leaf of *Bidens pilosa* Spanish Needles, Bur Marigold; or young leaf of *Aframomum melegueta* Guinea-grains; the undeveloped leaves of *Borassus aethiopum* Fan Palm and powdered leaves of *Piliostigma thonningii* (Opitipata). The roots of *Waltheria indica* (Sawai), bark of *Newbouldia laevis* (Sasanemasa), and a decoction of *Aspilia africanus* Haemorrhage Plant are recommended for pulmonary or internal haemorrhage; so are the leaves of *Lawsonia inermis* Henna (Jain, 1968) and the dried flowers of *Opuntia* species (Evans, 1983). Roots of *Blighia unijugata*

8. POTIONS AND MEDICINES

(Akyibiri) are cooked in palm soup to reduce the flow of blood in childbirth.

SUDORIFIC (see Diaphoretic)

SWELLINGS

Pounded leaves of *Datura metel* Hairy Thorn-Apple, Metel; *Guiera senegalensis, Adansonia digitata* Baobab, *Cassia occidentalis* Negro Coffee, *Premna quadrifolia* (Atantaba), *Cardiospermum grandiflorum* and *C. halicacabum* Heart Seed or Balloon Vine, *Ricinus communis* Castor Oil Plant, *Dissotis rotundifolia* (Adowaalie), *Tamarindus indica* Indian Tamarind, *Pulicaria crispa, Dichrostachys cinerea* (*glomerata*) Marabou Thorn, *Myrianthus arboreus* (Anyankoma), *Alstonia boonei* (Sinduro), and bruised, young leaves of *Aframomum melegueta* Guinea-grains with those of *Vismia guineensis* for inflammation of throat and tonsils. Leaves of *Commelina diffusa* pounded with seeds of *Leea guineensis* (Okatakyi) and *Piper nigrum* for swellings of the groin; whole plant of *Polygonum senegalense* pounded with natron as rub; scraped pulp of *Cucurbita pepo* Pumpkin, Vegetable Marrow and *C. maxima* Squash Gourd applied as poultice and pulp of young fruit of *Luffa cylindrica* Loofah Gourd, Vegetable Sponge applied as poultice. Paste of young shoots of *Albizia adianthifolia* (Pampena) with a snake-like gourd; leaf and crushed seed kernels of *Gossypium arboreum* Cotton as poultice; dried leaf of *Phytolacca dodecandra* (Ahoro) as bandage; leaf infusion of *Uvaria chamae* (Akotompotsen) as lotion and hot poultice or fomentation of leaves of *Trichodesma africanum* applied.

Pounded bark of *Gaertnera cooperi* as rub; powdered bark of *Sterculia rhinopetala* Sterculia Brown (Wawabimma) with oil; extract of bark and fruit of *Drypetes ivorensis*; ground bark extract of *Margaritaria discoidea* (*Phyllanthus discoideus*) (Pepea); bark of *Tetrorchidium didymostemon* (Aboagyedua) soaked in water or rum as purgative; pulverized bark of *Dichapetalum toxicarium* West African Ratbane (Ekum-nkura) as rub; bark decoction of *Erythrophleum suaveolens* (*guineense*) Ordeal Tree (Potrodom) as wash; macerated bark of *Pericopsis laxiflora* (*Afrormosia laxiflora*) Satinwood; bark ashes of *Milicia excelsa* (*Chlorophora excelsa*) Iroko (Odum) with palm oil; ground bark of *Treculia africana* var. *africana* African Breadfruit with oil and other plant parts and bark infusion of *Crossopteryx febrifuga* African Bark (Pakyisie).

Dried, ground roots of *Crateva adansonii* (*religiosa*) (Chelum Punga); pounded root of *Moringa oleifera* Horse-Radish Tree, Oil of Ben Tree with salt as poultice; root decoction of *Cochlospermum tinctorium* (Kokrosabia) as liniment; root-bark of *Entada abyssinica* (Sankasaa) as massage; pulverized root-bark of *Zanthoxylum xanthoxyloides* (*Fagara zanthoxyloides*) Candle Wood as poultice and root extract of *Allophylus africanus* (Hokple) taken internally. Ground, swollen roots of *Asparagus africanus* (Adedende) and *A. flagellaris* mixed with leaves of *Ricinus communis* Castor Oil Plant applied as poultice; root decoction of *Ziziphus mucronata* Buffalo Thorn taken internally and leaf paste applied externally for inflammation of the gland (*adenites*) and pounded roots of *Vernonia biafrae* (Hu) as poultice applied to the eyeballs for iritis.

Application of the heated stem of *Penianthus zenkeri;* swollen stems of *Cissus aralioides* (Asirimu); latex or scorched, pounded leaves of *Calotropis procera* Sodom Apple; latex of *Ficus asperifolia* Sandpaper Tree applied; pulverized fruit of *Combretum micranthum* (Landaga) with oil as ointment; pounded fruit and oil of *Momordica charantia* African Cucumber as dressing; crushed leaf and seeds of *Datura stramonium* Apple of Peru mixed with oil applied to allay pains; decoction of *Schwenckia americana* (Agyingyinsu) taken, also plant pounded with natron and water and rubbed locally with shea butter. Unspecified parts of *Monodora myristica* (Ayerew-amba) with Guinea-grains, *Machaerium lunatum* (*Drepanocarpus lunatus*) (Nkako), *Eriosema griseum* (Trindobaga), and *Byrsocarpus coccineus* (Awennade). A compress of root-bark or leaf of *Rauvolfia vomitoria* (Kakapenpen) with sap of plantain stem and sometimes lime-juice and spices is applied as counter-irritant.

SYPHILIS (see Venereal Diseases)

TAENIFUGE (see Anthelmintic)

TEETHING (see Dentition)

TETANUS

Leaf of *Cassia occidentalis* Negro Coffee as eyewash; root and leaf of *Paullinia pinnata* (Toa-ntini) mixed with shea butter as drink; decoction of root and leaf of *Ehretia cymosa* (Okosua) and crushed seeds of *Physostigma venenosum* Calabar Bean given hypodermically in acute cases. *Solanum aethiopicum* is used for cases after abortion.

TOOTHACHE

Bark decoction of *Allanblackia parvifolia* (*floribunda*) Tallow Tree (Sonkyi); *Vitex simplicifolia* (Abisa) as lotion, *Dialium guineense* Velvet Tamarind (Asenamba) with spices as mouthwash and gargle; *Erythrophleum africanum* African Black Wood (Bupunga) and *Tieghemella heckelii* (Baku) each as mouthwash and hot bark decoction of *Parkia biglobosa* (Duaga) as mouthwash and to steam mouth and throat. Bark infusion of *Musanga cecropioides* Umbrella Tree (Dwumma) as gargle; *Annona senegalensis* var. *senegalensis* Wild Custard Apple and *Lophira lanceolata* each as mouthwash, *Anacardium occidentale* Cashew Nut and *Mangifera indica* Mango each with leaf lotion as mouthwash to relieve, *Trema orientalis* (*guineensis*) (Sesea) with leaf and bark of *Pycnanthus angolensis* African Nutmeg (Otie). Inner bark of *Memecylon afzelii* (Otwe-ani) as poultice; powdered bark of *Anogeissus leiocarpus* (Sakanee) with that of *Terminalia* species rubbed on the gum to relieve; bark extract of *Harungana madagascariensis* (Okosoa); warm infusion of bark or leaf of *Piliostigma thonningii* (Opitipata) to relieve; bark of

8. POTIONS AND MEDICINES

193

Smeathmannia pubescens (Turunnua) and that of *Piptadeniastrum africanum* (Danhoma) as gargle; aromatic bark and leaf of *Zanthoxylum viride* (*Fagara viridis*) (Oyaanini) locally applied and outer part of stem of *Cocos nucifera* Coconut Palm applied.

Chewed root of *Waltheria indica* (Sawai) (Moro, 1984–5), *Monodora myristica* (Ayerew-amba), *Cnestis ferruginea* (Akitase), and root infusion of *Erythrina senegalensis* Coral Flowers; scraped root of *Newbouldia laevis* (Sasanemasa) with pepper for carious tooth; root or root-bark of *Nauclea latifolia* African Peach (Sukisia) as chewsticks; piece of root of *Fadogia agrestis* (Buruntirikwa) sucked and juice swallowed; pulverized root of *Zanthoxylum xanthoxyloides* (*Fagara zanthoxyloides*) Candle Wood with hot spices on carious tooth; decoction of root and leaf of *Clausena anisata* Mosquito Plant as mouthwash; peppermint-like root-wood of *Z. gilletii* (*F. macrophylla*) (Okuo) rubbed on gums to ease; powdered root of *Ximenia americana* Wild Olive rubbed on gums; chewed root of *Erythrina mildbraedii* (Nfona) or rhizome of *Zingiber officinale* Ginger (Kakadro) to relieve; root decoction of *Prosopis africana* (Sanga); powdered root or root-bark of *Acacia kamerunensis* (*pennata*) (Oguaben) in palm-wine and soaked on wool as plug for carious tooth, and powdered root of *Terminalia avicennioides* (Petni) rubbed on gum.

Pulverized leaves of *Nicotiana tabacum* and *N. rustica* Tobacco on carious tooth; dried leaves of *Palisota hirsuta* (Nzahuara) smoked like tobacco; soaked leaf of *Piper umbellatum* (Amuaha); leaf concoction of *Psychotria articulata*; powdered leaf or leaf decoction of *Tetracera potatoria* (Twihama); strongly scented leaf of *Psidium guajava* Guava chewed as relieve; crushed leaf of *Hymenocardia acida* (Sabrakyie) with lemon; decoction of leaf and bark of *Daniellia oliveri* African Copaiba Balsam Tree and mouthwash; hot leaf decoction of *Ficus glumosa* var. *glaberrima* (Galinziela) as mouthwash; warm leaf decoction of *Scleria boivinii*, *S. naumanniana*, and *S. depressa* as mouthwash to relieve; leaf decoction of *Maytenus senegalensis* (Kumakuafo) as mouthwash and gargle and *Heisteria parvifolia* (Sikakyia), *Clerodendrum capitatum* (Tromen) with inflorescence, and *Premna hispida* (Aunni) as mouthwash and gargle. Leaf-juice of *Portulaca oleracea* Purslane, Pigweed; *Spilanthes filicaulis* Para or Brazil Cress chewed as relief and powdered leaf placed in carious tooth; leaf juice of *Leea guineensis* (Okatakyi) as enema, *Motandra guineensis* (Amamfohae) as mouthwash and gum massage; leaf infusion of *Paullinia pinnata* (Toa-ntini); leaf of *Dodonaea viscosa* Switch Sorrel (Fomitsi) and pounded leaf of *Vernonia guineensis* applied to face.

Seeds of *Aframomum melegueta* Melegueta chewed with cola to relieve; roasted pulverized seed and latex of *Adansonia digitata* Baobab; latex of *Milicia excelsa* (*Chlorophora excelsa*) Iroko (Odum); *Calotropis procera* Sodom Apple, *Voacanga africana* (Ofuruma), and *Ficus vogeliana* (Opanto) each soaked in wool and applied to carious tooth; decoction of fruit of *Acacia nilotica* var. *tomentosa* with ginger as mouthwash; fruit of *Ziziphus mucronata* Buffalo Thorn chewed; unspecified parts of *Usteria guineensis* (Kwaemeko) and *Caesalpinia bonduc* Bonduc

(Oware-amba); and the caustic, toxic sap of male *Treculia africana* var. *africana* African Breadfruit (Brebretim) applied on cotton wool to carious tooth.

THROAT TROUBLES AND AFFECTIONS (see Bronchial Troubles)

THRUSH (INFESTATION OF THE TONGUE AND MOUTH LINING IN CHILDREN)

Leaf-juice of *Jatropha gossypiifolia* Physic Nut applied; bark of *Bridelia ferruginea* (Opam-fufuo) or *B. stenocarpa* (*micrantha*) (Opam) in preparations; root-pith of *Alchornea cordifolia* Christmas Bush (Gyamma); fruit-juice of *Citrus aurantiifolia* Lime as mouthwash and gargle; whole plant of *Euphorbia hirta* Australian Asthma Herb (Ayensu, 1978) and seed fat and probably leaf juice of *Pycnanthus angolensis* African Nutmeg (Otie).

TRACHOMA (see Eye Diseases)

TRYPANOSOMIASIS (see Sleeping Sickness)

TUBERCULOSUS (CONSUMPTION)

Whole plant of *Dissotis rotundifolia* (Adowa-alie) (Ayensu, 1978) and *Euphorbia hirta* Australian Asthma Herb (Ayensu, 1978); leaf decoction of *Centella asiatica* Indian Water Navelwort for certain cases (Jain, 1968); decoction of leafy tips of *Microglossa afzelii* (Pofiri) and the fruits of *Detarium senegalense* Tallow Tree as massage for cases of the spine. Bark decoction of *Nauclea latifolia* African Peach (Sukisia) alone or mixed with spices to relieve cough; the oil from *Allium sativum* Garlic known as 'sulphur oil' used in treatment and the bark of *Macaranga hurifolia* (Kpazina) chewed and swallowed or the powdered stem in palm wine as drink for patients; treatment being completed with a rub of the leaf on the back and chest.

TUMOURS

The latex of *Carica papaya* Pawpaw, *Ficus asperifolia* Sandpaper Tree, and *Milicia excelsa* (*Chlorophora excelsa*) Iroko (Odum); bruised leaf of *Hedranthera barteri*, *Byrsocarpus coccineus* (Awennade), *Ziziphus mucronata* Buffalo Thorn, and *Myrianthus arboreus* (Anyankoma); leaf decoction of *Dialium guineense* Velvet Tamarind, (Asenamba); boiled fruit of *Microdesmis puberula* (Fema); powdered bark of *Garcinia kola* Bitter Cola (Tweapea) and *Pachypodanthium staudtii* (Fale) with other ingredients.

ULCERS

Bark decoction of *Mammea africana* African Mammy Apple (Bompagya); *Harungana madagascariensis* (Okosoa), *Anogeissus leiocarpus* (Sakanee) to bathe, *Erythrophleum suaveolens* (*guineense*) Ordeal Tree (Potrodom) to bathe; *Acacia*

8. POTIONS AND MEDICINES

nilotica var. *tomentosa* as wash; *Khaya senegalensis* Dry-Zone Mahogany as lotion and *Pseudocedrela kotschyi* Dry-Zone Cedar as wash. Bark extract of *Enantia polycarpa* African Yellow Wood (Duasika) applied; powdered bark of *Terminalia ivorensis* (Emire) sprinkled on, *Trichilia monadelpha* (*heudelotii*) (Tanduro), *Maytenus senegalensis* (Kumakuafo), and *Hymenocardia acida* (Sabrakyie) or decoction as antiseptic; pulped bark of *Parkia biglobosa* (Duaga) with lemon-juice applied; bark infusion of *Ficus asperifolia* Sandpaper Tree as wash; pulverized bark and root of *Ximenia americana* Wild Olive; pulverized, dried, bitter, bark of *Manilkara multinervis* subsp. *lacera* African Pearwood; bark of *Lannea kerstingii* (Kobewu) with that of *L. velutina* (Sinsa) applied; bark of *Alstonia boonei* (Sinduro); crushed inner bark of *Kigelia acutifolia* and bark of *Spathodea campanulata* African Tulip Tree (Kokoanisua) as dressing.

Leaf of *Glyphaea brevis* (Foto) applied, *Grewia mollis* (Yualega), *Alchornea cordifolia* Christmas Bush (Gyamma) powdered; *Cassia obtusifolia* (*tora*) Foetid Cassia, *C. absus* Four-Leaved Henna, Black Grain pulverized as dressing; *Guibourtia copallifera* when young, *Lonchocarpus cyanescens* West African or Yoruba Indigo as poultice, *L. laxiflorus* (Nalenga), *Piliostigma thonningii* (Opitipata) powdered, *Blighia sapida* Akee Apple (Akye) crushed with salt, and *Tylophora conspicua* pounded and applied. Leaf of *Croton lobatus* (Akonansa) applied, *Torenia thouarsii* ground up with hot lime and applied after dressing; leaf pulp of *Sanseviera liberica* African Bowstring Hemp with other herbs or the expressed juice alone as ointment; crushed leaf of *Bryophyllum pinnatum* Resurrection Plant and that of *Ageratum conyzoides* Billy-goat Weed (Efumomoe) applied; ground whole plant of *Centella asiatica* Indian Water Navelwort (Evans, 1983); ground leaf of *Ananas comosus* Pineapple with copper or brass filings and palm oil applied and ashes of stem, leaf, and fruit-skin of *Musa paradisiaca* Plantain as dressing powder or stem core as lint; leaf infusion of *Lantana trifolia* as wash, *Rytigynia canthioides*, and powdered, young, green shoots of *Anthocleista nobilis* Cabbage Palm; strong decoction of fresh leaves of *Jasminum dichotomum* (Krampa) as lotion; leaf-juice of *Strophanthus gratus* (Omaatwa); leaf-ashes of *Gilbertiodendron limba* (Tetekon) with water applied; powdered leaf of *Mallotus oppositifolius* (Satadua) with shea butter as ointment; decoction of leafy stems of *Erythrococca anomala* as wash; powdered, dried, leaf and bark of *Copaifera salikounda* Bubinga (Entedua) with baked and powdered clay applied.

Powdered root of *Terminalia avicennioides* (Petni) sprinkled on; root and root-bark of *Ziziphus mauritiana* Indian Jujube powdered and mixed with oil; root-pulp infusion of *Strychnos spinosa* Kaffir Orange (Akankoa) taken internally as antiseptic; root bark and stem tips of *Strophanthus hispidus* Arrow Poison as drink, an enema, or in local application; charred roots and root-bark of *Caloptropis procera* Sodom Apple applied as ointment; scraped pulp of *Manihot esculenta* Cassava as poultice applied and boiled roots of *Boerhavia diffusa* and *B. repens* Hogweed as poultice applied.

Fruit ashes of *Morinda morindoides* as dressing; *Allium sativum* Galic pounded with natron and applied as dressing; oil from crushed fruits of *Panda oleosa*

(Kokroboba) with fat applied; juice of the roasted fruits of *Citrus aurantiifolia* Lime, *C. aurantium* Sour Orange, and *C. limon* Lemon applied and powdered pods of *Amblygonocarpus andongensis* (Nanzili) as dressing applied. The latex of *Adenium obesum* and astringent sap of *Pterocarpus erinaceus* Senegal Rose Wood Tree African Kino (Senya) applied. *Dracaena bicolor* (Gblieku) is poisonous and causes ulceration of the stomach if taken.

URETHRAL DISCHARGES AND TROUBLES (see Venereal Diseases)

URINARY COMPLAINTS (see Venereal Diseases)

VENEREAL DISEASES (SEXUALLY TRANSMITTED DISEASES)

Gonorrhoea

This is rated the fourth most common communicable disease in West Africa; together with other sexually transmitted diseases, it is identified as one of the major causes of menstrual troubles and infertility. Professor Oshoba, a Nigerian microbiologist, reports that 2 in every 20 women in West Africa suffer from gonococcal infections, while between 15 and 30 per cent of male adults are also affected.

Herbal treatment includes root or root-bark decoction of *Combretum mucronatum* (*smeathmannii*) with pepper, *Bridelia ferruginea* (Opam-fufuo) as diuretic, *Microdesmis puberula* (Fema), *Manniophyton fulvum* (Hunhun) with leafy twigs, *Afzelia africana* (Papao) with pepper, *Daniellia ogea* Gum Copal Tree, *Ostryderris stuhlmannii* (Kaman Godui), *Ficus elegans, Jatropha curcas* Physic Nut with leaves and natron, *Maytenus senegalensis* (Kumakuafo) or powdered root with salt, *Solanum incanum* Egg Plant (in small quantities) with Tamarind, natron and Bilma salt as drink, *Conocarpus erectus* Button Wood (Koka); *Paullinia pinnata* (Toa-ntini), *Leea guineensis* (Okatakyi) with pepper, *Lannea acida* (Kuntunkuri), *Citrus limon* Lemon, *Strophanthus gratus* (Omaatwa), *Fadogia agrestis* (Buruntirikwa), *Feretia apodanthera* (Bitinamusa), and *Vernonia amygdalina* Bitter Leaf (Bonwen) or *V. colorata* with that of *Rauvolfia vomitoria* (Kakapenpen). Root or root-bark infusion of *Moringa oleifera* Horse-Radish Tree, Oil of Ben Tree; *Securidaca longepedunculata* (Kpaliga), *Cassia sieberiana* African Laburnum as diuretic and *Mangifera indica* Mango. Root or powdered root taken internally of *Piper guineense* West African Pepper (Soro-wisa); *Saba florida* (Akontoma) with ingredients, and *Morinda lucida* Brimstone Tree. Root-juice of *Tetracera alnifolia* (Akotopa); steeped roots of *Cassia occidentalis* Negro Coffee and root-bark of *Zanthoxylum viride* (*Fagara viridis*) (Oyaanini) as enemas. Pounded roots of *Palisota hirsuta* (Nzahuara) taken with lime; corms of *Gladiolus daleni* (*psittacinus*), *G. klattianus, G. gregarius,* and *G. unguiculatus* Sword Lily, Corn-flag as purgative with *Colocynthis lanatus* Water Melon and onions as ingredients.

8. POTIONS AND MEDICINES

Decoction of cabbage or young leaf of *Elaeis guineensis* Oil Palm; whole plant decoction of *Euphorbia hirta* Australian Asthma Herb and for other urinogenital complaints (Jain, 1968); leaf decoction of *Pterygota macrocarpa* Pterygota, (Kyereye); *Sterculia tragacantha* African Tragacanth (Sofo); *Hibiscus rostellatus* with leafy twigs, *Macaranga heterophylla* (Opamkokoo), *Cassia alata* Ringworm Shrub with salt as purgative either as drink or enema, *Spondias mombin* Hog Plum, Ashanti Plum; *Spathodea campanulata* African Tulip (Kokoanisua); *Lankesteria elegans, Borreria stachydea* (Barungwini), decoction of twigs of *Capparis polymorpha* (Sansangwa), and decoction or fresh juice of *Ipomoea involucrata*. Cold infusion of *Scoparia dulcis* Sweetbroom Weed as repeated drink or *Stachytarpheta cayennensis* Brazilian Tea, Bastard Vervain; leaf infusion of *Cassia podocarpa* (Nsuduru) and *Parquetina nigrescens* (Aba-kamo); leaf extract of *Gouania longipetala* (Homabiri); leaf and bark of *Mitragyna ciliata* Abura, Poplar, (Subaha); pounded leaf of *Mussaenda afzelii* with lime; leaf of *Microglossa pyrifolia* (Asommerewa) or *Piper umbellatum* (Amuaha) as enema; young shoots of *Harungana madagascariensis* (Okosoa) chewed with cola; leaf of *Macaranga barteri* (Gyapam); leaf extract of *Sclerocarpus africanus* boiled with mutton and taken internally; pulverized, dried plant of *Trianthema pentandra* Horse Purslane taken with millet beer; infusion of *Cissus quadrangularis* Edible-Stemmed Vine, (Kotokoli) as purge, and water filtered through charcoal of *Zea mays* Maize sticks as prescription.

Bark decoction of *Ricinodendron heudelotii* (Wamma), *Bridelia atroviridis* (Asaraba) (Mandango and Bandole, 1985), *Dichrostachys cinerea (glomerata)* Marabou Thorn as wash, *Tetrapleura tetraptera* (Prekese) as beverage and enema, *Trichilia monadelpha (heudelotii)* (Tanduro) with pulp, *Anthocleista nobilis* Cabbage Palm as drink or bath or vapour bath, and *Mitragyna inermis* (Kukyafie) with leaves and sometimes roots and natron. Bark infusion of *Maesobotrya barteri* var. *sparsiflora* (Apotrewa), *Araliopsis tabouensis* (Emiahile), *Strychnos aculeata* with fruit of *Piper guineense* West African Pepper (Soro-wisa) in enema and liniment; *Nauclea diderrichii* (Kusia) and *Kigelia africana* Sausage Tree (Nufuten) boiled with red natron and Guinea-corn flower. Hot bark of *Petersianthus macrocarpus* (*Combretodendron macrocarpum*) Stinkwood Tree (Esia) applied; bark pulp of *Trichilia prieuriana* (Kakadikro) as enema; macerated bark of *Strophanthus preussii* (Dietwa) and *S. sarmentosus* (Adwokuma) as enema; bark of *Gardenia erubescens* (Dasuli) with others and bark of *Crossopteryx febrifuga* African Bark (Pakyisie). Unspecified part of *Microglossa afzelii* (Pofiri) as diuretic, *Phyllanthus muellerianus* (Potopoleboblo), *Combretum molle* (Gburega) (also for anuria), *Carica papaya* Pawpaw, *Trema orientalis (guineensis)* (Sesea), and *Alstonia boonei* (Sinduro). Decoction of *Merremia tridentata* subsp. *angustifolia* with natron, *Boerhavia diffusa* Hogweed as diuretic (Jain, 1968), *Hoslundia opposita* (Asifuaka) as purgative, and fruit of *Cyrtosperma senegalense* as ingredient in remedies; vapour of the tuberous roots of *Smilax kraussiana* (Kokora) (Mandago and Bandole, 1985); gum of *Daniellia oliveri* African Copaiba Balsam Tree with warm water; latex of young shoots of *Ficus capensis* (Nwadua) as drink; latex or gum of

Garcinia kola Bitter Cola (Tweapea) taken internally; fruit of *Hedranthera barteri*; fruit of *Colocynthis lanatus* (*vulgaris*) Water Melon cut and boiled with onions and other ingredients as decoction (also for leucorrhoea in women) and crushed *Secamone afzelii* (Kotohume) with palm nuts used in cooking patients' food. *Nymphaea* species Water Lily is also recommended for leucorrhoea (Evans, 1983).

Syphilis (including syphilitic sores or Chancre)

Bark of *Gardenia ternifolia* (Peteprebi), *Mammea africana* African Mammy Apple, (Bompagya); and *Trichilia emetica* subsp. *suberosa* (*roka*) (Kisiga); powdered bark of *Zanthoxylum gilletii* (*Fagara macrophylla*) (Okuo) with *Piper*, dried and eaten for cases of the throat; powdered bark of *Cola gigantea* var. *glabrescens* (Watapuo); bark of *Kigelia africana* Sausage Tree (Nufuten) boiled with natron and Guinea-corn; bark decoction of *Melia azedarach* Persian Lilac, Bead Tree and *Anogeissus leiocarpus* (Sakanee) as wash; bark of *Khaya senegalensis* Dry-Zone Mahogany sometimes with potash and decoction of *Hibiscus asper* as part treatment.

Dried roots of *Aristolochia albida* with natron; fresh roots of *Securidaca longepedunculata* (Kpaliga) with other ingredients as drink; powdered roots of *Terminalia glaucescens* (Ongo) with ingredients; root of *Grewia villosa* as ingredient; root decoction of *Acacia polyacantha* subsp. *campylacantha* African Catechu (Gorpila) with pulp of Baobab fruit taken internally; root decoction of *Newbouldia laevis* (Sasanemasa) and *Dichrostachys cinerea* (*glomerata*) Marabou Thorn; root of *Pseudarthria hookeri* var. *argrophylla* (Kwaheni), *Asparagus africanus* (Adedende), *A. flagellaris* and boiled root of *Solanum incanum* Egg Plant with sour milk and juice of fruit, sometimes with pounded leaves rubbed on the hands for infective eruptions. Powdered root of *Zanthoxylum xanthoxyloides* (*Fagara zanthoxyloides*) Candle Wood and *Eremospatha macrocarpa;* root decoction of *Boswellia dalzielii* Frankincense Tree (Kabona) with *Hibiscus sabdariffa* Roselle as drink; root decoction of *Feretia apodanthera* (Bitinamusa) as part of prescription for prevention or cure, and exposure of body (under clothing) to the vapour of boiled, tuberous roots of *Smilax kraussiana* (Kokora) (Mandango & Bandole, 1985).

Dried, pulverized leaf of *Strychnos spinosa* Kaffir Orange (Akankoa) as dressing and *Cassia absus* Four-Leaved Henna, Black Grain applied; leaf decoction of *Ziziphus mucronata* Buffalo Thorn; young shoots of *Morus mesozygia* (Wonton) as nasal drops; pulped leaf of *Desmodium velutinum* (Koheni-koko) as plaster; an infusion of *Cissus quadrangularis* Edible-Stemmed Vine (Kotokoli) as purge; *Portulaca oleracea* Purslane, Pigweed as ingredient in prescription; pounded *Polygonum senegalense* with natron applied to syphilitic sores; *Heliotropium ovalifolium* as a drastic purge and for local application to syphilitic ulcers; leaves of *Leonotis nepetifolia* var. *nepetifolia* or *L. nepetifolia* var. *africana* (Nyeddo) pounded with natron applied locally for swellings and ulcers supposed to be of syphilitic origin, and woody twigs of *Sterculia tragacantha* African

8. POTIONS AND MEDICINES   199

Tragacanth (Sofo) for chronic cases. A copious dosage of fermented liquid prepared from the pounded and macerated roots of *Calotropis procera* Sodom Apple to induce vomiting and diarrhoea to expel 'spawns' of syphilis; oil of *Carapa procera* Crabwood in small doses as expectorant; pulverized pods of *Acacia nilotica* var. *tomentosa*; pulverized fruits of *Combretum micranthum* (Landaga) with oil as ointment; bitter fruits of *Detarium senegalense* Tallow Tree; flowers of *Ipomoea asarifolia* boiled with beans and taken as remedy; essence of seeds of *Caloncoba glauca*; sap from the trunk of *Mangifera indica* Mango and sap of *Euphorbia laterifolia* (Kamfobarima) as a drastic purge; unspecified parts of *Okoubaka aubrevillei* (Odee), *Microdesmis puberula* (Fema), *Mallotus oppositifolius* (Satadua), *Chrozophora senegalensis* and *Petersianthus macrocarpus* (*Combretodendron macrocarpum*) Stinkwood Tree (Esia). *Waltheria indica* (Sawai) and *Bridelia ferruginea* (Opam-fufuo) are supposed to afford immunity to syphilis. The bark of the roots and stem tips of *Strophanthus hispidus* Arrow Poison are used as drink, an enema, or in local applications for serious cases of syphilis, bony syphilis, and heredo-syphilis.

Other Venereal diseases

Apart from gonorrhoea and syphilis there are about 23 other kinds of sexually transmitted diseases. These include urethral and urinary troubles and complaints, urethral discharges, genito-urinary troubles, urethral stricture, and venereal sores. Herbal treatment includes bark of *Chasmanthera dependens*, *Penianthus zenkeri* as enemas, *Syzygium guineense* var. *guineense* (Sunya) with root and leaf, *Acacia sieberiana* var. *villosa* (Kulgo) with root, and *Picralima nitida* (Ekuama). Bark decoction of *Mezoneuron benthamianus* (Akoobowerew) with root and leaf, *Pericopsis laxiflora* (*Afrormosia laxiflora*) Satinwood as wash, *Olax subscorpioidea* (Ahoohenedua) applied to cankers, *Zanthoxylum gilletii* (*Fagara macrophylla*) (Okuo) or macerated in palm-wine, and *Stereospermum kunthianum* (Sonontokwakofo) with natron.

Root decoction of *Oncoba spinosa* Snuff-box Tree (Asratoa) with leaf; *Ziziphus mauritiana* Indian Jujube, *Z. mucronata* Buffalo Thorn or a decoction of the leaves, *Nauclea latifolia* African Peach (Sukisia) in enemas as well as taken internally and *Cochlospermum tinctorium* (Kokrosabia) or infusion with other herbs. Rhizome decoction of *Cynodon dactylon* Bermuda or Bahama Grass (Boulos, 1983), root of *Calliandra portoricensis* (Nkabe) with pepper, *Erythrina senegalensis* Coral Flowers, *Ximenia americana* Wild Olive, *Adenium obesum* and ingredients boiled with Guinea-corn pap and blacksmith's slag as drink, *Rauvolfia vomitoria* (Kakapenpen) with the leaves, *Strophanthus hispidus* Arrow Poison in spirit, *Securinega virosa* (Nkanaa) with leaves, *Monodora brevipes* (Abotokuradua) used internally for urethral stricture, boiled roots of *Sansevieria liberica* African Bowstring Hemp, and root pulp or leafy tops of *Harrisonia abyssinica* (Penku) in hot water as drink or in enemas.

Leaf infusion of *Abutilon guineense* (Odonno-bea) with spices, *Rhodognaphalon* brevicuspe (*Bombax brevicuspe*) (Onyinakoben) with bark, and *Desmodium adscendens* var. *adscendens* (Akwamfanu) as wash. Leaf of *D. incanum* (*canum*), *Lippia multiflora* Gambian Tea Bush (Saa-nunum), *Byrsocarpus coccineus* (Awennade); macerated leaf of *Memecylon afzelii* (Otwe-ani) in water applied and leaf pulp of *Clerodendrum umbellatum* (Niabiri) as enema. Leaf decoction of *Machaerium lunatum* (*Drepanocarpus lunatus*) (Nkako), *Elytraria marginata, Lankesteria brevior, Borreria stachydea* (Barungwini), and whole plant of *Portulaca oleracea* Purslane, Pigweed with aromatic Labiate. Whole plant of *Corchorus aestuans* and the roots and *Pterocarpus erinaceus* Senegal Rose Wood Tree, African Kino as injection; whole plant of *Phyllanthus fraternus* subsp. *togoensis* (*niruri*) (Ombatoatshi), an infusion of *Euphorbia convolvuloides* taken orally and by rectal injection as an aperiant; an infusion of stigmas of *Zea mays* Maize and decoction of whole plant of *Indigofera hirsuta* Hairy Indigo (Moro, 1984–5) which is reported to be most effective in curing all venereal diseases. Further investigation into this plant is recommended in view of the high incidence of such diseases in West Africa.

Stem and root infusion of *Nymphaea lotus* Water Lily; root of *Momordica charantia* African Cucumber as ingredient and gum of *Sterculia setigera* (Pumpungo) and *Boswellia dalzielii* Frankincense Tree, (Kabona) as ingredients. Decoction of pods of *Acacia nilotica* var. *tomentosa*; seeds of *Abrus precatorius* Prayer Beads; fruits of *Lycopersicon esculentum* (*lycopersicum*) Tomato; boiled, immature fruits of *Ananas comosus* Pineapple; powdered leaves of *Baphia nitida* Camwood (Odwen) in food or palm-wine; thickened rhizomes of *Cissus cornifolia* (Sintanatora) with natron; stem decoction of *C. populnea* (Agyako) with natron and stems of *Alchornea cordifolia* Christmas Bush (Gyamma); steamy brew of leaf of *Citrus aurantiifolia* Lime under a blanket as vapour bath and an unspecified part of *Milicia excelsa* (*Chlorophora excelsa*) Iroko (Odum) as ingredient in mixture for hip-bath and *Maesopsis eminii* (Onwamdua) (Mandango and Bandole, 1985).

VERMIFUGE (see Anthelmintic)

VERTIGO

Leaf-juice of *Ficus vallis-choudae* (Aloma-bli) as drink or head ointment; juice of crushed leaf of *Hoslundia opposita* (Asifuaka) as eye drops; root-bark extract of *Rauvolfia vomitoria* (Kakapenpen) in eyes or infusion of young leaves rubbed on the face; cold infusion of seeds of *Copaifera salikounda* Bubinga (Entedua) as remedy; old leaves of *Leea guineensis* (Okatakyi) roasted and applied to head and pulp of crushed fruit of *Rutidea parviflora* (Akitankrua-nini) in boiled water.

VESICANT (RAISING BLISTERS)

The leaf-juice of *Clematis grandiflora, C. hirsuta*, and that of *Plumbago zeylanica* Ceylon Leadwort (Opapohwea) with lemon applied; the root of *P. indica* Red

8. POTIONS AND MEDICINES

Plumbago applied in leprous treatment and the root-latex of *Sapium grahamii* (Pampiga); pounded seeds of *Caesalpinia bonduc* Bonduc (Oware-amba); the seed oil of *Jatropha curcas* Physic Nut; the latex of *Euphorbia lateriflora* (Kamfobarima) and the shell-oil of the roasted seeds of *Anacardium occidentale* Cashew Nut (Atea).

VOMITIVE (see Emetic)

WALK EARLY, ENABLE CHILDREN TO

Decoction of roots or leaves of *Stylosanthes fruticosa* (*mucronata*) as daily bath for infants; leaf of *Gongronema latifolium* (Nsurogya) rubbed on joints; whole plant of *Cissus quadrangularis* Edible-Stemmed Vine (Kotokoli) with spices as enema; ground leaf of *Sabicea calycina* (Anansentoromahama) applied to limbs and a preparation of the fruits of *Cnestis ferruginea* (Akitase) and allied plants given to weak children. Leaf of *Launaea taraxacifolia* (*Lactuca taraxacifolia*) Wild Lettuce rubbed on limbs of children and decoction of *Sphaeranthus senegalensis* (also to strengthen them).

WHITLOW

Pounded leafy stems of *Opuntia* species Prickly Pear (Nkantonsoe) mixed with chewed palm kernels in a banana leaf and bandaged on the swelling to bring to a head; bark of *Sterculia tragacantha* African Tragacanth, (Sofo) mixed with palm kernels, and leaf of *Piper umbellatum* (Amuaha) with palm kernels to bring to a head. Crushed whole plant of *Portulaca oleracea* Purslane, Pigweed applied; fruit of *Momordica charantia* African Cucumber applied to festering finger and ground root of *Annona senegalensis* var. *senegalensis* Wild Custard Apple applied as paste (Moro, 1984–5).

WHOOPING COUGH (see Cough)

WOUNDS, SORES AND CUTS

The application of leaf juice, pounded leaf, or crushed leaf of the following plants: *Chenopodium ambrosioides* American or Indian Wormseed, Sweet Pigweed; *Maesobotrya barteri* var. *sparciflora* (Apotrewa), *Stictocardia beraviensis*, *Carica papaya* Pawpaw, *Glyphaea brevis* (Foto), *Flabellaria paniculata* (Okpoi), *Pupalia lappacea* (Akukuaba), *Blumea aurita* var. *aurita*, *Manihot esculenta* Cassava, *Margaritaria discoidea* (*Phyllanthus discoideus*) (Pepea), *Passiflora glabra* (*foetida*) Stinking Passion Flower, *Senecio biafrae* (*Crassocephalum biafrae*), *Luffa acutangula* (Akatong), *Parquetina nigrescens* (Aba-kamo), *Phyllanthus muellerianus* (Potopoleboblo), *Parinari curatellifolia* (Atena), *Tamarindus indica* Indian Tamarind, *Crotalaria goreensis*, *Desmodium incanum* (*canum*), *Lecaniodiscus cupanioides* (Dwindwere), *Strophanthus sarmentosus* (Adwokuma), *Tylophora conspicua*, *Plumbago zeylanica* Ceylon Leadwort, *Thunbergia chrysops*, *Ensete gilletii*

Wild Banana, *Cassia occidentalis* Negro Coffee, *Guibourtia copallifera*, *Entada africana* (Kaboya), *Gouania longipetala* (Homabiri), *Ziziphus spina-christi* var. *spina-christi*, *Lannea acida* (Kuntunkuri), *L. microcarpa*, *Mitragyna stipulosa* (Subaha-akoa), *Sabicea calycina* (Anansetoromahama), and *Clerodendrum volubile* (Kumamuno). Powdered or pulverized leaf of *Alchornea cordifolia* Christmas Bush (Gyamma); *Eriosema glomeratum* and *Crossopteryx febrifuga* African Bark, (Pakyisie); young leaf of *Musa paradisiaca* Plantain heated as dressing; leaves of some varieties of *Capsicum annuum* Pepper as dressing, *Euphorbia hirta* Australian Asthma Herb, *Asystasia gangetica* applied to wound after piercing lobe of ear; bruised leaves and flowers of *Aspilia africana* Haemorrhage Plant to cleanse surface of sores; leaf-juice of *Aneilema lanceolatum* subsp. *lanceolatum* and *A. lanceolatum* subsp. *subnudum* to cure water-sores.

Heated leaf of *Piper guineense* West African Pepper (Soro-wisa); dried leaf of *Phytolacca dodecandra* (Ahoro) as bandage; leaf decoction of *Caloncoba echinata* (Gorli), *Oncoba spinosa* Snuff-box Tree (Asratoa); *Erythrococca anomala*, *Lepisanthes senegalensis* (*Aphania senegalensis*) (Akisibaka), *Lantana camara* Wild Sage as lotion and that of *Phaulopsis barteri* as antiseptic for bathing. Leaf and crushed seed kernels of *Gossypium arboreum* Cotton; decoction of leafy twigs of *Sphenocentrum jollyanum* (Krakoo) with powdered bark; ashes of burned leaf of *Jatropha curcas* Physic Nut; powdered leaf of *Mallotus oppositifolius* (Satadua) with shea butter as ointment; pounded leaf and stem of *Cissus quadrangularis* Edible-Stemmed Vine, (Kotokoli); pulverized dried leaf of *Acacia kamerunensis* (*pennata*) (Oguaben) as dressing; astringent leaf and bark of *A. farnesiana* and *Anopyxis klaineana* (Kokote); dried and pulverized *Phaulopsis ciliata* (*falcisepala*) or *P. imbricata* used as dressing or fresh juice applied; leaf and fruits of *Solanum nigrum* (Nsusuabiri) applied; pounded leaf of *Desmodium adscendens* var. *adscendens* (Akwamfanu) with lime-juice as dressing and leaf and root of *Reissantia indica* (*Hippocratea indica*) applied. Pounded young leafy shoots of *Diospyros mespiliformis* West African Ebony applied as stimulating dressing; leaf-juice of *Psychotria peduncularis* (*Cephaelis peduncularis*) (Kwesidua); young leaf or bark of *Cordia vignei* (Tweneboa-akoa); leaf and bark decoction or bark of *Kigelia africana* Sausage Tree (Nufuten); crushed leaf and bark of *Markhamia lutea* (Sisimasa) or *M. tomentosa* (Tomboro) as paste with lime; bruised leaf and flower of *Spathodea campanulata* African Tulip (Kokoanisua); juice of fresh leaf and bark of *Stereospermum acuminatissimum* (Tokwa-kufuo); leaf of *Clerodendrum splendens* (Ekenyieya) as lotion and juice from young petioles of *Elaeis guineensis* Oil Palm.

Juice of root, stem, or leaf of *Uvaria chamae* (Akotompotsen); root shavings of *Penianthus zenkeri*; application of powdered, ground roots or root-bark of *Lophira lanceolata*, *Terminalia avicennioides* (Petni), *Waltheria indica* (Sawai), *Piliostigma thonningii* (Opitipata), *Lonchocarpus cyanescens* West African or Yoruba Indigo, *Zanthoxylum xanthoxyloides* (*Fagara zanthoxyloides*) Candle Wood, *Nauclea latifolia* African Peach (Sukisia), and *Saba senegalensis* (Sono-nantin). Pulped root of *Anogeissus leiocarpus* (Sakanee); root decoction of *Combretum micranthum* (Landaga) as wash, *Rhaphiostylis beninensis* (Akwakora gyahene) as lotion and

8. POTIONS AND MEDICINES

Ximenia americana Wild Olive as bath; root-bark of *Terminalia glaucescens* (Ongo) applied for the burning effect of iodine; lotions of root, bark, leaf, and seeds of *Pterocarpus santalinoides* (Hote) as wash; pounded root and root-bark of *Ziziphus mauritiana* Indian Jujube with oil; root-bark of *Paullinia pinnata* (Toa-ntini) as poultice; ground root of *Lannea kerstingii* (Kobewu) as hot poultice; root-bark of *Byrsocarpus coccineus* (Awennade) with *Piper* applied and a sprinkle of powdered roots of *Hoslundia opposita* (Asifuaka) with Guinea-grains and potassium nitrate.

Ground and roasted seeds of *Monodora myristica* (Ayerew-amba); pulverized pods and gum from bark of *Cassia sieberiana* African Laburnum as dressing; pounded fruit and oil of *Momordica charantia* African Cucumber as dressing; fruit of *Solanum anomalum* (Nsusoa) applied to sores on the ear; pulverized seeds of *Aframomum melegueta* Guinea-grains (Famu-wisa) applied; crushed fruits of *Maerua angolensis* (Pugodigo); acrid oil of seed of *Azadirachta indica* Neem Tree; oil of *Carapa procera* Crabwood and *Omphalocarpum elatum* (Timatibre) and fruit sap of *Rothmannia whitfieldii* (Sabobe).

Powdered stem and leaf of *Capparis polymorpha* (Sansangwa); bark extract of *Xylopia aethiopica* Ethiopian Pepper (Hwenetia) as ointment; powdered, ground, or macerated bark of *Combretum molle* (Gburega), *Parinari excelsa* Guinea Plum, (Ofam); *Detarium senegalense* Tallow Tree, *Urena lobata* Congo Jute, *Distemonanthus benthamianus* African Satinwood (Bonsamdua); *Erythrophleum suaveolens* (*guineense*) Ordeal Tree (Potrodom); *Parkia bicolor* (Asoma), *Maytenus senegalensis* (Kumakuafo), *Coula edulis* Gaboon Nut (Bodwue); *Alstonia boonei* (Sinduro), *Rauvolfia vomitoria* (Kakapenpen), and *Stereospermum kunthianum* (Sonontokwakofo). Bark decoction of *Terminalia ivorensis* (Emire) as lotion; pulped bark or bark decoction of *T. macroptera* (Kwatiri) and *Isoberlinia tomentosa* (*dalzielii*) (Kangkalaga) as wash; mucilaginous bark and leaf of *Grewia mollis* (Yualega); powdered bark and bark decoction of *Hymenocardia acida* (Sabrakyie); bark decoction of *Albizia ferruginea* (Awiemfosemina) as wash and *Pentaclethra macrophylla* Oil Bean Tree (Ataa) as lotion; pulped bark of *Parkia biglobosa* (Duaga) with lemon-juice; bark of *Prosopis africana* (Sanga) as dressing or lotion; bark lotion of *Ostryderris stuhlmannii* (Kaman Godui) as wash; bark infusion of *Ficus asperifolia* Sandpaper Tree as wash; ground bark of *F. dekdekena* (*thonningii*) (Gamperoga) mixed with gunpowder applied; pulped bark of *Lannea nigritana* var. *nigritana* (Sinsabgbetiliga) and pounded inner bark of *Cola lateritia* var. *maclaudi* (Watapuo-bere) with clay for cases in the nose. Bark decoction of *Trichilia monadelpha* (*heudelotii*) (Tanduro) and *Khaya senegalensis* Dry-Zone Mahogany as lotion; infusion of inner bark and leaf of *Diospyros gabunensis* Flint Bark Tree (Kusibiri) as antiseptic wash — the bruised leaves being applied as poultice; bark infusion of *Anthocleista nobilis* Cabbage Palm (Hohoroho) as poultice; crushed inner bark of *Kigelia acutifolia* and decoction of soft inner bark of *Newbouldia laevis* (Sasanemasa) as lotion.

Latex of *Garcinia kola* Bitter Cola (Tweapea) applied externally; grey galls of *Guiera senegalensis* applied; yellow resinous sap of *Garcinia smeathmannii* (*polyantha*) False Chewstick Tree (Bohwe) as dressing; resin of *Symphonia globulifera* (Ehureke); decoction of unspecified part of *Griffonia simplicifolia* (Kagya); sawdust of *Acacia macrothyrsa* produced by borers as dressing; oleo-resin of *Canarium schweinfurthii* Incense Tree applied as a substitute for gum mastic; latex from the stem and leaf of *Adenium obesum* applied; brownish-yellow latex of *Alafia multiflora* (Okum-adada) applied, also diluted and taken internally; latex of *Tabernaemontana crassa* (Pepae) for disinfecting or dressing, and young shoots of *Psilanthus mannii* (Gbomete). Leaves of *Spondias mombin* Hog Plum, Ashanti Plum with those of *Vitex doniana* Black Plum, (Afua) and *Terminalia avicennioides* (Petni) applied to prevent inflammation and perforated leaves of *Baphia nitida* Camwood (Odwen) as a receptacle for herbal remedies used as dressing.

The following plants are specific for circumcision wounds; pulverized bark of *Spondias mombin* Hog Plum, Ashanti Plum as dressing; crushed leaves of *Tamarindus indica* Indian Tamarind applied; sap of *Vismia guineensis* (Kosowanini); pounded and macerated bark of *Parinari excelsa* Guinea Plum (Ofam); ground bark of *Balanites aegyptiaca* Desert Date (Gongu); ointment of chewed bark of *Musanga cecropioides* Umbrella Tree (Dwumma) with palm oil and juice of *Ancistrocarpus densispinosus* and *Commelina diffusa*.

YAWS (FRAMBOESIA — AN INFECTIOUS TROPICAL SKIN DISEASE)

The fruits of *Omphalocarpum procerum* (Gyatofo-akongua); the bark decoction of *O. elatum* (Timatibre) and oil extract of seeds; bark decoction of *Terminalia laxiflora* and *Vernonia conferta* (Flakwa); the powdered or pulverized bark of *Trichilia monadelpha* (*heudelotii*) (Tanduro), *Albizia zygia* (Okoro), *Pausinystalia lane-poolei*, and *Cola gigantea* var. *glabrescens* (Watapuo). The oil of *Carapa procera* Crab Oil Tree, Crabwood; the lime-juice of *Citrus aurantiifolia* Lime with rust; the sap of *Thomandersia hensii* (*laurifolia*), latex and root bark of *Calotropis procera* Sodom Apple; and bark of *Santiria trimera* (Burkill, 1985). The hardened latex of *Alstonia boonei* (Sinduro), latex or leaf decoction of *Rauvolfia vomitoria* (Kakapenpen); powdered leaf of *Alchornea cordifolia* Christmas Bush (Gyamma); decoction of leafy stems of *Erythrococca anomala*, *Momordica charantia* African Cucumber taken internally and infusion as wash and powdered fruit with oil as dressing, and decoction of *Indigofera hirsuta* Hairy Indigo (Duke, 1981).

The crushed leaf of *Blighia sapida* Akee Apple (Akye) with salt; the leaves of *Ethulia conyzoides*, *Senecio abyssinicus*, *Emilia coccinea*, and leafy stems of *Heliotropium subulatum* (Burkill, 1985). The ground leaves of *Tylophora conspicua* heated and applied with pepper; the leaf-juice of *Indigofera dendroides*; the ground roots of *Lonchocarpus cyanescens* West African or Yoruba Indigo (Akase) and *Tetracera affinis*; the liquid of boiled roots of *Anacardium occidentale* Cashew Nut (Atea); the roots of *Carica papaya* Pawpaw ground with lime and the leaf of *Dalbergia saxatilis* (Nuodolega) to keep off flies. The leaves of *Cassia occidentalis* Negro Coffee used externally and internally; the pounded leaves of *Asystasia calycina* with lime-juice applied; the juice from *Solenostemon monostachys* (Sisiworodo) after scorching in fire; *Torenia thouarsii* ground up with hot lime applied after dressing; leaves of *Solanum nigrum* (Nsusuabiri) applied and *Brillantaisia lamium* or *B. nitens* (Guare-ansra) applied locally; fresh leaf of *Dissotis rotundifolia* (Adowa-alie) scorched by fire; leaves of *Launaea taraxacifolia* (*Lactuca taraxacifolia*) Wild Lettuce with ashes as rub; ground roots of *Boerhavia diffusa* or *B. repens* Hogweed; ground *Justicia flava* (Ntumenum) applied and scraped pulp of *Dioscorea* species applied. Leaf-spines of *Acanthus montanus* and leafy stems, root, and whole plant of *Momordica balsamina* (Burkill, 1985).

NINE
Poisons, Tannin, Dyes, etc.

In addition to the medicinal uses of plants, plant extracts are used for a variety of purposes. The healing properties of plants are often accepted with reservations or are disbelieved: their poisonous properties are, however, unquestionable. The Greek philosopher Socrates committed suicide with the fruits of *Conium maculatum* Poison Hemlock, a very poisonous plant in the carrot family Umbelliferae. In 1984 a housewife who mistakenly used the bark of *Erythrophleum suaveolens* (*guineense*) Ordeal Tree (Potrodom), instead of that of *Khaya senegalensis* Dry-Zone Mahogany, to prepare a concoction for fever died on admission to the Legon Hospital; earlier Irvine (1961) had reported an incidence of Potrodom poisoning of two men in Saltpond. The planting of *Khaya* and *Erythrophleum* together, as at Legon and Achimota, is rather dangerous and is to be discouraged. In a similar incident, also at Legon, a potential victim was lucky to be forewarned by a groundnut seller. The active principle in Potrodom is erythrophleguine (see Table 9.1), which acts as a local anaesthetic and cardiac inhibitor. It decreases respiration, and in overdoses there are symptoms of depression of the circulation, difficulty in breathing, vomiting, and convulsions, the latter resulting from a direct action on the medulla oblongata. *Mareya micrantha* (Odubrafo) and *Securidaca longepedunculata* (Kpaliga) are also well-known poisonous plants; the roots of the latter contain saponins and methyl salicylate. Other poisonous principles include tephrosine from *Tephrosia vogelii* Fish Poison Plant, abrine from *Abrus precatorius* Prayer Beads, and strychnine from *Strychnos nux-vomica* Nux Vomica, Poison Nut.

Ricinus communis Castor Oil Plant yields castor oil, but it can kill if as few as two seeds are swallowed. The active principle is a toxalbumin called ricin; the leaves containing a crystalline alkaloid called ricinine. It is important to appreciate the dangerous properties of such plants, and to observe the necessary precautions so as to prevent a possible accidental poisoning. An overdose of a medicinal preparation can be poisonous, while recognized toxins can be helpful in small doses. Fortunately, plant extracts as antidotes have the properties to neutralize poisons, minimize their effect, or induce vomiting in cases of accidental poisoning. There exist plant antidotes to arrow poison, snake venom, scorpion sting, spider bite, and other forms of general poisoning. It is advisable to seek medical help quickly in event of poisoning or consult a herbalist, rather than attempting self-medication.

Tanning and dyeing are important rural industries, especially in the Sudan

9. POISONS, TANNINS, DYES

Table 9.1 Common plant Poisons, their source and action

Species	Principle	Remarks
Abrus precatorius	abrine abralin	nerve poison and blood coagulant — human and stock
Acacia albida	prussic acid	stock poison
Ageratum conyzoides	hydrocyanic acid coumarin	poisonous to rabbits
Amanita species	muscarine cyclopeptides	human poison
Asclepias curassavica	asclepin	large doses fatal
Burkea africana	concentration of saponins	fish poison
Calotropis procera	calotropin	heart poison
Cassia absus	abrine chaksine	similar to abrine in *Abrus* depressant effect on respiration
Corchorus olitorius	corchorin	similar in action to strophantin
Crotalaria pallida *C. retusa*	cystine	poisonous to horses and birds
Croton tiglium	crotonin	human poison
Datura species	atropine or hyoscyamine daturine scopolamine	human poison
Dioscorea dumetorum	dioscorine dihydrodioscorine	human poison
Duranta repens	prussic acid	poisonous to man and livestock
Erythrophleum suaveolens	cassaidine nor-cassaidine cassaine cassaimin erythrophlamine erythrophleguine homophleine	human poison
Galerina species	cyclopeptides	human poison
Gloriosa simplex *G. superba*	superbine colchicine	poisonous to stock

Table 9.1 Cont'd.

Species	Principle	Remarks
Gyromitra species	monomethylhydrazine	human poison
Habropetalum dawei	phenol concentrations	fish poison
Hura crepitans	crepitin hitrine	poisonous to man and livestock
Khaya senegalensis	cailcedrin A cailcedrin B	poisonous to paramoecia
Lantana camara	lantanine	livestock poison
Lepiota species	cyclopeptides	human poison
Lonchocarpus sericeus	lonchocarpin	human poison
Luffa cylindrica	concentration of saponins	fish poison
Manihot esculenta	phaseolunatin	poisonous to man and livestock
Mansonia altissima	mansonin	cardiac poison
Mucuna pruriens var. *pruriens*	mucunain	burning, itching, vesication and stomatitis on contact with skin
Mundulea sericea	rotenone	fish poison
Nerium oleander	oleandrin neriin folinerin rosagenin cornerin pseudo-curanine rutin cortenerin	animal poison, e.g. cow and sheep
Oxalis corniculata	salts of oxalic acid	stock poison, e.g. sheep
Paullinia pinnata	concentration of saponins	fish poison
Pentaclethra macrophylla	pancine	fish poison; also for arrows
Phaseolus lunatus	limarin phaseolunatin	stock poison
Physostigma venenosum	esermine eseridine calabarine physovenine	general poison

9. POISONS, TANNINS, DYES

Table 9.1 Cont'd.

Species	Principle	Remarks
Securidaca longepedunculata	methyl salicylate	general poison
Sesbania sesban	concentration of saponins	fish poison
Solanum species	solanine	livestock and human poison
Sophora occidentalis	cystisine	insecticide
Spigelia anthelmia	spigeline	poisonous to stock
Strophanthus species	strophanthis sarmento-cymerin trigonelline	paralysis of the heart
Strychnos nux-vomica	brucine strychnine	general poison
Swartzia madagascariensis	concentration of saponins	poisonous to fish and schistosoma-carrying snails
Tephrosia vogelii	tephrosine	fish poison
Thevetia peruviana	thevetin	powerful heart poison
Vernonia amygdalina *V. nigritiana*	vernonine	heart action like digitaline, poisonous to mouse and dog
Ximenia americana	hydrocyanic acid	dangerous to livestock and human

and Guinea savanna countries of West Africa. These industries, which deal in basketry and related crafts and a variety of leather goods, use plant extracts as raw materials. *Acacia* and *Rhizophora* species yield a high proportion of tannin while *Indigofera* and *Lonchocarpus* species are famous for their indigo blue colour. The attractive colours that give a bright finish to the commercial Bolga baskets or the blend of warm-coloured leather goods, bags, hats, walking sticks, knife sheaths, etc. — are indicative of the standard of quality of the products of the tannin and dyeing industries. Dyes are also used for tattooing. Plants yield ink and saponin, and their ashes or potash are useful for local soap-making. (For plant exudates like latex, adulterants, bird-lime, coagulants, gum-yielding trees, resin-yielding trees, gum copal trees, and gutta-percha, see Chapter 1.)

Poisons

GENERAL

The following plants or plant parts are generally poisonous: the bark of *Andira inermis* Dog Almond in large doses, *Spondianthus preussii* var. *preussii* (Tweanka), *Trichilia emetica* subsp. *suberosa* (*roka*) (Kisiga), and *Albizia ferruginea* (Awiemfosamina) (including the leaf). The bark of *Napoleonaea vogelii* (*Napoleona vogelii*) (Obua) is deadly poisonous and that of *Pycnocoma cornuta* and *P. macrophylla* (Akofie-kofi) is used in palm-wine to discourage theft and to punish pilferers. Both leaf and fruit of *Duranta repens* Golden Dewdrop is reported by Storrs and Piearce (1982) to contain prussic acid, and deaths have been recorded of both humans and livestock from eating the plant. In humans the symptoms are sleepiness, high temperature, and finally convulsions. The leaf decoction of *Melia azedarach* Persian Lilac and also the fruits and seeds, 6 to 8 seeds being lethal; the bitter leaves and fruits of *Mareya micrantha* (Odubrafo); infusion of the leaves of *Dracaena bicolor* (Gblieku); the leaf of *Paullinia pinnata* (Toa-ntini), *Coula edulis* Gaboon Nut, (Bodwue), and the leafy twigs of *Datura metel* Metel and *D. stramonium* Apple of Peru together with the seeds — the principal alkaloid being hyoscyamine or atropine; the leaf of *Sapium ellipticum* (Tomi); the whole plant of *Heliotropium ovalifolium* which causes diarrhoea and vomiting, and that of *Pancratium trianthum* Pancratium Lily. *Secamone afzelii* (Kotohume) is suspected to be poisonous — when taken as a stomachic and purgative it results in vomiting and convulsions

Other toxic plants and plant extracts include the root of *Bersama abyssinica* subsp. *paullinioides* (Duantu), *Cocos nucifera* Coconut Palm, *Elaeis guineensis* Oil Palm, *Capparis corymbosa*, *C. polymorpha* (Sansangwa), *Dicranolepis persei* (Prahoma), some varieties of *Manihot esculenta* Cassava due to the presence of a cynogenetic glycoside called phaseolunatin, *Phyllanthus reticulatus* var. *reticulatus* and *P. reticulatus* var. *glaber* (Awobe) and the fruits; the pounded roots of *Adenium obesum*, *Rauvolfia vomitoria* (Kakapenpen), *Mimosa pudica* Sensitive Plant in large doses, *Gnidia kraussiana* (*Lasiosiphon kraussianus*) and the leaf (also to stock, fish, and for arrow poison); the woody roots of *Celtis wightii* (*brownii*) (Esafufuo), *Boscia salicifolia*, and *Buchholzia coriacea* (Esonobise); the root of *Maesa lanceolata*; the root-bark of *Psorospermum corymbiferum* var. *corymbiferum* and that of *Strophanthus gratus* (Maatwa); the roots of *Sapium grahamii* and *Cucumeropsis edulis* (Agushi); all parts of *Thevetia peruviana* 'Milk Bush', Exile Oil Plant, especially the roots due to a glycoside in the latex, closely allied to strophanthin, which acts as a heart poison; the tubers of *Euphorbia baga*, causing headache and drowsiness, *Gloriosa superba* and *G. simplex* Climbing Lily (due to an alkaloid, colchicine), and of *Tacca leontopetaloides* South Sea Arrowroot. The tuber of *Dioscorea dumetorum* contains two bitter, poisonous alkaloids dioscorine and dihydrodioscorine — the latter causing convulsions. In time of famine it is made safe by washing out the poison in running water (Storrs and Piearce, 1982).

Both the fruits and the latex of *Anthosthema aubryanum* (Kyirikesa) and *Jatropha multifida* Coral Plant, Spanish Physic Nut; the viscid sap of *J. gossypiifolia* Physic Nut, *Asclepias curassavica* Blood Flower — the active principle being asclepiadin, large doses of which cause death — and the sap of *Calotropis procera* Sodom Apple.

Other poisons are the fruit of *Lonchocarpus sericeus* Senegal Lilac, *Maerua angolensis* (Pugodigo), *Kigelia africana* Sausage Tree, (Nufuten); *Hunteria eburnea* (Kanwen-akoa) and *Ziziphus mucronata* Buffalo Thorn. The seeds of *Croton tiglium* Croton Oil Plant (a native of South-East Asia), *Abrus precatorius* Prayer Beads, the active principle being abrine, which has a coagulating effect on blood and poisons the nerves — in addition the seeds, like the roots contain abric acid; *Corchorus olitorius* Jews Mallow which contain a bitter, toxic glycoside, corchorin, similar in its action to strophanthin (Storrs and Piearce, 1982), *Jatropha curcas* Physic Nut, (Adadze); *Dichapetalum toxicarium* West African Ratbane, *Zanha golungensis*, *Argemone mexicana* Prickly Poppy in large quantities, *Ricinus communis* Castor Oil Plant, *Ostryderris stuhlmannii* (Kaman Godui), and *Lepisanthes senegalensis* (*Aphania senegalensis*) (Akisibaka); the seed aril or raphe of *Blighia sapida* Akee Apple if not perfectly ripe or if overripe; the fresh seeds of *Cassia occidentalis* Negro Coffee and those of *Anacardium occidentale* Cashew Nut (Atea); the seed of *Strychnos nux-vomica* Poison Nut, *S. innocua* subsp. *innocua* var. *pubescens* (Kampoye), *S. spinosa* Kaffir Orange (Akankoa); the germinating seeds of *Avicennia africana* Mangrove (Asoporo) if eaten raw or improperly cooked; the seeds of *Carapa procera* Crab Oil Tree — causing vomiting, *Crotalaria pallida* (*mucronata*) (Peagoro) and *C. retusa* Devil Bean both due to the presence of an alkaloid probably identical with cystisine, *Ximenia americana* Wild Olive and *Ipomoea nil* Morning Glory —being cathartic as a result of the presence of an acrid resin. The watery or alcoholic extract of *Ceiba pentandra* Silk Cotton Tree and its effect on the nerves; the hairs of *Bambusa vulgaris* Yellow and Green Striped Bamboo; the whole plant of *Trianthema pentandra* Horse Purslane —capable of causing death by acute nephritis; and unspecified parts of *Plumbago indica* Red Plumbago and *P. zeylanica* Ceylon Leadwort — both dangerous drugs.

Another group of plants to avoid is poisonous mushrooms. The safest way to tell whether mushrooms are edible is to ask an expert or an elder. Smith (1955) observes that tests on their ability to darken silver, the colour of the gills, or the manner in which the 'skin' peels off from the top of the cap are worthless. Storrs and Piearce (1982) add that kitchen 'tests' based on whether or not they are eaten by animals are also totally unreliable and highly dangerous. Mushrooms containing poisonous toxins include *Amanita* species, for example *A. phalloides* Death Cap; *Galerina* species and *Lepiota* species. The toxins are not destroyed by cooking.

TRIAL BY ORDEAL

This custom is now illegal and punishable by law. It is a tribal practice whereby accused persons or suspected criminals are made to swallow portions of plant toxin as a means to establishing their guilt or innocence. Those who drink and survive are held innocent; the guilty die. It thus serves as both trial and instant punishment. Plants used for their poisonous properties include the bark of *Erythrophleum sauveolens* (*guineense*) Ordeal Tree (Potrodom); *E. africanum* African Black Wood, *E. ivorense* Sasswood Tree, *Crossopteryx febrifuga* African Bark (Pakyisie); *Detarium senegalense* Tallow Tree (Takyikyiriwa) and the bark decoction of *Cyrtosperma senegalense* used to blind witches. The latex of *Calotropis procera* Sodom Apple, *Elaeophorbia drupifera* (Akane), *Euphorbia deightonii* (Sra); the sap of *Piptadeniastrum africanum* (Danhoma); the seeds of *Physostigma venenosum* Calabar Bean, Ordeal Bean; *Abrus precatorius* Prayer Beads; the pounded roots of *Adenium obesum* and *Bersama abyssinica* subsp. *paullinioides* (Duantu) and an unspecified part of *Andira inermis* Dog Almond and *Ekebergia senegalensis*.

ARROW POISON

Arrows (and sometimes hunting spears) are tipped with various poisonous plant extracts — some strong enough to stop an elephant. The poisonous extracts include the crushed seeds of *Strophanthus* species with the juice of *Colocasia*, *Aframomum*, or *Palisota*. The active principle is strophanthin, which causes death by paralysing the heart. *S. gratus* (Omaatwa), *S. hispidus* Arrow Poison, (Amamfohama); *S. preussii* (Dietwa) and *S. sarmentosus* (Adwokuma) are used. Many other plant parts serve as ingredients in the preparation. These include the leaves of *Ficus exasperata* (Nyankyeren), *Machaerium lunatum* (*Drepanocarpus lunatus*) (Nkako), *Maytenus senegalensis* (Kumakuafo), *Nicotiana tabacum* Tobacco, *N. rustica*, and *Microglossa pyrifolia* (Asommerewa); the bark of *Erythrophleum suaveolens* (*guineense*) Ordeal Tree (Potrodom); *Eriocoelum racemosum* (Onibonakokoo), *Khaya senegalensis* Dry-Zone Mahogany, *Pseudocedrela kotschyi* Dry-Zone Cedar, *Enantia polycarpa* African Yellow Wood, *Pachypodanthium staudtii* (Fale), *Buchholzia coriacea* (Esonobise), *Securidaca longepedunculata* (Kpaliga), and *Corynanthe pachyceras* (Pamprama); the bark and seeds of *Pentaclethra macrophylla* Oil Bean Tree; the bark and fruits of *Uapaca guineensis* Sugar Plum (Kuntan); the stems of *Adenia rumicifolia* var. *miegei* (*lobata*) (Peteha); the powdered bark of *Zanthoxylum gillettii* (*Fagara macrophylla*) (Okuo) with that of *Diospyros canaliculata* Flint Bark (Otwabere) and *Mansonia altissima* (Oprono). The seeds of *Tetrapleura tetraptera* (Prekese), *Raphia hookeri* Wine Palm (Adobe); *Tephrosia vogelii* Fish Poison Plant, *Dichapetalum toxicarium* West African Ratbane, and the fruit ashes of *Morinda morindoides*. Others are the roots of *Afzelia africana* (Papao), *Indigofera simplicifolia* (Nyagahe), *Nauclea latifolia*

African Peach (Sukisia); *Solanum incanum* Egg Plant crushed with that of *Amorphophallus dracontioides, Sapium grahamii* (Pampiga), *Pericopsis laxiflora* (*Afrormosia laxiflora*) Satinwood, *Asparagus africanus* (Adedende), *A. flagellaris* and *Capparis polymorpha* (Sansangwa). The tubers of *Gloriosa superba* and *G. simplex* Climbing Lily and unspecified parts of *Triclisia dictyophylla* (*gilletii*), *Acacia farnesiana*, and the stinging hairs of the fruits of *Mucuna pruriens* var. *pruriens* Cow Itch (Apea) serve as ingredients.

The latex of a number of plants is used more for adherents or adhesives than as ingredients. These are *Hunteria eburnea* (Kanwen-akoa), *Holarrhena floribunda* False Rubber Tree (Sese); *Funtumia africana* False Rubber Tree and *Alafia scandens* (Momonimo) with the juice of *Costus*; others are *Adenium obesum, Sapium ellipticum* (Tomi), *Euphorbia balsamifera* Balsam Spurge (Aguwa); *E. deightonii* (Sra), *E. lateriflora* (Kamfo-barima), *Harungana madagascariensis* (Okosoa), *Calotropis procera* Sodom Apple, *Landolphia owariensis* White Rubber Vine (Obowe); *Saba senegalensis* (Sono-nantin), *Ficus asperifolia* Sandpaper Tree, and *Elaeophorbia drupifera* (Akane) with the bark of *Parquetina nigrescens* (Aba-kamo).

PLANTS POISONOUS TO STOCK

The leaves of *Lantana camara* Wild Sage (due to the alkaloid lantanine), *Swartzia madagascariensis* Snake Bean, *Paullinia pinnata* (Toa-ntini) (to pigs), *Bersama abyssinica* subsp. *paullinioides* (Duantu), *Nerium oleander* Oleander (20 g of leaves being sufficient to kill a cow or horse and only 1–5 g to kill a sheep), *Dichapetalum madagascariensis* (*guineense*) (Antro), *Capparis corymbosa, Microglossa pyrifolia* (Asommerewa), *Vernonia amygdalina* Bitter Leaf (Bonwen); *V. colorata* (both to goats), *Rauvolfia vomitoria* (Kakapenpen), *Cryptostegia grandiflora, Ximenia americana* Wild Olive due to presence of hydrocyanic acid (cattle), *Spigelia anthelmia* Worm Weed, *Ficus exasperata* (Nyankyeren) (though eaten by elephants), *F. asperifolia* Sandpaper Tree, *Oxalis corniculata* due to salts of oxalic acid which cause fatalities in sheep and illness in humans, *Quassia undulata* (*Hannoa undulata*) (Kunmumi), *Sarcostemma viminale* (1–2 lb of fresh or dried plants killing sheep in a day or two), *Asclepias curassavica* Blood Flower, *Cassia alata* Ringworm Shrub, (Nsempii); *C. laevigata, C. siamea* (and the pods to pigs), *Erythrophleum africanum* African Black Wood (Bupunga); *E. sauveolens* (*guineense*) Ordeal Tree (Potrodom); *Catharanthus roseus* Madagascar Periwinkle and *Tephrosia vogelii* Fish Poison Plant. Some stock losses have been recorded as a result of eating the leaves of *Gloriosa superba* and *G. simplex* Climbing Lily (Storrs and Piearce, 1982). The poison produces nausea, vomiting, giddiness, and photophobia, followed by paralysis of the nervous system and finally, in severe cases, death by asphyxia. Leaves of *Melia azedarach* Persian Lilac, though considered to be good fodder in Madhya Pradesh and Bihar in India, are reported as being poisonous to stock in South Africa. The fruits of *Jaundea pinnata* are said to be lethal to sheep and goats in Kenya if eaten (Burkill, 1985).

Also poisonous to stock are the seeds of *Gossypium arboreum* Cotton (to pigs and dogs), *Dichapetalum toxicarium* West African Ratbane, and the fresh seeds of *Cassia occidentalis* Negro Coffee; the fruit of *Maerua angolensis* (Pugodigo) and *Spathodea campanulata* African Tulip Tree (Kokoanisua); and the nut of *Lepisanthes senegalensis* (*Alphania senegalensis*) (Akisibaka). The bark of *Bussea occidentalis* (Kotoprepre) mixed with maize is lethal to monkeys. Storrs and Piearce (1982) report that *Solanum incanum* Egg Plant, *S. indicum* subsp. *distichum* Children's Tomato, and *S. nigrum* (Nsusuabiri) contain a glycoalkaloid, solanine, which is moderately toxic. Livestock and humans have been poisoned. The fruit contains the most poison. Other poisonous plants are the bark and foliage of *Burkea africana* (Pinimo), the bark and roots of *Phytolacca dodecandra* (Ahoro) due to phytolaccotoxin and a saponin which causes vomiting and diarrhoea resulting in many deaths of livestock from eating the plant; and the root-bark of *Zanthoxylum xanthoxyloides* (*Fagara zanthoxyloides*) Candle Wood used subcutaneously on dogs.

The leaves and seeds of *Phaseolus lunatus* Lima Bean contain phaseolunatin and limarin, cyanogenetic glycosides. The coloured seeds are the most poisonous. It causes giddiness, vomiting, convulsions, and breathing difficulties. The animal dies in an hour or so. *Strychnos spinosa* Kaffir Orange, (Akankoa) is poisonous to guinea-pigs and the bark and seeds of *Mundulea sericea* (Gyamkawa) is said to kill crocodiles or drive them away. The roottubers of *Crassocephalum rubens* and *Gynura miniata*, which are said to have the same property as garlic, are also used in protecting against crocodiles. *Commelina diffusa* is reported to be harmful to sheep, causing foaming at the mouth and death if eaten in excess; *Ageratum conyzoides* Billy-goat Weed, (Efumomoe) to rabbits — the principle being hydrocyanic acid, coumarin, and an alkaloid; and the bulb of *Scadoxus multiflorus* (*Haemanthus multiflorus*) Blood Flower, Fire-ball Lily (Ngobo) is poisonous to pigs.

PLANTS POISONOUS TO RATS AND MICE

The leaves of *Vernonia amygdalina* Bitter Leaf (Bonwen) and *V. colorata* due to the presence of vernonine and administered subcutaneously — 10 g per kg of body weight proving 100 per cent fatal; the leaves of *Tephrosia vogelii* Fish Poison Plant (also poisonous to stock) mixed with food like groundnuts, *Microglossa pyrifolia* (Asommerewa), *Gliricidia sepium* Mother of Cocoa, *Dichapetalum madagascariense* (*guineense*) (Antro), and *Trema orientalis* (*guineensis*) (Sesea); the latex of *Elaeophorbia drupifera* (Akane) mixed with food; leafy stems and roots of *Machaerium lunatum* (Nkako) and *Pericopsis laxiflora* (*Afrormosia laxiflora*) Satinwood. The roots of *Zanthoxylum viride* (*Fagara viridis*) (Oyaanini) and of *Z. xanthoxyloides* (*F. zanthoxyloides*) Candle Wood (Kanto). The fruit of *Pierreodendron kerstingii* (Soitbia) mixed with boiled yams and palm oil; the seeds of *Physostigma venenosum* Calabar Bean, *Dichapetalum toxicarium* West African Ratbane (Ekum-nkura), and of *Jatropha curcas* Physic Nut mixed with palm oil. The bark of *Diospyros canaliculata* Flint Bark (Otwabere); *Erythrophleum suaveolens* (*guineense*) Ordeal Tree (Potrodom); *Spondianthus preussii* var. *preussii* (Tweanka) boiled with rice, *Amphimas pterocarpoides* (Yaya), *Securinega virosa* (Nkanaa), and an unspecified part of *Drypetes ivorensis*.

PLANTS POISONOUS TO BULINUS SNAILS (SCHISTOSOMIASIS-CARRYING)

Schistomiasis is a water-borne disease common in tropical Africa and among the rice-eating people of the East, where the flooded paddy-fields provide ideal conditions for continuous infection. The young stages are passed in water snails as secondary host and man is infected by the cercariae, through the skin mainly. Schistosome eggs have been found in Egyptian mummies of 3000 years ago. Vines and Rees (1968) estimate that over 200 million people are seriously affected. Plants used to control water snails include: the leaves of *Tephrosia vogelii* Fish Poison Plant — the active part being tephrosine; the leaves of *Swartzia madagascariensis* Snake Bean, and the fruit and bark of *Balanites aegyptiaca* Desert Date. The leaf-juice of *Millettia thonningii* (Sante) and *Agave sisalana* Sisal Hemp are reported lethal to water snails. The use of *Phytolacca dodecandra* (Ahoro) — called Endod in East Africa — as a molluscicide does not appear to be practised in West Africa. The berries are the most active part, but the leaves, stem and bark of the male plant seem to have higher molluscicidal activity than those of the female plant. In the U.S.A. researchers are working on a project to develop a soap from Endod to kill snails as a means of controlling the disease in Africa.

PLANTS POISONOUS TO WATER FLEAS AND PARAMOECIA

The fruits of *Balanites aegyptiaca* Desert Date and *B. wilsoniana* (Kurobow); the bark extract of *Securinega virosa* (Nkanaa) — 1 in 200 infusion being lethal in 3 hours; the bark of *Diospyros canaliculata* Flint Bark (Otwabere) and that of *Pseudocedrela kotschyi* Dry-Zone Cedar. The roots of *Zanthoxylum viride* (*Fagara viridis*) (Oyaanini) and the leaves of *Paullinia pinnata* (Toa-ntini) as a result of the saponin present — 1 in 500 infusion being lethal in an hour. *Cryptostegia grandiflora* is poisonous to vermin.

FISH POISONS

Plant toxins used to catch fish are the fruit of *Ziziphus mauritiana* Indian Jujube, *Acacia albida* (Gozanga), and *Cassia sieberiana* African Laburnum (with the roots); the powdered fruit shells of *Balanites aegyptiaca* Desert Date together with the roots and bark; the powdered seeds of *Annona muricata* Sour Sop, *A. squamosa* Sweet Sop, and *A. glauca* var. *glauca* (Mampihege); the fruits, seeds, bark, and young leaves of *Blighia welwitschii* (Akyekobiri); the seeds and oil from the pericarp of *Raphia hookeri* Wine Palm, (Adoka); the seeds of *Lagenaria breviflora* (*Adenopus breviflorus*) (Anuwatre) or *Cathormion altissimum* (Abobonkakyere); the fruit-husks of *Parkia biglobosa* (Duaga) (sometimes with the bark); the pounded pods and husks of *P. clappertoniana* West African Locust Bean; the seeds of *Strophanthus hispidus* Arrow Poison, *Tieghemella heckelii* (Baku) and *Croton tiglium* Croton Oil Plant; the fruits of *Diospyros piscatoria* (Benkyi); the fruit-pulp, seeds, and bark of *Strychnos aculeata*; the green fruits of *Picralima nitida* (Ekuama) and the dried, pulverized seeds of *Crinum zeylanicum* (*ornatum*).

Among the plants for poisoning fish, the most popular is perhaps *Tephrosia vogelii* Fish Poison Plant. It is also one of the most toxic — the active part being the toxin tephrosine, 15 per cent in the leaves and 3 per cent in the seeds —liquids as dilute as one part in 50 million being strong enough to paralyse or kill fish. *T. densiflora*, *T. nana*, *T. barbigera*, and *T. purpurea* are also used. Other fish poisons are the pounded leaves of *Eriosema glomeratum*, *E. psoraleoides*, *Ostryocarpus riparius* (also the hardened sap, bark, and roots), *Sesbania sericea* (*pubescens*) (Asrati), *Sophora occidentalis* (including the seeds) — the active principle being the alkaloid cystisine; the leaf-juice of *Paullinia pinnata* (Toa-ntini); the macerated leaves, twigs, and flowers of *Blighia unijugata* (Akyebiri); the leafy twigs of *Dodonaea viscosa* Switch Sorrel, (Fomitsi); *Bussea occidentalis* (Kotoprepre), *Acacia kamerunensis* (*pennata*) (Oguaben), and *Entada africana* (Koboya); the crushed leaves, sap, and stems of *Adenia rumicifolia* var. *miegei* (*lobata*) (Peteha); the roasted, pounded stems of *A. cissampeloides* (Akpeka); the latex of *Anthostema aubryanum* (Kyirikesa), *Elaeophorbia drupifera* (Akane), *Jatropha curcas* Physic Nut (Adadze); *Euphorbia poissonii* (Atroku) and *Hura crepitans* Sandbox Tree; the watery juice of *Culcasia scandens* (Otwa-tekyirema), *C. angolensis*, and the pounded leaves of *Ampelocissus multistriata* (Kodomba). Other fish poisons are the thickened, woody roots of *Eriosema griseum* (Trindobaga), either alone or with the fruits of *Balanites aegyptiaca* Desert Date; the crushed roots of *Swartzia madagascariensis* Snake Bean; the root-bark of *Zanthoxylum xanthoxyloides* (*Fagara zanthoxyloides*) Candle Wood (Kanto) and the pounded roots of *Adenium obesum*.

Other sources of fish poison are the whole plant of *Palisota hirsuta* (Nzahuara), *Adhatoda buchholzii*, *Vernonia macrocyanus*, *Rhinacanthus virens* var. *virens* (Kwaduko), and *Aizoon canariensis* mashed with shrimps; the seeds and wood of *Strophanthus gracilis*; the fruits *Luffa acutangula* (Akatong) or *L. cylindrica* Vegetable Sponge; the leaf of *Carica papaya* Pawpaw, leafy twigs of *Combretum nigricans* var. *elliotii*, and old leaf of *Xanthosoma mafaffa* Cocoyam; the whole plant and oil of *Cleome gynandra* (*Gynandropsis gynandra*) Cat's Whiskers, the latex of *Sarcostemma viminale*, the whole plant of *Lepidium sativum* The Common Cress and *Habropetalum dawei* as a result of unusually high concentration of a phenol (Burkill, 1985).

The bark of some trees is a source of fish poison. These are *Anacardium occidentale* Cashew Nut, *Millettia barteri*, *Amphimas pterocarpoides* (Yaya), *Pycnocoma cornuta* (with others), *Diospyros mespiliformis* West African Ebony (Okisibiri); *Pentaclethra macrophylla* Oil Bean Tree (with the seeds), *Pseudocedrela kotschyi* Dry-Zone Cedar (with the bitter gum), and *Mundulea sericea* (Gyamkawa) (together with the seeds), the toxic principle being rotenone. *Mundulea* is sometimes cultivated for the purpose, and it is considered more potent than *Tephrosia vogelii* Fish Poison Plant, and rather dangerous to humans since it kills rather than stupefies. Other useful barks are *Turraeanthus africanus* Avodire (with the leaves), *Securinega virosa* (Nkanaa), *Albizia coriaria* (Awiemfosemina-akoa), *Massularia acuminata* (Pobe) (with the leaves and fruits), and the orange-yellow bark of *Mammea africana* African Mammy Apple (Bompagya). The rest are the wood of *Thevetia peruviana* 'Milk Bush', Exile Oil Plant; the macerated stems of *Baphia capparidifolia* subsp. *polygalacea* (*polygalacea*) Walking-stick Camwood; the pounded whole plant of *Schwenckia americana* (Agyingyinsu), *Justicia extensa, J. laxa* and *Eremomastax speciosa* (*polysperma*) (Adubiri); the bark and fruits of *Burkea africana* (Pinimo); the bulb of *Scadoxus cinnabarinus* (*Haemanthus cinnabarinus*) and *S. multiflorus* subsp. *katerinae* (*H. multiflorus*) Blood Flower, Fire-Ball Lily; and the snuff of *Nicotiana tabacum* Tobacco Plant and *N. rustica* with food as bait.

Antidotes to poisons

GENERAL REMARKS

It is worth repeating the warning earlier that in the event of accidental poisoning it is advisable to consult a herbalist, rather than attempting self-medication. Antidotes include bark of *Pycnanthus angolensis* African Nutmeg (Otie) as enema; a decoction of *Combretum molle* (Gburega); a watery decoction of the bark and leaves of *Morinda lucida* Brimstone Tree (Kankroma) causing vomiting, excessive flow or urine, and profuse diarrhoea; the bark of *Ficus dekdekena* (*thonningii*) (Gamperoga) as ingredient; the powdered bark of *Khaya senegalensis* Dry-Zone Mahogany because of its emeto-purgative properties and the bark infusion of *Mitragyna stipulosa* (Subaha-akoa) taken internally or used as a bath. A decoction of the roots and leaves of *Tamarindus indica* Indian Tamarind with that of *Afzelia africana* (Papao) and *Ficus* species; an infusion of the root-pulp of *Strychnos spinosa* Kaffir Orange (Akankoa); the roots of *Vernonia amygdalina* Bitter Leaf (Bonwen) and those of *V. colorata*; the leaf decoction of *V. conferta* (Flakwa) with the bark of *Blighia sapida* Akee Apple (Akye) and the roots and leaves or the whole plant of *V. guineensis*; the root decoction of *Capparis polymorpha* (Sansangwa); the pods, bark, and root of *Swartzia madagascariensis* Snake Bean to induce vomiting and the rhizomes of *Zingiber officinale* Ginger with Guinea-grains to delay the action of poisons.

The seeds of *Physostigma venenosum* Calabar Bean are a specific antidote to strychnine and atropine poisoning. Other antidotes are the seeds of *Trichilia emetica* subsp. *suberosa* (*roka*) (Kisiga) and palm oil or the palm-kernel oil of *Elaeis guineensis* Oil Palm as drink; the leaf-juice or decoction of *Mareya micrantha* (Odubrafo) taken as a drink or enema; the whole plant of *Microdesmis puberula* (Fema) as ingredient and that of *Annona senegalensis* subsp. *onlotricha* (*arenaria*) (Aboboma); the leaf decoction of *Blighia sapida* Akee Apple (Akye) and that of *Byrsocarpus coccineus* (Awennade); the leaf-juice or decoction of *Ficus exasperata* (Nyankyeren); the juice of the leafy tips of *Tarenna thomasii*; a leaf decoction of *Spathodea campanulata* African Tulip Tree (Kokoanisua) and a decoction of the leaf-tips of *Diospyros mespiliformis* West African Ebony. The whole plant of *Ficus vallis-choudae* (Alomabli) and that of *Microglossa afzelii* (Pofiri) are useful as ingredients.

ANTIDOTE TO ARROW POISON

The powdered leaf of *Mitracarpus villosus* (*scaber*) applied to the wound and a decoction drunk; the leaf and berry of *Fadogia erythrophloea* chewed and applied and a decoction or infusion drunk to cause vomiting; a decoction of the leafy stem-tips of *Ficus platyphylla* Gutta-Percha Tree; the leaf of *Gardenia ternifolia* (Peteprebi) for bathing and as lotions and the thick, succulent leaves of *Aloe schweinfurthii* and *A. buettneri* (Sere-berebe) as ingredients. The roots of *Gymnema sylvestre* applied as a powder to the affected area and given as a decoction internally; the roots of *Securidaca longepedunculata* (Kpaliga) are used in a similar way. The powdered roots of *Cissus populnea* (Agyako) with the residue of *Hibiscus sabdariffa* Roselle after extracting the oil, then moistened and applied to the wound, and the root and bark of *Boswellia dalzielii* Frankincense Tree (Kabona) applied. To induce vomiting the fruit of *Solanum dasyphyllum*, the seed of *Garcinia kola* Bitter Cola (Tweapea or Minchingoro), and the young, soft seeds of *Detarium senegalense* Tallow Tree are taken. Others are the bark of *Annona senegalensis* var. *deltoides* (Batanga) applied, *Bridelia ferruginea* (Opamfufuo) chewed, applied to the wound, and sucked; the powdered bark of *Erythrophleum suaveolens* (*guineense*) Ordeal Tree (Potrodom) applied; the bark of *Isoberlinia tomentosa* (*dalzielii*) (Kangkalaga) applied and the bark of *Adansonia digitata* Baobab applied — the active principle being adansonin which is said to be an effective antidote. The latex of *Alstonia boonei* (Sinduro) and that of *Euphorbia hirta* Australian Asthma Herb are useful in assisting extractions of the head. The odorous smoke of *Stylosanthes fruticosa* (*mucronata*) produced in a pipe is blown on the wound as antidote.

ANTIDOTES TO SNAKE-VENOM

Dr Abayomi Sofora, Professor of Pharmacognosy at the University of Ife in Nigeria, observes that only 12 per cent of snakes in Africa are known to be poisonous. However, Cansdale (1961) reports that West Africa has at least ten kinds of snakes whose bites may easily prove fatal to a human being. These include the Gaboon Viper, Carpet Viper, Green Mambas, and Black Cobras or Spitting Cobras which must be reckoned among the most poisonous snakes in the world. There are plants remedial to snake venom. These are applied locally or taken internally or both. They include the whole plant of *Uraria picta* (Heowe) for the bite of the viper *Echis carinatus*; *Ipomoea alba* Prickly Ipomoea for general cases; the bitter *Lindernia diffusa* var. *diffusa*, *Macaranga heterophylla* (Opamkokoo) as ingredient, and *Paullinia pinnata* (Toa-ntini); *Gouania longipetala* (Homabiri) and *Vernonia guineensis* applied, while *Gnidia kraussiana* (*Lasiosiphon kraussianus*) is applied for the blistering properties; *Aerva javanica* is used internally; *Spilanthes filicaulis* Para or Brazil Cress chewed with others and swallowed in conjunction with local treatment of the bite; *Heliotropium strigosum* taken internally and applied externally and the whole plant of *Allium cepa* Onion and relatives applied externally.

The root-bark of *Annona senegalensis* var. *senegalensis* Wild Custard Apple, and that of *Entada abyssinica* (Sankasaa) eaten raw to induce vomiting; the roots of *Securidaca longepedunculata* (Kpaliga) used internally (after an incision has been made and a ligature applied) causing violent vomiting and purging, *Cissampelos mucronata* (Akuraso), *Alchornea cordifolia* Christmas Bush (Gyamma); *Acacia polyacantha* subsp. *campylacantha* African Catechu, *Mimosa pigra* Sensitive Plant (Kwedi); *Saba florida* (Akontoma), *Gymnema sylvestre* applied locally or taken internally, *Ocimum gratissimum* Tea Bush (Nunum) with others as an ingredient, and *Solanum aculeatissimum*; the powdered roots of *Capparis polymorpha* (Sansangwa), *Mareya micrantha* (Odubrafo), *Cochlospermum tinctorium* (Kokrosabia), and *Rauvolfia vomitoria* (Kakapenpen); the chewed, macerated roots of *Dichrostachys cinerea* (*glomerata*) Marabou Thorn (also for scorpion stings) and the macerated bark as an emetic; the root decoction of *Pericopsis laxiflora* (*Afrormosia laxiflora*) Satinwood as ingredient and that of *Parquetina nigrescens* (Aba-kamo) taken internally and powdered root applied to the bite; a cold infusion of the pulverized roots of *Cienfuegosia heteroclada* taken internally and the root scrapings of *Balanites aegyptiaca* Desert Date (Gongu) in water as drink to vomit poison (Moro, 1984-5).

The leaves of *Strophanthus gratus* (Maatwa) and *S. hispidus* Arrow Poison (Amamfohama) are specifically for the bite of the black-necked cobra *Naja nigricollis* and the leaf-juice of *Cissus populnea* (Agyako) applied to the blinding venom spat at the eyes. Other leaves used as antidote in general cases are *Cassia alata* Ringworm Shrub, *Sterculia tragacantha* African Tragacanth (Sofo); *Ficus sycomorus* (*gnaphalocarpa*) (Kankanga), *Bridelia ferruginea* (Opam-fufuo), *Olax subscorpioidea* (Ahoohenedua), *Ximenia americana* Wild Olive (also for other poisonous bites), *Clausena anisata* Mosquito Plant (Samanyobli); *Mikania chevalieri* (*cordata*) Climbing Hemp Weed, *Clerodendrum buchholzii* (Taasendua), *C. polycepalum*, and *C. umbellatum* (Niabiri). Other antidotes are the flowers of *Tacca leontopetaloides* South Sea Arrowroot as rub and the chewed gum of *Daniellia ogea* Gum Copal Tree (Hyedua) as ingredient; an infusion of the tubers of *Amorphophalus abyssinicus*, *A. aphyllus*, and *A. johnsonii* with other ingredients taken internally and applied locally; the corms of *Gladiolus daleni* (*psittacinus*), *G. klattianus*, *G. gregarius*, and *G. unguiculatus* collectively called Sword Lily or Corn-Flag as remedy; the bark of *Amblygonocarpus andongensis* (Nanzili); an infusion of the bark of *Bertiera racemosa* (Mantannua) and the bark and leaf decoction of *Kigelia africana* Sausage Tree (Nufuten) as vapour bath to soften wounds; the fruit and sometimes the roots of *Cnestis ferruginea* (Akitase); the shelled, unripe fruits or the roots of *Strychnos spinosa* Kaffir Orange (Akankoa); the ground seeds of *Azadirachta indica* Neem Tree applied and the ashes of *Cathormion altissimum* (Abobonkakyere) as ingredient.

PREVENTIVE MEASURES AGAINST SNAKE-BITE

Allium cepa Onions and relatives are supposed to repel snakes and are planted or scattered around houses; *Euphorbia convolvuloides* is swallowed after chewing, also mixed with saliva and added to the juice of *Calotropis procera* Sodom Apple and rubbed on the hands (also for scorpion stings); use of the whole plant of *Erythrococca anomala* as a rub; the ashes of the leaves of *Piliostigma thonningii* (Opitipata) rubbed into cuts on the body (the young leaves serving as a cure for dog-bite); wearing portions of the root-bark of *Securidaca longepedunculata* (Kpaliga); planting *Jatropha gossypiifolia* Physic Nut around homes; chewed roots of *Scoparia dulcis* Sweetbroom Weed with the juice of *Nicotiana tabacum* Tobacco is used to paralyse snakes. Also sprinkling the watery extract of the roots of *Datura metel* Hairy Thorn Apple, Metel on the floor is reported to keep away snakes. The leaves of *Hoslundia opposita* (Asifuaka) as ingredient with those of *Ocimum americanum* American Basil and other additives serve both as antidote and reportedly render immunity to snake-bite, while leaves of *Nauclea latifolia* African Peach (Sukisia) placed in the holes of snakes (and crocodiles) are said to be poisonous to these reptiles. *Swartzia madagascariensis* Snake Bean is reported to have a similar effect on crocodiles due to the presence of saponins.

ANTIDOTE TO SCORPION-STING

The dried roots of *Cissampelos owariensis* and *C. mucronata* (Akuraso) applied externally; the roots of *Aristolochia bracteolata* applied locally; the powdered roots of *Mareya micrantha* (Odubrafo) applied; the chewed, macerated roots of *Dichrostachys cinerea* (*glomerata*) Marabou Thorn applied and the roots of *Achyranthes aspera* macerated in water and applied to relieve pain. The latex of *Calotropis procera* Sodom Apple rubbed on the hands as both antidote and preventive; the crushed leaves or seeds of *Datura stramonium* Apple of Peru mixed with oil and applied externally; leaves of *Mikania chevalieri* (*cordata*) Climbing Hemp Weed; the chewed bark of *Commiphora africana* African

9. POISONS, TANNINS, DYES

Bdellium (Narga) with natron; the charred, pulverized bark of *Keetia hispida* (*Canthium hispidum*) (Homa-ben) with palm oil and a lotion from the leaves; the whole plant of *Heliotropium strigosum*, also *Schoenoplectus aureiglumis* (*Scirpus aureiglumis*) pounded in water and applied externally, and an unspecified part of *Annona senegalensis* subsp. *onlotricha* (*arenaria*) (Aboboma). *Euphorbia convolvuloides* is used more as a preventive than an antidote — a portion of the plant is swallowed after chewing and the rest with saliva is mixed with the latex of *Calotropis procera* Sodom Apple and rubbed on the hands (also for snake-bite); whole plant decoction of *Evolvulus alsinoides* is reported by Moro Ibrahim, a herbalist at Legon in Ghana, as protective to give immunity against stings, and the root of *Merremia tridentata* subsp. *angustifolia* eaten with bran is supposed to confer a prolonged immunity (a year or more) to scorpion-sting — provided one does not deliberately eat salt during the period (that is besides normal meals). For spider-bite a warm infusion of the leaves of *Ananas comosus* Pineapple is used, alternating the treatment with application of pieces of the fruit.

Tannin

Tanning is the process of converting hide into leather by steeping it in an infusion of crushed plant material or tan. The substance with this property is tannic acid or tannin. Leather tanning is an important rural industry in West Africa; mainly in the countries of the Sudan and Guinea savanna woodland where sheep and cattle rearing predominates. *Acacia* species are useful tannin-producing plants. They include the bark of *A. farnesiana* (23%) and the pods, *A. nilotica* var. *tomentosa* (18–23%) — the pods for dehairing, *A. albida* (Gozanga) (28%) also roots and pods (5%), *A. hockii* Shittim Wood (18–20%) and the pods, *A. kamerunensis* (*pennata*) (Oguaben), and *A. polyacantha* subsp. *campylacantha* African Catechu and the pods. Others are *Alchornea cordifolia* Christmas Bush (Gyamma) and the leaves; *Anogeissus leiocarpus* (Sakanee) — the wood-ashes for dehairing, *Bridelia stenocarpa* (*micrantha*) (Opam), *B. ferruginea* (Opam-fufuo) (3%), *Erythrophleum suaveolens* (*guineense*) Ordeal Tree (Potrodom), and *Hymenocardia acida* (Sabrakyie). Others are the bark of *Punica granatum* Pomegranate and fruit, *Tamarindus indica* Indian Tamarind (7.1%); *Ziziphus mauritiana* Indian Jujube and *Z. mucronata* Buffalo Thorn (12–15%), *Albizia coriaria* (Awiemfosemina-akoa), *A. lebbeck* East Indian Walnut (5–15%), *Anacardium occidentale* Cashew Nut (9–21%) and the leaf (23%), *Annona squamosa* Sweet Sop, *Avicennia africana* Mangrove (12.5%), *Bauhinia purpurea*, *B. rufescens* (Jinkiliza), *Burkea africana* (Pinimo) and the pods, and *Cassia alata* Ringworm Shrub, (Nsempii). Others are the bark of *C. auriculata*, *C. fistula* Indian Laburnum, Golden Shower (12–18%); *C. sieberiana* African Laburnum, *C. singueana*, *Casuarina equisetifolia* Whistling Pine (6–18%), *Chrysobalanus icaco* (*orbicularis*) (Abeble), *Conocarpus erectus* Button Wood (18%), *Entada pursaetha* Sea Bean and pods, *Eucalyptus citriodora*

Lemon-Scented Gum (12), *Eugenia jambos* Rose Apple (7–12%), and *E. uniflora* Pitanga Cherry (20% or more).

The list of tannin-producing plants includes the bark of *Ficus glumosa* var. *glaberrima* (Galinziela), *F. platyphylla* Gutta-Percha Tree, *Garcinia kola* Bitter Cola, (Tweapea), *G. mangostana* Mangosteen (an introduction from Indonesia), *Harungana madagascariensis* (Okosoa), *Hymenocardia acida* (Sabrakyie), *Khaya senegalensis* Dry-Zone Mahogany (10.2%), *Laguncularia racemosa* White Button Wood (Abin) (10–20%), *Mangifera indica* Mango (16–20%) the dried flowers (15%) and the seeds (8–9%), *Markhamia tomentosa* (Tomboro), *Moringa oleifera* Horse-Radish Tree, Oil of Ben Tree; *Parinari excelsa* Guinea Plum, (Ofam) and wood ashes, *Maranthes polyandra* (*P. polyandra*) (Abra-bese), *Parkia biglobosa* (Duaga), and *P. clappertoniana* West African Locust Bean and the pods. Others are the bark of *Margaritaria discoidea* (*Phyllanthus discoideus*) (Pepea), *Piliostigma reticulatum*, *P. thonningii* (Opitipata), *Pithecellobium dulce* Madras Thorn (12–37%), *Psidium guajava* Guava (11–30%), *Rhizophora* species Red Mangrove (about 30%), *Sclerocarya birrea* (Nanogba), *Spondias mombin* Hog Plum, Ashanti Plum, *Syzygium guineense* var. *guineense* (Sunya), *Heritiera utilis* (*Tarrietia utilis*) (Nyankom), *Terminalia catappa* Indian Almond (9–13%), *Thespesia populnea* (Frefi), and *Trichilia emetica* subsp. *suberosa* (*roka*) (Kisiga).

The stems of a number of plants yield tannin. These are *Cissus aralioides* (Asirimu), *Combretum nigricans* var. *elliotii*, *Mussaenda afzelii*, *Prosopis africana* (Sanga) (14–16%) and root-bark, *Pycnocoma cornuta*, *P. macrophylla* (Akofie-kofi), *Tetracera alnifolia* (Akotopa), *Calotropis procera* Sodom Apple — the macerated extract for dehairing hides, and *Uncaria talbotii* (Akoo-ano). Some leaves yield tannin. These are those of *Adansonia digitata* Baobab, the dried, young leaves of *Lawsonia inermis* Henna, and the leaf decoction of *Combretum micranthum* (Landaga). The leaves of *Ficus dicranostyla* are worked up in water and used for dehairing hides. Others are the pods or fruits of *Acacia sieberiana* var. *villosa* (Kulgo), *Lagenaria breviflora* (*Adenopus breviflorus*) (Anuwatre), *Cathormion altissimum* (Abobonkakyere), *Tetrapleura tetraptera* (Prekese), and the fruit shells of *Persea americana* Avocado Pear. Other sources of tannin are the seeds of *Cassia occidentalis* Negro Coffee, the powdery portions of the fruit fibres of *Cocos nucifera* Coconut Palm, the latex of *Jatropha curcas* Physic Nut, (Adaadze); the red resin (kino) from the dry sap of *Pterocarpus erinaceus* Senegal Rose Wood Tree, the juice of *Strophanthus preussii* (Dietwa), and the roots of *Securidaca longepedunculata* (Kpaliga). The genus *Erythrina* represented by *E. addisoniae* (Sorowa) contains tannin, and the wood of *Pericopsis elata* (*Afrormosia elata*) (Kokrodua) has some tannin. The bark of *Coccoloba uvifera* Sea-side Grape and of *Psidium guajava* is useful for tanning and dyeing.

Dyes

Plants yield many colours for dyeing a variety of articles. The colours are fixed with mordants — usually ashes of plant material or salt. In the forest zone plant

9. POISONS, TANNINS, DYES

dyes are used to dye raffia, *Edow*, for weaving mats, bags, and other raffia crafts. The most popular use of plant dyes in the forest zone is the preparation of saffron dyes from the bark of *Rhodognaphalon brevicuspe* (*Bombax brevicuspe*) (Onyinakobin) and that of *Lannea welwitschii* (Kumanini) for dyeing funeral cloths (*kuntunkuni*) and for stamping the *edinkra* signs and symbols. In the savanna zone plant dyes are employed mainly in the basket and leather industries.

INDIGO BLUE

Sources of this colour are the young leaves of *Lonchocarpus cyanescens* West Africa or Yoruba Indigo (Akase) for which the plant is sometimes cultivated; leaves of *L. laxiflorus* (Nalenga) and the leaves of several species of *Indigofera*. The richest source is from *I. arrecta* Natal or Java Indigo. Others are *I. tinctoria* True or Frank Indigo, *I. hirsuta* Hairy Indigo, *I. spicata*, and *I. suffruticosa* West Indian Indigo. The ashes of *Parkia biglobosa* (Duaga) and *Combretum glutinosum* (Nkunga) are used in the indigo industry. *Cycnium camporum* and *Striga hermonthica* (Wumlim) turn black on drying and are useful as mordant or as adjuvant with indigo dyeing to deepen the colour.

BLUE

This colour is obtainable from the fruits of *Cremaspora triflora* (Otu); the seeds as well as corollas of *Clitoria ternatea* Blue Pea as litmus substitute and for colouring boiled rice; the leaves of *Desmodium mauritianum*; and the leaves, flowers, and twigs of *Saba florida* (Akontoma). *Cassia obtusifolia* (*tora*) Foetid Cassia is also reported to yield a blue dye.

BLUE-BLACK

The following plants give a blue-black colour: *Buchnera leptostachya* (Dam-Pan); the bark of *Jatropha curcas* Physic Nut; the seeds of *Gardenia vogelii* usually used as a body-paint; the stems and leaves of *Mucuna flagellipes* (Tatwea) together with the fruits of *Alchornea cordifolia* Christmas Bush, (Gyamma); the pods and seeds of *Piliostigma thonningii* (Opitipata) and the juice of the fresh fruits of *Rothmannia whitfieldii* (Sabobe) with the dried and finely ground seeds, used both for staining the skin and for dyeing fabrics.

BROWN

Plants that yield this colour include the bark of *Khaya senegalensis* Dry-Zone Mahogany for camouflaging hunter's clothes, *Cola acuminata* Commercial Cola Nut Tree mixed with that of *Khaya*; *Pseudocedrela kotschyi* Dry-Zone Cedar for dyeing cloth, *Tamarindus indica* Indian Tamarind, *Entandrophragma angolense* Gedu Nohor, (Edinam); *Mimusops elengi* and that of *Leucaena leucocephala* (*glauca*) Leucaena usually for dyeing fishing tackle; the leaves of *Trema orientalis* (*guineensis*) (Sesea); the sap of *Maesa lanceolata* and the roots of

Terminalia laxiflora.

RED

The following plant parts give a red colour: the pods of *Acacia nilotica* var. *tomentosa*; the leaves of *Anthonotha macrophylla* (Totoro); *Lawsonia inermis* Henna for hands, feet, nails, and also for leather and wool; the leaves of *Tectona grandis* Teak and the leaf-sheaths of *Sorghum* species with natron for dyeing mats, fibres, and cloth; the leaves of *Strychnos afzelii* (Duapepere) are chewed to make the lips red; the red wood of *Haematoxylon campechianum* Logwood, *Adenanthera pavonina* Red Sandalwood, Bead Tree, and that of *Baphia nitida* Camwood (Odwen). This was exported in 1946 as Kam-dye chiefly to the United States and was priced in London at £5 to £8 per ton. The bark of *Avicennia africana* Mangrove and the latex and wood of *Trilepisium madagascariense* (*Bosqueia angolensis*) (Okure) for mats and clothing; the bark of *Cnestis ferruginea* (Akitase), *Psychotria psychotrioides*, and *Morinda lucida* Brimstone Tree. The genus *Dracaena*, especially *D. surculosa* var. *surculosa* (Mobia) yields a red colour; also the flowers and red-veined leaves and twigs of *Gossypium arboreum* Cotton; the roots and bark of *Phyllanthus reticulatus* var. *reticulatus* and *P. reticulatus* var. *glaber* (Awobe); the red resin, kino, from the stripped bark and stem of *Pterocarpus erinaceus* Senegal Rose Wood Tree, African Kino and *P. santalinoides* (Hote). The burnt bark of *Rhizophora* species Red Mangrove is mixed with soda to dye floors dark red, the boiled bark yielding a red dye for polishing wooden floors. The leaf-juice of *Jatropha curcas* Physic Nut stains red.

RED-BROWN

The bark of the following trees yields a red-brown colour: *Lannea kerstingii* (Kobewu), *L. velutina* (Sinsa), *L. welwitschii* (Kumanini), *Desplatzia dewevrei* (Wisamfia), *Trichilia monadelpha* (*heudelotii*) (Tanduro), *Rhodognaphalon brevicuspe* (*Bombax brevicuspe*) (Onyinakoben), and *B. buonopozense* Red-Flowered Silk Cotton Tree (Okuo) all for dyeing *edinkra* cloths; the bark and roots of *Vitex doniana* Black Plum (Afua) are also used, so are the leaves of *Bridelia ferruginea* (Opam-fufuo) boiled with rusty iron and some shea butter. The dye from the fresh fruits of *Cocos nucifera* Coconut Palm and the leaves of *Ipomoea batatas* Sweet Potato are used for fishermen's nets. The bark of *Erythrophleum suaveolens* (*guineense*) Ordeal Tree (Potrodom); *Ficus glumosa* var. *glaberrima* (Galinziela) and that of *Hymenocardia acida* (Sabrakyie) yield a red-brown dye for dyeing raffia; so do the roots of *Terminalia mollis* (Kwarili) and *T. laxiflora*.

CRIMSON

This dye is obtained by steeping the flowers of *Mirabilis jalapa* Pride of Peru in water. It is used in China for colouring cakes and jellies made from seaweeds.

9. POISONS, TANNINS, DYES

YELLOW

The leaves of *Anacardium occidentale* Cashew Tree, *Anogeissus leiocarpus* (Sakanee) both yield a yellow colour for dyeing tanned skins; the rinds of ripe fruits of *Borassus aethiopum* Fan Palm for dyeing mats; the roots of *Cochlospermum planchonii* and *C. tinctorium* (Kokrosabia) for dyeing fabric, leather, mats, and to colour shea butter; the bark of *Craterispermum laurinum* to dye cloth; the seeds of *Cassia obtusifolia* (*tora*) Foetid Cassia sometimes used with indigo in dyeing; the stems of *Dracaena perrottetii* (Opetentong) and *D. mannii* (Kesene); the bark of *Casuarina equisetifolia* Whistling Pine; and an extract of the bark and wood of *Enantia polycarpa* African Yellow Wood, (Duasika). The late flowers of *Gossypium arboreum* Cotton yield a yellow colour — also the sap of *Harungana madagascariensis* (Okosoa); the bark of *Mitragyna inermis* (Kukyafie) and that of *Mangifera indica* Mango; the roots of *Morinda lucida* Brimstone Tree (Kankroma); the bark, root, and wood of *Nauclea latifolia* African Peach, (Sukisia) and the thickened rhizomes of *Stylochiton lancifolius* and *S. hypogaeus*. The bark of *Terminalia ivorensis* (Emire) yields a yellowish-red pigment for dyeing clothing and the fruit and bark of *Punica granatum* Pomegranate a yellow colouring body.

ORANGE

The colour is obtainable from the seeds of *Bixa orellana* Anatto and used for dyeing silk, cotton, and foods like butter. The seeds of *Lonchocarpus sericeus* Senegal Lilac give a yellowish-orange dye and *Tephrosia purpurea* an orange-brown colour.

PURPLE

The correct combination of red and blue dyes produces purple. A purplish dye is obtained from the bark of *Bridelia ferruginea* (Opam-fufuo) with an alum mordant while the ground bark, wood, and roots of *Pterocarpus erinaceus* Senegal Rose Wood Tree gives a dark purple colour when shea butter or palm oil is rubbed on the material.

GREEN

The colour is usually obtained by mixing yellow dyes with indigo — a secret trade of certain Hausa families. The roots of *Cochlospermum tinctorium* (Kokrosabia) is one such plant used with indigo. The young leaves of *Morinda lucida* Brimstone Tree (Kankroma) together with those of *Lonchocarpus* yield a green dye for mats. Also a cold infusion of the pods of *Tamarindus indica* Indian Tamarind is used with brass, copper filings, and local ammonium chloride to dye leather green.

GREY

The colour is obtained from the leaves of *Jatropha curcas* Physic Nut (Adadze);

the fruits of the Mango tree and the bark of *Ziziphus mauritiana* Indian Jujube.

CINNAMON

A fast, cinnamon-coloured dye is obtained from the bark of *Ziziphus mauritiana* Indian Jujube in Kenya.

BLACK

This dye is obtained from the seed of *Acacia nilotica* var. *tomentosa*, *Dictyandra arbrescens* (Kwaku asra), *Gardenia erubescens* (Dasuli), *Rothmannia urcelliformis* (Obohwe), and the roasted seeds of *Piliostigma thonningii* (Opitipata); the fruit of *Alchornea cordifolia* Christmas Bush (Gyamma) for dyeing fabrics, pottery, calabashes and leather; the fruit of *Kigelia africana* Sausage Tree (Nufuten); *Rothmannia longiflora* (Samankube) and *Hyphaene thebaica* Dum Palm for dyeing leather; the bark of *Acacia farnesiana*, *Syzygium rowlandii* (Asibenyanya) with lime-juice for dyeing cloth; the leaf-juice of *Combretum mucronatum* (*smeathmannii*); the leaf of *Clerodendrum splendens* (Ekenyieya), *C. umbellatum* (Niabiri), *Physostigma venenosum* Calabar Bean for dyeing wool, *Rothmannia hispida* (Tukobo) with the fruits (often applied to the skin in south-east Nigeria), *Griffonia simplicifolia* (Kagya), *Terminalia catappa* Indian Almond, and *T. macroptera* (Kwatiri). The rest are the stems of *Phyllanthus muellerianus* (Potopoleboblo) for dyeing fibres; the roots and bark of *P. reticulatus* var. *reticulatus* and *P. reticulatus* var. *glaber* (Awobe); the stems and leaves of *Mucuna flagellipes* (Tatwea), *M. pruriens* var. *pruriens* Cow Itch (Apea), and *M. sloanei* Horse-Eye Bean (Samante) all with the fruits of *Alchornea cordifolia* Christmas Bush. The leaf-juice of *Jatropha curcas* Physic Nut stains linen an indelible black and a decoction of *Ipomoea asarifolia* also stains cloth black.

Mordants

These serve to fix colours. They include the wood ashes of *Albizia adianthifolia* (Pampena), *Alchornea cordifolia* Christmas Bush (Gyamma); *Erythrina mildbraedii* (Nfona), *Anthocleista vogelii* (Awora-bontodee), *Balanites aegyptiaca* Desert Date, *Vitellaria paradoxa* (*Butyrospermum paradoxum* subsp. *parkii*) Shea Butter Tree, *Combretum glutinosum* (Nkunga), *Cussonia arborea* (*barteri*) (Saaborofere), *Ficus asperifolia* Sandpaper Tree, *Parkia biglobosa* (Duaga), *Spondias mombin* Hog Plum, Ashanti Plum, and *Anogeissus leiocarpus* (Sakanee); the ashes of wood and fruit of *Sterculia tragacantha* African Tragacanth (Sofo); the ashes from the nuts of *Borassus aethiopum* Fan Palm; the pod ashes of *Ceiba pentandra* Silk Cotton Tree, *Pentaclethra macrophylla* Oil Bean Tree, and *Piliostigma thonningii* (Opitipata); the ashes from the root and branches of *Jatropha curcas* Physic Nut; the ashes of the fruits of *Parinari excelsa* Guinea Plum (Ofam) and a cold infusion of the pods of *Tamarindus indica* Indian Tamarind with their seeds. The fruits of *Blighia sapida* Akee Apple are used in fixing colours. Common salt and natron are useful as ingredients in colour fixing.

Ink

Writing or marking ink is made by mixing carbon with a gum in water. The carbon is obtained by scraping the exterior of cooking pots where soot accumulates from the burning. Plant sources of writing and marking ink are the pods of *Acacia nilotica* var. *tomentosa*, *A. farnesiana*, and *Cathormion altissimum* (Abobonkakyere) (red colour); the fruits of *Desplatzia subericarpa* (Esonowisamfie-bere), *Eugenia coronata* (Kraku), *Securinega virosa* (Nkanaa), and *Solanum nigrum* (Nsusuabiri). Samples of Nsusuabiri used in Angola, reportedly, have retained the purple colour for many years. The charcoal of *Balanites aegyptiaca* Desert Date (Gongu) makes ink; also the latex of *Trilepisium madagascariense* (*Bosqueia angolensis*) (Okure) (red indelible) and the juice of the fruits of *Opuntia* species Prickly Pear (Nkantonsoe) as ingredient in red ink. Others are the seed extract of *Rothmannia whitfieldii* (Sabobe); the young leaves or fruits or bark of *Vitex doniana* Black Plum, (Afua) and *V. fosteri* (Otwentorowa). The gum of a number of plants is an ingredient in the preparation of ink, especially Malam's ink. They include *Acacia dudgeoni* (Gosei), *A. polyacantha* subsp. *campylacantha* African Catechu, (Gorpila); *A. senegal* Gum Arabic of commerce, *Anogeissus leiocarpus* (Sakanee), and *Combretum nigricans* var. *elliotii*. *Boswellia dalzielii* Frankincense Tree, (Kabona) is an ingredient with Black Plum in Malam's ink. The gum of *Sclerocarya birrea* (Nanogba) is mixed with soot and water and used as ink.

Tattooing

This involves the puncturing of the skin with patterns, designs or symbols, and names and rubbing in a pigment to produce a permanent mark. Plants used are the juicy green leaves of *Eclipta alba* (*prostrata*) rubbed over the skin after puncturing to produce an indelible deep bluish-black; the juicy roots of *Cassia alata* Ringworm Shrub (also for tribal or facial markings), and the roots of *C. podocarpa* (Nsuduru). The ground roots of *Sapium grahamii* (Pampiga) with a little water produces red or black marks on the face causing swelling and ultimately tattoo-like marks. The gum of *Daniellia ogea* Gum Copal Tree (Hyedua) burned and mixed with soot and oil and the carbon obtained from burning the gum of *Canarium schweinfurthii* Incense Tree is used for tribal markings; also the ground fruits and seeds of *Coffea ebracteolata* (Nhwesono); the seed extract of *Rothmannia whitfieldii* (Sabobe) and the fruits of *R. longiflora* (Samankube) for blue-black markings. The fresh juice of the fruits of *Anacardium occidentale* Cashew Nut is used for tribal markings and that of *Sorindeia warneckei* (Akpokpoe) gives a bluish colour, also does *S. grandifolia*.

Saponin-producing Plants

Some plants have the property of producing saponin which lathers in water and is capable of use for some washing. They are the bark of *Adenanthera pavonina* Red Sandalwood, Bead Tree; *Albizia lebbeck* East Indian Walnut, *Cathormion altissimum* (Abobonkakyere), *Cylicodiscus gabunensis* African Greenheart, (Denya); *Piptadeniastrum africanum* (Danhoma), *Psorospermum corymbiferum* var. *corymbiferum* when boiled, *Samanea dinklagei, Tetrorchidium didymostemon* (Aboagyedua), and *Vitex doniana* Black Plum, (Afua). Others are the leaves of *Albizia coriaria* (Awiemfosamina-akoa), *A. ferruginea* (Awiemfosamina), *Vitellaria paradoxa* (*Butyrospermum paradoxum* subsp. *parkii*) Shea Butter Tree, *Dracaena mannii* (Kesene), *Grewia bicolor*, and *Xylia evansii* (Samantawa); also the fruits of *Balanites aegyptiaca* Desert Date, Soap-berry Tree; *Blighia sapida* Akee Apple (Akye), *Piliostigma thonningii* (Opitipata), *P. reticulatum*, and *Tetrapleura tetraptera* (Prekese). The seeds of *Lepisanthes senegalensis* (*Aphania senegalensis*) (Akisibaka), *Cassia sieberiana* African Laburnum, *Moringa oleifera* Horse-Radish Tree, *Securidaca longepedunculata* (Kpaliga) (or the root bark) also yield saponin. The rest are the root of *Cassia nodosa* Pink Cassia and the bark, and *Olax subscorpioidea* (Ahoohenedua), the stems of *Gouania longipetala* (Homabiri), *Phytolacca dodecandra* (Ahoro), and the sap of *Jatropha curcas* Physic Nut.

Sources of Potash for Soap-making

Soap-making is an important rural industry in West Africa especially among women, either individually or in organized groups. The materials are all locally obtained, and consist of plant ashes and oils from palm nuts and kernels and coconut. To make the soap more detergent and antiseptic dried leaves of *Carica papaya* Pawpaw are used (Boateng, 1970) and soap so manufactured is said to cure skin diseases. The source of potash includes the wood-ashes of *Acacia albida* (Gozanga), *Afraegle paniculata* (Obuobi), *Annona senegalensis* var. *senegalensis* Wild Custard Apple, *Anthocleista nobilis* Cabbage Palm, *A. vogelii* (Awora-bontodee), *Ceiba pentandra* Silk Cotton Tree, *Cocos nucifera* Coconut Palm, *Cussonia arborea* (*barteri*) (Saa-borofere), and the wood and ashes of *Dracaena arborea* (Ntonme) and *Jatropha curcas* Physic Nut. The wood-ashes of *Erythrina mildbraedii* (Nfona), *Ficus vogeliana* (Opanto), *Gardenia ternifolia* (Peteprebi), *Isoberlinia doka* (Sapelaga), *Musanga cecropioides* Umbrella Tree (Dwumma), *Myrianthus arboreus* (Anyankoma), *Oxyanthus unilocularis* (Kwaetawa), *Parkia biglobosa* (Duaga), and *P. clappertoniana* West African Locust Bean. The wood-ashes of *Ricinodendron heudelotii* (Wamma), *Scottellia klaineana* var. *klaineana* (*coriacea*) Odoko, (Lakpa), *Sterculia rhinopetala* Sterculia Brown, (Wawabimma); and *S. tragacantha* African Tragacanth (Sofo) (and the fruit-ashes); and the wood-ashes of *Vitex doniana* Black Plum (Afua). The seed, shell, and bark-ashes of *Adansonia digitata* Baobab, the bark-ashes of *Afzelia africana* (Papao), and the burned, dried, pounded bark of *Mammea africana* African Mammy Apple (Bompagya). The dried husks and seeds of *Blighia sapida* Akee Apple (Akye); the pod-ashes of *Bussea occidentalis* (Kotoprepre), *Calpocalyx brevibracteatus* (Atrotre), *Theobroma cacao* Cocoa, *Prosopis africana* (Sanga), *Samanea dinklagei*, *Tetrapleura tetraptera* (Prekese), and *Xylia evansii* (Samanta). The pericarp ashes of *Caloncoba echinata* (Gorli), the pods and wood-ashes of *Entada abyssinica* (Sankasaa); the pod and bark-ashes of *Piptadeniastrum africanum* (Danhoma) and the ashes of the flowering stalks and leaves of *Elaeis guineensis* Oil Palm. The ashes of the fronds of *Raphia hookeri* Wine Palm; the leaf-ashes of *Trema orientalis* (*guineensis*) (Sesea), and the ashes of *Triumfetta cordifolia* Burweed, *T. rhomboidea* Burweed (Petekuku), and *Vernonia conferta* (Flakwa).

Aromatic Plants

Plants used for their aroma include the seeds of *Aframomum melegueta* Melegueta (Famu-wisa) and *Monodora myristica* (Ayerew-amba); the flowers of *Acacia farnesiana*, *Beilschmiedia mannii* Spicy Cedar, *Dalbergia heudelotii* (Akasia), *Jasminum sambac* (introduced — for scenting tea), *Lawsonia inermis* Henna, *Lecaniodiscus cupanioides* (Dwindwera), *Macrosphyra longistyla* (Zetitsui), *Plumeria rubra* var. *acutifolia* Red Frangipani, Temple Flower; and *Tetrapleura tetraptera* (Prekese); the fermented seed-pod of *Vanilla planifolia* (cultivated for the purpose) and the burnt bark of *Terminalia macroptera* (Kwatiri); the leaves of *Haumaniastrum lilacinum* (Taga), *Strychnos afzelii* (Duapepere), *Chenopodium ambrosioides* American or Indian Wormseed, *Hyptis suaveolens* Bush Tea-Bush, *Lippia multiflora* Gambian Tea Bush (Saa-nunum), *Lantana camara* Wild Sage, *Annona senegalensis* var. *senegalensis* Wild Custard Apple, *Leucas deflexa*, *Steganotaenia araliacea* (Pienwogo), and *Platostoma africanum*; the bark of *Pachypodanthium staudtii* (Fale), *Oxyanthus speciosus* (Korantema), *Canarium schweinfurthii* Incense Tree (and oleo-resin), and *Cinnamomum zeylanicum* Cinnamon.

Other aromatic sources are the essential oil of *Hoslundia opposita* (Asifuaka), *Cymbopogon citratus* Lemon Grass (and leaves), *C. giganteus* var. *giganteus* (and inflorescence), and the wood-smoke of *Combretum ghasalense* (Atena); the rhizomes of *Cyperus articulatus*, *C. digitatus* subsp. *auricomus* var. *auricomus*, *C. maculatus*, *C. rotundus* Nut Grass, *Kyllinga tenuifolia*, *K. erecta*, *K. pumila*, *K. squamulata*, and *Vetiveria zizanioides* Vetiver (cultivated for the purpose), the resin of *Commiphora africana* African Bdellium (Narga); *C. dalzielii* African Myrrh, *C. pedunculata*, *Copaifera salikounda* Bubinga, (Entedua); *Daniellia oliveri* African Copaiba Balsam Tree, *D. thurifera* Niger Copal Tree, and *Boswellia dalzielii* Frankincense Tree (Kabona).

Drugs

A drug is a substance used alone or as ingredient in medicine and for narcotic effect (see Table 9.2).

Table 9.2 Common plant-derived drugs and their effects

Name	Chemical or Botanical Name	Source	Classification	Medical Use	Effects Sought	Long-term/Heavy-use Symptoms
Marijuana	*Cannabis sativa*	Hemp	Relaxant Euphoriant Hallucinogen (in high doses)		Relaxation Increased euphoria or perception	Possible psychological addiction Possible lung, memory, or sexual damage
Alcohol	Ethanol or Ethyl alcohol	Palm Trees, Sugarcane, Pineapple, other fruits and grain	Sedative Hypnotic	Solvent Antiseptic Dietary	Sense alteration Anxiety reduction and Sociability	Cirrhosis Toxic psychosis Neurologic damage Addiction
Nicotine	*Nicotiana tabacum* *N. rustica*	Tobacco	Stimulant Sedative	Sedative Emetic	Calmness Sociability	Emphysema Lung, mouth, and throat cancer Cardiovascular damage Loss of appetite Addiction
Caffeine	*Coffea* spp. *Camellia sinensis* *Theobroma cacao* *Cola* spp	Coffee Tea Cocoa Kola Nuts	Stimulant	Stimulant	Alertness Sociability	Jitteriness Mild addiction

9. POISONS, TANNINS, DYES

Other plant drugs include the smoked, dried leaves of *Palisota hirsuta* (Nzahuara), the seeds of *Datura metel* Metel, Hairy Thorn Apple and *D. innoxia*, and the starchy roots of *Ipomoea pes-caprae* subsp. *brasiliensis* Beach Convolvulus, Goat's Foot Convolvulus. (See also Chapter 8.)

Essential Oils

Plants that yield essential oil include *Cinnamomum camphora* Japanese Camphor Tree, *C. zeylanicum* Cinnamon, *Cananga odorata* Perfume Tree, *Hoslundia opposita* (Asifuaka), *Ocimum gratissimum* Tea Bush, (Nunum); *Piper guineense* West African Pepper (Soro-wisa); *Vertiveria zizanioides* Vetiver, *Cymbopogon citratus* Lemon Grass, *C. nardus* Citronella, *Cyperus articulatus*, and *Aframomum melegueta* Melegueta (Famu-wisa). Essential oils are useful in the perfumery industry.

Other Oils

Plants that contain inedible but useful oils include *Licania elaeosperma* (Kokorobe) (substitute for linseed oil), *Afzelia africana* (Papao), *Allanblackia parviflora* (*floribunda*) Tallow Tree (Sonkyi) (suitable for soap-making); *Azadirachta indica* Neem Tree (hair oil, etc.), *Caloncoba echinata* (Gorli) and *C. glauca* (for soap and candle-making), *Chrysophyllum delevoyi* (*albidum*) White Star Apple (for soap-making), *Quassia undulata* (*Hannoa undulata*) (Kunmuni) (for soap-making), *Irvingia gabonensis* Wild Mango, (Abesebuo) (for soap-making), *Jatropha curcas* Physic Nut (for various purposes including soap and candles), *Parinari curatellifolia* (Atena) (for paints and varnishes), *Pentadesma butyraceum* (*butyracea*) Tallow Tree (for candle-making and margarine manufacture), *Pycnanthus angolensis* African Nutmeg (Otie) (for soap-making), *Ricinodendron heudelotii* (Wamma) (for varnish and soft soap), *Trichilia emetica* subsp. *suberosa* (*roka*) (Kisiga) (for soap-making), and *Caesalpinia bonduc* Bonduc (Oware-amba) (fatty oil). (For edible oils, see Chapter 2.)

TEN
Amenity Landscaping and Gardening

The natural beauty of plants is as obvious to the casual observer as it is to the professional botanist. The appreciation of the beauty of decorative plants, and man's desire to live in beautiful environments, have resulted in the introduction of many exotics to alien countries. Beautiful surroundings are healthy surroundings. The aesthetic values of trees are obvious. A forest scene, an avenue, stands in a courtyard, or a few trees that shade a cottage make all the difference. The effectiveness of the presence of trees is best appreciated by making a visual comparison between an estate surrounded by a garden of trees and one without. A third aspect is the shade, protection, and shelter given by trees. In the tropics and sub-tropics especially, shade-bearing trees are useful in residential areas, public places, and the countryside. Effective landscaping relies on the beauty, scenery, and shading provided by plants and trees in and around residential areas, offices, public buildings, churches, mosques, and along roadsides, on forest boundaries, in fields, parks, open markets, and cemeteries.

The flora of West Africa includes a wide range of decorative plants worth cultivating. A few like *Spathodea campanulata* African Tulip (Kokoanisua); *Ruspolia hypocrateriformis* and *Millettia thonningii* (Sante) are being cultivated while others have been introduced to tropical America and Asia. The majority still grow wild, however. Many decorative plants, especially annuals, have also been introduced from other tropical and sub-tropical countries into West Africa. A plant like *Catharanthus roseus* Madagascar Periwinkle has since naturalized and often grows wild. In Ghana, the Department of Parks and Gardens is responsible for the development of the landscape with the planting of decorative plants in all the regional capitals and major towns. There are a few botanical gardens in the country. The biggest and oldest, established in 1890, is the Aburi Botanical Gardens situated on the Akwapim Range some 36 km north-east of Accra at an elevation above sea level of 462 m (1500 ft) at the highest point. The garden covers 64 hectares (160 acres), but only 12 hectares have been developed. There are botanical gardens at each of the three universities: University of Ghana, Legon; University of Science and Technology, Kumasi; and the University of Cape Coast. Such gardens are very useful for visitors who wish to see the shape and form of plants.

As avenues, shade trees, hedges, lawns, forests, and woodlands — whether in the countryside, residential areas, or industrial centres — plants effectively contribute, in no small measure, to the scenery.

Avenues

An avenue is a wide, tree-lined street. The trees are evenly spaced along both sides of the street in pure stands, usually, or sometimes in mixed stands. Avenues give shade, improve the scenery, and add to the aesthetic value of the landscape. Decorative and evergreen trees make the best avenues. Trees with deep but moderate root-systems are preferred to those with big surface roots. Some streets — Neem Avenue, Palm Avenue, and Indian Cedar Avenue, for example — are named after the trees that line them. Suitable trees for avenues include *Albizia lebbeck* East Indian Walnut, *Anogeissus leiocarpus* (Sakanee), *Blighia sapida* Akee Apple (Akye) (also as shade tree); *B. unijugata* (Akyebiri) (also as shade tree), *Burkea africana* (Pinimo), *Cedrela mexicana* Mexican Cedar (also for boundary planting), *Delonix regia* Flamboyante, Flame Tree (decorative, with bright red flowers — yellow-flowered varieties occur); *Ficus dekdekena* (*thonningii*) (Gamperoga), *F. lutea* (*vogelii*) (Ofonto), *Hildegardia barteri* (Akyere), *Khaya senegalensis* Dry-Zone Mahogany (often heavily debarked for medicinal purposes), *Lonchocarpus sericeus* Senegal Lilac (also as shade tree), *Mangifera indica* Mango (also for boundary planting), *Millettia thonningii* (Sante) (decorative, with the bloom of purple flowers at the beginning of the first rains), *Parkia clappertoniana* West Africa Locust Bean, *Peltophorum pterocarpum* Copper Red, *Pentaclethra macrophylla* Oil Bean Tree (Ataa); *Tamarindus indica* Indian Tamarind (also as shade tree), *Tectona grandis* Teak, *Trichilia emetica* var. *suberosa* (*roka*) (Kisiga), *Samanea saman* Rain Tree, *Erythrophleum suaveolens* (*guineense*) Ordeal Tree (Potrodom); *Hura crepitans* Sandbox Tree and *Spathodea campanulata* African Tulip (Kokoanisua). Introduced trees used for avenues include *Roystonia regia* (*Ereodoxa regia*) Royal Palm, *Tabebuia pentaphylla* Roble Blanco, *Millingtonia hortensis*, *Azadirachta indica* Neem Tree, *Michelia champaca,* and *Casuarina equisetifolia* Whistling Pine.

Shade Trees

In the tropics especially, shade trees offer cover from the scorching sun, and are usefully employed in recreational areas, market centres, and school yards. Preferably, evergreen trees with wide crowns are selected. Shade trees include *Acacia sieberiana* var. *villosa* (Kulgo), *Adenanthera pavonina* Red Sandalwood, Bead Tree (also decorative); *Balanites aegyptiaca* Desert Date, *Berlinia grandiflora* (Tetekono) (decorative and very beautiful when in full bloom), *Cola gigantea* var. *glabrescens* (Watapuo), *Cordia millenii* Drum Tree, *C. platythyrsa* (Tweneboa), *Lecaniodiscus cupanioides* (Dwindwera) (decorative), *Melia azedarach* Persian Lilac, and *Morus mesozygia* (Wonton). A number of *Ficus* species are useful as shade trees. These are *F. glumosa* var. *glaberrima* (Galinziela) (also for live fence), *F. ingens* var. *ingens* (Kunkwiya) (also for live fence), *F. lyrata* (and for avenues), *F. ovata* (also for live fence), and *F. platyphylla* Gutta-Percha Tree. Climbing plants like *Vitis vinifera* Vine Plant,

Argyreia speciosa Woolly Morning Glory, a native of India, Java, and China, and *Ipomoea* and *Combretum* species are cultivated over sheds to provide shade.

Hedges

A hedge is a fence of bushes and serves as boundary, screen, and noise breaker. A well-cut hedge improves the beauty of a garden and of a property's façade. Hedge plants include *Acacia farnesiana* (decorative when in flower), *Acalypha wilkesiana* (with variegated leaves), *Bauhinia rufescens* (Jinkiliza) (also as ornamental shrub), *B. tomentosa* Napoleon's Hat (decorative), *Bixa orellana* Anatto (decorative), *Breynia nivosa* Snow Bush, *Bridelia stenocarpa* (*micrantha*) (Opam) (or boundary plant), *Caesalpinia pulcherrima* Pride of Barbados, Flower Fence (decorative and impenetrable due to the thorns); *Calophyllum inophyllum* Alexandra Laurel (also a useful windbreak), *Carissa edulis* (Botsu) (thorny and forms a dense, impenetrable hedge), *Cassia bicapsularis, C. laevigata, Codiaeum variegatum* Garden Crotons, *Dodonaea viscosa* Switch Sorrel (Fomitsi); *Duranta repens* Golden Dewdrop (most decorative with the hanging, yellow fruits — varieties with variegated leaves occur), *Eugenia coronata* (Kraku) (introduced to the United States by Dr Dalziel as ornamental hedge plant), *E. leonensis, E. uniflora* Pitanga Cherry (introduced from Brazil and famous for its sharp, sweet fruits), *Euphorbia lateriflora* (Kamfo-barima) (planted as a 'charm' by farmers to protect their crops), *Ficus leprieuri* (Amangyedua) (also as shade tree), and *Flacourtia flavescens* Niger Plum (impenetrable due to the thorns). Other plants used for hedges are *Haematoxylon campechianum* Logwood (decorative), *Jatropha curcas* Physic Nut (usually as hedge plant in farms and villages), *J. gossypiifolia, Lantana camara* Prickly Lantana, Wild Sage; *Lawsonia inermis* Henna, *Leucaena leucocephala* (*glauca*) Leucaena, Wild Tamarind (also as a windbreak), *Chionanthus mannii* (*Linociera mannii*) (Akokotua), *Maesa lanceolata, Malpighia glabra* Barbados Cherry, *Oxytenanthera abyssinica* Bamboo (also for windbreaks), *Parkinsonia aculeata* Jerusalem Thorn (effective when combined with Pride of Barbados), *Pisonia aculeata* (Mpintinko), *Pithecellobium dulce* Madras Thorn, *Securinega virosa* (Nkanaa), *Steganotaenia araliacea* (Pienwogo), *Tecoma stans* Trumpet Bush, Yellow Bells, *Thevetia peruviana* 'Milk Bush', Exile Oil Plant; *Thunbergia erecta* (decorative), *Ximenia americana* Wild Olive, *Ziziphus abyssinica* (Larukluror) (stems as hedge to exclude wild beast), *Z. mauritiana* Indian Jujube (also to exclude wild beasts), and *Dracaena arborea* (Ntonme) (boundary fence and planted around shrines).

Lawns

Lawns serve as recreational grounds, games fields, and picnic sites. Suitable grasses for lawns, football fields, and golf courses include *Axonopus*

10. AMENITY LANDSCAPING

compressus, Chrysopogon aciculatus Love Grass, *Digitaria debilis, D. ciliaris, D. horizontalis,* and *D. longiflora*; others are *Paspalum conjugatum* Sour Grass (in forest areas) and *P. vaginatum; Stenotaphrum secundatum* Buffalo Grass, *Panicum parvifolium,* and *P. laxum.* By far the most widely used lawn grass in West Africa is *Cynodon dactylon* Bahama Grass, Bermuda Grass. Other grasses suitable for lawns are *Bothriochloa bladhii, Chrysochloa hindsii, Dactyloctenium aegyptium, Eleusine indica, Monocymbium ceresiiforme, Perotis hildebrandtii,*and *Poa* species, an introduced grass. A few prostrate dicots also make good lawns, alone or mixed with the grasses above. These are *Alysicarpus rugosus, Desmodium triflorum, Indigofera kerstingii, I. spicata, Evolvulus alsinoides, E. nummularius, Cassia rotundifolia, Gomphrena celosioides, Oxalis corniculata, Portulaca grandifolia, P. quadrifida, P. foliosa, Richardia brasiliensis, Sesuvium portulacastum, Tribulus terrestris* Devil's Thorn, and *Zornia latifolia.*

Water Plants

Decorative water plants are *Nymphaea micrantha, N. lotus* Water Lily, *N. maculata, Nelumbo nucifera* (introduced), *Typha domingensis (australis)* Bulrush, *Thalia welwitschii, Crinum natans, Ipomoea aquatica,* and *Neptunia oleracea. Ceratophyllum demersum, Vallisneria aethiopica, Anubias* species such as *A. minima, Lagarosiphon hydrilloides,* and *Elodea canadensis* American Water Weed are used as decoration in home aquariums. *Eichhornia crassipes* Water Hyacinth (an introduction) is a decorative weed. As Water Hyacinth has become a notorious water weed in other parts of the continent, its introduction into West Africa poses a problem (see Chapter 11). *Ceratopteris cornuta* Water Fern and *Marsilea* species are decorative aquatic ferns. Water plants contribute to the beauty of ponds, aquariums, dams, and canals.

Decorative Plants

A number of plants, from herbs to trees, are of value for decorative purposes. These range from ferns, orchids, climbers, shrubs, to small and big trees.

FERNS

Decorative ferns include *Lycopodiella cernua (Lycopodium cernuum), Selaginella vogelii S. myosurus, Marattia fraxinea, Dicranopteris linearis (Gleichenia linearis), Lygodium smithianum, Microlepia speluncae, Pteridium aquilinum, Adiantum confine, A. incisum, A. philippense, A. mettenii (soboliferum), A. tetraphyllum* var. *vogelii (vogelii), Doryopteris concolor* var. *kirkii (kirkii), D. concolor* var. *nicklesii (nicklesii), Pityrogramma calomelanos* Silver Fern (a New World species), *Pteris intricata, P. preussii, P. pteridoides, P. tripartita, Platycerum elephantotis (angolense), P. stemaria, Arthropteris monocarpa, A. palisotii, Davallia chaerophylloides, Nephrolepis davallioides, Asplenium akimense, A. buettneri, A. cuneatum, A. diplazisorum, A. dregeanum, A. formosum, A. megalura, A. preussii,*

A. protensum, A. schnellii, A. stuhlmannii, A. unilaterale, Pneumatopteris afra (*Cyclosorus afer*), *Pseudophegopteris cruciata* (*Thelypteris cruciata*), *Diplazium hylophilum, Bolbitis heudelotii, Triplophyllum vogelii* (*Ctenitis lanigera*), *T. pilosissimum* (*C. pilosissima*), *T. protensum* (*C. protensa*), *T. securidiforme* (*C. securidiformis*), *Lastreopsis nigritiana,* (*Ctenitis pubigera*), *L. subsimilis* (*C. subsimilis*), *L. vogelii* (*C. subcoriacea*), and *Tectaria angelicifolia.* (See also under Water Plants above.)

ORCHIDS

Orchids are among the most decorative and fascinating members of the plant kingdom. They also represent one of the largest groups of higher plants. Decorative ground orchids include *Corymborkis corymbis* (*corymbosa*), *Habenaria buettneriana, H. buntingii, H. filicornis, H. gabonensis, H. genuflexa, H. huillensis, H. macrandra, H. procera, H. zambesina, Hetaeria occidentalis, Liparis nervosa* (*guineensis* and *rufina*), *Nervilia adolphii, N. fuerstenbergiana, N. kotschyi, N. reniformis, N. umbrosa, Malaxis maclaudii, Manniella gustavii, Platylepis glandulosa* (introduced to Kew), and *Zeuxine elongata. Eulophia* species are easily transported and propagated by the tuberous underground stems; the decorative species include *E. cristata, E. alta, E. buettneri, E. angolensis, E. euglossa, E. gracilis, E. juncifolia, E. guineensis, E. horsfallii* (with big flowers), *E. orthoplectra, E. odontoglossa, E. warneckeana, E. sordida, E. cucullata* (*dilecta*), and *E. flavopurpurea.* Other ground orchids are *Oeceoclades maculata* (*Eulophidium maculatum*), *O. saundersiana* (*E. saundersianum*), and *O. latifolia* (*E. latifolium*). Orchids are in great demand by both domestic and commercial horticulturists in European countries, Britain, and the United States. Exports of orchids exceed the export of the other decorative plants collectively. This demand is a threat to some of the endangered orchids like *Plectrelminthus caudatus* and *Ancistrochilus rothschildianus.* The largest threat, however, is the continuous destruction of the tropical rain forests — the natural habitat of many orchids — mainly for timber and for agriculture.

A large number of epiphytic orchis are very decorative. They include *Aerangis biloba, A. calantha, Summerhayesia laurentii* (*A. laurentii*), *Ancistrochilus rothschildianus, A. thomsonianus, Ancistrorrhynchus apitatus, A. cephalotes, Angraecum birrimense, A. chevalieri, A. multinominatum, Ansellia africana, Bolusiella batesii, B. imbricata, B. talbotii, Bulbophyllum barbigerum, B. falcatum* var. *bufo* (*bufo*), *B. calyptratum, B. cocoinum, B. colubrinum, B. scaberulum* (*congolanum*), *B. saltatorium* var. *albociliatum* (*distans*), *B. falcatum, B. pumilum* (*flavidum*), *B. calyptratum* var. *graminifolium* (*graminifolium*), *B. imbricatum* (*linderi*), *B. lupulinum, B.magnibracteatum, B. maximum, B. nigritianum, B. schinzianum* var. *phaeopogon* (*phaeopogon*), *B. pipio, B. purpureorhachis* (with coiled inflorescence), *B. recurvum, B. falcatum* var. *velutinum* (*rhizophorae*), *B. saltatorium, B. falcatum* var. *velutinum* (*velutinum*), and *B. oreonastes* (*zenkerianum*). Many species of *Bulbophyllum* have been introduced to Kew. Other decorative orchids are *Calyptrochilum christyanum, Chamaeangis* species,

Cheirostylis divina, Calyptrochilum emarginatum, Cyrtorchis arcuata, C. aschersonii, C. hamata, C. monteiroae, C. ringens, Diaphananthe bidens, D. curvata , D. laxiflora, D. laticalcar, D. pellucida, D. quintasii, D. rutila, D. sarcorhynchoides, Eurychone rothschildiana, Graphorkis lurida, Listrostachys pertusa, Microcoelia caespitosa, M. dahomeensis, M. macrorrhynchia, Plectrelminthus caudatus, Podangis dactyloceras, Polystachya adansoniae, P. affinis, P. dolichophylla, P. fractiflexa, P. fusiformis, P. galeata, P. golungensis, P. inconspicua, P. laxiflora, P. monolenis, P.mukandaensis, P. paniculata, P. polychaete, P. ramulosa, P. reflexa, P. subulata, P. tessellata, Rangaeris muscicola, R. rhipsalisocia, Solenangis scandens, S. clavata, Tridactyle anthomaniaca, T. armeniaca, T. bicaudata, T. crassifolia, T. gentilii, T. tridentata, Vanilla africana, V. crenulata, and *V. ramosa. Vanilla* species are climbing orchids.

HERBACEOUS PLANTS, CLIMBERS, SHRUBS AND SMALL TREES

These constitute the largest proportion of decorative plants, and are made up of *Acalypha* species, *Acridocarpus alternifolius* (worth cultivating), *A. smeathmannii* (Alasaayo) (beautiful both in flower and fruit, worth cultivating), *Haumanniastrum lilacinum* (Taga), *Adenium obesum, Agelaea obliqua* (Apose), *A. trifolia, Alafia scandens* (Momonimo) (worth cultivating for the handsome flowers), *Sherbournea bignoniiflora* (Kyerebeteni) and *S. calycina* (both worth cultivating), *Baissea multiflora, Barleria oenotheroides* (introduced into Europe as ornamental), *Bauhinia monandra* Pink Bauhinia, *B. purpurea, Byrsocarpus coccineus* (Awennade), *Hedranthera barteri* (introduced into Europe as decorative plant), *Caloncoba gilgiana* (Kotowhiri) (introduced to Cuba), *Keetia hispida (Canthium hispidum)* (Homa-ben), *Cassia alata* Ringworm Shrub, *C. auriculata, Cissus aralioides* (Asirimu), *Clappertonia ficifolia* (Nwohwea) (worth cultivating for the beautiful purple flowers), *Clematis grandiflora,* and *C. hirsuta.* Many *Clerodendrum* species are decorative: these are *C. buchholzii* (Taasendua), *C. capitatum* (Tromen), *C. formicarum, C. japonicum, C. splendens* (Ekenyieya) (introduced to European greenhouses), *C. umbellatum* (Niabiri), *C. violaceum* (worth cultivating), and *C. volubile* (Kumamuno) (worth cultivating).

Others are *Cnestis ferruginea* (Akitase) (the fruits), *Cochlospermum tinctorium* (Kokrosabia) (worth cultivating), and *Cordia rothii.* Several species of *Combretum* are decorative — these include *C. collinum* subsp. *binderianum (binderianum)* (Domapowa), *C. grandiflorum* (Ohwiremnini), *C. paniculatum* (Omeha), *C. platypterum* (Ohwirem) (worth cultivating) *C. racemosum* (Wota), and *C. tarquense* (both worth cultivating); *Cordyline terminalis, Cremaspora triflora* (Otu), *Crotalaria pallida (mucronata)* (Peagoro), *C. verrucosa, Croton zambesicus,* (Dodwatu), *Dalbergia saxatilis* (Nuodolega) (worth introducing into cultivation), *Datura suaveolens, Dicranolepis grandiflora* (Wankya) (worth cultivating), *D. persei* (Prahoma), and *Drypetes floribunda* (Bedibesa) (worth cultivating). *Dracaena* species are often decorative and include *D. arborea*

(Ntonme), *D. adamii, D. bicolor, D. camerooniana, D. elliotii, D. mannii* (Opetentong), *D. phrynioides, D. scoparia* (Kotwe-ake), *D. smithii, D. surculosa* var. *capitata*, and *D. surculosa* var. *surculosa* (Mobia) (both introduced to Europe), *D. ovata* and *D.* species (with yellow-spotted leaves, worth cultivating and introduced into Missouri Botanical Garden and Kew); *Encephalartos barteri* Ghost or Hosanna Palm, *Erythrina mildbraedii* (Nfona), *E. senegalensis* Coral Flowers, *Euadenia eminens* (Dinsinkoro), *Euclinia longiflora* (Anyonfonbokowa) (introduced to Kew, worth cultivating), *Feretia apondanthera* (Bitinamusa), *Gaertnera paniculata* (Pitapo), *Didymosalpinx abbeokutae* (introduced to Kew), *Gardenia ternifolia* (Peteprebi), *G. vogelii, Heisteria parvifolia* (Sikakyia), *Hugonia planchonii* (worth cultivating), *Jasminum dichotomum* (Krampa) (introduced to Florida as Gold Coast Jasmine), *Leea guineensis* (Okatakyi), *Macrosphyra longistyla* (Zetitsui), *Mussaenda afzelii* (worth cultivating), *M. chippii, M. elegans* (Damaram), *M. erythrophylla* Ashanti Blood, and *Aganope leucobotrya* (*Ostryderris leucobotrya*) (Anso dua anso huma).

Many species of *Ochna* are very decorative such as *O. staudtii* (*kibbiensis*) (Kwaasiwa), *O. multiflora, O. ovata, O. rhizomatosa, O. schweinfurthiana* (Bibie) and species of *Ouratea* such as *O. affinis* (Ananse Don), *O. calophylla* (Opunini), *Campylospermum flavum* (*O. flava*) (Epebegai) (cultivated in the Royal Botanic Garden, Edinburgh), *C. glaberrimum* (*O. glaberrima*), *Rhabdophyllum affine* subsp. *myrioneurum* (*O. myrioneura*) (Duabogo), *C. sulcatum* (*O. sulcata*), *C. reticulatum* (*O. reticulata*), *C. schoenleinianum* (*O. schoenleiniana*), and *C. vogelii* (*O. vogelii*). Others are *Phyllocosmus chippii* (*Ochthocosmus chippii*), *Duparquetia orchidacea* (Pikeabo), *Oncinotis glabrata, O. gracilis*, and *O. nitida; Oncoba brachyanthera, O. spinosa* Snuff-Box Tree, (Asratoa); *Oxyanthus speciosus* (Korantema), *O. subpunctatus* (Aburubiri-anwa), and *O. tubiflorus; Pachycarpus lineolatus, Pararistolochia goldieana* (introduced to Britain), *Parquetina nigrescens* (Aba-kamo), *Pleiocarpa mutica* (Kanwen) (grown as a stove plant in Europe), *Plumbago zeylanica* Ceylon Leadwort, *Premna hispida* (Aunni), *P. quadrifolia* (Atantaba) (worth cultivating), *Calycobolus heudelotii, Protea madienensis* var. *elliotii* (*elliotii*) (worth cultivating), *Pseuderanthemum ludovicianum* (cultivated at Kew), *P. tunicatum, Quisqualis indica* Rangoon Creeper, *Raphia hookeri* Wine Palm (decorative as interior potted plant), *Ritchiea reflexa* (Aayerebi) *Rothmannia longiflora* (Saman-kube), *R. whitfieldii* (Sabobe) (introduced to Kew for the large, showy flowers), *Ruspolia hypocrateriformis, Securidaca longepedunculata* (Kpaliga), and *Sericostachys scandens.*

One of the poisonous plants, *Strophanthus*, also has very decorative flowers and is often cultivated for the purpose. The species are *S. gratus* (Maatwa), *S. hispidus* Arrow Poison, and *S. sarmentosus* (Adwokuma) (introduced to Florida and Cuba); other decoratives are *Tacazzea apiculata* (the decorative flowers are also considered edible), *Tephrosia vogelii* Fish Poison Plant, *Thespesia populnea* (Frefi), *Thunbergia chrysops* (worth cultivating and introduced to Missouri Gardens), *Uncaria africana, U. talbotii* (Akoo-ano), *Usteria guineensis*

10. AMENITY LANDSCAPING

(Kwaemeko), *Turraea heterophylla* (Ahunanya nkwa), and *Vernonia nigritiana* (Gyakuruwa); *Aniseia martinicensis, Bonamia thumbergiana, Cuscuta australis* (semi-parasite), *Evolvulus alsinoides, Hewittia sublobata, Ipomoea acuminata, I. aitonii, I. argentaurata,* and *I. asarifolia.* Many other species of *Ipomoea* are decorative; they are *I. cairica* (also food for rabbits), *I. coptica, I. eriocarpa, I. involucrata, I. mauritiana,* (Disinkoro), *I. heterotricha, I. hederifolia, I. nil* Morning Glory, *I. ochracea, I. pes-caprae* subsp. *brasiliensis* Beach Convolvulus, *I. quamoclit, I. rubens, I. stonolifera, I. triloba, I. tuba,* and *I. verbascoidea*; other decorative plants are *Merremia dissecta, M. hederacea, M. kentrocaulos, M. pinnata, M. pterygocaulis, M. tridentata* subsp. *angustifolia, M. tuberosa,* and *M. umbellata*; *Calonyction aculeatum, Jacquemontia ovalifolia, Lepistemon owariensis, Operculina macrocarpa* (Abia), *Stictocardia beraviensis, Dicellandra barteri, Dissotis amplexicaullis, D. entii, D. grandiflora, D. irvingiana, D. perkinsiae, D. rotundifolia* (Adowa-alie), and *D. tubulosa* (*Osbeckia tubulosa*); others are *Melastomastrum capitatum, M. theifolium, Spathandra blakeoides* (*Memecylon blakeoides*) (Kyekyereantena), *Preussiella kamerunensis* (*chevalieri*), *Tristemma coronatum, T. hirtum, T. incompletum* (Anidan), *Triaspis odorata* (Kwaemu Sabrakyee), and *T. stipulata.*

Decorative species of *Hibiscus* include *H. articulatus, H. articulatus* var. *glabrescens, H. cannabinus* Kenaf, Fibre Plant; *H. congestiflorus, H. gourmania, H. lunariifolius, H. manihot, H. panduriformis, H. squamosus, H. tiliaceus* (Nwohwea), *H. surattensis, H. vitifolius,* and *Abelmoschus moschatus* (*H. abelmoschus*); the rest of the decorative plants are *Azanza garckeana, Cienfuegosia digitata, Pavonia urens* var. *glabrescens, Gladiolus daleni* (*psittacinus*), *G. klattianus, G. gregarius, G. unguiculatus* Sword Lily, Corn-Flag; *Pancratium trianthum* Pancratium Lily, *P. hirtum, Scadoxus cinnabarinus* (*Haemanthus cinnabarinus*), *S. multiflorus* subsp. *katerinae* (*H. multiflorus*) Fire-Ball Lily, Blood Flower (introduced to Europe); *Gloriosa suberba, G. simplex* Climbing Lily, *Amorphophallus aphyllus* (introduced to Europe), *Aframomum longiscapum, Asparagus racemosus, A. warneckei* (both introduced to Europe), *Palisota barteri* (also hedge plant worth cultivating for the inflorescence of white flowers), *Ensete gilletii* Wild Banana, *Gladiolus aequinoctialis* var. *aequinoctialis* (*Acidanthera aequinoctialis*), *Brillantaisia patula, Crossandra guineensis,* and *Rhynchosia buettneri.*

In addition to the indigenous decorative plants above, exotics have been introduced to our area. These include *Allamanda cathartica* Yellow Allamanda (sometimes weedy when it escapes cultivation), *A. neriifolia, Aristolochia elegans* Dutchman's Pipe (native of Brazil), *A. ridicula, Asclepias curassavica* Red Head, Blood Flower (native of tropical America); *Cassia ligustrina, C. multijuga, Clerodendrum thomsonae* Bleeding Heart, *Codiaeum variegatum* Garden Croton, *Cordia sebestena* Geiger Tree, *Cryptostegia grandiflora, Ixora coccinea, Jasminum multiflorum* and *J. sambac* (both native of tropical Asia), *Jatropha multifida* Coral Plant, Spanish Physic Nut and *J. podagrica* (both of tropical American origin), *Catharanthus roseus* Madagascar Periwinkle, *Nerium oleander*

Oleander (a native of Asia Minor), *Panax* species, *Pandanus utilis* and *P. veitchi* Screw Pine, *Plumbago capensis* Blue Plumbago (native of South Africa), *Plumeria rubra* var. *acutifolia* Red Frangipani, Temple Flower; *Punica granatum* Pomegranate, *Solanum seaforthianum* Small-Flowered Potato Creeper (native of West Indies) *S. wrightii* Potato Tree (native of Bolivia), *Tecomaria capensis* Red Tecoma (native of S. Africa), *Thespesia lampas* (Native of tropical Asia), *Thunbergia alata* Black-Eyed Susan, *T. fragrans* (native of India), and *T. grandiflora* (native of Bengal). Others are *Bougainvillea glabra*, *B. spectabilis*, *Beaumontia grandiflora* Nepal Trumpet Flower, *Quassia amara* Surinam Quassia, *Caesalpinia pulcherrima* Pride of Barbados, *Holmskioldia sanguinea* Parasol Flower (native of north India), *Lagerstroemia speciosa* Queen Grape Myrtle (native of India, China, and Australia), *L. indica* Crepe Myrtle, *Thryallis glauca* (*Galphinia glauca*) Showers of Gold (native of Central America), *Sanchezia nobilis* (native of Ecuador), *Odontonema strictum* (native of Central America), *Russelia equisetiformis* Antigua Heath (native of tropical America), *Jacobinia carnea* (native of Brazil), *Graptophyllum pictus*, *Crossandra* species, *Barleria cristata* (native of India), *Pseuderanthemum* species, *Angelonia salicariifolia* (native of North and South America and West Indies), *Scindapsus aureus*, *Murraya paniculata* Orange Jessamine (native of India), *Petrea volubilis* Queens' Wreath (native of Cuba and Brazil), *Clytostoma callistegioides* (native of southern Brazil and Argentina), *Nicolaia elatior* (*Phaeomeria magnifica*) Torch Ginger, and *Erythrina indica*. Others are *Coleus*, *Monstera*, *Philodendron*, *Ceropegia*, *Poinsettia*, *Cordyline*, and the following potted and indoor plants: *Heliconia*, *Oxalis*, *Fittonia*, *Aglaonema*, *Anthurium*, *Alocasia*, *Caladium*, *Dieffenbackia*, *Stapelia*, *Begonia*, *Billbergia*, *Cryptanthus*, *Cephalocereus*, *Tradescantia*, *Kohleria*, *Calathea*, *Peperomia*, *Hydrangea*, *Pellionea*, *Alpinia*, and *Kaempferia*.

BIG TREES

Decorative trees include *Afzelia africana* (Papao), *Bombax buonopozense* Red-Flowered Silk Cotton Tree, and many *Cassia* species: *C. fistula* Golden Shower (with the hanging, yellow flowers — native of India), *C. grandis* (native of Indonesia), *C. nodosa* Pink Cassia (native of South-East Asia), and *C. sieberiana* African Laburnum (very decorative with the hanging inflorescence); others are *Ficus elastica* Rubber Plant, *F. abutilifolia*, *F. ottoniifolia* (also for shade), and *F. umbellata* (Gyedua); *Guaiacum officinale* Lignum Vitae (native of tropical America), *Holarrhena floribunda* False Rubber Tree, *Homalium letestui* (Esononankoroma) (worth cultivating), *Hymenostegia afzelii* (Takorowa), *H. gracilipes*, *Jacaranda mimosifolia* Blue Jacaranda, *Lonchocarpus griffonianus* (Senyana), *L. laxiflorus* (Nalenga), *Lophira lanceolata*, *Majidea fosteri* (Ankyewa) (opened fruits), *Monodora brevipes* (Abotokuradua), *M. myristica* (Ayerewamba), *M. crispata* (worth cultivating), *M. tenuifolia* (Motokuradua), *Newbouldia laevis* (Sasanemasa), *Sclerocarya birrea* (Nanogba), *Stereospermum acuminatissimum* (Tokwa-kufuo) (worth cultivating), *S. kunthianum*

10. AMENITY LANDSCAPING

(Sonontokwakofo) (worth growing as decorative tree), and *Symphonia globulifera* (Ehureke).

Introduced decorative palms include *Washingtonia filifera, Verschaffeltia splendida, Veitchia merrillii, Sabal mexicana, Aiphane caryotifolia, A. corallina, Neodypsis decaryii, Cryosophila warscewiezii, Hyophorbe vershaffeltii, Ravenala madagascariensis* Traveller's Palm, *Chrysalidocarpus lutescens*, and *Caryota* species.

ELEVEN
Weeds

Weeds are able to thrive and grow vigorously in relatively poor or marginal soils, successfully competing with others and adapting to adverse conditions. The life-cycle is generally very short, producing many seeds that have a fairly long period of viability, and germinate easily; and they are equally easily dispersed by wind, water, or animal agent. This ease of dispersal accounts for the rapid manner in which weeds spread and establish themselves; it also explains the relative difficulty in eliminating an infestation or controlling them. The accidental introduction of *Chromolaena odorata* (*Eupatorium odoratum*) Siam Weed, and the resultant devastation caused, is testament to those facts. Another factor is the prolific, easy regeneration from seeds, stolons, or other vegetative parts.

Any cultivated plant that escapes its designated area of cultivation may be termed a weed. *Catharanthus roseus* Madagascar Periwinkle now common in Ghana, along the coasts of Ada to the east and Axim to the west, and *Ruellia tuberosa* are two such examples. Weeds have been described as plants growing where they are not wanted — a description which technically can fit any plant however useful economically.

Although weeds may be of use in some way or other, the disadvantages outweigh any advantages. They are a particular nuisance when established —the labour force required and the cost of herbicides employed for their control being immense. An estimated four-fifths of the farming work in developing countries is spent on weed control alone. Weed control is the main concern of Weed Research Organizations. In our area, the application of plant hormones and the use of poisons, herbicides, and arboricides are restricted to research stations. High initial costs combined with the complexities of import exchange rates mean that very few farmers have access to these forms of weed control.

The application of herbicides has adverse effects on the fertility of the land. Excessive use affects the phytotoxicity of the soil and reduces yield. The influence of herbicides on the yield is dependent on the soil moisture. Carson (1978) observed that during a relatively dry period in 1976, the 2.0 (kg.a.i./ha) rate of herbicidal application depressed yields, while the same rate produced optimum yields in 1977 when very wet periods preceded application. Problems with herbicides also include toxicities to some crop plants.

Biological control of weeds as practised in other countries has not yet been tried on a large scale in West Africa — at best it has only been proposed and is under consideration. There is the possibility that insects used for biological control will attack food plants and cash crops. Thus a proposed solution could become a potential problem. For example, Schroder (1976) reports that in Uganda surplus *Teleonemia scrupulosa* (a bio-control for *Lantana*) attacked *Sesamum indicum* Sesame, Beniseed and caused distortion and loss of yield through folding and oviposition damage. Bennett and Rao (1968) found that the Cerambycid *Plagiohammus spinipennis*, a stem-borer (another bio-control for *Lantana*) oviposits in captivity on two species of *Cordia*. A successful biological control is the cheapest means of controlling weeds; but by far the most common and widely used means of control in West Africa is physical or mechanical — principally manual.

The presence of weeds in cash or food farms retards the development of these products and causes a considerable reduction in yield. Carson (1978) reports that 30–60 per cent or more reduction is possible. Weeds smother and choke out food crops, flowers, and vegetables — a nuisance to gardeners, backyard farmers, and commercial and peasant farmers generally.

Some weeds can, however, prove advantageous. For example *Panicum maximum* Guinea Grass and *Leucaena leucocephala* (*glauca*) Leucaena are useful fodder. Dr Vietmeyer of the Smithsonian Institution reports that cattle feeding on Leucaena show weight gains comparable to those of cattle feeding on the best pastures anywhere. Weeds also bind the soil, preventing erosion; and they may be cut and utilized as raw materials for compost.

Noxious Weeds

CYPERUS ROTUNDUS

Nut-grass, Coco-grass is a widely distributed common weed in over 100 tropical countries and a pest in cultivated fields, flower and vegetable gardens, and backyard farms. Nut-grass is the world's worst weed (Holm and others, 1977). It prefers sandy fields. Its ability to propagate readily from seeds, tubers, or rhizomes enables it to regenerate and spread quickly and easily, making its control difficult. An effective method of control is by digging it out and destroying the rhizomes before the flowering spikes mature. The chances of a successful biological control of sedges are slender in view of the fact that they are allied to graminaceous crops. However, insects associated with Nut-grass are being especially investigated in Pakistan. Schroder (1976) reports that the life-history of the most important phytophagous insects is being studied, and feeding and oviposition tests have been carried out with *Bactra* species, *Athesapeuta cyperi*, and others. Terry (1983) reports that temporary suppression of *C. rotundus* is possible with several herbicides, but glyphosate can give an almost complete kill when applied to vigorous shoots which are connected to an intact system of rhizomes and tubers. The slightly fragrant roots are edible and used for various medicinal purposes (see

Chapter 8). The tubers are eaten by pigs, while the leaves serve as fodder.

CHROMOLAENA ODORATA (EUPATORIUM ODORATUM)

Siam Weed is a native of the West Indies and continental America from Florida to Paraguay. It is an annual composite, growing to a height of 2 metres. Like *Vernonia biafrae* (Hu) it may assume a climbing habit. In 1960, C. Simmonds observed that Siam Weed had become a serious problem in Nigeria. It was first observed in Ghana some nine years later by C. Parker. *Chromolaena*, first introduced into the University of Ghana Botanical Gardens, later spread beyond the institute's perimeters and into neighbouring land. Now it is widespread in all the forest areas of Ghana as far north as Berekum in the Brong Ahafo Region. Siam Weed occurs also in the Ivory Coast and Cameroun. It is a pest weed in both new and abandoned farms, growing in open sunlight and partial shade. It does not tolerate thick shade and disappears under dense canopy. This is an effective control. In Ghana, Siam Weed is popularly called Akyeampong (or Busia Weed around Kisi near Cape Coast). It is particularly troublesome in oil palm and cassava plantations. Slashing with a cutlass is partially effective in the dry season and before the plants fruit; otherwise the roots regrow and mature seeds quickly germinate.

Chemical control using Gramoxone applied at a ratio of 0.5 km/acre and 2,4-D at a ratio of 2.5 lb/acre to eradicate Siam Weed has been tried in India and Nigeria respectively. However, weeds killed in this way quickly re-establish themselves from roots and seeds. The weed is also susceptible to the herbicides atrazine, diuron, and glyphosate. Hall, Kumar, and Enti (1972) observed the green aphid *Aphis spiraecola* attacking the growing young leaves of the plant. Affected leaves rolled and distorted. It was observed that damage to the plants was so severe that affected shoots failed to bear any flowers. However, the aphids were themselves vulnerable to the larvae of *Parogus borbonicus*, and the effective control of Siam Weed by aphids is hindered in this way. The weevil *Apion brunneonigrum*, whose larvae develop and feed on the florets of Siam Weed, has been introduced into Nigeria as a biological control. The weed is reported to be an indicator of good soil. The leaves are styptic and they repel mosquitoes when fresh — this latter property is worth investigating and developing.

IMPERATA CYLINDRICA VAR. AFRICANA

Lalang Grass is a perennial grass of wide distribution throughout tropical areas, South Africa, and in Madagascar. It also occurs in tropical countries like Malaya, Sri Lanka, India, and Australia. It is a well-known pest to cultivation and a notorious weed in the savanna zone. It is quite difficult to control or eradicate Lalang, for it quickly sprouts from the rhizomes after being slashed with cutlasses or burned. *Anogeissus leiocarpus* (Sakanee) has been suggested as a form of control, but it is slow growing. *Gmelina arborea* and *Lantana camara* Wild Sage have also been suggested, but *Lantana* presents its own problems (see below). Cutting back Lalang at intervals apparently starves the rhizomes to check growth; and where it is left alone for eight to ten years it tends to die out and is gradually replaced by other perennial grasses. Swaine and others (1979) observed at Botianor on the Accra plains in Ghana that the flood of January 1977 killed *Imperata* below the maximum flood level, and thus noted an intolerance to flooding by the lagoon water. This is a possible method of control. Lalang is susceptible to the herbicides dalapon and glyphosate. Though it thrives well in moist depressions it is fairly drought-tolerant. It is of little value as fodder, even when young, as the shoots prick and hurt the muzzles of animals. The leaves are useful as thatch, for stuffing mattresses, and as material for mat-making, cowry bags, baskets, and plates. The flossy inflorescence is useful for stuffing pillows and cushions. It is medicinal (see Chapter 8)

LANTANA CAMARA

Camara, Prickly Lantana, Wild Sage is a native of tropical America, but naturalized in many warm countries and widely distributed in West Africa. Camara was initially introduced as a decorative plant. It is easily raised from seeds or cuttings and grows freely and vigorously in almost all conditions. Consequently, it easily escapes its cultivation area. As an escapee, *Lantana* is a pest. Schroder (1976) records that it is an abundant and noxious weed in peasant agriculture in many parts of the Central and Eastern Regions of Ghana; and it was found to be most abundant south of Koforidua, where it forms dense stands covering several hundred square miles. The thorny, scrambling stems which form dense thickets makes it practically impenetrable by man. Slashing is laborious and time-consuming; and the problem is compounded by the ease with which cut stems quickly regenerate from suckers. Burning fresh *Lantana* has proved equally ineffective.

Biological control of Camara was started in 1902 by the Government of Hawaii. Schroder (1976) suggests that *Teleonemia elata* and *Leptobyrsa decora* might be useful insects for biological control. Other insects suggested are *Uroplata girardi* and *Octotoma scabripennis*, beetles from Central and South America. Other possible insects are *Syngamia haemorrhoidalis* and *Diastema tigris*. The plant forms a useful hedge and the fruits are edible and aromatic. It is medicinal (see Chapter 8).

PISTIA STRATIOTES

Water Lettuce is a free-floating aquatic plant widely distributed in the tropics and sub-tropics (Texas, Florida, and the Gulf Coast). It propagates mainly vegetatively by stolons and has numerous fibrous root systems sometimes reaching the bed in shallow water. It may form immense colonies several kilometres long. Schroder (1976) reports that it prefers slightly acidic (pH 6.5–7.0) water and a nutritious bed with plenty of organic matter. After the Volta Lake in Ghana was formed in 1964 large colonies of Water Lettuce were observed. Okali and Hall (1974) observed that in some shelter bays, especially the upper bays of the rivers Afram, Pawnpawn, and Dayi, these mats attained a total area of about 350 km^2, covering the water surface completely and obstructing water transport and fishing. On the Barikese Dam near Kumasi, approximately 400 hectares (that is about three-quarters of the total surface of the reservoir) was covered by *Pistia*. In 1971 large colonies were reported on the Weija Dam near Accra. Luckily all these three initial outbursts of *Pistia* were followed by annual die-back which greatly reduced the sizes of the colonies. Pettet and Pettet (1970) have shown that the annual die-back was associated with a viral infection which reached epidemic proportions during the wet season; and they suggest that the partial disappearance of *Pistia* on the Volta Lake may have been caused by the same disease. Large colonies of *Pistia* have also been recorded on the Niger and other rivers in West Africa.

In 1971, Pierce observed the colonization of *Pistia* by *Oxycaryum cubense* (*Scirpus cubensis*) on the Volta Lake and suggested that the leafy canopy of *O. cubense* eventually kills the *Pistia* because it shades it, the resulting pure *O. cubense* mat being unstable and often sinking. Okali and Hall (1974) studied this observation and concluded that *Pistia* declined on all the occasions where this *Oxycaryum cubense* – *Pistia* association occurred. They observed also that *Pistia* sudds from which *O. cubense* was absent did not show a similar decline. The study also established that the decline of *Pistia* following invasion by *O. cubense* did not lead, as had earlier been suggested, to sinking of the sudd.

Schroder (1976) suggests the introduction of the insects *Paulinia acuminata* and *Samea multiplicalis* for biological control trials of *Pistia* in Ghana. The plant serves as a breeding site for mosquitoes. It is eaten in eastern Sudan in time of famine and in northern Nigeria it is used to feed ostriches. It is cultivated in China to feed pigs. *Pistia* is medicinal (see Chapter 8).

EICHHORNIA CRASSIPES

Water Hyacinth, a very notorious aquatic weed has of late been introduced into Ghana as a decorative water plant (despite phytosanitary regulations to the contrary) and escapees are gradually spreading. The Ministry of Agriculture, the Volta River Authority, the Environmental Protection Council and the Institute of Aquatic Biology are making a joint effort to control and irradicate it. The weed has been a serious problem in parts of Southern Nigeria since 1983.

Common Weeds

A list of common weeds includes the following:

HERBACEOUS

These are *Acanthospermum hispidum* Star Bur, *Achyranthes aspera*, *A. indica* (*obtusifolia*) — a new weed in Legon, Ghana, not previously recorded for West Africa (Abbiw, 1985), *Ageratum conyzoides* Billy-goat Weed, *Alternanthera pungens* (*repens*), *Alysicarpus rugosus*, *Amaranthus spinosus* Prickly Amaranth, *A. viridis* Wild or Green Amaranth, *Aneilema beniniense* (in the forest), *Aspilia africana* Haemorrhage Plant, *A. bussei*, *Asystasia calycina*, *A. gangetica*, *Bidens pilosa* Bur Marigold, *Blumea perrottetiana*, *Boerhavia* species Hogweed, *Borreria* species, *Bryophyllum pinnatum* Resurrection Plant, *Celosia leptostachya*, *Centrosema plumieri*, *Chrysanthemum procumbens*, *Cleome rutidosperma* (*ciliata*) Cat's Whiskers, *C. viscosa* Wild Mustard, *Clitoria ternatea* Blue Pea (also decorative), *Commelina* species, *Conyza aegyptiaca*, *Corchorus olitorius* Jew's Marrow, Long-fruited Jute; *Croton lobatus*, *Dicoma tomentosa*, *D. sessiliflora*, *Elytraria* species, *Emilia coccinea*, *E. sonchifolia*, and *Erigeron floribundus*. Others are *Euphorbia cyathophora* (*heterophylla*), *E. hirta* Australian Asthma Herb, *E. prostrata*, *Gomphrena celosioides*, *Hyptis suaveolens* Bush Tea-Bush, *Ipomoea asarifolia*, *I. mauritiana* (Disinkoro), *I. pes-caprae* subsp. *brasiliensis* Beach Convolvulus, *Launaea taraxacifolia* (*Lactuca taraxacifolia*) Wild Lettuce, *L. capensis*, *Laggera heudelotii*, *Lindernia diffusa* var. *diffusa*, *Melanthera elliptica*, *M. scandens*, *Mitracarpus villosus* (*scaber*), *Mollugo nudicaulis*, *Micrococca mercurialis*, *Nelsonia canescens*, *Hedyotis corymbosa* (*Oldenlandia corymbosa*), *Oxalis corniculata*, *Mikania chevalieri* (*cordata* var. *chevalieri*) Climbing Hemp Weed, *Acalypha ciliata*, and *Canscora decussata*. The rest are *Passiflora glabra* (*foetida*) Stinking Passion Flower, *Pergularia daemia*, *Phyllanthus niruri* var. *amarus* (*amarus*), *Physalis angulata* (Totototo), *P. micrantha*, *Piper umbellatum* (Amuaha) (in cocoa farms), *Platostoma africanum*, *Portulaca oleracea* Purslane, Pigweed; *P. quadrifida*, *Priva lappulacea*, *Pupalia lappacea* (Akukuaba), *Schrankia leptocarpa*, *Scoparia dulcis* Sweetbroom Weed, *Sesuvium portulacastrum* Seaside Purslane, *Solanum nigrum* (Nsusuabiri) (also a weed in Britain), *Spermacoce verticillata* (*Borreria vertcillata*), *Spigelia anthelmia* Worm Weed, *Stachytarpheta cayennensis* Brazilian Tea, Bastard Vervain; *S. indica*, *Synedrela nodiflora*, *Talinum triangulare* Water Leaf, Wild Spinach; *Trianthema portulacastrum*, *Tribulus terrestris* Devil's Thorn, *Tridax procumbens*, *Vernonia cinerea*, *V. perrottetii*, and *Zornia latifolia*.

WOODY OR SHRUBBY

These are *Abutilon mauritianum* (Nwaha), *Alchornea cordifolia* Christmas Bush, (Gyamma); *Calotropis procera* Sodom Apple, *Cassia obtusifolia* (*tora*) Foetid Cassia, *C. mimosoides* (Langrinduo), *C. occidentalis* Negro Coffee, *C. rotundifolia*, *Corchorus olitorius* Jew's Marrow, Long-fruited Jute; *C. tridens* (Sanvoa), *Crotalaria* species, *Desmodium ramosissimum*, *Dichrostachys cinerea* (*glomerata*) Marabou Thorn, *Grewia carpinifolia* (Ntanta), *Icacina oliviformis* (*senegalensis*) False Yam, *Jatropha gossypiifolia* Physic Nut, and *Justicia flava*.

The rest are *Leea guineensis* (Okatakyi), *Leucaena leucocephala* (*glauca*) Leucaena, *Malvastrum coromandelianum*, *Melochia* species, *Phyllanthus reticulatus* var. *reticulatus*, *Pouzolzia guineensis* (in forest zone), *Securinega virosa* (Nkanaa), *Sida acuta*, *S. rhombifolia*, *S. ovata*, *Solanum torvum* (Saman-ntoroba), *S. erianthum* (often a serious problem), *Tephrosia* species, *Triumfetta rhomboidea* (Petekuku), *T. cordifolia* Burweed (Ekuba); *Urena lobata* Congo Jute, *Vernonia amygdalina* Bitter Leaf, (Bowen); *V. colorata*, *Waltheria indica* (Sawai), and *Wissadula amplissima*.

GRASSES AND SEDGES

These are *Axonopus compressus*, *Bothriochloa bladhii*, *Brachiaria distichophylla*, *B. lata*, *Centotheca lappacea* (in plantations and forest), *Cenchrus biflorus* (pest on cultivated land), *C. echinatus*, *Chloris barbata*, *Cyperus difformis*, *C. distans*, *C. iria*, *C. tuberosus*, *Dactyloctenium aegyptium*, *Digitaria ciliaris*, *D. horizontalis*, *D. longiflora*, *Echinochloa colona* Jungle Rice, *Eleusine indica*, *Eragrostis tenella*, *Heteropogon contortus* Spear Grass, *Ischaemum longiflora*, *Kylinga squamulata*, *Mariscus alternifolius*, *M. flabelliformis*, *Oryza barthii* Wild Rice, *O. longistaminata* (both in rice fields), *Panicum maximum* Guinea Grass, *P. repens*, *Paspalum conjugatum* Sour Grass, *P. vaginatum* (in rice fields), *P. scrobiculatum* (*orbiculare*) Bastard or Ditch Millet, *Pennisetum subangustum*, *Pseudobrachiaria deflexa* (*Brachiaria deflexa*), *Remirea maritima* (along sea shore), *Rhynchelytrum repens*, *Rottboellia exaltata*, *Setaria barbata*, *S. pallide-fusca* Cat's Tail Grass, *S. verticillata* Rough Bristle Grass, *S. longiseta*, *Sporobolus pyramidalis* Rat's Tail Grass, *S. robusta*, and *Streptogyna crinita* (a harmful weed in agriculture).

FERNS

These are *Adiantum philippense*, *A. tetraphyllum* var. *vogelii* (*vogelii*), *Asplenium* species, *Triplophyllum protensum* (*Ctenites protensa*), *Dicranopteris linearis* (*Gleichenia linearis*) (also decorative), *Lycopodiella cernua* (*Lycopodium cernuum*) (also decorative), *Lonchitis currori*, *L. ruducta*, *Marattia fraxinea*, *Microlepia speluncae*, *Nephrolepis biserrata*, *Pneumatopteris afra* (*Cyclosorus afer*) (Mmeyaa) (in cocoa farms), *Pteridium aquilinum* Bracken, *Pteris atrovirens*, *P. burtonii*, and *P. vittata* (on and around gold-mine slag heaps in Ghana).

PARASITIC WEEDS

These are *Cassytha filiformis*, *Cuscuta australis*, *Phragmanthera incana*, *P. nigritana*, *Tapinanthus bangwensis*, *T. globiferus* Mistletoe, and *Striga hermonthica* (Wumlim) and *S. gesnerioides*. *Striga* is a problem in the Sahel and Sudan savanna areas of West Africa where it parasitizes mainly on sorghum and bulrush or pearl millet — important cereals in the area. Other crops attacked are upland rice, maize, cowpea, groundnuts and sugar-cane. Crops such as cotton can stimulate germination of *Striga* seeds without themselves becoming parasitized. Leakey and Wills (1977) observe that suitable rotations, therefore, offer the opportunity to reduce *Striga* damage on sorghum by planting this crop after one such as cotton. Foliar applications of the herbicides linuron and ametryne kill *Striga* at application rates which are well tolerated by other plants, including sorghum. Granular applications of 2,4-D and MCPA have also been used. Removal and burning of *Striga* plants before they produce seeds and improving the soil fertility also reduce the severity of this weed. (See also Biological Control of Weeds, below.)

SEMI-AQUATIC WEEDS

These include *Alternanthera sessilis, Eclipta alba (prostrata), Ethulia conyzoides, Glinus oppositifolius, Heliotropium indicum* Indian Heliotrope, *Hydrolea glabra, Indigofera paniculata, Lindernia diffusa* var. *diffusa, Ludwigia erecta, L. hyssopifolia, L. octovalvis (Jussiaea suffruticosa), Mimosa pigra* Sensitive Plant, *Mitragyna inermis* (Kukyafie), *Pentodon pentandrus* (Buburanya), *Polygonum lanigerum, P. senegalense, Sphenoclea zeylanica,* and *Thalia welwitschii*. The grasses and sedges are *Andropogon gayanus* var. *gayanus, Cyperus articulatus, Echinochloa pyramidalis* Antelope Grass, *E. stagnina, Fimbristylis dichotoma, Leersia hexandra* Rice Grass, *Leptochloa caerulescens, Paspalum scrobiculatum (orbiculare)* Bastard or Ditch Millet, *Phragmites karka* Common Reed, *Sacciolepis africana, Torulinum odoratum, Urochloa mutica (Brachiaria mutica),* and *Coix lacryma-jobi* Job's Tears.

AQUATIC WEEDS

These are *Ceratophyllum demersum, Cyperus papyrus* (especially around Lake Chad), *Ipomoea aquatica, Ludwigia leptocarpa, L. stolonifera, Neptunia oleracea, Nymphaea lotus* and *N. micrantha* both Water Lilies; *Potamogeton schweinfurthii, P. octandrus, Salvinia nymphellula,* and the sedge *Oxycaryum cubense (Scirpus cubensis)*. Others are *Typha domingensis (australis)* Bulrush, Cat-tail and *Vossia cuspidata*. The rest are *Azolla pinnata* var. *africana (africana)* and the Duckweeds: *Lemna perpusilla (paucicostata), Spirodela polyrhiza,* and *Wolffia arrhiza*. Aquatic ferns are *Ceratopteris cornuta* Water Fern and *Cyclosorus striatus*. *Cressa cretica* is a weed of sandy places near the sea.

Weed Control

CHEMICAL CONTROL

Chemicals used in weed control as chemical contact and translocated weed-killers are obtainable under various trade names. Primagram is a maize weed-killer, Avirosan a rice weed-killer, and Gesapax is a specific weed-killer for both pineapple and sugar-cane. Primagram and Avirosan are pre-emergent weedicides sprayed at a concentration of 1.5 litres in 90 litres of water; Gesapax is sprayed at 2 litres per acre after planting. Other weedicides are Gramoxone, Gramurone, and Bellater. Weedicides may be pre-emergent or post-emergent. Arboricides include sodium chlorate, sodium sulphamate, and sodium arsenite which is perhaps the commonest of the three. A 10 per cent aqueous solution has been found satisfactory — that is about one kilogram in about ten litres of water — the solution being poured into a continuous girdle round the tree. Sodium arsenite is a cumulative poison, therefore preventive measures are necessary when using it. Plant hormones like 2,4-D (2,4-dichloro-phenoxyacetic acid) and 2,4,5-T (2,4,5-trichlorophenoxyacetic acid) are being used as arboricides.

BIOLOGICAL CONTROL

Growing leguminous plants like *Pueraria phaseoloides* as cover crops helps to check and suppress other weeds. Cover crops include *Mucuna* species, *Desmodium gangeticum* var. *gangeticum*, *Crotalaria pallida* (*mucronata*) (Peagoro), *Macroptilium atropurpureus* (*Phaseolus atropurpureus*), and *M. lathyroides* (*P. lathyroides*). Negro Coffee is said to be inimical to *Striga*. The cultivation of *Cassia occidentalis* Negro Coffee for a year in fields affected by *Striga hermonthica* causes the latter to disappear. Light-demanding weeds are intolerant of shade, for example *Gmelina arborea* has been suggested to shade out *Imperata cylindrica* Lalang Grass. For the biological control of weeds by insects, see above. Grazing is another means of controlling weeds biologically.

MECHANICAL CONTROL

This is the traditional method of weed control — by the use of either the cutlass, for slashing and cutting, or the hoe. Axes are used for any hardwoods that are too big to be cut with a cutlass. Manual weeding (and burning) is the most widespread form of weed control by the peasant farmer and the backyard gardener in West Africa and most developing countries. The inability of traditional farming methods to cover large expanses of the countryside is principally attributable to the laboriously slow method of manual weeding; many peasant food farms average about three acres or less per farmer, very few exceed five acres. Weeding takes about 80 per cent of the traditional farmer's working hours: that is, he spends four times the number of hours on weeding than on planting and harvesting together. The relationship between the problem of manual weed control, the effects of weeds on crops (see above), and the importance of the peasant farmer to the national agricultural output (see Chapter 2) is worth examining.

The amount of labour invested in weeding is enormous, not only for the peasant farmer, but also for field workers. Of the 300 workers at the Agricultural Research Station at Kade in Ghana, one-third are engaged on hand-weeding throughout the year. The Curator of the Grounds and Gardens of the University of Ghana, Legon, reports that weeding is a major problem. The total labour force is engaged in the control of weeds throughout the year, assisted by temporary workers employed during the rainy season when weeds are most abundant. This labour force is in addition to a mechanical slasher. The use of mechanical slashers and lawn mowers for weeding is exclusive to some institutions and wealthy homes respectively.

Advantages of Weeds

Many weeds serve as a source of fodder, a good example being *Melanthera scandens*. Weeds also bind the soil and prevent erosion and provide the basis for compost and mulch. Weeds such as *Imperata cylindrica* Lalang are useful as thatching material and as straw for mattresses and cushions. Dried grass and *Dichrostachys cinerea* (*glomerata*) Marabou Thorn can be utilized as fuel, while *Cleome viscosa* and *Aspilia africana* Haemorrhage Plant are medicinal (see also Chapter 8). Some weeds like *Talinum triangulare* Water Leaf, Wild Spinach; *Launaea taraxacifolia* (*Lactuca taraxacifolia*) Wild Lettuce, and *Emilia* species are useful as pot-herbs, spinach, and vegetables. Weeds can be employed as cover crops to shade out and suppress other weeds. Others like *Cassia obtusifolia* (*tora*) Foetid Cassia yield a yellow and blue dye, *Mitragyna inermis* (Kukyafie) a yellow dye, *Leucaena leucocephala* (*glauca*) Leucaena a brown dye, and the latex of *Calotropis procera* Sodom Apple rubbed on the hands prevents scorpion stings (see Chapter 9). Similarly, *Uraria picta* (Heowe) is an antidote to the venom of the viper, while the latex of *Euphorbia hirta* Australian Asthma Herb assists in the extraction of poisonous arrows.

TWELVE
Plants and Soil Nutrients

Plants influence the soil's structure by a combination of the penetrating and binding action of the roots (see Chapter 1) and the addition of organic matter; and its chemical composition by the recycling of soil minerals and the fixing of free atmospheric nitrogen into organic nitrogen-containing compounds. Soil minerals taken up by the plants during absorption are stored in the branches, twigs, leaves, flowers, and fruits. By the breakdown of the organic matter of the plant material into manure, the minerals are released back into the soil to be reutilized by the plants — a recycle.

Manure

An efficient and useful method of disposal for garden and farm wastes and weeds is to heap them in a corner of the plot as raw materials for compost or humus. Any form of organic matter — fresh and dead leaves, straw, lawn cuttings, weeds, sawdust — is suitable. Cutting them up is not essential, though this provides a greater surface area to speed up microbial action. Taylor (1962) notes that in the higher-rainfall areas heaps are usually preferred because pits may become flooded and thus many of the necessary microorganisms may be killed. In dry regions the heaps need to be watered as the material is added. If there is sufficient space, several heaps may be made to facilitate easy turning and reduce the distance across which both raw materials and compost have to be carried. The larger the heap, the greater the concentration of heat favourable to the oxidation of the organic matter by the metabolic activities of various saprophytic fungi and bacteria, and therefore the quicker the breakdown of the fallen leaves and vegetable matter to form humus. Temperatures up to 70°C for larger heaps are possible while smaller ones usually register around 40°C, the outer layers of organic matter acting as insulating material. The period of complete composting varies from 2 to 6 weeks, depending on the size of heap, the amount of water added, and the aeration. The high temperatures are also useful in that pathogenic bacteria, cysts, eggs, other plant parasites, and weed seeds are destroyed. Where optimum conditions of heat and moisture prevail, the breakdown can be rapid. However, in exceptionally favourable conditions, there is complete oxidation of the organic matter and it disappears. Meyer and Anderson (1965) report that soils with low humus content may result from a sparse contribution of organic matter, as in desert or semi-desert regions, or from a rapid oxidation of organic material which prevents accumulation of humus even when the supply of organic residues to the soil is large; this latter condition occurs in many tropical and sub-tropical soils.

Plants require various selective minerals — mainly calcium, potassium, magnesium, nitrogen, and phosphorus — for their proper growth and development. Also chemical compounds like carbonates, silicates, and phosphates obtainable from burnt trees are needed for plant nutrition. Being immobile, they gain these minerals from their surrounding soil matrix. Water is absorbed by the root hairs, conducted through the stem, and deposited through the leaves as water vapour during transpiration. Varying quantities of many chemical compounds are dissolved in the soil water. These chemical compounds originate principally from the chemical weatherings or dissolution of rock particles, the decomposition of organic matter as a result of the activities of micro-organisms, and the reactions between the roots of plants and the soil constituents. The soil thus looses its mineral contents gradually to the plants which store these in the roots, stems, and other parts named above. This is particularly the case in tropical forests, where the bulk of nutrients are present in the trees rather than in the soil (see Chapter 1). Another means by which tropical soils loose their nutrients (particularly carbonates of calcium and magnesium) is by leaching, the washing down of soluble salts from the upper layers of the soil by the heavy tropical rains.

The preparation and effective application of manure to the soil, therefore, is a means of replenishing the lost mineral contents and plant food to enrich it again. In theory it is possible to use the same piece of land over and over again by this method. The practice of crop rotation in addition to the application of manure further helps in maintaining the soil fertility, since different crops have different mineral requirements and therefore make different demands on the soil. In the forest, protected timber land, sacred grove, or other uncultivated land, the rate of nutrient absorption by plants is proportional to the rate at which fallen leaves and debris return their absorbed minerals to the soil. The two systems of shifting cultivation and extended bush-fallow are practical applications of this principle — to replenish the soil with lost minerals and organic matter. But for man's agricultural activities, there would be a natural balance which would automatically ensure and maintain the soil quality almost indefinitely.

MANURE FROM THE FOREST FLOOR AND REFUSE DUMPS

Leaf litter, twigs, rotten branches, and vegetative matter in various stages of decomposition may be collected directly from the forest floor as humus or manure. This compost material is usually only about four inches or so deep and is easily recognizable by its darkish colour as compared with the reddish-brown or yellowish soil beneath. To obtain the full benefits from the use of forest-floor manure only organic material that is completely decomposed is worth using; if compost is added to the soil before the composting process is complete, it may actually rob the soil of nitrogen. The leaves of *Musanga cecropioides* Umbrella Tree (Dwumma); *Myrianthus arboreus* (Anyankoma), *Adansonia digitata* Baobab, *Azadirachta indica* Neem Tree, *Parkia clappertoniana* West African Locust Bean (Daudawa), and *Samanea saman* Rain Tree particularly form a thick layer of humus to protect and improve the soil. Manure may also be obtained directly from refuse dumps or incinerators. As the refuse piles up higher and higher, only the top parts receive attention and are burned or carried away. The lower parts which are neglected are acted upon by the necessary decomposing agents to form manure. Such manure contains animal remains and other household wastes in addition to plant remains.

MANURE FROM FARM WASTES

In addition to the use of garden wastes to prepare manure, animal wastes (which are mainly plant material in a processed form) may be used to prepare farmyard manure. Droppings of sheep, goats, cattle, and poultry are collected and dug into the soil. Cow dung is an important means of improving soil fertility in West Africa — especially in the Sudan and Guinea savanna where cattle rearing is predominant. Droppings can also be added to the soil and turned over or ploughed over with it during the preparation of the land for planting. This is the usual method of applying farmyard manure.

MULCHING

The practice of burning crop residues and vegetable wastes either to prepare the land for farming or to get rid of them, or both, is a waste of valuable organic material. Firstly, the residues are the only natural source of cheap plant nutrients and organic matter, and secondly, the burning completely destroys all the nitrogen and micro-organisms in the soil. Crop residues can be used to advantage by employing them for mulching. This is the practice of placing straw or dead leaves round plants or seedlings to reduce attack by weeds and prevent soil erosion and excessive evaporation. It also increases the effectiveness of any rainfall. Mulching material eventually decomposes with the rains to provide manure.

GREEN MANURE

Another method of maintaining the fertility of the soil is to work the crop residue together with the cover crops into the soil as green manure to recycle the plant nutrients. This method is more effective immediately after harvesting, for the plants are attacked by termites during the dry season and their excreta in turn improves the soil fertility. The following plants are recommended as useful for green manure: *Cassia kirkii* var. *guineensis*, *Mimosa pudica* Sensitive Plant, *Crotalaria pallida* (*mucronata*) (Peagoro), *Dichrostachys*

12. PLANTS AND SOIL NUTRIENTS

cinerea (*glomerata*) Marabou Thorn, *Erythrina senegalensis* Coral Flowers, *Indigofera arrecta* Natal or Java Indica, *I. tinctoria* True or Frank Indigo, *Sesbania pachycarpa* (*bispinosa*), *S. sesban* Egyptian Sesban (Tingkwanga); *S. sericea* (*pubescens*) (Asratsi), *S. sudanica* subsp. *occidentalis* (*dalzielii*), *Tephrosia platycarpa* (Buruguni), and *T. purpurea*. Others are *Desmodium triflorum* and *Phaseolus lunatus* Butter Bean. The above are all leguminous plants, that is plants in the bean family. The seed cake of *Lophira alata* Red Ironwood, (Kaku); *Trichilia emetica* subsp. *suberosa* (*roka*) (Kisiga), and *Ricinodendron heudelotii* (Wamma) are useful sources of green manure. Cover crops include *Pueraria phaseoloides*, a native of tropical Asia, *Ipomoea involucrata*, *Passiflora glabra* (*foetida*) Stinking Passion Flower, *Dissotis rotundifolia* (Adowa-alie), and *Mikania chevalieri* (*cordata* var. *chevalieri*) Climbing Hemp-weed. *Indigofera hirsuta* Hairy Indigo is a valuable cover crop used extensively for soil improvement. Duke (1981) gives the manurial constituents of the leaves as nitrogen, phosphoric acid, potash, and lime. The grass *Melinis minutiflora* var. *minutiflora* Stink Grass is useful as a cover and mulch and to suppress other weeds.

Artificial manures or fertilizers — that is mineral salts like superphosphates, manufactured or mined — may be applied to supplement the natural manure. The Ghana Cocoa Research Institute (CRIG) has successfully completed research into the use of cocoa husks for the manufacture of fertilizer (see Chapter 3).

ADVANTAGES OF ORGANIC MANURE

Organic manure has the advantage of being less expensive than imported fertilizers. It might even involve no cost except labour. Since it is prepared and used on the farm or garden there are hardly any transportation costs either. The practice of using organic manure, therefore, saves the government the financial strain of importing chemical fertilizers, thus making funds available for other essentials that cannot readily be manufactured locally. The substantial reduction in the size of the original material or plant waste, makes composting an efficient means of waste disposal. In complete decomposition the plant material may be reduced to one-fifth the original volume.

EFFECTS ON THE SOIL

As indicated above, the application of manure to the soil returns the soil nutrients taken up by the plants. Physically, the humus in the manure favours a looser soil structure and better aeration. It conditions the soil, making sandy ground form crumbs and increasing the water-holding capacity. In clay soils it reduces the cohesiveness making the ground more pervious to water. Although chemical analyses show that the value of compost as a fertilizer is limited, the effective application of this abundant natural resource ensures higher yields. Up to 2½ times more produce per acre or hectare is possible.

Nitrogen Fixation

Another means by which plants improve the nutrients in the soil is by fixing nitrogen. Nitrogen fixation is a process by which the free nitrogen in the air (about 78 per cent of the earth's atmosphere) is converted into organic nitrogen-containing compounds in the soil, in which form it can be utilized by plants. This is of importance economically, since a shortage of usable nitrogen is often a major limiting factor in plant growth. The actual fixation is carried out by certain bacteria which live in symbiosis with some leguminous plants (in their root nodules) and some blue-green algae. The former process is the most important in terms of total nitrogen fixation. The legume provides a convenient anaerobic site from where the bacteria produces the necessary enzyme for fixing nitrogen.

Of the several nitrogen-fixing bacteria, the commonest is *Rhizobium* which invades the root nodules of plants in the family Leguminosae. The extra nitrogen released by these leguminous plants into the soil becomes available to other crops. This beneficial effect is the basis for intercropping plants in the bean family with others — a practice used by both simple, traditional farmers and scientific agriculturists. In addition, the leguminous plants are used as fodder or for preparing compost or ploughed back into the soil as green manure (see above). Raven, Evert, and Curtis (1976) report that a good crop of alfalfa that is ploughed back into the soil may add as much as 450 kilograms of nitrogen per hectare. Certain other groups of plants establish partnership with symbiotic bacteria, mostly actinomycetes; but by far the largest group of plants responsible for nitrogen fixation are leguminous — the bacteria partnership being basically *Rhizobium*.

Examples of leguminous plants commonly used for their nitrogen-fixing properties, in addition to those listed above as green manure, are *Cajanus cajan* Pigeonpea, *C. kerstingii*, *Arachis hypogaea* Groundnut, *Glycine max* Soya Bean, *Macroptilium atropurpureus* (*Phaseolus atropurpureus*), and *Leucaena leucocephala* (*glauca*) Leucaena, Wild Tamarind. Rows of Pigeonpea are usually planted in food farms primarily for the nitrogen-fixing property. Intercropping* groundnuts or *Vigna subterranea* Bambara Groundnut (Aboboe) with maize or alternating them is equally common. It is reported that plots left weedy with *Crotalaria retusa* Devil Bean, another leguminous plant, yield highly when planted with maize. Leucaena is of special interest. Dr Vietmeyer of the Smithsonian Institution in Washington reports that Leucaena plants fix their own nitrogen from the air in the soil, making it possible for them to grow in nitrogen-poor soils and reducing the need for expensive fertilizers. Furthermore, the report adds that Leucaena foliage, rich in nitrogen, harvested and placed around nearby maize plants, has produced maize yields approaching those obtained using commercial fertilizers. Leucaena serves also as fuel-wood and fodder.

Azolla pinnata var. *africana* (*africana*) is a small free-floating fern in fresh,

12. PLANTS AND SOIL NUTRIENTS

sheltered water, often forming a continuous mat. Living in symbiosis with *Azolla* is *Anabaena*. The possibility of the living *Azolla*–Anabaena association releasing appreciable quantities of nitrogen compounds into the soil is being exploited in rice farming and with other crops grown on flooded land, such as *Colocasia esculenta* Taro. *Azolla* is used as a nitrogen fertilizer in rice fields in China and Vietnam — and is being recommended to rice growers in West Africa. It is hoped that the project will lead to increased rice production in the region. Four basic *Azolla* green manuring systems have been developed:

1 Monocropping of *Azolla* in rice fields — the crop being buried by trampling and ploughing before the rice is transplanted. This system ensures a good nitrogen supply for the initial development of the rice but has to be repeated twice or thrice more during the maturation stage.

2 Intercropping of *Azolla*, that is *Azolla* is cultivated together with the rice — the system providing nitrogen during the maturation stage of the rice but not at the development stage.

3 The combination system (monocropped and intercropped *Azolla*) — the system providing nitrogen during both the development and the maturation stage of the crop.

4 Monocropping of *Azolla* outside the rice field — for a continuous stock of *Azolla*, both fresh and composted and available for use in the field before planting the rice.

Azolla has fixed 70–100 kg of nitrogen ha^{-1} per rice crop in the Philippines. In addition *Azolla* is grown as feed for fish, chickens, ducks, pigs, and ruminants (and possibly man) and as a feed stock for biomethane digesters.

Nitrogen fixation in wet rice is also attributable to a wide range of blue-green algae which live in the water above the soil surface. Examples are *Tolypothrix*, *Nostoc*, *Schizothrix*, and *Calothrix*. Norman, Pearson, and Searle (1984) report that the annual amount of nitrogen fixed can exceed 70 kg ha^{-1}. Other fixing agents in wet rice are heterotrophic rhizosphere organisms and non-rhizosphere heterophs.

Nitrogen fixation by leguminous plants is an important part of the nitrogen cycle — the processes involved in the continuous circulation of nitrogen in various forms through plants, the soil, and animals by the action of living organisms and other agencies like lightning.

Soil Indicators

Traditional peasant farmers have, by experience, associated some plants with particular types of soils. The presence of pioneer trees like *Harungana madagascariensis* (Okosoa) and *Musanga cecropioides* Umbrella Tree are associated with secondary forest; while *Corynanthe pachyceras* (Pamprama) and *Nesogordonia papaverifera* (Danta) indicate high forest, and therefore good farming soil. *Raphia* species or *Anthocleista vogelii* (Awora-bontodee) are indicative of swampy areas and the presence of *Suaeda monoica* and *Epaltes gariepina* as indicative of saline soils (Burkill, 1985).

Plants reported to be indicators of good soil include *Aneilema beniniense, Commelina diffusa, Polyspatha paniculata* (Burkill, 1985), wild species of *Aframomum* such as *A. daniellii*; *Thaumatococcus daniellii* Katemfe, *Hilleria latifolia, Trema orientalis (guineensis)* (Sesea), *Nelsonia canescens, Momordica foetida* (Sopropo), *M. charantia* African cucumber (Nyanya) *Geophila* species, and the grass *Olyra latifolia*. However, the presence of *Paspalum conjugatum*, a forest grass popularly known as '*Asamoa nkwanta*', indicates poor soil.

Hildegardia barteri (Akyere) is indicative of rocky soil, *Calotropis procera* Sodom Apple of exhausted soil (Burkill, 1985), and *Drosera indica* of acidic soil (Burkill, 1985).

Bibliography and References

ABBIW, D.K. (1985) *Achyranthes indica* (L.) Mill. (*A. obtusifolia*) — a new weed in Ghana. (mimeographed).

—, (1988) The seed structure of *Nauclea pobeguinii*. Monogr. Syst. Bot. Missouri Bot. Gard. 25: 497–98.

ABEBE, W. (1984) Traditional pharmaceutical practice in Gondar region, northwestern Ethiopia. Journal of Ethnopharmacology 11: 33–47.

ADAMS, C.D. and ALSTON, A.H.G. (1955) *A list of the Gold Coast Pteridophyta*. Bulletin of the British Museum (Natural History) Vol. 1, No. 6.

—, and BAKER, M.G. (1962) Weeds of cultivation and grazing Land. In: J.B. WILLS (editor) *Agriculture and land use in Ghana*. Oxford University Press. London.

ADAMS, D.T. (1940) *An elementary geography of the Gold Coast*. University of London Press.

ADAMS, J.G. (1971) Some uses of plants by the Manon of Liberia (Nimba Mountains). Journal Agric. Trop. Bot. Appl. 18(9/10): 372–378.

ADDAE-MENSAH, I. and SOFOWORA, E.A. (1958) Constituents of *Fagara tessmannii* Pharm. Bull. 6: 300.

ADESINA, S.K. and ETTE, E.I. (1982) The isolation and identification of anticonvulsant agents from *Clausena anisata* and *Afraegle paniculata*. Fitoterapia 53: 63–66.

ADOMAKO, D. (1975) A review of researches into the commercial utilisation of Cocoa by-products, with particular reference to the prospects in Ghana. Ghana Cocoa Marketing Board Newsletter, No. 61.

ADUONUM, K. (1981) *Atenteben Tutor*. University of Cape Coast.

ADU-TUTU, M. and others (1979) Chewing stick usage in southern Ghana. Economic Botany 33(3): 320–328.

AIKMAN, L. (1974) Nature's gifts to medicine. National Geographic.

AINSLIE, J.R. (1937) A list of plants used in native medicine in Nigeria. Imperial Forestry Institute, Oxford, Institute Paper 7. (mimeographed).

AKE-ASSI, L. (1985) Flore de la Côte d'Ivoire: étude descriptive et biogéographique, avec quelques notes ethnobotanique. Monographs in Systematic Botany from the Missouri Botanical Garden 25: 305–307.

—, (1985) Plantes médicinales: quelques légumineuses utilisées en médecine de tradition africaine en Côte d'Ivoire. Monographs in Systematic Botany from the Missouri Botanical Garden 25: 309–313.

—, (1985) Personal communication.

AKEHURST, B.C. (1981) *Tobacco*. Longmans. London and New York.

AKOBUNDO, I.O. (Editor) (1980) Weeds and their control in the humid and subhumid tropics. Proc. Series No. 3 of a conference held at the International Institute of Tropical Agriculture, Ibadan. July 1978.

AKOBUNDU, I.O. & AGYAKWA, C.W. (1987) *A Handbook of West African Weeds*. IITA, Nigeria.

AKUBUE, P.I. and MITTAL, G.C. (1982) Clinical evaluation of a traditional herbal practice in Nigeria: a preliminary report. Journal of Ethnopharmacology 6(3): 355–359.

ALSTON, A.H.G. (1959) *The ferns and fern-allies of West Tropical Africa*. Crown Agents for Overseas Governments and Administration.

AMARTEIFIO, E. and others (1969) *A recipe book for Ghana Schools*. Macmillan. London.

AMICO, A. (1977) Medicinal Plants of Southern Zambesia. Fitoterapia 48: 101–135.

—, (1980) Plants used in traditional medicine in lower Zambesia. Journal of Ethnopharmacology 2(1): 33–34.

AMOAH, G.K.A. (Editor) (1962) *The Aburi Garden*. Ministry of Information, Ghana.

AMPOFO, O. (1977) Plants That Heal. In: *World Health* (The magazine of the World Health Organization) 26: 28–30.

—, (1984) *First Aid in Plant Medicine*. Presby Press, Accra. Ghana.

ANON (1937) *Draft of First Descriptive Check-list of the Gold Coast*. Imperial Forestry Institute, Oxford.

—, (1963) Journal of African Timber & Plywood (Ghana) Limited (A.T.P.) Samreboi, A.T.P.

—, (1970) *Learn about musical instruments*. McGraw-Hill Far Eastern Publishers. Singapore.

—, (1970) *New Facts about Marijuana*. Ambassador College Press. Pasadena, California.

—, (1979) *Timbers of the World*. The Construction Press. England.

—, (1984) Leaf protein production at Kpone Bawaleshie. (mimeographed).

—, 'GRAMOXONE' A revolutionary herbicide for use in tropical crops. Focus Journal: A 'Plant Protection' Information Service.

—, (1987) Report of the Task Force on the Development of Oil Palm Industry in Ghana.

ARDAYFIO-SCHANDORF, E. (Editor) (1983) Bulletin of the Ghana Geographical Association Vol. 1. (mimeographed).

ARE, L.A. and GWYNE-JONES, D.R.G. (1974) *Cacao in West Africa*. Oxford University Press. Ibadan.

ATA, J.K.B.A. (1977) New sources of vegetable fats and oils. (mimeographed).

AUBREVILLE, A. (1959) *La Flore Forestière de la Côte d'Ivoire*. 3 Vols. Centre Technique Forestière Tropical. Nogent-sur-Marne.

AVUMATSODO, S.K. (1984–5) Personal communication.

AYENSU, E.S. (1978) *Medicinal Plants of West Africa*. Reference Publications. Michigan.

—, (Editor) (1980) *Jungles*. Jonathan Cape, London.

—, and BENTUM, A. (1974) *Commercial Timbers of West Africa*. Smithsonian Institution Press, Washington.

—, and COURSEY, D.G. (1972) Guinea Yams — The botany, ethnobotany, use and possible future of yams in West Africa. Economic Botany 26(4): 301–318

—, and others (1975) *Underexploited Tropical Plants with Promising Economic Value*. National Academy of Sciences. Washington.

—, and others (1980) *Firewood Crops, Shrubs and Tree Species for Energy Production*. National Academy of Sciences. Washington.

BAIDOE, J.F. (1976) Yield regulation in the high forest of Ghana. Ghana Forestry Journal 2: 22–27.

BAILEY, L.H. (1924) *Manual of Cultivated Plants*. Macmillan. New York.

BEDIGIAN, D. and HARLAN, J.R. (1983) Nuba agriculture and ethnobotany, with particular reference to sesame and sorghum. Economic Botany 37(4): 384–395.

BENNEH, G. (1974) Environment and Development in West Africa. Ghana Academy of Arts and Sciences.

BENNET, F.D. and RAO, V.P. (1968) Distribution of an introduced weed *Eupatorium odoratum* Linn. (Compositae) in Asia and Africa and possibilities of its Biological Control. Pest Articles and News Summaries.

BENSON, L. (1959) *Plant Classification*. D.C. Heath. Lexington, Massachusetts.

BERHAUT, J. (1954) *Flore du Sénégal*. Librairie Clairafrique. Dakar.

BERTHERAND, E.L. (1886) Acclimatation. Flore medicale de l'Afrique occidentale. Algerie.

BISACRE, M., CARLISLE, R., ROBERTSON, D. and RUCK, J. (Editors) (1984) *Illustrated Encyclopedia of Plants*. Marshall Cavendish.

BOATENG, E.A. (1960) *A Geography of Ghana*. Cambridge University Press.

BOATENG, S. (1970) Soap-making in Ghana. In: The Gliksten Journal 10(6): 23.

BOULOS, L. (1983) *Medicinal Plants of North Africa.* Reference Publications. Michigan.

BOUQUET, A. (1972) Plantes médicinales du Congo-Brazzaville. Travaux et Documents de l'O.R.S.T.O.M., No. 13, Office de la Recherche Scientifique et Technique Outre-Mer, Paris.

—, and DEBRAY, M. (1974) Plantes médicinales de la Côte d'Ivoire. Office de la Recherche Scientifique et Technique Outre-Mer., 1, Paris.

BOUTELJE, J.B. (1980) *Encyclopedia of world timbers.* Swedish Forest Products Research Laboratory.

BRITWUM, S.P.K. (1976) Natural and artificial regeneration practices in the high forest of Ghana. Ghana Forestry Journal 2: 45–49.

BRUCE, R.K. (Chairman) and others (1977) *An Environmental Studies Programme for Schools in Ghana. Our Country II.* Ministry of Education, Ghana.

BRUNEL, J.F., HIEPKO, P. and SCHOLZ, H. (1984) *Flore analytique du Togo. Phañerogames.* Eschborn.

BURGER, W.C. (1967) *Families of Flowering Plants in Ethiopia.* Oklahoma State University Press. Stillwater, Oklahoma, U.S.A.

BURKILL, H.M. (1985) *Useful Plants of West Tropical Africa* Vol. 1, (A–D). Royal Botanic Gardens, Kew.

CANSDALE, G.S. (1961) *West African Snakes.* Longmans. London

CARSON, A.G. (1978) Integrated Chemical Control of Weeds with Handweeding in Rainfed Cotton in Northern Ghana. (mimeographed).

CERES (July–Aug 1980) No. 76 (Vol. 13, No. 4) FAO Review on Agriculture and Development.

—, (Sep–Oct 1980) No. 77 (Vol. 13, No. 5).

—, (Nov–Dec 1980) No. 78 (Vol. 13, No. 6).

—, (May–June 1981) No. 81 (Vol. 14, No. 3).

—, (Jan–Feb 1983) No. 91 (Vol. 16, No. 1).

—, (May–June 1983) No. 93 (Vol. 16, No. 3).

—, (July–Aug 1983) No. 94 (Vol. 16, No. 4).

CHAKRAVARTY, H.L. (1968) *Cucurbitaceae* of Ghana. Bull. Inst. fond. Afrique noire 30 ser. A.: 400–468.

—, (1974) *Flora of Cape Coast* Part 1, P.K. Chakraborty, 15/4, N.N. Ghosh Lane, Calcutta 40.

CHANDRA, S. (Editor) (1984) *Edible aroids.* Clarendon Press. Oxford.

CHEREMISINOFF, N.P. (1980) *Wood for Energy Production.* Ann Arbor Science Publishers.

CHILD, R. (1974) *Coconuts.* Longmans. London.

CHIPP, T.F. (1913) *A list of the trees, shrubs and climbers of the Gold Coast, Ashanti and the Northern Territories.* London.

—, (1922) *The Forest Officers' Handbook of the Gold Coast, Ashanti and the Northern Territories.* Crown Agents for the Colonies. London.

CLARK, J.D. and STEMLER, A. (1975) Early domesticated *Sorghum* from Central Sudan. Nature Vol. 254, No. 5501.

CLARKE, J. (Editor) (1985) Maff Bulletin Vol. 29, No. 5. Maff Publication.

CLAYTON, W.D. (1966) *A Key To Nigerian Grasses.* Samaru Research Bulletin No. 1. The Government Printer. Kaduna.

CLERK, G.C. (1974) *Crops and their diseases in Ghana.* Ghana Publishing Corporation.

CONSTANTINE, A. Jr. (Revised HOBBS, H.J.) (1975) *Know Your Wood.* Charles Scribner's Sons. New York.

COURIERS, THE (July–Aug 1982) No. 74. Journal of African–Caribbean–Pacific–European Community.

COURSEY, D.G. and BOOTH, R.H. (1972) The Post-Harvest Phytopathology of Perishable Tropical Produce. Review of Plant Pathology 51: 751–765.

CURTIS, D.L. (1965) *Sorghum* in West Africa. Samaru Research Bulletin No. 59. The Government Printer. Kaduna.

CUTLER, D.F. (1968) Detection of adulterants in tobacco papers using Anoptral Contrast Microscopy. Microscopy 31: 29–32.

DALZIEL, J.M. (1937) *The Useful Plants of West Tropical Africa.* The Crown Agents. London.

De WET, J.M.J. and HARLAN, J.R. (1971) The Origin and Domestication of *Sorghum bicolor.* Economic Botany 25(2): 128–135

DIRAR, H.A. (1984) Kawal, meat substitute from fermented *Cassia obtusifolia* leaves. Economic Botany 38(3): 342–349.

DODU, S.R.A. (1972) *Our Heritage – the Traditional Medicine of Mankind.* Ghana Universities Press. Accra.

DOKOSI, O.B. (1969) Some herbs used in the traditional system of healing diseases in Ghana — 1. Ghana Journal of Science 9(2): 119–130.

—, (1982–85) Personal communication.

DOKU, E.V. (1966) Cultivated Cassava Varieties in Ghana. Ghana Journal of Science 6(3 & 4): 74–86.

—, (1966) Root Crops in Ghana. Ghana Journal of Science 6(1 & 2): 15–36.

—, (1970) Viability in local and exotic varieties of Cowpea (*Vigna unguiculata* (L.) Walp.) in Ghana. Ghana Journal of Agric. Sci. 3: 139–143.

—, and KARIKARI, S.K. (1971) Bambara Groundnut. Economic Botany Vol. 25, No. 3.

DUDGEON, G.C. (1922) *The Agricultural and Forest Products of British West Africa*. John Murray. London.

DUKE, J.A. (1981) *Handbook of Legumes of World Economic Importance*. Plenum Press. New York.

EDEN, T. (1958) *Tea*. Longman, Green and Co.

ENTI, A.A. (1975) Distribution and Ecology of *Thaumatococcus daniellii* (Benn.) Benth. (mimeographed).

—, (1979) Notes on the collection of *Heliotropium indicum* Linn. in Ghana. (mimeographed).

—, (1979) Notes on *Synsepalum dulcificum* (The Miraculous Berry). (mimeographed).

—, (1982–85) Personal communication.

von EPENHUIJSEN (1974) *Growing native vegetables in Nigeria*. Food and Agriculture Organization of the United Nations. Rome.

EVANS, F.J. and others (1983) *British Herbal Pharmacopoeia*. British Herbal Medicine Association.

EVERARD, B. and MORLEY, B.D. (1970) *Wild Flowers of the World*. Ebury Press and Michael Joseph. London.

FANSHAWE, D.B. (1972) The Bamboo, *Oxytenanthera abyssinica*, its Ecology, Silviculture and Utilization. Kirkia 8(2): 157–166.

FELLOWS, L.E. (1970) 5-Hydroxy-L-Tryptophan, 5-Hydroxytryptamine and L-Tryptophan-5-Hydroxylase in *Griffonia simplicifolia*. Phytochemistry 9: 2389–2396.

—, (1985) The sugar-shaped weapons of plants. New Scientist No. 1469: 40–41.

—, and others (1985) Clues to new strategies of pest control from the chemicals of wild plants. (mimeographed).

—, and others (1986) Polyhydroxy Plant Alkaloids as Glycosidase Inhibitors and Their Possible Ecological Role. American Chemical Society. Symposium ser. 296: 72–78.

FISH, F. and WATERMAN, P.G. (1971) A note on the Chloroform-soluble alkaloids of *Fagara macrophylla*. Journal of Pharmacy 23 suppl.: 67–68.

—, and —, (1971) Alkaloids from two Nigerian species of *Fagara*. Journal of Pharmacy, 23 suppl.: 1325–1355.

—, and —, (1972) Methanol-soluble quarternary alkaloids from African *Fagara* species. Phytochemistry 11: 3007-3014.

—, and —, (1972) Rutaceae-Lirioresinol-B-dimethyl ether from the bark of *Fagara leprieurii*. Phytochemical Reports 11: 1527-1528.

FOGGIE, A. (1962) Natural regeneration in the humid tropical forest. Fifth Wildlife & Forestry Congr. 3: 1941-1946.

—, and PIASECKI, B. (1962) Timber, fuel and minor forest produce. In: *Agriculture and land use in Ghana*. (WILLS, J.B. — Editor) 236-251. Oxford University Press. London.

FOLEY, G. and MOSS, P. (1983) *Improved Cooking Stoves in Developing Countries*. Earthscan.

—, and TIMBERLAKE, L. (1984) *Stoves and Trees*. Earthscan.

FOSTER, W.H. and MUNDY, E.J. (1961) Forage species in Northern Nigeria. Samaru Research Bulletin No. 14. The Government Printer, Kaduna.

FOWLER, F.G. and FOWLER, H.W. (1953) *The Pocket Oxford Dictionary*. (4th Edition). Clarendon Press. Oxford.

GBILE, Z.O. and SOLADOYE, M.O. (1985) Plants in traditional medicine in West Africa. Monographs in Systematic Botany from Missouri Botanical Garden 25: 343-349.

GEREAU, R.E. (1985) Insect Control. AETFAT Congress Herbarium Curation Workshop. Missouri Botanical Garden.

GHANA Forestry Department Annual Report (1963-72) Forestry Division, Accra.

GHANA Seven-year Development Plan (1963/64-69/70) Government Printing Department. Accra, Ghana.

GHANA Cocoa Marketing Board at Work. (1977) Public Relations Department of the Cocoa Marketing Board. 6th Edition.

GHANA Cotton Co. Ltd. (personal communication).

GLEDHILL, D. (1972) *West African Trees*. Longmans. London.

GODIN, V.J. and SPENSLEY, P.C. (1971) *Oils and Oilseeds*. The Tropical Products Institute (mimeographed).

GOFFE, T. (1978) *XYZ of Musical Instruments*. Transworld Publishers.

GREENSILL, T.M. (1959) *The Food Crops*. Evans Brothers.

—, (1959) *The School Vegetable Garden*. Evans Brothers.

GREENWAY, P.G. (1941) Gum, resinous and mucilaginous plants in East Africa. East African Agricultural Journal 6: 241-250.

—, (1941) Bowstring hemp of *Sansevieria* fibre. East African Agricultural Journal 7: 96-97.

—, (1941) Dyeing and tanning plants in East Africa. Bull. Imp. Inst. 39: 222–245.

—, (1942) Wild rubber in East Africa. East African Agricultural Journal 8: 224–227.

—, (1942) A use for saponins and some possible local sources. East African Agricultural Journal 9: 85–86.

GROULEZ, J. and WOOD, P.J. (1985) *Terminalia superba*. Commonwealth Forestry Institute, Oxford.

GRUBBER, G.J.H. (1977) (Edited by TINDAL, H.D. and WILLIAMS, J.T.) *Tropical Vegetables and their Genetic Resources*. IBPGR Publication.

GUILLARMOD, A.J. (1971) *Flora of Lesotho (Basutoland)*. Verlag Von J. Cramer.

GUILLEBAUD, C.W. (1958) *An economic survey of the sisal industry of Tanganyika*. The Tanganyika Sisal Growers Association and James Nisbet. England.

GUNN, C.R. (1984) *Fruit and seeds of genera in the subfamily Mimosoideae (Fabaceae)*. United States Department of Agriculture.

—, and DENNIS, J.V. (1976) *World guide to tropical drift seeds and fruits*. Quadrangle/The New York Times Book Co.

HALL, D.O., BARNARD, G.W. and MOSS, P.A. (1982) *Biomass for energy in the developing countries*. Pergamon Press.

HALL, J.B. (1973) *Malvastrum corchorifolium* (Desr.) Britton in Ghana. The Nigerian Field 38(4): 189–192.

—, (1980) New and little-known species of *Placodiscus* (*Sapindaceae*) in West Africa. Adansonia, Ser. 2.20(3): 287–295.

—, ABBIW, D.K. and ENTI, A.A. (1980) List of Indigenous Trees found in the Forest Zone of Ghana. (mimeographed).

—, and JENIK, J. (1967) A New Weed for West Africa: *Ruellia tuberosa* Linn. 32(4): 188–191.

—, and —, (1968) Contribution towards the classification of savanna in Ghana. Bulletin de l'I.F.A.N. 30(1): 84–99.

—, KUMAR, R. and ENTI, A.A. (1972) The obnoxious weed *Eupatorium odoratum* (Compositae) in Ghana. Ghana Journal of Agric. Sci. 5: 75–78.

—, and OKALI, D.U.U. (1976) Vegetation Report — IITA Watershed Study. Ghana University Press.

—, PIERCE, P. and LAWSON, G. (1971) *Common plants of the Volta Lake*. Department of Botany, University of Ghana, Legon.

—, and SIAW, D.E.K.A. (1980) The varieties of *Grewia carpinifolia* Juss. (*Tiliaceae*). Adansonia, Ser. 2.20(3): 339–347.

—, and SWAINE, M.D. (1974) *Classification and ecology of forests in Ghana*. Ghana University Press.

—, and —, (1976) *Classification and ecology of closed-canopy Forest in Ghana*. Blackwell Scientific Publications.

—, and -, (1981) *Distribution and ecology of vascular plants in a tropical rain forest. Forest Vegetation in Ghana*. Dr W. Junk Publishers.

—, and WATERMAN, P.G. (Edited KUNKEL, G.) (1979) Some contributions of phytochemistry to the classification of African *Rutaceae*. *Taxonomic aspects of African economic Botany* pp. 105–108. Las Palmas de Gran Canaria.

HARPER, E. (1969) Some trees to look out for. The Gliksten Journal 10(3): 19.

—, (1970) Wood has a special appeal. The Gliksten Journal 10(6): 23.

HECKEL, E. (1903) Catalogue alphabétique raisonné des plantes médicale et toxiques de Madagascar avec leur emploi indigène. Annales Mus. Colon. Marseille II (sec. 2, Vol. 1), fasc. 2: 59–204.

—, (1910) Les plantes utiles de Madagascar. Annales Mus. Colon. 18 (sec. 2, Vol. 8): 5: 372.

HEDIN, L. (1930) *Étude sur la forêt et les bois du Cameroun sous mandat français*. Paris.

HENRY, T.A. (1939) *The plant alkaloids*. Edition 3. Churchill. London.

HEPPER, F.N. (1966) The duckweeds of West Africa. Nigerian Field 31(1): 18–21.

—, (1967) The identity of Grains-of-paradise and Melegueta pepper (*Aframomum, Zingiberaceae*) in West Africa. Kew Bulletin 21(1): 129–137.

—, (1970) Bambara groundnut. Field Crop Abstracts 23: 1–6.

—, (1976) *The West African Herbaria of Isert and Thonning*. Bentham-Moxon Trust, Kew.

—, (1985) A visit to a lake stilt village in Ghana. Nigerian Field 49: 45–51.

—, and NEATE, F. (1971) *Plant collectors in West Africa*. Utrecht.

HEYWOOD, V.H. and CHANT, S.R. (Editors) (1982) *Popular Encyclopedia of Plants*. Cambridge University Press.

HOLLAND, J.H. (1922) *The useful plants of Nigeria*. His Majesty's Stationery Office, London.

HOLM, Le Roy G. and others (1977) *The world's worst weeds – distribution and biology*. University Press of Hawaii.

—, (1979) *A geographical atlas of world weeds*. John Wiley and Sons. New York

HOPKINS, B. (1974) *Forest and Savanna*. Heinemann. Ibadan.

HOSSAIN, MOSHARRAF and HALL, J.B. (1969) *A field key to the trees of Mole Game Reserve, Damongo, Ghana.* Institute of African Studies. University of Ghana. Legon.

HOWES, F.N. (1974) *Useful and everyday plants.* Cambridge University Press.

HUTCHINSON, J. and DALZIEL, J.M. (1927-36) *Flora of West Tropical Africa.* 3 Vols. Revision by R.W.J. KEAY, Vol. 1, Part 1 (1954), Part 2 (1958); Vol. II by F.N. HEPPER (Editor) (1963); and Vol. III by F.N. HEPPER (Editor) Part I (1968) Part II (1972). Crown Agents. London.

HUYNH, K.-L. (1984) Étude des *Pandanus* (*Pandanaceae*) d'Afrique occidentale (Première partie). Bull. Mus. Nat. Hist. Nat., Paris, 4e sér., 3, section B, Adansonia No. 1: 37-55.

HYMOWITZ, T. and BOYD, J. (1977) Origin, ethnobotany and agricultural potential of the Winged Bean — *Psophocarpus tetragonolobus.* Economic Botany 31: 180-188.

INGLETT, G.E. and MAY, J.F. (1968) Tropical Plants with unusual taste properties. Economic Botany 22(4): 326-331.

IRVINE, F.R. (1930) *Plants of the Gold Coast.* Oxford University Press.

—, (1931) *West African Botany.* Oxford University Press.

—, (1956) The edible cultivated and semi-cultivated leaves of West Africa. Dr W. Junk. Den Haag.

—, (1961) *Woody plants of Ghana.* Oxford University Press.

—, (1969) *West African Crops.* Oxford University Press.

IVENS, G.W.K., MOODY and EGUNJOBI, J.K. (1978) *West African Weeds.* Oxford University Press. Ibadan.

JAIN, S.K. (1968) *Medicinal Plants.* National Book Trust, India.

JOHN, D.M. (1986) *The inland waters of tropical West Africa. An Introduction and Botanical Review.* E. Schweizerbart'sche Verlagsbuchhandung Stuttgart.

—, and ASARE, S.O. (1975) A Preliminary Study of the Variations in Yield and Properties of Phycocolloids from Ghanaian Seaweeds. Marine Biology 30: 325-330.

JUDD, B.I. (1979) *Handbook of Tropical Forage Grasses.* Garland S.T.P.M. Press.

JUDD, W.J. and others (1977) BS II Lectin: A Second Hemagglutinin Isolated from *Bandeiraea simplicifolia* Seeds with affinity for type III Polyagglutinable Red Cells. Vox Sanguinis 33: 246-251.

KEAY, R.W.J. (1949) *An outline of Nigerian vegetation.* Government Printer. Lagos.

—, ONOCHIE, C.F.A. and STANFIELD, D.P. (1960) *Nigerian trees Vol. 1.* Federal Department of Forestry, Ibadan, Nigeria.

—, —, and —, (1964) *Nigerian trees Vol. II*. Department of Forestry Research. Ibadan.

KENNEDY, J.D. (1936) *Forest flora of Southern Nigeria*. Government Printer.

KERHARO, J. (1969) *Cassia* in the Senegalese pharmacopoeia use in traditional medicine, chemistry, and pharmacology. Bull. de la Société Medicale d'Afrique Noire de Lanque Francaise 14: 1019.

KNORR, K.E. (1945) *World rubber and its regulation*. Stanford University Press. California.

KOLI, S.E. (1973) *A guide to cotton cultivation in Ghana*. Ghana Publishing Corporation.

KOKWARO, J.O. (1976) *Medicinal plants of East Africa*. East African Literature Bureau.

KROGSGAARD-LARSEN, P., CHRISTENSEN, S.B. and KOFOD, H. (Editors) (1984) *Natural products and drug development*. Munksgaard. Copenhagen.

KRUG, C.A. and De POERCK, R.A. (1968) *World coffee survey*. FAO of UN.

KUNKEL, G. (1965) *The trees of Liberia*. BLV Bayerischer Landwirtschaftsverlag.

KURANCHIE, P.A. (1971) Model farm plans for tobacco farmers in Ejura and Wenchi areas. (mimeographed).

LANE-POOLE, C.E. (1916) *A list of trees, shrubs, herbs and climbers of Sierra Leone*. Government Printing Office. Freetown.

LAWES, D.A. (1964) A new cultivation technique in tropical Africa. Samaru Research Bulletin No. 32. The Government Printer. Kaduna.

LAWRENCE, G.H.M. (1963) *Taxonomy of vascular plants*. The MacMillan Company. New York.

LAWSON, G.W. (1967) Sudd formation on Volta Lake. Bulletin de l'I.F.A.N. 29: 1–4.

—, and JOHN, D.M. (1982) *The marine algae and coastal environment of tropical West Africa*. J. Cramer.

LEAKEY, C.L.A. and WILLS, J.B. (Editors) (1977) *Food Crops of the Lowland Tropics*. Oxford.

LELY, H.V. (1925) *The useful trees of Northern Nigeria*. Crown Agents. London.

LEMMA, A., HEYNEMAN, D. and SILANGWA, S.M. (1984) *Phytolacca dodecandra* (Endod). Tycooly International Publishing Limited. Dublin.

LEWIS, Memory P.F.E. (1980) Chewing Sponges for Teeth Cleaning. J. Prev. Dent. 6: 75–80.

—, (1980) Plants Used for Teeth Cleaning throughout the World. J. Prev. Dent. 6: 61–70.

LEWIS, W.H. (1980) Plants Used as Chewing Sticks. J. Prev. Dent. 6: 71–73.

—, and ELVIS-LEWIS, Memory P.F.E. (1977) *Medical botany*. John Wiley and Sons.

LIEBERMAN, D. and others (1979) Seed dispersal by baboons in the Shai Hills, Ghana J. Ecology 60: 65–75.

LOCK, J.M. and HALL, J.B. (1973) Three new species of *Aframomum* K. Schum. (*Zingiberaceae*) from Ghana. Kew Bull. 28: 441–449.

—, —, and ABBIW, D.K. (1977) The cultivation of Melegueta Pepper (*Aframomum melegueta*) in Ghana. Economic Botany 31(3): 321–29.

LONGMAN, K.A. and JENIK, J. (1974) *Tropical forest and its environment*. Longman. London.

LOWE, J. and STANFIELD, D.P. (1974) *The flora of Nigeria. Sedges*. Ibadan University Press.

McGINNIES, W.G., GOLDMAN, B.J. & PAYLORE, P. (1971) *Food, fiber and the arid lands*. University of Arizona Press.

MANDANGO, M.A. and BANDOLE, M.B. (1988) Contribution à la connaissance des plantes médicinales des Turumbu de la zone de Basoko (Zaire). Monographs in Systematic Botany from Missouri Botanical Garden 25: 373–384.

MANU, C.K. (1984) Report of the Department of Game and Wildlife in combating drought and desertification in Ghana. (mimeographed).

MARTIN, F.W. (Editor) (1984) *Handbook of tropical food crops*. CRC Press, Inc. Boca Raton, Florida.

von MAYDELL, H.-J. (1983) *Arbres et arbustes du Sahel* (Leurs caractéristiques et leurs utilisation). Eschborn.

MEYER, B.S. and ANDERSON, D.B. (1965) *Plant physiology*. D. Van Nostrand.

MILLER, T.B., RAINS, A.B. and THORPE, R.J. (1964) The nutritive value and agronomic aspects of some fodders in Northern Nigeria — III. Hays and dried crop residues. Samaru Research Bulletin No. 39. The Government Printer. Kaduna.

MIRACLE, M.P. (1966) *Maize in tropical Africa*. The University of Wisconsin Press.

MORO, IBRAHIM (1984–5) Personal communication.

MORTON, J.K. (1957) Sand-dune formation on a tropical shore. J. Ecol. 45: 495–497.

—, (1961) *West African lilies and orchids*. Longman. London.

—, (1966) The Commelinaceae of West Africa: A biosystematic survey. J. Linn. Soc. (Bot.), 60(382): 167–221.

—, (1972) Phytogeography of the West African mountains. In: (VALENTINE, D.N.) (Editor) *Taxonomy, phytogeography and evolution*. Academic Press. London.

—, (1981) Introgression in West African Orchids of the genus *Eulophia*. Separata Bol. Soc. Brot., Ser. 2, 53(2): 1437–1458.

MUGERA, G.M. (1970) Toxic and medicinal plants of East Africa. Bull. of Epizootic Diseases of Africa 1; 2; 18: 377–387, 389–403.

NEWTON, L.E. (1979) Chaya, a food plant new to West Africa. Appropriate Technology Vol. 6, No. 3.

NORMAN, M.J.T., PEARSON, C.J. and SEARLE, P.G.E. (1984) *The Ecology of Tropical Food Crops*. Cambridge University Press.

NYE, P.H. (1957) Some prospects for subsistence agriculture in West Africa. J. W. Afr. Sci. Ass. 3: 91–95.

OBENG-ASAMOA, E.K. (1977) A limnological study of the Afram arm of Volta Lake. Hydrobiologia 55(3): 257–264.

OGUAKWA, J.U. (1980) Plants used in traditional medicine in West Africa. Journal of Ethnopharmacology 2(1): 29–31.

OKALI, D.U.U. and ATTIONU, R.H. (1974) The quantities of some nutrient elements in *Pistia stratiotes* L. from the Volta Lake. Ghana Jul. Agric-Science 7: 203–208.

—, and HALL, J.B. (1974) Colonization of *Pistia stratiotes* L. mats by *Scirpus cubensis* Poeppig & Kunth on the Volta Lake. Ghana University Press.

—, — (1977) A structural and floristic analysis of woody fallow vegetation near Ibadan, Nigeria. J. Ecology 67(1): 322–346.

OKIGBO, B.N. (1980) *Plants and food in Igbo culture and civilization*. Government Printer, Owerri, Nigeria.

OKOLI, B.E. (1984) Wild and cultivated cucurbits in Nigeria. Economic Botany 38(3): 350–357.

—, and MGBEOGU, C.M. (1983) Fluted pumpkin, *Telfairia occidentalis*: West African vegetable crop. Economic Botany 37(2): 145–149.

OLDEMAN, R.A. (1977) L'architecture de la forêt Guyanaise. Mem. ORSTOM, 73: 1–204.

—, (Editor) (1982) *Tropical hardwood utilization, practice and prospects*. D.W. Junk. The Hague.

OLIVER-BEVER, B. (1983) Medicinal plants in tropical West Africa. II. Plants acting on the nervous system. Journal of Ethnopharmacology 7(1): 1–93.

ONWUEME, I.C. (1978) *The tropical tuber crops*. John Wiley and Sons.

OOI, J.B. (Editor) (1983) *Natural Resources in Tropical Countries*. Singapore University Press.

OPEKE, L.K. (1982) *Tropical tree crops*. John Wiley and Sons. New York.

OSAFO, D.M. (1968) Some Views on Agricultural Development in Ghana. Insight and Opinion Vol. 3, No. 3. Accra Catholic Press.

PALGRAVE, O.H.C. (1957) *Trees of Central Africa*. National Publications Trust. Rhodesia & Nyasaland.

PAPADAKIS, J. (1965) *Crop ecology survey in West Africa*. FAO. Rome.

PAPERNA, I. (1969) Aquatic weeds, snails and transmission of bilharzia in the new man-made Volta Lake, Ghana. Bulletin de l'I.F.A.N. 31: 840–841.

PARDY, A.A. (1951–55) *Notes on indigenous trees and shrubs of S. Rhodesia*. The Art Printing Works Limited, Salisbury.

PARKER, C. (1978) Internal Report No. 129. Weed Research Organisation.

PENNINGTON, T.D. and STYLES, B.T. (1975) A generic monograph of the *Meliaceae*. Blumea 22: 419–540.

—, and —, (1981) *Flora Neotropica – Meliaceae*. The New York Botanical Garden.

PENZER, B.A. (1920) *Cotton in British West Africa*. The Federation of British Industries.

PERDUE, R. (U.S. Agric. Res. Services) (1985) Personal communication.

PETTET, A. and PETTET, S.J. (1970) Biological control of *Pistia stratiotes* L. in Western State, Nigeria. Nature Vol. 226, No. 282.

PIERCE, P.C. (1971) Aquatic Weed Development, Impact and Control at Volta Lake 1967–71. USAID Volta Lake Technical Assistance Project (641.11.190.028) 90 pp. (mimeographed).

POBÉGUIN, H. (1906) *Flore de la Guinée Française*. Augustin Challamel, Paris.

PORTÈRES, R. (1976) African cereals. In. J.R. Harim et al. *Origin of African plant domestication* pp. 409–52. Mouton. The Hague & Paris.

PURSEGLOVE, J.W. (1968) *Tropical Crops* Vol. 1 (Dicotyledons). Longmans.

—, (1972) *Tropical Crops* Vol. II (Monocotyledons). Longmans.

—, BROWN, E.C., GREEN, C.L. and ROBBINS, S.R.J. (1981) *Spices* Vol. II. Longmans.

QUARCOO, A.K. (1968) The visual Arts of Ghana. A discipline and a dimension of sociological studies. Insight and Opinion Vol. 3, No. 3. Accra Catholic Press.

QUIGLEY, F. and HALL, J.B. (1979) *Dioscorea* L. (*Dioscoreaceae*)in Ghana. Bulletin de l'I.F.A.N. 41, ser. A, No. 3: 490–504.

RACHIE, K.O. (Chairman) and others (1979) *Tropical legumes: resources for the future*. National Academy of Sciences, Washington D.C.

RAVEN, P.H., EVERT, R.F. and CURTIS, H. (1976) *Biology of plants* (Second Edition). Worth Publishers Inc.

RICHARDS, O.W. and DAVIES, R.G. (1977) *Imms' general textbook of entomology* Vol. II. Chapman and Hall.

RICHARDS, P.W. (1952) *The tropical rain forest*. Cambridge University Press. London.

—, (1955) The secondary succession in the tropical rain forest. Sci. prog. 50: 45–57.

RIDSDALE, C.E. (1975) A synopsis of the African and Madagascan *Rubiaceae -Naucleeae*. Blumea 22(3): 541–553.

RODIN, R.J. (1985) *The ethnobotany of the Kwanyama Ovambos*. Missouri Botanical Garden.

ROSE INNES, R. (1977) *A manual of Ghana grasses*. Overseas Development Administration, Land Resources Division, Surbiton, U.K.

ROWSON, J.M. (1969) Nostrums and medicinal plants of West Africa. Annales Pharmaceutiques Françaises 27: 439–448.

RUSSELL, T.A. (1955) The kola of Nigeria and the Cameroons. Tropical Agric. 32: 210–240.

—, (1965) The Raphia palms of West Africa. Kew Bull. 19(2): 173–196.

SANFORD, W.W. (1968) Distribution of epiphytic orchids in a semi-deciduous tropical forest in Southern Nigeria. J. Ecol. 56: 697–705.

SAUNDERS, H.N. (1958) *A handbook of West African flowers*. Oxford University Press.

SCHRODER, D. (1976) Noxious weeds in West African Commonwealth Countries (especially in Ghana). Commonwealth Institute of Biological Control Report.

SEGERBACK, L.B. (1983) *Orchids of Nigeria*. A.A. Balkema, Rotterdam.

SEIBIRE, A. (1899) *Les plantes utiles du Senegal: plantes indigenes, plantes exotiques*. Paris.

SEIGLER, D.S. (Editor) (1977) *Crop Resources*. Academic Press. New York.

SHALES, J.W. (Editor) (1969) The Importance of Wood in Gunmaking. The Gliksten Journal 10(3): 9–10.

—, (1970) Hardwoods used in England team coach by Plaxtons. The Gliksten Journal 10(3): 21.

SINGH, S., VRIES, J. de, HULLEY, J.C.L. and YEUNG, P. (1977) *Coffee, tea and cocoa.* World Bank Staff Occasional Papers No. 22.

SINGHA, S.C. (1965) *Medicinal plants of Nigeria.* M.O.E. Publ., Enugu, Nigeria.

SKERMAN, P.J. (1977) *Tropical forage legumes.* FAO Plant Production and Protection Series No. 2.

SMART, J. (1976) *Tropical pulses.* Longman.

SMITH, G.M. (1955) *Cryptogamic botany, algae and fungi* Vol. 1. 2nd Edition. McGraw-Hill.

SOFOWORA, A. (1982) *Medicinal plants and traditional medicine in Africa.* John Wiley.

STANFIELD, D.P. (1970) *The Flora of Nigeria. Grasses.* Ibadan University Press.

STONE, R.H. and COZENS, A.B. (1958) *Biology for tropical schools.* Longman, Green.

STORRS, A.E.G. and PIEARCE, G.D. (1982) *Don't eat these* (A guide to some local poisonous plants). Forestry Department, Ndola, Zambia.

STORY, R. (1958) Some plants used by the Bushmen in obtaining food and water. Botanical Survey of South Africa, Mem. 30.

STYLES, B.T. (1986) *Infraspecific classification of wild and cultivated plants.* Clarendon Press. Oxford.

SWAINE, M.D. and HALL, J.B. (1974) Ecology and Conservation of Upland Forests in Ghana. Proc. Ghana Scope's Conf. Environ. Dev. West Afr., 151–158.

—, and —, (1983) Early Succession on Cleared Forest Land in Ghana. J. Ecology 71: 601–627.

—, and others (1979) Zonation of a Coastal Grassland in Ghana, West Africa. Folia Geobot. Phytotax., Praha 14: 11–27.

TAILOR, W.I. and FARNSWORTH, N.R. (Editors) (1975) *The Catharanthus Alkaloids.* Marcel Dekker, Inc. New York.

TAYLOR, C.J. (1952) The vegetation zones of the Gold Coast. Bull. Gold Coast. For. Dep. 4: 1–12.

—, (1959) *Synecology and silviculture in Ghana.* Thomas Nelson.

—, (1962) *Tropical forestry with particular reference to West Africa.* Oxford University Press.

TERRY, P.J. (1983) *Some common crop weeds of West Africa and their control.* USAID.

THOMPSON, H.N. (1910) Gold Coast: report on forests. Colon. Rep. Miscell. 66: 1–238.

THORPE, R.J. (1964) Cereal-legume silage mixtures for the Northern Guinea Zone, Nigeria. Samaru Research Bulletin No. 34. The Government Printer. Kaduna.

TINDALL, H.D. (1983) *Vegetables in the tropics*. Macmillan Press. London.

TUFUOR, K. (1976) Prospects for increasing forest productivity in West African high forest-through forest genetics and tree improvement. Ghana For. J. 2: 15-19.

UNESCO, (1978) Tropical Forest Ecosystems. Natural Resources Report. 15. Paris.

—, (1981) The coastal ecosystems of West Africa Coastal Lagoons, Estuaries, and Mangroves. Workshop Report, Dakar: 11-15 June, 1979.

VALENTINE, P. (1980) Ein Besuch bei Kameruner Medizinleuten. Ethnomedizin 6(1-4): 9-18.

VAUGHAN, J.C. (1970) *The structure and utilization of oil seeds*. Chapman and Hall.

VERDCOURT, B. and TRUMP, E.C. (1969) *Common poisonous plants of East Africa*. Collins.

VICKERY, M.L. and VICKERY, B. (1979) *Plant products of tropical Africa*. MacMillan International College Editions.

VIETMEYER, N. (1984) *Leucaena: promising forage and tree crop for the tropics*. National Academy Press. Washington, D.C.

VINES, A.E. and REES, N. (1968) *Plant and animal biology* Vol. 1. Pitman.

VISSER, L.E. (1975) Medicinal plants of the Ivory Coast: ethnobotanical study of medicinal and culinary uses of wild plants by the Ando of the Ivory Coast (Western Africa). Meded Landbouwhogesch Wageningen 75(15): 1-79.

VOORHOEVE, A.G. (1965) *Liberian high forest trees*. Centre for Agricultural Publications and Documentation. Wageningen.

WALKER, A. and SILLANS, P. (1961) *Les plantes utiles du Gabon*. Paul Lechevalier. Paris.

WALTERS, P.R., MACFARLANE, N. and SPENLEY, P.C. (1979) *Jojoba: an assessment of prospects*. Tropical Products Institute. London.

WATERMAN, P.G. (1975) Decarine from the bark of *Zanthoxylum viride*. Phytochemistry 14: 843-844.

WATSON, J.W. and WAREHAM, A.K. (Editors) (1963) *West African secondary school atlas*. Thomas Nelson.

WHITE, F. (1978) The taxonomy, ecology and chorology of African Ebenaceae 1. The Guineo-Congolian species. Bull. Jard. Bot. Nat. Belg. 48: 245-358.

—, (1983) *The vegetation of Africa.* UNESCO.

WICKENS, G.E. (1983) The baobab — Africa's upside-down tree. Kew Bulletin 37: 173–209.

WICKHAM, C. (1981) *Common plants as natural remedies.* Frederick Muller. London.

WILDEMAN, E. de, (1935) A propos de médicaments indigènes congolais. Mem. Sect. Sci. Nat. ed Med. Inst. Roy. Colon. Belge Collect., n-8, v.3, fasc. 3: 1–127.

—, (1938) Sur des plantes médicinales ou utiles du Mayumbe (Congo Belge) d'après des notes du WELLENS, R.P. (1891-1924). Mem. Sect. Sci. Nat. ed Med. Inst. Roy. Colon. Belge Collect., n-8, v.6, fasc. 4: 1–97.

WIK, O. (1977) *Wood stoves.* Alaska Northwest Publishing Co.

WILLIAMS, C.N., CHEW, W.Y. and RAJARATNAM, J.H. (1980) *Trees and field crops of the wetter regions of the tropics.* Longmans.

WILLIS, J.C. (1973) *A Dictionary of the flowering plants and ferns.* Cambridge University Press.

WILLS, J.B. (Editor) (1962) *Agriculture and land use in Ghana.* Oxford University Press.

WOME, B. (1982) Febrifuge and antimalarial plants from Kisangani, Zaire. Bull. Soc. Roy. Bot. Belg. 115(2): 243–250.

YANNEY EWUSIE, J. (1968) Preliminary studies on the phenology of some woody species of Ghana. Ghana Journal of Science 8: 126–150.

—, SHARMA, B.B. and SIEBER, M. (1974) A note on the occurrence of *Nymphaea guineensis* Schum. & Thonn. in Ghana. Ghana Journal of Science 14(1): 69–70.

YOUNG, L.Y. (Editor-Documentalist) (1978) *Bambara Groundnut.* International Grain Legume Information Centre.

ZEVEN, A.C. (1967) *The semi-wild oil palm and its industry in Africa.* Centre for Agricultural Publications and Documentation. Wageningen.

ZOBERI, M.H. (1972) *Tropical macrofungi.* London.

—, (1973) Some edible mushrooms from Nigeria. The Nigerian Field, 38, No. 2: 81–90.

LOCAL NAMES (GHANAIAN)

Page references to the species may be found by consulting the Index of Scientific Names, p. 288.

Aayerebi *Ritchiea reflexa*
Aba-kamo *Parquetina nigrescens*
Abeble *Chrysobalanus icaco*
Abesebuo *Irvingia gabonensis*
Abia *Operculina macrocarpa*
Abin *Laguncularia racemosa*
Abisa *Vitex simplicifolia*
Aboagyedua *Tetrorchidium didymostemon*
Aboboe *Vigna subterranea*
Aboboma *Annona senegalensis* ssp. *onlotricha*
Abobonkakyere *Cathormion altissimum*
Abokodidua *Castanola paradoxa*
Abontore *Salacia stuhlmanniana*
Abotesima *Soyauxia grandifolia*
Abotoasebie *Pentadesma butyraceum*
Abotokuradua *Monodora brevipes*
Abrabesi *Maranthes polyandra*
Abruma *Coelocaryon oxycarpum*
Abuana *Ophiobotrys zenkeri*
Aburubiri-anwa *Oxyanthus subpunctatus*
Adadima *Annona glabra*
Adadze *Jatropha curcas*
Adedende *Asparagus africanus*
Adobe *Raphia hookeri*
Adoka *Raphia hookeri*
Adowa-alie *Dissotis rotundifolia*
Aduba *Argocoffeopsis rupestris*
Adubiri *Eremomastax speciosa*
Adwea *Dacryodes klaineana*
Adwokule *Sabicea africana*
Adwokuma *Strophanthus sarmentosus*
Afena *Strombosia glaucescens* var. *lucida*
Afena-akoa *Leptaulus daphnoides*
Afiafia *Carpolobia alba*
Aframsua *Pachystela brevipes*
Afua *Vitex doniana*
Agbana *Uvaria doeringii*
Agushie *Cucumeropsis edulis*
Aguwa *Euphorbia balsamifera*
Agyahere *Protomegabaria stapfiana*
Agyako *Cissus populnea*
Agyngyinsu *Schwenckia americana*
Ahaemete *Millettia irvinei*
Aheneba-nsatea *Diospyros barteri*
Ahensaw *Momordica angustifolia*
Ahoohenedua *Olax subscorpioidea*
Ahoro *Phytolacca dodecandra*
Ahunanyankwa *Turraea heterophylla*
Ahunanyankwa *Turraea heterophylla*
Ahyehyew-nsa *Urera mannii*
Akane *Elaeophorbia drupifera*
Akankoa *Strychnos spinosa*
Akasaa *Chrysophyllum delevoyi*
Akase *Lonochocarpus cyanescens*
Akasie *Dalbergia heudelotii*
Akatong *Luffa acutangula*
Akisibaka *Lepisanthes senegalensis*
Akitankrua-nini *Rutidea parviflora*
Akitase *Cnestis ferruginea*

Akladepka *Simirestis welwitschii*
Ako *Diospyros tricolor*
Akofie-kofi *Pycnocoma macrophylla*
Akokotua *Chionanthus mannii*
Akonansa *Croton lobatus*
Akontoma *Saba florida*
Akoo-ano *Uncaria talbotii*
Akoobowerew *Mezoneuron benthamianus*
Akotompo *Uvaria ovata*
Akotompotsen *Uvaria chamae*
Akotopa *Tetracera alnifolia*
Akpeka *Adenia cissampeloides*
Akpokpoe *Sorindeia warneckei*
Akukuaba *Pupalia lappacea*
Akuraso *Cissampelos mucronata*
Akwakora-gyahene *Rhaphiostylis beninensis*
Akwamfanu *Desmodium adscendens* var. *adscendens*
Akye *Blighia sapida*
Akyebiri *Blighia unijugata*
Akyekobiri *Blighia welwitschii*
Akyere *Hildegardia barteri*
Akyirinian *Diospyros monbuttensis*
Alari *Xylopia staudtii*
Alasaayo *Acridocarpus smeathmannii*
Aloma-bli *Ficus vallis-choudae*
Amamfohae *Motandra guineensis*
Amamfohama *Strophanthus hispidus*
Amandidua *Hymenodictyon floribundum*
Amangyedua *Ficus leprieuri*
Amponimpo *Solanum anguivi*
Amuaha *Piper umbellatum*
Amugui *Flacourtia flavescens*
Anaku *Trichoscypha arborea*
Ananse Don *Ouratea affinis*
Anansedodowa *Cola millenii*
Anansentoromahama *Sabicea calycina*
Ananta *Cynometra ananta*
Anidan *Tristemma incompletum*
Ankumabaka *Uvariastrum pierreanum*
Ankyewa *Majidea fosteri*
Ankye-wobiri *Placodiscus pseudostipularis*
Anmada *Enneastemon vogelii*
Anokye-hyedua *Guibourtia ehie*
Anso dua anso hama *Aganope leucobotrya*
Antro *Dichapetalum madagascariense*
Anuwatre *Lagenaria breviflora*
Anyankoma *Myrianthus arboreus*
Anyankom-nini *Myrianthus libericus*
Anyen-enyiwa *Abrus precatorius*
Anyofon-bokowa *Euclinia longiflora*
Apana *Capparis erythrocarpos*
Apapaye *Turraeanthus africanus*
Apazina *Macaranga heudelotii*
Apea *Mucuna pruriens* var. *pruriens*
Apenkwa *Entandrophragma cylindricum*
Aporo *Kalanchoe integra* var. *crenata*
Apose *Agelaea trifolia*
Aposin *Psychotria subobliqua*

277

Apotrewa *Maesobotrya barteri* var. *sparsiflora*
Aprim *Scytopetalum tieghemii*
Aprokuma *Antrocaryon micraster*
Asaa *Synsepalum dulcificum*
Asamoa nkwanta *Paspalum conjugatum*
Asaraba *Bridelia atroviridis*
Asenamba *Dialium guineense*
Asibenyanya *Syzygium rowlandii*
Asifuaka *Hoslundia opposita*
Asimba *Xylopia quintasii*
Asipiriwa *Justicia insularis*
Asirimu *Cissus aralioides*
Asoma *Parkia bicolor*
Asomanini *Swartzia fistuloides*
Asommerewa *Microglossa pyrifolia*
Asonsom *Urena lobata*
Asoporo *Avicennia africana*
Asrati *Sesbania sericea*
Asratoa *Oncoba spinosa*
Asumpa *Anonidium mannii*
Ataa *Pentaclethra macrophylla*
Ataaba *Spondias mombin*
Atabene *Chrysophyllum beguei*, *C. perpulchrum*
Atadwe *Cyperus esculentus*
Atantaba *Premna quadrifolia*
Atea *Anacardium occidentale*
Atena *Combretum ghaselense*, *Parinari curatellifolia*
Atobe-gyaso *Rinorea subintegrifolia*
Atoe *Anchomanes difformis*
Atoko *Sorghum bicolor*
Atroku *Euphorbia poissonii*
Atropo *Solanum macrocarpon*
Atrotre *Calpocalyx brevibracteatus*
Atwea *Diospyros viridicans*
Aunni *Premna hispida*
Avunle *Syzygium guineense* var. *littorale*
Awendade *Dialium dinklagei*
Awennade *Byrsocarpus coccineus*
Awiemfosemina *Albizia ferruginea*
Awiemfosemina-akoa *Albizia coriaria*
Awobe *Phyllanthus reticulatus* var. *glaber*
Awohanya *Manotes longiflora*
Awora *Marantochloa* species
Awora-bontodee *Anthocleista vogelii*
Awororo *Solanum aethiopicum*
Awuram-asie *Thaumatococcus daniellii*
Awuruku *Strephonema pseudocola*
Ayemtudua *Aptandra zenkeri*
Ayerew-amba *Monodora myristica*
Ayike *Laccosperma secundiflorum*
Bakapembe *Pleioceras barteri*
Baku *Tieghemella heckelii*
Bakunini *Afrosersalisia afzelii*
Balinyiri *Centaurea perrottetii*
Bambra *Ficus mucuso*
Bameha *Aerva tomentosa*
Banfa-banfa *Crassocephalum rubens*
Bangama *Myrianthus serratus*
Bankye *Manihot esculenta*
Barungwini *Borreria stachydea*
Batanga *Annona senegalensis* var. *deltoides*
Batatwene *Cercestis afzelii*
Ba-Udiga *Bridelia scleroneura*
Bebenga *Cucurligo pilosa*
Bedibesa *Drypetes floribunda*

Bediwonua *Canarium schweinfurthii*
Benaduru *Leptadenia hastata*
Benkasa *Opilia celtidifolia*
Benkyi *Diospyros piscatoria*
Berekankum *Manilkara obovata*
Bese *Cola nitida*
Bibie *Ochna schweinfurthiana*
Bisi *Hunteria elliotii*
Bitinamusa *Feretia apodanthera*
Bli *Vitex rivularis*
Blohunyi *Ficus polita*
Bobiri *Chytranthus carneus*
Boboe *Cynometra vogelii*
Bodwe *Ongokea gore*
Bodwue *Coula edulis*
Boeboe *Jacquemontia tamnifolia*
Bohwe *Garcinia smeathmannii*
Bomene *Ancylobotrys scandens*
Bompagya *Mammea africana*
Bonsa-dua *Alsodeiopsis staudtii*
Bonsamdua *Distemonanthus benthamianus*
Bonwen *Vernonia amygdalina*
Botsu *Carissa edulis*
Brebretim *Treculia africana* var. *africana*
Buburanya *Pentodon pentandrus*
Bunglale *Monechma ciliatum*
Bungu *Ceratotheca sesamoides*
Bupungu *Erythrophleum africanum*
Buruguni *Tephrosia platycarpa*
Buruntirikwa *Fadogia agrestis*
Chelum Punga *Crateva adansonii*
Chenchen-dibiga *Kohautia senegalense*
Chinchapula *Combretum collinum* ssp. *hypopilinum*
Dafa *Lasiodiscus mannii* var. *chevalieri*
Damaram *Mussaenda elegans*
Damaramma *Mussaenda erythrophylla*
Dam-Pan *Buchnera leptostachya*
Danhoma *Piptadeniastrum africanum*
Danhomanua *Aubrevillea kerstingii*
Danta *Nesogordonia papaverifera*
Dasuli *Gardenia erubescens*
Daudawa *Parkia clappertoniana*
Demmere *Calamus deeratus*
Denya *Cylocodiscus gabunensis*
Dietwa *Strophanthus preussii*
Dila *Boscia senegalensis*
Dinsinkoro *Euadenia eminens*
Disinkoro *Ipomoea mauritiana*
Dodwatu *Croton-zambesicus*
Domapowa *Combretum collinum* ssp. *binderianum*
Duabiri *Greenwayodendron oliveri*
Duabogo *Rhabdophyllum affine* ssp. *myrioneurum*
Duade *Craterispermum caudatum*
Duaga *Parkia biglobosa*
Duantu *Bersama abyssinica* ssp. *paullinioides*
Duapepere *Strychnos afzelii*
Duapompo *Omphalocarpum ahia*
Duasika *Enantia polycarpa*
Duatadwe *Chrysophyllum pruniforme*
Dubini *Khaya ivorensis*
Dubinibiri *Lovoa trichilioides*
Dweraba *Crescentia cujete*
Dwindwera *Lecaniodiscus cupanioides*

LOCAL NAMES

Dwombobre *Xylopia acutifolia*
Dwumma *Musanga cecropioides*
Dziiball *Artocarpus communis*
Dzogbedzro *Uapaca togoensis*
Edinam *Entandrophragma angolensis*
Efumomoe *Ageratum conyzoides*
Ehureke *Symphonia globulifera*
E-kariga *Psorospermum corymbiferum* var. *kerstingii*
Ekenyieya *Clerodendrum splendens*
Ekuama *Picralima nitida*
Ekuba *Triumfetta cordifolia*
Ekum-nkura *Dichapetalum toxicarium*
Eme *Ocimum basilicum*
Emiahile *Araliopsis tabouensis*
Emire *Terminalia ivorensis*
Enotipsi *Ampelocissus grantii*
Entedua *Copaifera salikounda*
Epebegai *Campylospermum flavum*
Esa *Celtis mildbraedii*
Esa-fufuo *Celtis wightii*
Esa-kokoo *Celtis zenkeri*
Esakosua *Celtis adolfi-friderici*
Esia *Petersianthus macrocarpus*
Esono-bise *Buchholzia coriacea*
Esononankoroma *Homalium letestui*
Esonowisamfie-bere *Desplatzia subericarpa*
Etwa prada *Hexalobus crispiflorus*
Ewio *Pennisetum americanum*
Eyitro *Hygrophila auriculata*
Fafaraha *Malacantha alnifolia*
Fale *Pachypodanthium staudtii*
Famu-wisa *Aframomum melegueta*
Fan *Talinum triangulare*
Fema *Microdesmis puberula*
Fetefele *Pellegriniodendron diphyllum*
Fetefre *Discoglypremna caloneura*
Flakwa *Vernonia conferta*
Foba *Voacanga thouarsii*
Folie *Dichapetalum pallidum*
Fomitsi *Dodonaea fiscosa*
Fonto *Ficus lutea*
Fotie *Hannoa klaineana*
Foto *Glyphaea brevis*
Frefi *Thespesia populnea*
Funtum *Funtumia elastica*
Galinziela *Ficus glumosa* var. *glaberrima*
Gamperoga *Ficus dekdekena*
Gblieku *Dracaena bicolor*
Gblitso *Diospyros abyssinica*
Gbomete *Psilanthus mannii*
Gburega *Combretum molle*
Gongu *Balanites aegyptiaca*
Gorli *Caloncoba echinata*
Corpila *Acacia polycantha* ssp. *campylacantha*
Gosei *Acacia dudgeoni*
Gowuraga *Acacia gourmaensis*
Gozanga *Acacia albida*
Guare-ansra *Brillantaisia nitens*
Gwodei *Hippocratea apocynoides* ssp. *guineensis*
Gyakuruwa *Vernonia nigritiana*
Gyambobre *Xylopia parviflora*
Gyamkawa *Mundulea sericea*
Gyamma *Alchornea cordifolia*
Gyapam *Macaranga barteri*
Gyatofo-Akongua *Omphalocarpum procerum*

Gyedua *Ficus umbellata*
Hama-kyereben *Salacia debilis*
Heowe *Uraria picta*
Hohoroho *Anthocleista nobilis*
Hokple *Allophyllus africanus*
Homa-ben *Canthium hispidum*
Homabiri *Gouania longipetala*
Homafuntum *Alafia* species
Hote *Pterocarpus santaloides*
Hu *Vernonia biafrae*
Humatarakwa *Santaloides afzelii*
Hunhon *Laportea aestuans*
Hunhun *Manniophyton fulvum*
Hwenetia *Xylopia aethiopica*
Hyedua *Daniellia ogea*
Jinkiliza *Bauhinia rufescens*
Kabona *Boswellia dalzielii*
Kaboya *Entada africana*
Kabuga *Digitaria exilis*
Kadze *Tiliacora dielsiana*
Kagya *Griffonia simplicifolia*
Kakadikro *Trichilia prieuriana*
Kakaduro *Zingiber officinale*
Kakali *Pleiocarpa bicarpellata*
Kakapenpen *Rauvolfia vomitoria*
Kakapimbe *Pleiocarpa pycnantha* var. *tubicina*
Kaku *Lophira alata*
Kaman Godui *Ostryderris stuhlmannii*
Kamfo-barima *Euphorbia lateriflora*
Kampoye *Strychnos innocua* ssp. *innocua* var. *pubescens*
Kangkalaga *Isoberlinia tomentosa*
Kankanga *Ficus sycomorus*
Kankroma *Morinda lucida*
Kanto *Zanthoxylum xanthoxyloides*
Kanwen *Pleiocarpa mutica*
Kanwen-akoa *Hunteria eburnea*
Katawani *Pseudospondias microcarpa* var. *microcarpa*
Katopa *Tiliocora funifera*
Katrika *Drypetes gilgiana*
Keklengbe *Longocarpus sericeus*
Kesene *Dracaena mannii*
Kisiga *Trichilia emetica* ssp. *suberosa*
Kobewu *Lannea kerstingii*
Kodia *Chaetacme aristida*
Kodomba *Ampelocissus multistriata*
Koheni-koko *Desmodium velutinum*
Koka *Conocarpus erectus*
Koklotade *Philoxerus vermicularis*
Kokoanisua *Spathodea campanulata*
Kokofobene *Schrebera arborea*
Kokoo *Triplotaxis stellulifera*
Kokora *Smilax kraussiana*
Kokoro *Aeglopsis chevalieri*
Kokorobe *Licania elaeosperma*
Kokote *Anopyxis klaineana*
Kokroboba *Panda oleosa*
Kokrodua *Pericopsis elata*
Kokrosabia *Cochlospermum tinctorium*
Kokyi *Cyperus articulatus*
Konkon sibere *Ataenidia conferta*
Kookoo *Colocasia esculenta*
Korantema *Oxyanthus speciosus*
Kosowanini *Vismia guineensis*
Kotamenyati *Allophylus spicatus*

Koto-haban *Sarcophrynium brachystachys*
Kotohume *Secamone afzelii*
Kotokoli *Cissus quadrangularis*
Kotoprepre *Bussea occidentalis*
Kotowhiri *Caloncoba gilgiana*
Kotwe-ake *Dracaena scoparia*
Kpaliga *Securidaca longepedunculata*
Kpazina *Macaranga hurifolia*
Kpeteple *Talbotiella gentii*
Kpitikpiti *Capparis brassii*
Kpleng *Salacia staudtiana* var. *leonensis*
Kpoploti *Antidesma laciniatum* var. *laciniatum*
Krakoo *Sphenocentrum jollyanum*
Kraku *Eugenia coronata*
Krampa *Jasminum dichotomum*
Krobonko *Telfairia occidentalis*
Kroma *Klainedoxa gabonensis* var. *oblongifolia*
Kronko *Pavetta corymbosa*
Kruba *Khaya grandifoliola*
Krumben *Khaya anthotheca*
Kukyafie *Mitragyna inermis*
Kulgo *Acacia sieberiana* var. *villosa*
Kultia *Syzygium guineense* var. *macrocarpum*
Kumakuafo *Maytenus senegalensis*
Kumamuno *Clerodentrum volubile*
Kumanini *Lannea welwitschii*
Kunkiwiya *Ficus ingens* var. *ingens*
Kunmuni *Quassia undulata*
Kuntammiri *Uapaca corbisieri*
Kuntan *Uapaca guineensis*
Kuntan-akoa *Uapaca heudelotii*
Kuntunkuri *Lannea acida*
Kurobow *Balanites wilsoniana*
Kurubeta *Pseudocedrela kotschyi*
Kusia *Nauclea diderrichii*
Kusibiri *Diospyros gabundensis*
Kwaasiwa *Ochna staudtii*
Kwabohoro *Guarea cedrata*
Kwaduko *Rhinacanthus virens* var. *virens*
Kwadwuma *Guarea thompsonii*
Kwaebrofre *Cussonia kirkii* var. *kirkii*
Kwaeheni *Pseudarthria hookeri* var. *argrophylla*
Kwaemeko *Usteria guineensis*
Kwaemu-Sabrakyee *Triaspis odorata*
Kwae-susua *Psydrax parviflora*
Kwaetawa *Oxyanthus unilicularis*
Kwaginyanga *Combretum fragrans*
Kwakenya *Tricalysia reticulata*
Kwaku asra *Dictyandra arbrescens*
Kwanga *Daniellia thurifera*
Kwarili *Terminalia mollis*
Kwatafo-mpaboa *Berlinia occidentalis*
Kwatiri *Terminalia macroptera*
Kwedi *Mimosa pigra*
Kwesidua *Psychotria peduncularis*
Kwinabuka *Maranthes robusta*
Kwonkwia *Ficus iteophylla*
Kyekyereantena *Spathandra blakeoides*
Kyeneboa *Cordia senegalensis*
Kyenkyen *Antiaris toxicaria*
Kyerebeteni *Sherbournea bignoniiflora*
Kyereye *Pterygota macrocarpa*
Kyirikasa *Anthosthema aurbyanum*
Lakpa *Scottellia klaineana* var. *klaineana*
Landaga *Combretum micranthum*
Langrinduo *Cassia mimosoides*

Laruklukor *Ziziphus abyssinica*
Mampihege *Annona glauca* var. *glauca*
Mankani *Xanthosoma mafaffa*
Mantannua *Bertiera racemosa*
Minchingoro *Garcinia kola*
Mmeyaa *Pneumatopteris afra*
Mobia *Dracaena surculosa* var. *surculosa*
Momonimo *Alafia scandens*
Momorehemo *Alafia barteri*
Motokuradua *Monodora tenuifolia*
Mpatowansoe *Alternanthera pungens*
Mpawu *Cleidion gabonicum*, *Rinorea oblongifolia*
Mpepea *Antidesmia venosum*
Mpintinko *Pisonia aculeata*
Mpupua *Cyathula prostrata*
Mu *Barleria opaca*
Mumue *Salacia whytei*
Mutuo *Stachyanthus occidentalis*
Nakwa *Holoptelea grandis*
Nalenga *Lonchocarpus laxiflorus*
Nanogba *Sclerocarya birrea*
Nanzili *Amblygonocarpus andongensis*
Narga *Commiphora africana*
Nasia *Ozoroa reticulaa*
Nfona *Erythrina mildbraedii*
Ngalenge *Crinum zeylanicum*
Ngo ne nkyene *Cleistopholis patens*
Ngobo *Scadoxus multiflorus*
Nhwesono *Coffea ebracteolata*
Niabiri *Clerodendrum umbellatum*
Nkabe *Calliandra portoricensis*
Nkako *Machaerium lunatum*
Nkanaa *Securinega virosa*
Nkantonsoe *Opuntia* species
Nkatie *Paropsia adenostegia*
Nkunga *Combretum glutinosum*
Nkyewodue *Celosia argentea*
Nnanfuro *Bequaertiodendron oblanceolatum*
Noto *Loeseneriella africana*
Nsempii *Cassia alata*
Nserewedua *Rutidea glabra*
Nsokodua *Garcinia afzelii, G. epunctata*
Nsoto *Scaphopetalum amoenum*
Nsuduru *Cassia podocarpa*
Nsurogya *Gongronema latifolium*
Nsusoa *Solanum anomalum*
Nsusuabiri *Solanum americanum (nigrum)*
Ntanta *Grewia carpinifolia*
Nte *Dioclea reflexa*
Ntehama *Dioclea reflexa*
Ntentrema *Marantochloa cuspidata*
Ntetekon *Monanthotaxis foliosa*
Nton *Pandanus abbiwii*
Ntonme *Dracaena arborea*
Ntonto *Momordica cissoides*
Ntoroba-banyin *Solanum melongena*
Ntoropo *Solanum macrocarpon, S. melongena*
Ntroba *S. aethiopicum*
Ntum *Eclipta alba*
Ntumenum *Justicia flava*
Ntwea *Loeseneriella rowlandii*
Ntwibo *Colocasia esculenta*
Nufu-nufu *Bequaertiodendron magalismontanum*
Nufuten *Kigelia africana*
Nunum *Ocimum gratissimum*
Nuodolega *Dalbergia saxatilis*

LOCAL NAMES

Nwadua *Ficus capensis*
Nwaha *Abutilon mauritianum*
Nwohwea *Clappertonia ficifolia, Hibiscus tiliaceus*
Nya *Neocarya macrophylla*
Nyagahe *Indigofera simplicifolia*
Nyamele-buruma *Vitex micrantha*
Nyamrem *Croton penduliflorus*
Nyankom *Heritiera utilis*
Nyankyeren *Ficus exasperata*
Nyeddo *Leonotis nepetifolia* var. *africana*
Nyenyanke *Pavetta crassipes*
Nzahuara *Palisota hirsuta*
Oba *Xylopiastrum villosum*
Obohwe *Rothmannia urcelliformis*
Obowe *Landolphia owariensis*
Obua *Napoleonaea vogelii*
Obuobi *Aeglopsis paniculata, Afraegle paniculata*
Odee *Okoubaka aubrevillei*
Odom *Erythrophleum ivorense*
Odonno-bea *Abutilon guineense*
Odubrafo *Mareya micrantha*
Odum *Milicia excelsa*
Odum-nua *Milicia regia*
Odwen *Baphia nitida*
Odwenkobiri *Baphia pubescens*
Ofam *Parinari excelsa*
Ofram *Terminalia superba*
Ofuruma *Voacanga africana*
Oguaben *Acacia kamerunensis*
Ogyafam *Indigofera macrophylla*
Ohaa *Sterculia oblonga*
Ohwirem *Combretum platypterum*
Ohwirem-nini *Combretum grandiflorum*
Okatakyi *Leea guineensis*
Okosibiri *Diospyros mespiliformis*
Okoli Awotso *Ochna afzelii*
Okoro *Albizia zygia*
Okosoa *Harungana madagascariensis*
Okusua *Ehretia cymosa*
Okpoi *Flabellaria paniculata*
Okum-adada *Alafia multiflora*
Okuo *Bombax buonopozense, Zanthoxylum gilletii*
Okure *Trilepisium madagascariensis*
Okyini *Ehretia trachyphylla*
Omaatwa *Strophanthus gratus*
Ombatoatshi *Phyllanthus fraternus* ssp. *togoensis*
Omeha *Combretum paniculatum*
Omenewa *Diospyros kamerunensis*
Omenewabere *Diospyros heudelotii*
Ongo *Terminalia glaucescens*
Onibona *Eriocoelum pungens*
Onibonakokoo *Eriocoelum racemosum*
Onwam-dua *maesopsis eminii*
Onyina *Ceiba pentandra*
Onyina-koben *Rhodognaphalon brevicuspe*
Opahanini *Drypetes afzelii*
Opam *Bridelia stenocarpa*
Opam-fufuo *Bridelia feruginea*
Opamkokoo *Macaranga heterophylla*
Opanto *Ficus vogeliana*
Opapohwea *Plumbago zeylanica*
Ope-igbo(?) *Dracaena mannii*
Opetentong *Dracaena perrottetii*
Opitipata *Piliostigma thonningii*
Oprono *Mansonia altissima*

281

Opunini *Ouratea calophylla*
Osurosuso *Urera obovata*
Otie *Pycnanthus angolensis*
Otu *Cremaspora triflora*
Otwabere *Diospyros canaliculatus*
Otwa-tekyirema *Culcasia scandens*
Otwe-ani *Memecylon afzelii*
Otwentorowa *Vitex fosteri*
Otweto *Diospyros soubreana*
Otwewa *Carpolobia lutea*
Oware-amba *Caesalpinia bonduc*
Owebiribi *Teclea verdoorniana*
Oyaa *Zanthoxylum leprieurii*
Oyaabere *Zanthoxylum chevalieri*
Oyaanini *Zanthoxylum viride*
Pakyisie *Crossopteryx febrifuga*
Pampena *Albizia adianthifolia*
Pampiga *Sapium grahamii*
Pamprama *Coryananthe pachyceras*
Papao *Afzelia africana*
Papaonua *Afzelia bella* var. *glacior*
Peaba *Hyptis pectinata*
Peagoro *Crotalaria pallida*
Peteku-nsoe *Acanthospermum hispidum*
Pempen *Landolphia heudelotii*
Penku *Harrisonia abyssinica*
Pepae *Tabernaemontana crassa*
Pepea *Margaritaria discoidea*
Pepesia *Erythroxylum mannii*
Peteha *Adenia rumicifolia* var. *miegei*
Petekuku *Triumfetta rhomboidea*
Peteprebi *Gardenia ternifolia*
Petni *Terminalia avicennioides*
Peytuba *Combretum sericeum*
Pienwogo *Steganotaenia araliacea*
Pikeabo *Duparquetia orchidacea*
Pinimo *Burkea africana*
Pitapo *Gaertnera paniculata*
Pobe *Massularia acuminata*
Pofiri *Microglossa afzelii*
Potoke *Deinbollia grandifolia*
Potopoleboblo *Phyllanthus muellerianus*
Potrodom *Erythrophleum suaveolens*
Prahoma *Dicranolepis persei*
Prekese *Tetrapleura tetraptera*
Pugodigo *Maerua angolensis*
Pumpune *Landolphia hirsuta*
Pumpungo *Sterculia setigera*
Punini *Maranthes glabra*
Punum *Casearia barteri*
Saa-borofere *Cussonia arborea*
Saa-nunum *Lippia multiflora*
Sabobe *Rothmannia whitfieldii*
Sabrakyie *Hymenocardia acida*
Sakanee *Anogeissus leiocarpus*
Samankube *Rothmannia longiflora*
Saman-ntoroba *Solanum torvum*
Samanta *Xylia evansii*
Samante *Mucuna sloanei*
Samanyobli *Clausena anisata*
Samfena *Aningeria altissima*
Samparanga *Celtis integrifolia*
Sanbrim *Aulacocalyx jasminiflora*
Sanga *Prosopis africana*
Sankasaa *Entada abyssinica*
Sansammuro *Claoxylon hexandrum*

Sansangwa *Capparis polymorpha*
Sante *Millettia thonningii*
Santom *Ipomoea batatas*
Sanvoa *Corchorus tridens*
Sanya *Daniellia oliveri*
Sanzamulike *Diospyros sanza-minika*
Sapelaga *Isoberlinia doka*
Sasanemasa *Newbouldia laevis*
Satadua *Mallotus oppositifolius*
Sawai *Waltheria indica*
Se ngo se bari *Faurea speciosa*
Selena *Millettia zechiana*
Senyana *Lonchocarpus griffonianus*
Sere-berebe *Aloe buettneri*
Sese *Holarrhena floribunda*
Sese-dua *Christiana africana*
Sesea *Trema orientalis*
Sibere *Marantochloa leucantha*
Sikakyia *Heisteria parviflora*
Silikawakawle *Neostenanthera hamata*
Sinduro *Alstonia boonei*
Sinsa *Lannea velutina*
Sinsabgbetiliga *Lannea nigritana* var. *nigritana*
Sintanatora *Cissus cornifolia*
Siresireke *Platostoma africanum*
Sisimasa *Markhamia lutea*
Sisiworodo *Solenostemon monostachys*
Sofo *Sterculia tragacantha*
Soitbia *Pierreodendron kerstingii*
Sonkyi *Allanblackia parviflora*
Sono-nantin *Saba senegalensis*
Sonontokwakofo *Stereospermum kunthianum*
Sopropo *Momordica foetida*
Sorowa *Erythrina addisoniae*
Soro-wisa *Piper guineenes*
Sra *Euphorbia deightonii*
Subaha *Mitragyna ciliata*
Subaha-akoa *Mitragyna stipulosa*
Sugugwa *Marantochloa purpurea*
Sukisia *Nauclea latifolia*
Sukuruwa *Byttneria catalpifolia* ssp. *africana*
Sukusia *Nauclea pobeguinii*
Sunya *Syzygium guineense* var. *guineense*
Supuwa *Vitex grandifolia*
Susanfo *Tiliacora dinklagei*
Taasendua *Clerodendrum buchholzii*
Tadatso *Combretum zenkeri*
Taga *Haumaniastrum lilacinum*
Takorowa *Hymensostegia afzelii*

Takyikyiriwa *Detarium senegalense*
Tanamfre *Cola chlamydantha*
Tanduro *Trichilia monadelpha*
Tanuronini *Trichilia tessmannii*
Tatwea *Mucuna flagellipes*
Tetekon *Gilbertiodendron limba*
Tetekono *Berlinia grandiflora*
Tetia-dupon *Psydrax subcordata*
Tiabutuo *Sacoglottis gabonensis*
Timatibre *Omphalocarpum elatum*
Tingkwanga *Sesbania sesban*
Toa-ntini *Paullinia pinnata*
Tokwa-kufuo *Stereospermum acuminatissimum*
Tomboro *Markhamia tomentosa*
Tomi *Sapium ellipticum*
Totoro *Anthonotha macrophylla*
Totoronini *Anthonotha fragrans*
Totototo *Physalis angulata*
Trindobaga *Eriosema griseum*
Tromen *Clerodendrum capitatum*
Tukobo *Rothmannia hispida*
Tulingi *Parinari congensis*
Tungbo *Cordia myxa*
Turunnua *Smeathmannia pubescens*
Tweanka *Spondianthus preussii* var. *preussii*
Tweapea *Garcinia kola*
Tweapeakoa *Garcinia gnetoides*
Tweneboa *Cordia platythyrsa*
Tweneboa-akoa *Cordia vignei*
Twihama *Tetracera potatoria*
Wamma *Ricinodendron heudelotii*
Wankya *Dicranolepis grandiflora*
Watapuo *Cola gigantea* var. *glabrescens*
Watapuobere *cola lateritia* var. *maclaudi*
Wawa *Triplochiton scleroxylon*
Wawabimma *Sterculia rhinopetala*
We-ana *Cola heterophylla*
Wisamfia *Desplatzia dewevrei*
Wisuboni *Octoknema borealis*
Wonane *Ruspolia hypocrateriformis*
Wonton *Morus mesozygia*
Wota *Combretum racemosum, Dalbergia hostilis*
Woteegbogbo *Deinbollia pinnata*
Wumlim *Striga hermonthica*
Yaya *Amphimas pterocarpoides*
Yisa *Celtis africana*
Yualega *Grewia mollis*
Zetitsui *Macrosphyra longistyla*
Zinzam *Sesamum radiatum*

COMMON NAMES (ENGLISH)

Page references to the species may be found by consulting the Index of Scientific Names, p. 288

Abura *Mitragyna ciliata*
Abyssinian Coffee *Coffea arabica*
Afara *Terminalia superba*
African Bark *Crossopteryx febrifuga*
    Bdellium *Commiphora africana*
    Black Wood *Dalbergia* species, *Erythrophleum africanum*
    Bowstring Hemp *Sansevieria liberica*
    Bread Fruit *Treculia africana* var. *africana*
    Catechu *Acacia polyacantha* ssp. *campylacantha*
    Cherry Orange *Citropsis articulata*
    Copaiba Balsam Tree *Daniellia oliveri*
    Cucumber *Momordica charantia*
    Greenheart *Cylicodiscus gabunensis*
    Kino *Pterocarpus erinaceus*
    Laburnum *Cassia sieberiana*
    Mahogany *Khaya ivorensis*
    Mammy Apple *Mammea africana*
    Myrrh *Commiphora dalzielii*
    Nutmeg *Pycnanthus angolensis*
    Peach *Nauclea latifolia*
    Pearwood *Manilkara multinervis* ssp. *lacera*
    Satinwood *Distemonanthus benthamianus*
    Tragacanth *Sterculia tragacantha*
    Tulip Tree *Spathodea campanulata*
    Walnut *Coula edulis, Lovoa trichilioides*
    Yam Beans *Sphenostylis stenocarpa*
    Yellow Wood *Enantia polycarpa*
Akee Apple *Blighia sapida*
Alexandra Laurel *Calophyllum inophyllum*
American Basil *Ocimum americanum*
    Water Weed *Elodea canadensis*
    Wormseed *Chenopodium ambrosioides*
Anatto *Bixa orellana*
Antelope Grass *Echinochloa pyramidalis*
Antigua Heath *Russelia equisetiformis*
Apple of Peru *Datura stramonium*
Arabian Coffee *Coffea arabica*
Arrow Poison *Strophanthus hispidus*
Ashanti Blood *Mussaenda erythrophylla*
    Plum *Spondias mombin*
Asiatic Rice *Oryza sativa*
Assyrian Plum *Cordia myxa*
Aubergine *Solanum melongena*
Australian Asthma Herb *Euphorbia hirta*
Avocado Pear *Persea americana*
Avodire *Turraeanthus africanus*
Bahama Grass *Cynodon dactylon*
Balloon Vine *Cardiospermum grandiflorum*, *C. halicacabum*
Balsam Spurge *Euphorbia balsamifera*
Bambara Groundnut *Vigna subterranea*
Bamboo *Bambusa vulgaris, Oxytenanthera abyssinica*
Banana *Musa paradisiaca* var. *sapientum*
Baobab *Adansonia digitata*
Bark Cloth Tree *Antiaris toxicaria*
Barbados Cherry *Malpighia glabra*

Bastard Millet *Paspalum scrobiculatum*
    Vervain *Stachytarpheta cayennensis*
Beach Convolvulus *Ipomoea pes-caprae* ssp. *brasiliensis*
Bead Tree *Adenanthera pavonina, Melia azedarach*
Belembe *Xanthosoma brasiliense*
Beniseed *Sesamum indicum*
Bermuda Grass *Cynodon dactylon*
Bilimbing *Averrhoa bilimbi*
Billy-goat Weed *Ageratum conyzoides*
Bitter Cola *Cola nitida, Garcinia kola*
    Leaf *Vernonia amygdalina*
Black Grain *Cassia absus*
    gram *Phaseolus mungo*
    Guarea *Guarea thompsonii*
    Plum *Vitex doniana*
    -eyed Susan *Thunbergia alata*
Bleeding Heart *Clerodendrum thomsonae*
Blood Flower *Asclepias curassavica, Scadoxus cinnabarinus* ssp. *katerinae, S. multiflorus*
    Plum *Haematostaphis barteri*
Blue Jacaranda *Jacaranda mimosifolia*
    Pea *Clitoria ternatea*
    Plumbago *Plumbago capensis*
Bonduc *Caesalpinia bonduc*
Bottle Gourd *Lagenaria siceraria*
Brazil Cress *Spilanthes filicaulis*
    Nut *Bertholetia excelsa*
Brazilian Coffee *Coffea arabica*
    Tea *Stachytarpheta cayennensis*
Breadfruit *Artocarpus communis*
Broad-leaved Mahogany *Khaya grandifoliola*
Broccoli *Brassica oleracea* var. *italica*
Bubinga *Copaifera salikounda*
Buffalo Grass *Stenotaphrum secundatum*
    Thorn *Ziziphus mucronata*
Buffel Grass *Setaria chevalieri*
Bulrush *Typha domingensis*
    Millet *Pennisetum americanum*
Bur Marigold *Bidens pilosa*
Burweed *Triumfetta cordifolia, T. rhomboidea*
Bush Apple *Heinsia crinita*
    Tea-bush *Hyptis suaveolens*
Butter Bean *Phaseolus lunatus*
Button Wood *Conocarpus erectus*
Cabbage Palm *Anthocleista nobilis*
Calabash *Lagenaria siceraria*
    Tree *Crescentia cujete*
Calabar Bean *Physostigma venenosum*
Camara *Lantana camara*
Cameroons Mountain Rubber Vine *Landolphia landolphioides*
Camwood *Baphia nitida*
Candle Wood *Zanthoxylum xanthoxyloides*
Candollei *Entandrophragma candollei*
Carambola *Averrhoa carambola*
Carrot *Daucus carota*
Cashaw Bean *Prosopis chilensis*

Cashew Nut *Anacardium occidentale*
Castor Oil Plant *Ricinus communis*
Cassava *Manihot esculenta*
Cat-tail *Typha domingensis*
Cat's Tail Grass *Setaria pallide-fusca*
    Whiskers *Cleome gynandra, C. rutidosperma*
Cauliflower *Brassica oleracea* var. *botrytis*
Ceara Rubber *Manihot glaziovii*
Ceylon Leadwort *Plumbago zeylanica*
    Olive *Elaeocarpus serratus*
    Spinach *Basella alba*
Chaya *Cnidoscolus aconitifolius*
Chayote *Sechium edule*
Cherimola *Annona cherimola*
Children's Tomato *Solanum indicum* ssp. *distichum*
Christmas Bush *Alchornea cordifolia*
Cinnamon *Cinnamomum zeylanicum*
Citron *Citrus medica*
Citronella *Cymbopogon nardus*
Climbing Hemp Weed *Mikania chevalieri*
    Lily *Gloriosa simplex, G. superba*
Clove *Eugenia caryophyllata*
Cocaine Plant *Erythroxylum coca*
Cockspur *Pilsonia aculeata*
Coco Plum *Chrysobalanus icaco* ssp. *atacorensis*
Cocoa *Theobroma cacao*
Coconut Palm *Cocos nucifera*
Cocoyam *Xanthosoma mafaffa*
Coffee Senna *Cassia occidentalis*
Cola Nut *Cola nitida*
Commercial Cola Nut Tree *Cola acuminata*
Common Cress *Lepidium sativum*
    Reed *Phragmites karka*
Congo Jute *Urena lobata*
Copper Pod *Peltophorum pterocarpum*
Coral Flowers *Erythrina senegalensis*
    Plant *Jatropha multifida*
Corn *Zea mays*
    -flag *Gladiolus daleni, G. gregarius, G. klattianus, G. unguiculata*
Cotton *Gossypium arboreum*
Cow Itch *Mucuna pruriens* var. *pruriens*
Cowpea *Vigna unguiculata*
Crab Oil Tree *Carapa procera*
Crab's Eyes *Abrus precatorius*
Crabwood *Carapa procera*
Crepe Myrtle *Lagerstroemia indica*
Croton Oil Plant *Croton tiglium*
Cucumber *Cucumis edulis*
Custard Apple *Annona reticulata*
Date Palm *Phoenix dactylifera*
Death Cap *Amanita phalloides*
Desert Date *Balanites aegyptiaca*
Devil Bean *Cortalaria retusa*
Devil's Thorn *Tribulus terrestris*
Ditch Millet *Paspalum scrobiculatum*
Dog Almond *Andira inermis*
Drum Tree *Cola millenii*
Dry-zone Cedar *Pseudocedrella kotschyi*
    Mahogany *Khaya senegalensis*
Duckweed *Lemna perpusilla, Spirodela polyrhiza, Wolffia arrhiza*
Dum Palm *Hyphaene thebaica*
Dutchman's Pipe *Aristolochia elegans*
Earth Almond *Cyperus esculentus*

East African Copal *Trachylobium verrucosum*
    Indian Walnut *Albizia lebbeck*
Ebony *Diospyros* species
Eddoes *Colocasia esculenta*
Edible-stemmed Vine *Cissus quadrangularis*
Egg Plant *Garcinia xanthochymus, Solanum aethiopicum*
Egyptian Mimosa *Acacia nilotica* var. *adansonii*
    Sesban *Sesbania sesban*
Elephant Grass *Pennisetum purpureum*
Elo *Xylopia quintasii*
Elo Pubescent *Xylopiastrum villosum*
English Plum *Spondias cytherea*
Ethiopian Pepper *Xylopia aethiopica*
Exile Oil Plant *Thevetia peruviana*
False Chewstick Tree *Garcinia smeathmannii*
    Rubber Tree *Funtumia africana, Holarrhena floribunda*
    Thistle *Acanthus montanus*
    Yam *Icacina olivaeformis*
Fan Palm *Borassus aethiopum*
Fever Plant *Ocimum gratissimum*
Fibre Plant *Hibiscus cannabinus*
Fig-Leaved Cola *Cola caricifolia*
Fire-ball Lily *Scadoxus cinnabarinus* ssp. *katerinae, S. multiflorus*
Fish Poison Plant *Tephrosia vogelii*
Flamboyante *Delonix regia*
Flame Tree *Delonix regia*
Flint Bark *Diospyros canaliculata*
Flint Bark Tree *Diospyros gabunensis*
Flower Fence *Caesalpinia pulcherrima*
Fluted Pumpkin *Telfairia occidentalis*
Foetid Cassia *Cassia obtusidolia*
Four-Leaved Henna *Cassia absus*
Frafra Potato *Solenostemon rotundifolius*
Frank Indigo *Indigofera tinctoria*
Frankincense Tree *Boswellia dalzielii, Daniellia thurifera*
Gaboon Ebony *Diospyros viridicans*
    Nut *Coula edulis*
Gambian Tea Bush *Lippia multiflora*
Garden Crotons *Codiaeum variegatum*
    Hibiscus *Hibiscus rosa-sinensis*
Garlic *Allium sativum*
Gedu Nohor *Entandrophragma angolense*
Geiger Tree *Cordia sebestena*
Ghost Palm *Encephalartos barteri*
Giant Granadilla *Passiflora quadrangularis*
Ginger *Zingiber officinale*
    Lily *Costus afer*
Goat's Foot Convolvulus *Ipomoea pes-caprae* ssp. *brasiliensis*
Golden Dewdrop *Duranta repens*
    Shower *Cassia fistula*
    Timothy Grass *Setaria sphacelata*
Governor's Plum *Flacourtia indica*
Grains of Paradise *Aframomum melegueta*
Grapefruit *Citrus paradisi*
Greater Yam *Dioscorea alata*
Green Amaranth *Amaranthus viridis*
    and Yellow Striped Bamboo *Bambusa vulgaris*
    gram *Phaseolus aureus*
Groundnut *Arachis hypogaea*
Guava *Psidium guajava*

COMMON NAMES

Guinea Corn *Sorghum bicolor, S.* species
   Grains *Aframomum melegueta*
   Grass *Panicum maximum*
   Potato *Dioscoreophyllum cumminsii*
   Plum *Parinari excelsa*
   Yam *Dioscorea cayenensis* ssp. *cayenensis*
Gum Arabic *Acacia senegal*
   Copal Tree *Daniellia ogea*
Gutta-Percha Tree *Ficus platyphylla*
Haemorrhage Plant *Aspilia africana*
Hairy Indigo *Indigofera hirsuta*
   Thorn Apple *Datura metel*
Hausa Potato *Solenostemon rotundifolius*
Heart Seed *Cardiospermum grandiflorum, C. halicacabum*
Henna *Lawsonia inermis*
Hog Plum *Spondias mombin*
   weed *Boerhavia diffusa, B. repens*
Honey Berry *Melicoccus bijugatus*
Honda Berry *Dillenia indica*
Hope Beetle *Xylotripes gideon*
Horse-eye Bean *Mucuna sloanei*
Horse Purslane *Trianthema penandra*
Horse-radish Tree *Moringa oleifera*
Horse Tamarind *Leucaena leucocephala*
Hosanna Palm *Encephalartos barteri*
Hungry Rice *Digitaria debilis*
Incense Tree *Canarium schweinfurthii*
Indian Almond *Terminalia catappa*
   Heliotrope *Heliotropium indicum*
   Jujuba *Ziziphus jujuba*
   Jujube *Z. mauritianus*
   Laburnum *Cassia fistula*
   Shot *Canna indica*
   Spinach *Basella alba*
   Tamarind *Tamarindus indica*
   Water Navelwort *Centella asiatica*
   Wormseed *Chenopodium ambrosioides*
Irish Potato *Solanum tuberosum*
Iroko *Milicia excelsa*
Jack Fruit *Artocarpus heterophyllus*
Jambolan *Syzygium cumini*
Japanese Camphor Tree *Cinnamomum camphora*
Java Almond *Canarium commune*
   Indigo *Indigofera arrecta*
   Plum *Syzygium cumini*
Jequé Rubber Tree *Manihot dichotoma*
Jerusalem Thorn *Parkinsonia aculeata*
Jew's Marrow *Corchorus olitorius*
Job's Tears *Coix lacryma-jobi*
Jojoba *Simmondsia chinensis*
Jungle Rice *Echinochloa colona*
Kaffir Orange *Strychnos spinosa*
Kamerun Grass *Sorghum arundinaceum*
Katemfe *Thaumatococcus daniellii*
Kenaf *Hibiscus cannabinus*
Khaki Weed *Alternanthera pungens*
Kola Nut *Cola nitida*
Lablab Bean *Lablab niger*
Lalang *Imperata cylindrica* var. *africana*
Landa *Erythroxylum mannii*
Laurel-leaved Cola *Cola laurifolia*
Lebbeck *Albizia lebbeck*
Lemon *Citrus limon*
   Grass *Cymbopogon citratus*

-scented Gum *Eucalyptus citriodora*
Lesser Yam *Dioscorea esculenta*
Lettuce *Launaea sativa*
Leucaena *Leucaena leucocephala*
Liberian Coffee *Coffea liberica* var. *liberica*
Lignum Vitae *Guaiacum officinale*
Lima Bean *Phaseolus lunatus*
Lime *Citrus aurantiifolia*
Logwood *Haematoxylon campechianum*
Long-fruited Jute *Corchorus olitorius*
Loofah Gourd *Luffa cylindrica*
Love Grass *Chrysopogon aciculatus*
Madagascar Periwinkle *Catharanthus roseus*
Madras Thorn *Pithecellobium dulce*
Mahogany *Khaya* species
Maize *Zea mays*
Malay Apple *Eugenia malaccensis*
Male Bamboo *Dendrocalamus strictus*
Mandarin *Citrus reticulata*
Mango *Mangifera indica*
Mangosteen *Garcinia mangostana*
Mangrove *Avicennia africana*
Marabou Thorn *Dichrostachys cinerea*
Marble Vine *Dioclea reflexa*
Marcella *Grangea maderaspatana*
Marijuana *Cannabis sativa*
Marsh Corkwood *Annona glabra*
Mauritius Grass *Uruchloa mutica*
Melegueta *Aframomum melegueta*
Metel *Datura metel*
Mexican Cedar *Cedrela mexicana*
   Poppy *Argemone mexicana*
Milk Bush *Thevetia peruviana*
Miraculous Berry *Synsepalum dulcificum*
Mission Grass *Pennisetum polystachion*
Mistletoe *Phragmanthera* species, *Tapinanthus bangwensis, T. globiferus*
Monkey Akee *Eriocoelum kerstingii*
   Apple *Annona glabra*
   Cola *Carapa procera, Cola caricifolia*
   Pot *Lecythis zabucajo*
Monrovian Coffee *Coffea liberica* var. *dewevrei*
Moon Flower *Ipomoea alba*
Morning Glory *Ipomoea nil*
Mosquito Plant *Clausena anisata*
Mother of Cocoa *Gliricidia sepium*
Napoleon's Hat *Bauhinia tomentosa*
Narrow-Leaved Coffee *Coffea stenophylla*
Natal Indigo *Indigofera arrecta*
Neem Tree *Azadirachta indica*
Negro Coffee *Cassia occidentalis*
Nepal Trumpet Flower *Beaumontia grandiflora*
Nettle Tree *Celtis integrifolia*
Nicker Nut *Caesalpinia bonduc*
Niger Copal Tree *Daniellia thurifera*
   Plum *Flacourtia flavescens*
Nut Grass *Cyperus rotundus*
Nutmeg *Myristica fragrans*
Nux vomica *Strychnos nux-vomica*
Odoko *Scottellia klaineana* var. *klaineana*
Oil Bean Tree *Pentaclethra macrophylla*
   of Ben Tree *Moringa oleifera*
   Palm *Elaeis guineensis*
Okra, Okro *Abelmoschus esculentus*
Oleander *Nerium oleander*
Omu *Entandrophragma candollei*

Onion *Allium cepa*
Orange-barked Terminalia *Holoptelea grandis*
Orange Jessamine *Murraya paniculata*
Ordeal Bean *Physostigma venenosum*
    Tree *Erythrophleum suaveolens*
Pancratium Lily *Pancratium trianthum*
Para Cress *Spilanthes filicaulis*
    Grass *Urochloa mutica*
    Rubber Tree *Hevea brasiliensis*
Parasol Flower *Holmskioldia sanguinea*
Passion Fruit *Passiflora edulis*
Pawpaw *Carica papaya*
Peanut *Arachis hypogaea*
Pearl Millet *Pennisetum americanum*
Pepper *Capsicum annuum* (also *Piper*)
Perfume Tree *Cananga odorata*
Persian Lilac *Melia azedarach*
Physic Nut *Jatropha curcas, J. gossypiifolia*
Pigeonpea *Cajanus cajan*
Pigweed *Portulaca oleracea*
Pimento Grass *Stenotaphrum secundatum*
Pineapple *Ananas comosus*
Pines *Pinus* species
Pink Cassia *Cassia nodosa*
    Bauhinia *Bauhinia monandra*
Pitanga Cherry *Eugenia uniflora*
Plantain *Musa paradisiaca*
Poison Hemlock *Conium maculatum*
    Nut *Strychnos nux-vomica*
Pomegranate *Punica granatum*
Pommelo *Citrus grandis* Poplar *Mitragyna ciliata*
Poppy (opium) *Papaver somniferum*
Potato Tree *Solanum wrightii*
    Yam *Dioscorea bulbifera*
Prayer Beads *Abrus precatorius*
Prickly Amaranth *Amaranthus spinosus*
    Ipomoea *Ipomoea alba*
    Lantana *Lantana camara*
    Pear *Opuntia* species
    Poppy *Argemone mexicana*
Pride of Barbados *Caesalpinia pulcherrima*
    of Peru *Mirabilis jalapa*
Pterygota *Pterygota macrocarpa*
Pumpkin *Cucurbita pepo*
Purslane *Portulaca oleracea*
Pyrethrum *Chrysanthemum cinerariifolium*
Queen Grape Myrtle *Lagerstroemia speciosa*
Queen's Wreath *Petrea volubilis*
Rain Tree *Samana saman*
Rambuttan *Nephelium lappaceum*
Rangoon Creeper *Quisqualis indica*
Rat's Tail Grass *Sporobolus pyramidalis*
Rattan Palm *Laccosperma opacum*
Red Head *Asclepias curassavica*
    Frangipani *Plumeria rubra* var. *acutifolia*
    Ironwood *Lophira alata*
    Mangrove *Rhizophora racemosa, R.* species
    Plumbago *Plumbago indica*
    Sandalwood *Adenanthera pavonina*
    Sorrel *Hibiscus sabdariffa*
    Tecoma *Tecomaria capensis*
    -Flowered Silk Cotton Tree *Bombax buonopozense*
Resurrection Plant *Bryophyllum pinnatum*
Rhinoceros Beetle *Oryctes monoceros*

Rhodesian Timothy *Setaria sphacelata*
Rice Grass *Leersia hexandra*
Ringworm Shrub *Cassia alata*
Rio Nunez Coffee *Coffea canephora*
Roble Blanco *Tabebuia pentaphylla*
Rose Apple *Eugenia jambos*
Roselle *Hibiscus sabdariffa*
Rough Brittle Grass *Setaria verticillata*
Royal Palm *Roystonia regia*
Rubber Plant *Ficus elastica*
Rush Nut *Cyperus esculentus*
Salaga Potato *Solenostemon rotundifolius*
Salt and Oil Tree *Cleistopholis patens*
Sandbox Tree *Hura crepitans*
Sandpaper Tree *Ficus asperifolia*
Sapele *Entandrophragma cylindricum*
Sapistan Plum *Cordia myxa*
Sapodilla Plum *Achras zapota*
Sasswood Tree *Erythrophleum ivorense*
Satinwood *Pericopsis laxiflora*
Sausage Tree *Kigelia africana*
Scented Guarea *Guarea cedrata*
Screw Pine *Pandanus abbiwii, P. utilis, P. veitchi*
Sea Bean *Entada pursaetha*
Seaside Grape *Coccoloba uvifera*
Seaside Purslane *Sesuvium portulacastrum*
Senegal Lilac *Lonchocarpus sericeus*
    Rose Wood Tree *Pterocarpus erinaceus*
Sensitive Plant *Mimosa pigra, M. pudica*
Sesame *Sesamum indicum*
Seville Orange *Citrus aurantium*
Shaddock *Citrus grandis*
Shallot *Allium ascalonicum*
Shama Millet *Echinochloa colona*
Shea Butter Tree, Shea Nut Tree *Vitellaria paradoxa*
Shittim Wood *Acacia hockii*
Shoe Flower *Hibiscus rosa-sinensis*
Showers of Gold *Thryallis glauca*
Siam Weed *Chromolaena odorata*
Sierra Leone (Upland) Coffee *Coffea stenopylla*
Silk Cotton Tree *Ceiba pentandra*
Silver Fern *Pityrogramma calomelanos*
Sisal Hemp *Agave sisalana*
Small-Flowered Potato Creeper *Solanum seaforthianum*
Snake Bean *Swartzia madagascariensis*
    Gourd *Trichosanthes cucumerina* var. *anguina*
Snow Bush *Breynia nivosa*
Snuff-box tree *Oncoba spinosa*
Soap-berry Tree *Balanites aegyptiaca*
Sodom Apple *Caloptropis procera*
Sour Grass *Paspalum conjugatum*
    Orange *Citrus aurantium*
    Sop *Annona muricata*
South Sea Arrowroot *Tacca leontopetaloides*
Soya Bean *Glycine max*
Spanish Needles *Bidens pilosa*
    Physic Nut *Jatropha multifida*
Spear Grass *Heteropogon contortus*
Spice Tree *Xylopia aethiopica*
Spicy Cedar *Beilschmiedia mannii*
Spiked Millet *Pennisetum americanum*
Spiny Amaranth *Amaranthus spinosus*
Squash Gourd *Cucurbita maxima*

COMMON NAMES

Star Apple *Chrysophyllum africanum, C. cainito*
    Bur *Acanthospermum hispidum*
    Thistle *Centaurea perrottetii, C. praecox*
Sterculia Brown *Sterculia rhinopetala*
Stink Grass *Eragrostis cilianensis, Melinis minutiflora* var. *minutiflora*
Stinking Passion Flower *Passiflora glabra*
Stinkwood Tree *Petersianthus macrocarpus*
Strawberry Guava *Psidium cattleianum*
Sugar-cane *Saccharum officinarum*
Sugar Plum *Uapaca guineensis*
Surinam Quassia *Quassia amara*
Sweet Orange *Citrus sinensis*
    Pigweed *Chenopodium ambroisioides*
    Potato *Ipomoea batatas*
    Sop *Annona squamosa*
Sweetbroom Weed *Scoparia dulcis*
Switch Sorrel *Dodonaea viscosa*
Sword Bean *Canavalia ensiformis*
    Lily *Gladiolus daleni, G. gregarius, G. klattianus, G. unguiculata*
Tahitian Taro *Xanthosoma brasiliense*
Tallow Tree *Allanblackia parviflora, Detarium senegalense, Pentadesma butyraceum*
Tangerine *Citrus reticulata*
Tannia *Xanthosoma mafaffa*
Taro *Colocasia esculenta*
Tea *Camellia sinensis*
    Bush *Ocimum gratissimum*
Teak *Tectona grandis*
Temple Flower *Plumeria rubra* var. *acutifolia*
Ten-Month Yam *Dioscorea alata*
Tiger Nut *Cyperus esculentus*
Tobacco *Nicotiana rustica, N. tabacum*
Tomato *Lycopersicon esculentum*
Torch Ginger *Nicolaia elatior*
Traveller's Palm *Ravenala madagascariensis*
True Indigo *Indigofera tinctoria*
Trumpet Bush *Tecoma stans*
Turmeric *Curcuma domestica*
Turnip *Brassica campestris* var. *rapifera*
Umbrella Tree *Musanga cecropioides*
Unscented Mahogany *Entandrophragma candollei*
Upland Rice *Oryza glaberrima*
Utile *Entandrophragma utile*
Vegetable Marrow *Cucurbita pepo*
    Sponge *Luffa cylindrica*
Velvet Tamarind *Dialium guineense*
Vetiver *Vetiveria zizanioides*
Vine Plant *Vitis vinifera*
Walking Stick Camwood *Baphia capparidifolia* ssp. *polygalacea*
Water Fern *Ceratopteris cornuta*
    Grass *Urochloa mutica*
    Hyacinth *Eichhornia crassipes*
    Leaf *Talinum triangulare*
    Lettuce *Pistia stratiotes*
    Lily *Nymphaea lotus*
    Melon *Colocynthis lanatus*
    Yam *Dioscorea alata*
Weaver Bird *Quelea quelea*
West African Cedar *Entandrophragma cylindricum*
    Ebony *Diospyros mespiliformis*
    Indigo *Lonchocarpus cyanescens*
    Locust Bean *Parkia clappertoniana*
    Pepper *Piper guineense*
    Ratbane *Dichapetalum toxicarum*
    Rubber Plant *Funtumia elastica*
    Serendipity Berry *Dioscoreophyllum cumminsii*
West Indian Indigo *Indigofera suffruticosa*
    Raspberry *Rubus* species
Wheat *Triticum aestivum*
Whistling Pine *Casuarina equisetifolia*
White Button Wood *Laguncularia racemosa*
    Cabbage *Brassica oleracea* var. *capitata*
    Guinea Yam *Dioscorea cayenensis* ssp. *rotundata*
    Mahogany *Khaya anthotheca*
    Pumpkin *Lagenaria siceraria*
    Rubber Vine *Landolphia owariensis*
    Star Apple *Chrysophyllum delevoyi*
Wild Amaranth *Amaranthus viridis*
    Banana *Ensete gilletii*
    Custard Apple *Annona senegalensis* var. *senegalensis*
    Date Palm *Phoenix reclinata*
    Lettuce *Launaea taraxacifolia*
    Mango *Irvingia gabonensis*
    Mustard *Cleome viscosa*
    Olive *Ximenia americana*
    Rice *Oryza barthii*
    Sage *Lantana camara*
    Spinach *Talinum triangulare*
    Tamarind *Leucaena leucocephala*
    Tobacco *Nicotiana rustica*
    Yam *Dioscorea lecardii, D. praehensilis*
Wine Palm *Raphia hookeri, R.* species
Winged Bean *Psophocarpus tetragonolobus*
    Yam *Dioscorea alata*
Witchweed *Striga* species
Woolly Morning Glory *Argyreia speciosa*
Worm Weed *Spigelia anthelmia*
Yang *Dipterocarpus* species
Yellow Allamanda *Allamanda cathartica*
    Bells *Tecoma stans*
    Sterculia *Sterculia oblonga*
    -flowered Tobacco *Nicotiana rustica*
Yoruba Ebony *Diospyros monbuttensis*
    Indigo *Lonchocarpus cyanescens*

INDEX TO SCIENTIFIC NAMES

Abelmoschus esculentus (Hibiscus esculentus) 38, 65
   moschatus (H. abelmoschus) 239
Abrus precatorius 111, 113, 127, 129, 132, 133, 134, 136, 138, 139, 140, 142, 148, 151, 152, 158, 162, 178, 180, 186, 188, 190, 200, 206, 207, 211, 212
   pulchellus 152
Abutilon guineense 142, 146, 148, 156, 176, 186, 199
   mauritianum 142, 247
Acacia albida 14, 46, 53, 97, 108, 141, 151, 157, 169, 171, 177, 207, 215, 221, 228
   ataxantha 106
   dudgeoni 14, 53, 171, 227
   farnesiana 13, 14, 53, 105, 145, 183, 202, 213, 221, 226, 227, 229, 234
   gourmaensis 13, 53, 142
   hockii 14, 53, 130, 144, 158, 165, 170, 221
   kamerunensis (pennata) 8, 53, 112, 129, 154, 183, 189, 193, 202, 216, 221
   macrothyrsa 178, 204
   nilotica var. adansonii 53
   nilotica var. tomentosa 9, 14, 53, 79, 86, 102, 105, 108, 138, 142, 145, 153, 156, 158, 171, 184, 190, 193, 194, 199, 200, 221, 224, 226, 227
   pentagona 8, 138, 163, Plate 10
   polyacantha ssp. campylacantha 14, 49, 102, 175, 189, 198, 219, 221, 227
   senegal 15, 53, 77, 227
   seyal 55, 77, 171
   sieberiana var. villosa 13, 14, 53, 79, 91, 104, 105, 110, 128, 133, 138, 158, 163, 188, 199, 222, 233
   species 11, 104, 209, 221
Acalypha ciliata 247
   wilkesiana 234
   species 237
Acanthospermum hispidum 171, 247
Acanthus montanus 141, 205
Achras zapota 46
Achyranthes aspera 220, 247
   indica (obtusifolia) 247
Acidanthera aequinoctialis (see Gladiolus aequinoctialis)
Acridocarpus alternifolius 237
   smeathmannii 139, 143, 152, 155, 164, 184, 188, 237
Acroceras amplectens 55
   zizanioides 55
Adansonia digitata 10, 11, 12, 14, 41, 51, 53, 69, 111, 147, 149, 155, 159, 173, 185, 191, 193, 218, 222, 228, 254, Plate 13
Adenanthera pavonina 15, 91, 114, 224, 228, 233
Adenia cissampeloides 10, 216
   rumicifolia var. miegei (lobata) 10, 130, 135, 136, 148, 154, 166, 176, 178, 182, 189, 212, 216

Adenium obesum 13, 196, 199, 204, 210, 212, 213, 216, 237
   somalense 10
Adenodolichos paniculatus 79, 88, 187
Adenopus breviflorus (see Lagenaria breviflora)
Adhatoda buchholzii 216
   robusta 125
Adiantum confine 235
   incisum 235
   mettenii (soboliferum) 235
   philippense 235, 248
   tetraphyllum var. vogelii (vogelii) 235, 248
Adina microcephala
   (see Breonadia salicina)
Aeglopsis chevalieri 43, 53
   paniculata 169
Aeolanthus pubescens 41, 154
Aerangis biloba 236
   calantha 236
   laurentii (see Summerhayesia laurentii)
Aerva javanica 219
   tomentosa 40
Afraegle paniculata 11, 14, 15, 43, 53, 110, 114, 119, 180, 188, 228
Aframomum daniellii 49, 258
   latifolium 156
   longiscapum 239
   melegueta 51, 52, 75, 119, 125, 127, 136, 141, 147, 150, 156, 158, 162, 190, 191, 193, 203, 229, 231
   sulcatum 49, 125, 127
   species 146, 212, 258
Afrosersalisia afzelii 46, 101
Afrormosia (see Pericopsis)
Afzelia africana 6, 41, 53, 78, 83, 84, 86, 87, 88, 92, 97, 108, 110, 148, 151, 157, 162, 163, 172, 175, 179, 185, 188, 196, 212, 217, 228, 231, 240
   bela var. glacier 6, 91, 114, 183
Aganope leucobotrya (Ostryoderris leucobotrya) 238
Agave sisalana 77, 115, 215
Agelaea obliqua 8, 237
   trifolia 8, 81, 237
Ageratum conyzoides 143, 144, 150, 152, 156, 177, 185, 187, 195, 207, 214, 247
Aglaonema species 240
Aiphane caryotifolia 241
   corallina 241
Aizoon canariensis 216
Alafia barteri 10, 14, 157
   multiflora 204
   scandens 13, 182, 213, 237
   species 14
Albizia adianthifolia 6, 15, 101, 105, 106, 128, 152, 167, 174, 180, 183, 187, 191, 226
   chevalieri 41
   coriaria 91, 217, 221, 228
   ferruginea 6, 83, 144, 178, 203, 210, 228

INDEX TO SCIENTIFIC NAMES

lebbeck 13, 15, 53, 91, 97, 221, 228, 233
zygia 6, 11, 15, 41, 93, 98, 101, 130, 142, 153, 154, 180, 181, 184, 188, 204
species 97
Alchornea cordifolia 126, 130, 135, 139, 145, 147, 150, 152, 155, 160, 168, 170, 171, 174, 176, 181, 182, 187, 188, 194, 195, 200, 202, 204, 219, 221, 223, 226, 247
floribunda 51, 129
species 126, 156
Allamanda cathartica 239
neriifolia 239
Allanblackia parviflora (floribunda) 6, 15, 87, 88, 126, 144, 192, 231
Allium ascalonicum 38, 155
cepa 38, 219, 220
sativum 128, 131, 132, 135, 137, 139, 141, 142, 181, 183, 194, 195
Allophylus africanus 9, 96, 127, 139, 146, 154, 157, 161, 169, 176, 190, 191
spicatus 146
Allopteropsis paniculata 55
semialata 55
Alocasia species 240
Aloe buettneri 159, 218, Plate 13
schweinfurthii 159, 218
Alpinia species 240
Alsodeiopsis staudtii 46
Astonia boonei 14, 84, 91, 107, 119, 127, 132, 157, 167, 170, 179, 181, 191, 197, 203, 204, 218, Plate 8
Alternanthera maritima 21
nodiflora 134, 189
pungens (repens) 125, 145, 155, 170, 175, 177, 247
sessilis 134, 249
Alysicarpus rugosus 55, 253, 247
Amanita phalloides 211
species 207, 211
Amaranthus caudatus 148
hybridus ssp. incurvatus 40
spinosus 55, 144, 176, 247
viridis 40, 55, 134, 247
species 40
Amblygonocarpus andongensis 196, 220
Amorphophallus abyssinicus 169, 220
aphyllus 31, 220, 239
dracontioides 31, 213
johnsonii 220
Ampelocissus gracilipes 12, 46
grantii 134, 154, 161, 170, 183
multistriata 216
Amphimas pterocarpoides 15, 127, 128, 145, 215, 217
Anabaena limnetia 58
species 257
Anacardium occidentale 13, 15, 16, 41, 45, 51, 52, 53, 86, 102, 108, 119, 145, 148, 158, 165, 176, 184, 186, 192, 201, 204, 211, 217, 221, 225, 227
Anadelphia afzeliana 82
leptocoma 82
pumila 82
trispiculata 82
Ananas comosus 44, 127, 133, 148, 149, 173, 195, 200, 221

Anchomanes difformis 31, 152
Ancistrocarpus densispinosus 204
Ancistrochilus rothschildianus 236
thomsonianus 236
Ancistrocladus abbreviatus 125
Ancistrophyllum (see Laccosperma)
Ancistrorrhynchus capitatus 236
cephalotes 236
Acylobotrys amoena (Landolphia amoena) 11, 13, 47
scandens 13, 46
Andira inermis 119, 128, 210, 212
Andropogon africanus 82
ascinodis 82
canaliculatus 82
curvifolius 55, 82
fastigiatus 55, 82
gayanus var. bisquamulatus 55, 82, 89, 112, 249, Plate 8
gayanus var. squamulatus 55, 82, 89, 112
gayanus var. tridentatus 55, 82, 89, 112
incanellus 55, 82
macrophyllus 82, 89
perligulatus 82
pseudapricus 55, 81, 82
pteropholis 82, 89
schrensis 55
tectorum 55, 82, 89, 112
tenuiberbis 82, 89
species 82
Androsiphonia (see Paropsia)
Aneilema beniniense 177, 247, 258
lanceolatum ssp. lanceolatum 153, 202
lanceolatum ssp. subnudum 202
Angelonia salicariifolia 240
Angraecum birrimense 236
chevalieri 236
multinominatum 236
species 130, 185
Aningeria altissima 83, 86
robusta 46, 91, 113
Aniseia martinicensis 239
Ankistrodesmus falcatus 57
Annona cherimola 45
glabra 45
glauca var. glauca 164, 215
muricata 45, 51, 145, 154, 164, 184, 215
reticulata 45
senegalensis ssp. deltoides 218
senegalensis ssp. onlotricha (arenaria) 45, 141, 142, 147, 180, 182, 218, 221
senegalensis ssp. senegalensis 41, 45, 53, 137, 145, 147, 150, 159, 163, 164, 170, 185, 189, 192, 201, 219, 228, 229
squamosa 45, 119, 164, 215, 221
species 45
Anogeissus leiocarpus 15, 53, 80, 84, 95, 97, 137, 141, 145, 156, 166, 168, 172, 192, 194, 198, 202, 221, 225, 226, 227, 233, 244
Anonidium mannii 46
Anopyxis klaineana 6, 78, 87, 91, 98, 183, 202
Ansellia africana 236
species 130, 185
Anthocleista nobilis 119, 128, 140, 170, 174, 176, 189, 195, 197, 203, 228
vogelii 138, 226, 228, 257

species 172
Anthonotha fragrans 15
  macrophylla 141, 169, 224
Anthosthema aubryanum 133, 179, 211, 216
Anthurium species 240
Antiaris toxicaria (africana and welwitschii) 6, 14, 86, 93, 104, 109, 113, 135, 169, 170, 171
Antidesma laciniatum var. laciniatum 126, 130
  membranaceum 126
  venosum 15, 41, 46, 80, 127, 160, 166, 188
Antrocaryon micraster 46, 51, 53, 84, 91, 189
Anubias minima 57, 235
  species 57, 235
Aphania (see Lepisanthes)
Aphanostylis mannii 13
Aphis spiraecola 244
Apion brunneonigrum 244
Apocynaceae 13
Aporrhiza urophylla 80, 84
Aptandra zenkeri 180
Arachis hypogaea 31, 256
Araliopsis tabouensis 197
Argemone mexicana 147, 150, 152, 159, 165, 169, 177, 179, 211
Argocoffeopsis afzelii 65
  rupestris 65
  subcordata 65
Argyreia speciosa 234
Aristida adscensionis 55, 82, 112
  hordeacea 56
  kerstingii 56
  mutabilis 56, 82
  recta 56
  sieberiana 56, 82
Aristolochia albida 131, 159, 198
  bracteolata 220
  elegans 239
  indica 164, 173, 189
  ridicula 239
Artabotrys thomsonii 12
Arthropteris monocarpa 235
  palisotii 235
Artocarpus communis 46, 53, 113
  heterophyllus 46
Asclepias curassavica 13, 115, 151, 180, 207, 211, 213, 239
  species 175
Asparagus africanus 80, 148, 159, 160, 191, 198, 213
  flagellaris 31, 80, 148, 159, 160, 191, 198, 213
  racemosus 49, 239
  warneckei 239
Aspergillus niger 27, 38
Aspilia africana 55, 81, 142, 151, 161, 169, 188, 190, 202, 247, 251
  bussei 247
Asplenium akimense 235
  buettneri 235
  cuneatum 235
  diplazisorum 235
  dregeanum 235
  formosum 235
  megalura 236
  preussii 236

protensum 236
schnellii 236
stuhlmannii 236
unilaterale 236
species 248
Asystasia calycina 143, 204, 247
  gangetica 169, 202, 247
Ataenidia conferta 81, 143
Athesapeuta cyperi 243
Atroxima afzeliana 46
Aubrevillea kerstingii 87
  platycarpa 87
Aulacocalyx jasminiflora 9, 88
Auricularia species 40
Averrhoa bilimbi 46
  carambola 46
Avicennia africana 49, 97, 165, 166, 183, 211, 221, 224, Plate 7
Axonopus compressus 55, 234, 248
  flexuosus 56, 82
Azadirachta indica 9, 15, 53, 84, 95, 96, 97, 111, 128, 154, 164, 183, 203, 220, 231, 233, 254, Plate 7
Azanza garckeana 239
Azolla pinnata var. africana (africana) 57, 249, 256
  species 257

Bacillus species 27
Bactra species 243
Baissea multiflora 237
Balanites aegyptiaca 11, 12, 15, 41, 51, 53, 102, 105, 108, 114, 125, 128, 133, 143, 163, 178, 183, 204, 215, 216, 219, 226, 227, 228, 233
  wilsoniana 16, 49, 215
Bambusa vulgaris 53, 79, 84, 110, 211
Baphia capparidifolia ssp. polygalacea (polygalacea) 219
  nitida 9, 104, 108, 119, 137, 151, 158, 168, 178, 182, 184, 187, 190, 200, 204, 224
  pubescens 108
Barleria oenotheroides 237
  cristata 240
  opaca 41, 176
  species 187
Basella alba 40
Basilicum polystachyon 161
Bauhinia monandra 237
  purpurea 15, 41, 221, 237
  rufescens 10, 46, 53, 145, 152, 156, 171, 221, 234
  tomentosa 234
  species 158
Beaumontia grandiflora 240
Beckeropsis laxior (see Pennisetum laxior)
  uniseta (see P. unisetum)
Begonia species 240
Beilschmiedia mannii 46, 52, 53, 131, 135, 162, 182, 229
Bequaertiodendron megalismontanum 46
  oblanceolatum 46
Berlinia confusa 149, 158, 178
  grandiflora 15, 91, 179, 233
  occidentalis 85, 86
Bersama abyssinica ssp. paullinioides 80, 128, 131, 172, 178, 189, 210, 212, 213

INDEX TO SCIENTIFIC NAMES

Bertholetia excelsa 46
Bertiera racemosa 143, 190, 220
Bidens pilosa 41, 149, 190, 247
Billbergia species 240
Bixa orellana 133, 147, 156, 225, 234, Plate 14
Blepharis linariifolia 55
Blighia sapida 53, 91, 102, 140, 152, 161, 163, 166, 168, 185, 195, 204, 211, 217, 218, 226, 228, 229, 233
    unijugata 11, 98, 150, 157, 179, 190, 216, 233
    welwitschii 11, 52, 98, 109, 216, Plate 7
Blumea aurita var. aurita 145, 152, 165, 177, 190, 201
    aurita var. foliosa 157, 184
    mollis 157
    perrottetiana 247
Boerhavia diffusa 55, 57, 127, 132, 134, 136, 140, 149, 150, 154, 159, 162, 163, 168, 174, 185, 195, 197, 205
    repens 55, 57, 132, 134, 140, 150, 154, 159, 174, 185, 195, 205
    species 247
Bolbitis heudelotii 236
Bolusiella batesii 236
    imbricata 236
    talbotii 236
Bombax brevicuspe (see Rhodognaphalon brevicuspe)
    buonopozense 16, 41, 51, 53, 81, 107, 115, 143, 170, 173, 224, 240
    species 115
Bonamia thunbergiana 239
Borassus aethiopum 15, 49, 50, 51, 78, 80, 81, 83, 86, 87, 88, 107, 109, 111, 113, 114, 116, 132, 135, 190, 225, 226, Plate 8
Borreria species 247
    stachydea 169, 174, 197, 200
    verticillata (see Spermacoce verticillata)
Boscia salicifolia 41, 210
    senegalensis 46, 53, 65
Bosqueia (see Trilepisium)
Boswellia dalziellii 16, 131, 150, 157, 158, 165, 176, 180, 198, 200, 218, 227, 229
Bothriochloa bladhii 235, 248
Botryodiploda theobromae 30
Botrytis cineria 38
    species 38
Bougainvillea glabra 240
    spectabilis 240
    species 96
Brachiaria brachylopha 56
    brizantha 55
    deflexa (see Pseudobrachiaria deflexa)
    distachyoides 56
    distichophylla 25, 55, 248
    falcifera 25, 55
    jubata 25, 55
    lata 82, 248
    mutica (see Urochloa mutica)
    plantaginea 56
    stigmatisata 25, 56
    xantholeuca 25, 55
Brassica campestris var. rapifera 41
    oleracea var. botrytis 41
    oleracea var. capitata 41

oleracea var. italica 41
Breonadia salicina (Adina microcephala) 9, 86, 91, 119, 145
Breynia nivosa 234
Bridelia atroviridis 130, 148, 177, 197
    ferruginea 80, 81, 133, 144, 145, 148, 158, 162, 175, 180, 184, 194, 196, 199, 218, 219, 221, 224, 225
    grandis 170
    scleroneura 46, 126, 157
    stenocarpa (micrantha) 46, 78, 79, 88, 126, 141, 184, 189, 194, 221, 234
Brillantaisia lamium 144, 149, 154, 205
    nitens 144, 149, 154, 205
    patula 187, 239
Bryophyllum pinnatum 134, 142, 147, 149, 151, 160, 187, 195, 247
Buchholzia coriacea 46, 132, 134, 139, 142, 149, 162, 167, 175, 210, 212
Buchnera leptostachya 223
Bulbophyllum barbigerum 236
    bufo (see B. falcatum var. bufo)
    calyptratum 236
    calyptratum var. graminifolium (graminifolium) 236
    cocoinum 236
    colubrinum 236
    congolanum (see B. scaberulum)
    distans (see B. saltatorium var. albociliatum)
    falcatum 236
    falcatum var. bufo (bufo) 236
    falcatum var. velutinum (velutinum) 236
    flavidum (see B. pumilum)
    graminifolium (see B. calyptratum var. graminifolium)
    imbricatum (linderi) 236
    linderi (see B. imbricatum)
    lupilinum 236
    magnibracteatum 236
    maximum 236
    nigritianum 236
    oreonastes (zenkerianum) 236
    phaeopogon (see B. schizianum var. phaeopogon)
    pipio 236
    pumilum var. flavidum 236
    purpureorhachia 236
    recurvum 236
    rhizophorae (see B. falcatum var. velutinum)
    saltatorium 236
    saltatorium var. albociliatum (distans) 236
    scaberulum (congolanum) 236
    schinzianum var. phaeopogon (phaeopogon) 236
    velutinum (see B. falcatum var. velutinum)
    zenkerianum (see B. oreonastes)
    species 130, 185, 236
Bulbostylis metralis 115
    pilosa 115
Bulinus species 215
Burkea africana 9, 13, 15, 53, 86, 91, 102, 104, 207, 214, 217, 221, 233
Burseraceae 16

Bussea occidentalis 84, 97, 162, 214, 216, 229
Butyrospermum (see Vitellaria)
Byrsocarpus coccineus 96, 149, 151, 163, 164, 167, 173, 186, 192, 194, 200, 203, 218, 237
Byttneria catalpifolia ssp. africana 12

Cadaba farinosa 41, 54, 139, 142, 156
Caesalpinia bonduc 16, 114, 119, 129, 141, 151, 155, 166, 175, 186, 187, 190, 193, 201, 231
    pulcherrima 234, 240
Cajanus cajan 31, 32, 54, 163, 173, 256, Plate 2
    kerstingii 256
Caladium species 240
Calamus deeratus 41, 49, 80, 81, 87, 89, 92, 111, 112, 116
Calathea species 240
Calliandra portoricensis 126, 161, 172, 178, 186, 199
Caloncoba echinata 50, 51, 105, 134, 164, 167, 171, 202, 229, 231
    gilgiana 46, 80, 237
    glauca 46, 199, 231
Calonyction aculeatum 239
Calophyllum inophyllum 53, 234
Calothrix 257
Calotropis procera 10, 12, 77, 81, 102, 106, 115, 120, 125, 126, 128, 130, 132, 133, 134, 137, 140, 142, 146, 148, 153, 159, 160, 164, 165, 168, 170, 171, 178, 182, 186, 187, 192, 193, 195, 199, 204, 207, 211, 212, 213, 220, 221, 222, 247, 251, 258
Calpocalyx brevibracteatus 49, 80, 87, 105, 189, 229
Calvatia species 40
Calycobolus heudelotii 238
Calycosiphonia macrochlamys (Coffea macrochlamys) 65
    spathicalyx (Coffea spathicalyx) 65
Calyptrochilum christyanum 236
    emarginatum 236
Camellia sinensis 50, 230
Campylospermum flavum (Ouratea flava) 48, 238
    glaberrimum (O. glaberrima) 238
    reticulatum (O. reticulata) 126, 154, 238
    schoenleinianum (O. schoenleiniana) 238
    sulcatum (O. sulcata) 238
    vogelii (O. vogelii) 238
Cananga odorata 231
Canarium commune 46
    schweinfurthii 6, 16, 46, 53, 83, 84, 91, 108, 114, 120, 137, 140, 141, 144, 165, 167, 177, 204, 227, 229
Canavalia ensiformis 31, 33, 114
Canna indica 113, 153, 154
Cannabis sativa 126, 127, 130, 230
Canscora decussata 247
Canthium hispidum (see Keetia hispida)
    horizontale (see Psydrax horizontalis)
    subcordatum (see P. subcordata)
    venosum (see P. venosa)
    vulgare (see P. parviflora)
Capparis brassii 127
    corymbosa 41, 46, 130, 210, 213
    erythrocarpos 46, 160, 161, 169
    polymorpha 54, 152, 197, 203, 210, 213, 217, 219
Capsicum annuum 35, 125, 141, 162, 165, 181, 183, 202, Plate 2
    frutescens 35
    species 130
Carapa procera 16, 78, 81, 91, 98, 104, 109, 120, 129, 133, 135, 137, 142, 150, 153, 156, 168, 179, 181, 182, 185, 187, 189, 199, 203, 204, 211
Caratocystis fibriata 29
Carcospora citrullina 39
Cardiospermum grandiflorum 114, 131, 143, 152, 166, 173, 178, 184, 191
    halicacabum 40, 114, 131, 143, 152, 166, 173, 178, 191, Plate 10
Carica papaya 41, 45, 49, 120, 125, 161, 163, 176, 189, 190, 194, 197, 201, 204, 216, 228
Carissa edulis 46, 52, 54, 96, 129, 172, 234
Carpolobia alba 9, 47, 79, 129, 180, 188
    lutea 9, 12, 79, 157, 162, 188
    species 105, 110
Caryota species 241
Casearia barteri 9
Cassia absus 120, 151, 182, 195, 198, 207
    alata 13, 120, 128, 132, 146, 149, 168, 171, 178, 183, 184, 197, 213, 219, 221, 227, 237
    auriculata 221, 237
    bicapsularis 234
    fistula 15, 179, 221, 240
    grandis 240
    kirkii var. guineensis 47, 79, 143, 254
    laevigata 154, 213, 234
    ligustrina 239
    mimosoides 50, 54, 140, 247
    multijuga 239
    nigricans 50, 128, 131, 155, 178, 186
    nodosa 41, 178, 228, 240
    obtusifolia (tora) 41, 54, 65, 120, 120, 142, 158, 167, 178, 183, 195, 223, 225, 247, 251
    occidentalis 65, 120, 128, 132, 133, 135, 138, 144, 145, 148, 149, 151, 153, 155, 159, 160, 163, 164, 167, 168, 170, 171, 175, 180, 182, 183, 186, 187, 188, 191, 192, 196, 202, 204, 211, 214, 222, 247, 250
    prodocarpa 117, 159, 178, 197, 227
    rotundifolia 235, 247
    siamea 54, 95, 96, 97, 213
    sieberiana 8, 85, 91, 105, 108, 127, 129, 131, 148, 163, 166, 171, 175, 176, 188, 196, 203, 215, 221, 228, 240
    singueana 152, 154, 221
    sophera 178
    species 240
Cassipourea barteri 97
Cassytha filiformis 248
Castanola paradoxa 9, 41
Casuarina equisetifolia 83, 97, 105, 221, 225, 233
Catharanthus roseus 120, 124, 136, 144, 213, 232, 239, 242, Plate 10
Cathormion altissimum 47, 159, 188, 216, 220, 222, 227, 228
Cedrela mexicana 107, 120, 233
Ceiba pentandra 13, 16, 41, 49, 53, 54, 107, 108, 115, 116, 139, 146, 150, 157, 160, 171, 182, 211, 226, 228

INDEX TO SCIENTIFIC NAMES 293

species 115
Celastraceae 10, 80
Celosia argentea 40, 55, 127, 145, 147, 184
    leptostachya (laxa and trigyna) 127, 130, 134, 162, 169, 181, 183, 247
Celtis adolfi-friderici 6, 84, 85
    africana 54, 105, 152, 157, 161
    integrifolia 10, 41, 47, 54, 112, 128, 161, 169, 173, 181
    mildbraedii 7, 11, 84, 85, 87, 91, 93, 94
    wightii (brownii) 11, 83, 210
    zenkeri 7, 11, 84, 85, 94
    species 11, 79, 88
Cenchrus biflorus 25, 56, 129, 248
    ciliaris 56
    echinatus 56, 248
    setigerus 56
Centaurea perrottetii 131, 177
    praecox 196
Centella asiatica 125, 171, 181, 182, 194, 195
Centotheca lappacea 56, 248
Centrosema plumieri 247
Cephaelis peduncularis (see Psychotria peduncularis)
Cephalocereus species 240
Cephalostigma (see Wahlenbergia)
Ceratocystis fimbriata 29
    paradoxa 30
Ceratophyllum demersum 57, 235, 249
Ceratopteris cornuta 235, 249
Ceratostomella paradoxa 44
Ceratotheca sesamoides 41, 66, 77, 81, 152
Cercestis afzelii 80
Cercospora caricae 45
    coffeicola 65
    elaedis 69
    musae 45
    nicotianae 74
    species 33
Ceropegia species 240
Chaetacme aristata 96, 139, 178
Chamaeangis species 236
Chasmanthera dependens 157, 187, 199
Chasmopodium caudatum 56, 89, 106
Cheirostylis divina 237
Chenopodium ambrosioides 127, 201, 229
    murale 41
Chionanthus mannii (Linociera mannii) 234
Chloris barbata 56, 248
    gayana 56
    pilosa 56
    prieuri 56
    pycnotrix 56
    robusta 82, 89
Chlorophora (see Milicia)
Christiana africana 10, 41, 104, 107
Chromolaena odorata (Eupatorium odoratum) 164, 242, 244, Plate 16
    species 244
Chrozophora senegalensis 54, 147, 174, 182, 199
Chrysalidocarpus lutescens 241
Chrysanthellum indicum var. afroamericanum 162
Chrysanthemum cinerariifolium 77
    procumbens 247

Chrysobalanus icaco (ellipticus and orbicularis) 12, 97, 147, 221
    icaco ssp. atacorensis 12
Chrysochloa hindsii 235
Chrysophyllum africanum 11, 14
    beguei 114
    cainito 46
    delevoyi (albidum) 11, 14, 53, 91, 105, 114, 231
    perpulchrum 16, 46
    pruniforme 46
    welwitschii (see Donella welwitschii)
Chrysopogon aciculatus 235
Chrysops species 153
Chytranthus carneus (villiger) 180
Cienfuegosia digitata 239
    heteroclada 219
Cinnamomum camphora 231
    zeylanicum 50, 52, 136, 137, 145, 229, 231
Cissampelos mucronata 111, 126, 219, 220
    owariensis 126, 220
Cissus aralioides 47, 134, 192, 222, 237
    arguta 47
    cornifolia 47, 200
    polyantha 152
    populnea 12, 13, 80, 128, 129, 153, 158, 168, 170, 175, 200, 218, 219
    producta 162
    quadrangularis 10, 131, 158, 187, 197, 198, 201, 202
Citropsis articulata 43
Citrus aurantiifolia 43, 50, 77, 128, 129, 131, 136, 147, 150, 153, 156, 161, 163, 165, 168, 183, 186, 194, 196, 200, 204
    aurantium 43, 104, 182, 196
    grandis 43
    limon 43, 140, 147, 196
    medica 43
    paradisi 43
    reticulata 43
    sinensis 43, Plate 2
    species 43, 44, 50, 52
Claoxylon hexandrum 180
Clappertonia ficifolia 10, 237
Clausena anisata 96, 120, 228, 135, 152, 161, 162, 164, 166, 173, 176, 177, 181, 184, 188, 190, 193, 219
Cleidion gabonicum 9
Cleistopholis patens 12, 41, 107, 112, 114, 128, 154, 163, 185
Clematis grandiflora 200, 237
    hirsuta 139, 140, 155, 171, 183, 200, 237
Cleome gynandra (Gyanandropsis gynandra) 40, 128, 149, 216
    rutidosperma (ciliata) 247
    viscosa 149, 247, 251
Clerodendrum buchholzii 138, 181, 219, 237
    capitatum 133, 155, 163, 175, 188, 189, 193, 237
    formicarum 237
    japonicum 237
    polycephalum 219
    splendens 187, 202, 226, 237
    thomsonae 239
    umbellatum 139, 169, 200, 219, 226, 237
    violaceum 237

volubile 172, 175, 202, 237
species 237
Clitoria ternatea 54, 149, 150, 179, 223, 247
Closterium kutzingii 57
Clytosoma callistegioides 240
Cnestis ferruginea 8, 12, 106, 126, 129, 139, 146, 152, 155, 161, 174, 178, 184, 193, 201, 220, 224, 237, Plate 15
Cnidoscolus aconitifolius 40
Coccoloba uvifera 46, 222
Cocculus pendulus 155
Cochlospermum planchonii 52, 115, 225, Pl. 8
    tinctorium 52, 54, 80, 115, 135, 140, 152, 155, 163, 166, 168, 169, 171, 173, 176, 177, 187, 188, 189, 191, 199, 219, 225, 237
Cocos nucifera 10, 13, 41, 49, 50, 67, 79, 81, 88, 96, 98, 101, 111, 112, 113, 114, 115, 149, 193, 210, 222, 224, 228, Plate 5
Codiaeum variegatum 234, 239
Coelocaryon oxycarpum 179
Coffea afzelii (see Argocoffeopsis afzelii)
    arabica 64, 65, 120
    canephora 65, 120, Plate 5
    ebracteolata 159, 227
    liberica var. liberica 65, 177
    liberica var. dewevrei 65
    macrochlamys (see Calycosiphonia macrochlamys)
    rupestris (see Argocoffeopsis rupestris)
    spathicalyx (see Calycosiphonia spathicalyx)
    stenophylla 65
    subcordata (see Argocoffeopsis subcordata)
    species 64, 230
Coix lacryma-jobi 56, 113, 117, 153, 249
Cola acuminata 72, 79, 120, 130, 138, 140, 141, 147, 151, 223
    buntingii 11, 80
    caricifolia 73, 106, 152, 176, 178, 185, 186
    chlamydantha 11, 73
    digitata 80
    gigantea var. glabrescens 10, 15, 73, 81, 98, 198, 233
    heterophylla 73
    lateritia var. maclaudi 73, 89, 110, 134, 203
    laurifolia 9, 10, 54, 105, 147
    millenii 104
    nitida 51, 72, 85, 91, 98, 104, 120, Plate 5
    species 72, 230
Coleus species 240
Colletotrichum species 38
Colocasia esculenta 29, 40, 257, Plate 2
    species 212
Colocynthis lanatus (vulgaris) 39, 53, 65, 196, 198, Plate 2
Combretodendron (see Petersianthus)
Combretum aculeatum 171, 178
    collinum ssp. binderianum (binderianum) 15, 237
    collinum ssp. hypopilinum (hypopilanum) 15, 178
    fragrans (ghasalense) 15, 79
    ghasalense 52, 229
    glutinosum 79, 105, 223, 226
    grandiflorum 237
    micranthum 102, 111, 120, 128, 131, 134, 138, 139, 148, 151, 154, 158, 172, 181, 192, 199, 202, 222
    molle 15, 80, 135, 144, 156, 169, 189, 197, 203, 217
    mucronatum (smeathmannii) 138, 159, 196, 226
    nigricans var. elliotii 15, 97, 216, 222, 227
    paniculatum 42, 105, 237
    platypterum 42, 154, 237
    racemosum 52, 128, 237, Plate 15
    sericeum 15
    tarquense 237
    zenkeri 146, 174
    species 96, 234, 237
Commelina diffusa 152, 191, 204, 214, 258
    erecta ssp. erecta 57
    forskalaei 57
    species 55, 247
Commicarpus plumbagineus 54
Commiphora africana 114, 131, 153, 164, 220, 229
    dalzielii 229
    pedunculata 16, 47, 229
Compositae 55
Conium maculatum 206
Conocarpus erectus 87, 97, 137, 154, 190, 196, 221
Convolvulaceae 20
Conyza aegyptiaca 247
Copaifera salikounda 15, 91, 93, 120, 195, 200, 229
Corchorus aestuans 40, 200
    olitorius 42, 54, 207, 211, 247
    tridens 42, 54, 247
    species 10, 40, 71
Cordia africana 47
    millenii 81, 91, 107, 109, 128, 132, 139, 141, 167, 183, 233
    myxa 10, 14, 15, 47, 135, 138, 142, 185
    platythyrsa 81, 107, 109, 117, 233
    rothii 10, 47, 54, 97, 106, 237
    sebestena 239
    senegalensis 132
    vignei 177, 181, 202
    species 243
Cordyline fruticosa 52
    terminalis 237
    species 240
Corticium rolfsii 29
    solani 32
Corymborkis corymbis (corymbosa) 236
Corynanthe pachyceras 51, 94, 101, 108, 120, 124, 127, 130, 141, 157, 170, 212, 257
Costus afer 12, 142, 181, 185
    schlechteri 144
    species 213
Coula edulis 11, 53, 80, 85, 86, 87, 88, 91, 101, 127, 131, 144, 187, 203, 210
Crassocephalum biafrae (see Senecio biafrae)
    crepioides 41, 131, 161
    rubens 41, 151, 214
Craterispermum caudatum 9
    laurinum 9, 78, 81, 128, 130, 137, 154, 167, 225

Crateva adansonii (religiosa) 15, 42, 47, 54, 107, 156, 160, 170, 189, 191
Cremaspora triflora 51, 98, 223, 237
Crescentia cujete 13, 49, 105, 108, 110, 117
Cressa cretica 249
Crinum natans 235
    zeylanicum (ornatum) 128, 159, 178, 182, 216
Crossandra guineensis 146, 239
    species 240
Crossopteryx febrifuga 97, 108, 120, 128, 130, 137, 138, 139, 141, 145, 156, 158, 179, 182, 183, 184, 187, 189, 191, 197, 202, 212, Plate 13
Crotalaria goreensis 210
    ochroleuca 42
    pallida (falcata and mucronata) 117, 207, 211, 237, 250, 254, 207, 211, 256
    retusa 121, 133, 137, 140, 154, 207, 211, 256
    verrucosa 237
    species 247
Croton lobatus 57, 160, 195, 247
    membranaceum 148
    penduliflorus 154
    tiglium 207, 211, 216
    zambesicus 80, 140, 145, 154, 237
Cryosophila warsceweizii 241
Cryptanthus species 240
Crypostegia grandiflora 10, 213, 215, 239
Ctenitis lanigera (see Triplophyllum vogelii)
    pilosissima (see T. pilosissimum)
    protensa (see T. protensum)
    pubigera (see Lastreopsis nigritiana)
    securidiformis (see T. securidiforme)
    subcoriacea (see L. vogelii)
    subsimilis (see L. subsimilis)
Ctenium canescens 82
    elegans 82
    newtonii 82
    villosum 82
Cucumeropsis edulis 39, 53, 210
Cucumis edulis 169
    melo var. agrestis 53
Cucurbita maxima 39, 108, 110, 117, 127, 134, 160, 175, 187, 191
    pepo 39, 110, 127, 134, 160, 175, 187, 191
Cucurbitaceae 39
Culcasia angolensis 216
    scandens 188
Curculigo pilosa 178, 216
Curcuma domestica 52, 129
Cuscuta australis 178, 239, 248
Cussonia arborea (barteri) 15, 140, 174, 177, 226, 228
    kirkii var. kirkii (bancoensis) 108, 110
Cuviera nigrescens 108
Cyathula prostrata 126, 129, 143, 147, 162, 182, 184
Cyclosorus afer (see Pneumatopteris afra)
    striatus 249
Cycnium camporum 223
Cylicodiscus gabunensis 7, 84, 86, 87, 91, 101, 161, 180, 189, 228
Cymbopogon citratus 20, 50, 52, 154, 164, 167, 229, 231
    giganteus var. giganteus 52, 56, 82, 89, 112, 142, 154, 160, 167, 229
    nardus 231
    schoenanthus ssp. proximus 129, 156
    schoenanthus ssp. schoenanthus 129, 156
Cynodon dactylon 20, 55, 133, 144, 148, 173, 199, 235
Cynometra ananta 7, 86
    vogelii 15, 54
Cyperus articulatus 112, 142, 143, 161, 229, 231, 249, Plate 9
    difformis 248
    digitatus ssp. auricomus var. auricomus 229
    distans 31, 248
    esculentus 30, 66, 129, 169
    haspan 50
    iria 248
    maculatus 229
    papyrus 249
    rotundus 55, 142, 165, 173, 174, 229, 243
    sphacelatus 55
    tuberosus 248
Cyrtorchis arcuata 237
    aschersonii 237
    hamata 237
    monteiroae 237
    ringens 237
Cyrtosperma senegalense 50, 147, 197, 212, Plate 13

Dacryodes klaineana 11, 98, 108
Dactyloctenium aegyptium 25, 56, 235, 248
Dalbergia heudelotii 229
    hostilis 105, 157
    saxatilis 104, 105, 135, 185, 204, 237
    sissoo 97
    species 116
Dalbergiella welwitschii 128, 134, 135, 178
Daniellia ogea 16, 93, 180, 196, 220, 227
    oliveri 12, 16, 42, 54, 83, 91, 98, 102, 107, 108, 135, 140, 143, 155, 162, 166, 167, 170, 174, 180, 193, 197, 229
    thurifera 14, 15, 16, 92, 142, 183, 229
Datura innoxia 8, 231
    metel 134, 153, 165, 174, 181, 191, 210, 220, 231
    stramonium 132, 164, 192, 210, 220
    suaveolens 237
    species 207
Daucus carota 41
Davallia chaerophylloides 235
Deightoniella torulosa 30
Deinbollia grandifolia 47, 80, 153, 168
    pinnata 11, 96, 130, 132, 135, 166
Delonix regia 15, 96, 233
Dendrocalamus strictus 42, 54, 102, 105, 111, 112
Dennettia tripetala 125
Desmodium adscendens var. adscendens 128, 132, 138, 140, 142, 145, 177, 200, 202
    gangeticum var. gangeticum 132, 137, 138, 145, 148, 157, 159, 250
    incanum (canum) 145, 149, 153, 152, 190, 200, 201
    mauritianum 223

ramosissimum 247
triflorum 55, 235, 255
velutinum 54, 133, 146, 157, 160, 166, 171, 172, 198
Desplatzia chrysochlamys 47
dewevrei 47, 224
subericarpa 15, 47, 227
Detarium microcarpum 47
senegalense 16, 47, 54, 88, 114, 127, 132, 133, 138, 146, 155, 165, 167, 169, 190, 194, 199, 203, 212, 218
Dialium dinklagei 110, 139, 142, 147, 166, 168, 179, 180
guineense 9, 11, 15, 51, 95, 102, 104, 114, 132, 137, 138, 147, 155, 169, 175, 176, 186, 189, 192, 194, Plate 10
Diaphananthe bidens 237
curvata 237
laxicalcar 237
laxiflora 237
pellucida 237
quintasii 237
rutila 237
sarcorhynchoides 237
Diaporthe batatis 29
Diastema tigris 245
Dicellandra barteri 239
Dichapetalum madagascariensis (guineense) 9, 11, 110, 167, 213, 214
pallidum 145
toxicarium 15, 154, 191, 211, 212, 214, 215
Dichrostachys cinera (glomerata) 15, 47, 54, 96, 102, 106, 111, 112, 128, 131, 134, 145, 148, 150, 160, 171, 188, 191, 197, 198, 219, 220, 247, 251, 254
Dicoma sessiliflora 247
tomentosa 247
Dicranolepis grandiflora 237
persei 150, 169, 210
Dicranopteris linearis (Gleichenia linearis) 235, 248
Dictyandra arbrescens 226
Dictyophleba leonensis 14, 47
macrophylla 13
Didymella lycopersici 35
Didymosalpinx abbeokutae 134, 238
Dieffenbachia species 240
Digitaria ciliaris 25, 55, 235, 248
debilis 25, 55, 235
exilis 23, 24, 56, 81
gayana 56, 111
horizontalis 25, 55, 235, 248
longiflora 235, 248
Diheteropogon amplectens var. amplectens 82
amplectens var. catangensis 56
hagerupii 56, 82
Dillenia indica 46
Dinophora spenneroides 47, 142
Dioclea reflexa 15, 114, 165
Dioncophyllum (see Habropetalum)
Dioscorea alata 27, 28
bulbifera 27, 28
cayenensis ssp. cayenensis 27, 28
cayenensis ssp. rotundata 27, 28
dumetorum 27, 207, 210

esculenta 27, 28
hirtiflora 28
lecardii 27
praehensilis 27
smilacifolia 28
species 27, 187, 205
Dioscoreophyllum cumminsii 12, 31
Diospyros abyssinica 80, 108
barteri 9, 11
canaliculata 11, 121, 171, 212, 215
elliotii 9
gabunensis 80, 110, 203
heudelotii 9, 11
kamerunensis 11, 80, 87, 104
mespiliformis 11, 13, 15, 54, 91, 104, 105, 107, 109, 121, 125, 141, 146, 149, 154, 167, 190, 202, 217, 218
monbutensis 170
piscatoria 216
sanza-minika 7, 11, 87, 88, 110
soubreana 11, 190
tricolor 9, 11
viridicans (kekemi) 11, 172
species 11, 101, 105, 116
Diplazium hylophylum 236
Dipterocarpus species 85
Discoglypremna caloneura 109, 110, 135
Dissotis amplexicaulis 239
entii 239
grandiflora 239
irvingiana 239
perkinsiae 239
rotundifolia 138, 142, 152, 155, 181, 185, 191, 194, 205, 239, 255, Plate 15
tubulosa (Osbeckia tubulosa) 239
Distemonanthus benthamianus 7, 78, 81, 83, 84, 85, 87, 88, 92, 93, 98, 107, 183, 203
Dodonaea viscosa 21, 54, 80, 97, 105, 157, 170, 180, 187, 193, 216, 234
Donella welwitschii (Chrysophyllum welwitschii) 114, 142
Doryopteris concolor var. kirkii (kirkii) 235
concolor var. nicklesii (nicklesii) 235
Dracaena adamii 238
arborea 228, 234, 237
bicolor 183, 196, 210, 238
camerooniana 238
elliotii 238
fragrans 10, 156, 179, 180, 188
mannii 10, 42, 154, 225, 228, 238
ovata 238
perrottetii 225
phrynioides 238
scoparia 238
smithii 238
surculosa var. capitata 238
surculosa var. surculosa 129, 224, 238
species 224, 237, 238
Drepanocarpus (see Machaerium)
Drosera indica 258
Drypetes afzelii 15, 80
chevalieri 115, 137, 147, 184, 186
floribunda 9, 47, 97, 110, 237
gilgiana 47
ivorensis 47, 108, 134, 191, 215
Duparquetia orchidacea 10, 125, 150, 238,

INDEX TO SCIENTIFIC NAMES

Plate 10
Duranta repens 207, 210, 234
Dyschoriste perrottetii 153

Echinochloa colona 25, 56, 248
    crus-pavonis 56, 82, 89
    pyramidalis 25, 50, 56, 82, 249
    stagnina 25, 56, 82, 249
    species 25
Echinops longifolius 153
Echis carinatus 219
Eclipta alba (prostrata) 177, 227, 249
Ectadiopsis oblongifolia 47, 139
Ehretia cymosa 9, 47, 141, 146, 155, 161, 178, 192
    trachyphylla 105
Eichhornia crassipes 235, 246
Ekebergia senegalensis 140, 161, 184, 186, 212
Elaeis guineensis 49, 50, 68, 79, 81, 88, 96, 98, 111, 113, 114, 116, 135, 143, 161, 169, 174, 183, 197, 202, 218, 229, Plate 4, 5
Elaeocarpus serratus 46
Elaeophorbia drupifera 133, 142, 159, 179, 182, 212, 213, 214, 216
Eleusine indica 21, 25, 55, 128, 135, 145, 148, 155, 157, 171, 174, 235, 248
    species 25
Elionurus elegans 56
    hirtifolius 56
Elodea canadensis 57, 235
Elytraria marginata 142, 200
    species 247
Emilia coccinea 41, 151, 204, 247
    sonchifolia 41, 157, 186, 247
    species 251
Enantia polycarpa 117, 121, 129, 171, 195, 212, 225
Encephalartos barteri 238
Enneastemon foliosus (see Monanthotaxis foliosa)
    vogelii 47
Ensete gilletii 113, 173, 201, 239
Entada abyssinica 54, 135, 145, 151, 154, 186, 191, 219, 221, 229
    africana 10, 15, 54, 112, 125, 136, 139, 156, 188, 202, 216
    pursaetha 66, 80, 112, 114, 116, 127, 156
Entandrophragma angolense 5, 6, 91, 93, 223
    candollei 6, 84, 93
    cylindricum 5, 6, 84, 91, 92, 93, 109, Plate 1
    utile 5, 6, 84, 86, 91, 109
    species 7, 108
Epaltes gariepina 257
Eragrostis aspera 56, 82, 115
    atrovirens 56, 115
    barteri 56, 82
    blepharostachya 56
    chalarothyrsos 56, 82
    cilianensis 25, 56, 82, 112
    ciliaris 26, 56, 80, 82, 113, 115
    cylindriflora 56, 115
    domingensis 56, 82, 115
    egregia 56, 82, 115
    gangetica 26, 56, 80, 82, 113
    namaquensis var. diplachnoides 56, 82, 115

    namaquensis var. namaquensis 56, 82, 115
    pilosa 26, 56, 80, 82
    pobeguinii 56
    scotelliana 56, 115
    squamata 56, 82, 115
    tenella 248
    species 56, 82, 115
Eragrostis tremula 26, 56, 80, 82, 113
    turgida 56, 82, 115
    welwitschii 56
Eremomastax speciosa (polysperma) 217
Eremospatha hookeri 80
    macrocarpa 81, 87, 89, 92, 111, 112, 198
Erigeron floribundus 247
Eriochloa fatmensis (nubica) 26, 55
    meyerana 56
Eriocoelum kerstingii 104
    pungens 80, 83
    racemosum 80, 83, 130, 212
Eriosema glomeratum 42, 134, 153, 177, 202, 216
    griseum 128, 192, 216
    psoraleoides 152, 162, 165, 183, 184, 188, 216
    species 129
Erwinia carotovora 35
Erythrina addisoniae 222
    fusca 114
    indica 240
    mildbraedii 130, 193, 226, 228, 238
    senegalensis 54, 114, 130, 135, 145, 168, 169, 171, 177, 181, 182, 187, 193, 199, 238, 255
    species 222
Erythrococca anomala 127, 133, 134, 149, 160, 181, 186, 195, 202, 204, 220
Erythrophleum africanum 15, 177, 192, 212, 213
    ivorense 7, 87, 88, 101, 212
    suaveolens (guineense) 7, 16, 84, 86, 87, 88, 91, 95, 127, 138, 150, 153, 162, 165, 166, 179, 180, 183, 184, 186, 191, 194, 203, 206, 207, 212, 213, 215, 218, 221, 224, 233, Plate 14
    species 121, 206
Erythroxylum coca 50, 121
    mannii 91, 116, 175
Ethulia conyzoides 126, 161, 178, 204, 249
Euadenia eminens 47, 130, 149, 153, 238
    trifoliolata 42, 129, 149, 152, 17
Eucalyptus citriodora 221
    species 97
Euclinia longiflora 47, 238
Eudorina elegans 57
Eugenia caryophyllata 52
    coronata 9, 227, 234
    jambos 13, 46, 222
    leonensis 115, 234
    malaccensis 46
    uniflora 46, 222, 234
Eulophia alta 236
    angolensis 236
    buettneri 236
    cristata 236
    cucullata 236

dilecta 236
euglossa 236
flavopurpurea 236
gracilis 236
guineensis 236
horsfallii 236
juncifolia 236
odontoglossa 236
orthoplectra 236
sordida 236
warneckeana 236
species 130, 185, 236
Eulophodium (see Oeceoclades)
Eupatorium (see Chromolaena)
Euphorbia baga 210
  balsamifera 54, 213, Plate 13
  convolvuloides 138, 170, 185, 200, 220, 221
  cyathophora (heterophylla) 247
  deightonii 212, 213
  hirta 128, 132, 135, 137, 151, 153, 162, 170, 173, 177, 194, 197, 202, 218, 247, 251
  lateriflora 159, 165, 168, 179, 182, 199, 201, 213, 234
  possonii 216
  prostrata 247
  species 77
Euphorbiaceae 13, 14
Eurychone rothschildiana 237
Evolvulus alsinoides 132, 136, 147, 165, 221, 235, 239
  nummularius 20, 235

Fadogia agrestis 130, 133, 146, 182, 188, 193, 196
  cienkowskii 42
  erythrophloea 218
Fagara (see Zanthoxylum)
Faurea speciosa 11, 79, 102
Feretia apodanthera 41, 171, 196, 198, 238
Ficus abutilifolia 240
  asperifolia 125, 136, 142, 156, 161, 173, 188, 192, 194, 195, 203, 213, 226
  barteri 47
  capensis (or sur) 47, 108, 130, 136, 140, 146, 148, 162, 170, 172, 175, 182, 186, 188, 197
  capreifolia 47
  congensis 47
  dekdekena (thonningii) 14, 113, 139, 186, 203, 217, 233
  dicranostyla 222
  elastica 14, 176, 240
  elegans 47, 176, 196
  eriobotryoides 47
  exasperata 128, 139, 142, 152, 158, 190, 212, 213, 218
  glumosa var. glaberrima 14, 42, 47, 113, 139, 222, 224, 233
  ingens var. ingens 10, 14, 47, 54, 233
  iteophylla 47, 54, 171
  leprieuri 14, 159, 234
  lutea (vogelii) 14, 47, 139, 145, 189, 233
  lyrata 233
  mucuso 49

  natalensis 133
  ottoniifolia 240
  ovata 14, 113, 233
  platyphylla 14, 16, 47, 97, 113, 218, 222, 233
  polita 47, 113
  sagittifolia 141, 150, 190
  sur (under capensis)
  sycomorus (gnaphalocarpa) 16, 42, 47, 51, 54, 137, 141, 219
  umbellata 47, 108, 240
  urceolaris 47
  vallis-choudae 42, 47, 91, 113, 136, 158, 168, 200, 218
  vogeliana 47, 109, 113, 125, 150, 168, 171, 193, 228
  species 47, 113, 156, 185, 217, 233
Fimbristylis dichotoma 249
  species 112
Fittonia species 240
Flabellaria paniculata 80, 201
Flacourtia flavescens 11, 138, 147, 180, 234, Plate 10
  indica 46
Fleurya (see Laportea)
Fomes species 27
Fragillaria species 57
Fuirena umbellata 50, 55
Funtumia africana 14, 75, 107, 115, 148, 164, 178, 213
  elastica 13, 75, 107, 115, 164, 176
  species 98
Fusarium oxysporum f. cubense 29, 45

Gaertnera cooperi 191
  paniculata 238
Galerina species 207, 211
Galphinia (see Thryallis)
Ganoderma lucidum 69
Garcinia afzelii 9, 46, 129, 146
  epunctata 9
  gnetoides 177
  kola 9, 15, 46, 73, 128, 135, 138, 140, 162, 179, 183, 194, 198, 203, 218, 222
  mangostana 46, 222
  smeathmannii (polyantha) 46, 203
  xanthochymus 46
  species 9, 159
Gardenia erubescens 51, 105, 110, 129, 197, 226
  nitida 187
  ternifolia 104, 132, 156, 166, 171, 172, 180, 181, 198, 218, 228, 238
  vogelii 223, 238
Geophila species 258
Gigartina acicularis 58
Gilbertiodendron limba 84, 155, 195
Gladiolus aequinoctialis var. aequinoctialis (Acidanthera aequinoctialis) 239
  daleni (psittacinus) 31, 147, 153, 178, 196, 220, 239
  gregarius 31, 147, 153, 178, 196, 220, 239
  klattianus 31, 147, 153, 178, 196, 220, 239
  unguiculatus 31, 147, 153, 178, 196, 220, 239

INDEX TO SCIENTIFIC NAMES 299

Gleichenia (see Dicranopteris)
Glinus oppositifolius 249
Gliricidia sepium 21, 54, 214
Globimetula braunii 42, 67
Gloriosa simplex 165, 175, 207, 210, 213, 239
    superba 125, 165, 175, 207, 210, 213, 239
Gluema ivorensis 14
Glycine max 256
Glyphaea brevis 106, 129, 131, 138, 146, 168, 176, 178, 185, 188, 189, 195, 201
Gmelina arborea 8, 54, 95, 97, 245, 250
Gnidia kraussiana (Lasiosiphon kraussianus) 210, 219
Gomphrena celosioides 164, 235, 247
Gongronema latifolium 9, 129, 140, 179, 189, 201
Gossypium arboreum 53, 69, 125, 133, 146, 162, 173, 191, 202, 214, 224, 225
    barbadense 69
    herbaceum 69, 174
    hirsutum 69
    species 69
Gouania longipetala 130, 152, 154, 161, 163, 187, 197, 202, 219, 228
Gracilaria dentata 58
Grangea maderaspatana 149
Graphorkis lurida 237
Graptophyllum pictus 240
Grateloupia filicina 58
Greenwayodendron oliveri (Polyalthia oliveri) 80, 105, 128, 188
Grewia barteri 10, 47
    bicolor 10, 47, 166, 175, 228
    carpinifolia 11, 42, 47, 96, 111, 114, 129, 158, 165, 166, 169, 173, 247
    carpinifolia var. rowlandii 47
    carpinifolia var. hiernia 47
    mollis 10, 47, 49, 105, 106, 115, 140, 155, 185, 195, 203
    venusta 47, 105
    villosa 10, 47, 105, 198
    species 47
Griffonia simplicifolia 9, 54, 96, 111, 124, 129, 130, 133, 146, 165, 177, 204, 226, Plate 11
Guaiacum officinale 240
Guarea cedrata 6, 85, 91, 93
    thompsonii 6, 86, 91
Guibourtia copallifera 16, 195, 202
    ehie 6, 16
Guiera senegalensis 8, 54, 131, 137, 138, 146, 147, 151, 158, 159, 160, 163, 170, 172, 181, 188, 191, 203
Gymnema sylvestre 12, 125, 144, 218, 219
Gyanandropsis gynandra (see Cleome gynandra)
    pentaphylla 160
    speciosa 141
Gynerium saccharum 79
Gynura miniata 214
Gyromitra species 208

Habenaria buettneriana 236
    buntingii 236
    filicornis 236
    gabonensis 236
    genuflexa 236
    huillensis 236
    macrandra 236
    procera 236
    zambesina 236
    species 130, 185
Habropetalum dawei (Dioncophyllum dawei) 208, 216
Hackelochloa granularis 55
Haemanthus (see Scadoxus)
Haematostaphis barteri 15, 47, 185
Haematoxylon campechianum 13, 224, 234
Halopegia azurea 81
Hannoa klaineana 83, 84, 140, 156
    undulata (see Quassia undulata)
Harrisonia abyssinica 108, 129, 146, 199
Harungana madagascariensis 10, 47, 80, 83, 127, 131, 132, 138, 143, 144, 150, 151, 158, 163, 167, 168, 169, 171, 174, 179, 183, 184, 188, 190, 192, 194, 197, 213, 222, 225, 257
Haumaniastrum lilacinum 41, 160, 229
Hedranthera barteri 128, 148, 194, 198, 237
Hedyotis corymbosa (Oldenlandia corymbosa) 129, 147, 247
Heeria (see Ozoroa)
Heinsia crinita 42, 47, 105, 143, 161, 163, 164, 181
Heisteria parvifolia 47, 98, 105, 106, 142, 182, 188, 193, 238
Heliconia species 240
Helicotylenchus multinctus 29, 32, 39
Helictonema velutina (Hippocratea velutina) 10, 156, 160
Heliotropium indicum 124, 125, 127, 152, 154, 249
    ovalifolium 150, 198, 210
    strigosum 219, 221
    subulatum 204
Heritiera utilis (Tarrietia utilis) 6, 49, 81, 85, 86, 87, 91, 109, 130, 135, 222
Hetaeria occidentalis 236
Heteropogon contortus 56, 82, 113, 115, 248
Heteropteris leona 155, 160, 165
Hevea brasiliensis 13, 74
Hewittia sublobata 239
Hexalobus crispiflorus 47, 104, 109
    monopetalus var. monopetalus 47, 79, 105, 139, 147
    monopetalus var. parvifolius 47
Hibiscus abelmoschus (see Abelmoschus moschatus)
    articulatus 239
    articulatus var. glabrescens 239
    asper 198
    cannabinus 10, 71, 239
    congestiflorus 239
    esculentus (see Abelmoschus esculentus)
    gourmania 239
    lunariifolius 42, 57, 127, 159, 239
    manihot 239
    micranthus 115
    penduliformis 239
    rosa-sinensis 9, 42
    rostellatus 42, 116, 197
    subdariffa 10, 51, 71, 179, 184, 198, 218

squamosus 239
surattensis 239
tiliaceus 42, 54, 104, 239
vitifolius 239
species 239
Hildegardia barteri 10, 15, 81, 110, 112, 116, 233, 258, Plate 15
Hilleria latifolia 136, 258
Hippocratea africana (see Loeseneriella africana)
   apocynoides ssp. guineensis (guineensis) 10
   indica (see Reissantia indica)
   rowlandii (see L. rowlandii)
   velutina (see Helictonema velutina)
   welwitschii (see Simirestis welwitschii)
Holarrhena floribunda 14, 107, 121, 144, 157, 167, 183, 190, 213, 240
Holmskioldia sanguinea 240
Holoptelea grandis 7, 84, 110, 128, 179, 181
Homalium letestui 80, 240
Hoslundia opposita 47, 54, 129, 138, 139, 141, 148, 152, 166, 267, 180, 183, 186, 188, 197, 200, 203, 220, 229, 231
Hugonia planchonii 142, 238
Hunteria eburnea 47, 80, 110, 125, 131, 157, 171, 211, 213
   elliotii 104
   umbellata 80, 105, 126, 174
Hura crepitans 133, 179, 208, 216, 233
Hydrangea species 240
Hydrolea glabra 249
Hygrophila auriculata 50, 143, 154, 188
   spinosa 50
Hymenocardia acida 11, 47, 105, 129, 135, 139, 140, 146, 148, 152, 155, 156, 162, 181, 193, 195, 203, 221, 222, 224
Hymenodictyon floribundum 14, 106
Hymenostegia afzelii 97, 104, 142, 176, 240
   gracilipes 240
Hyophorbe vershaffeltii 241
Hyparrhenia cyanescens 56, 82, 89
   familiaris 56, 82
   glabriuscula 56, 82
   involucrata var. breviseta 56, 82
   involucrata var. involucrata 56, 89
   mutica 56
   nyassae 56, 82
   rudis 56, 82, 89
   rufa 56, 81, 82, 89
   smithiana var. major 82, 89
   subplumosa 56, 82
   welwitschii 56, 82, 89
Hyperthelia dissoluta 56, 82, 113
Hyphaene thebaica 49, 50, 54, 79, 81, 88, 111, 113, 114, 115, 160, 164, 226
Hypnea flagelliformis 58
   musciformis 58
Hypolytrum purpurascens 112
Hypselodelphis poggeana 116
   violacea 116
Hyptis pectinata 41, 137, 154, 168
   spicigera 138, 161, 165
   suaveolens 50, 52, 55, 134, 139, 154, 160, 188, 229, 247

Icacina olivaeformis (senegalensis) 31, 49, 164, 247
Imperata cylindrica var. africana 82, 113, 115, 148, 155, 190, 244, 250, 251
   species 245
Indigofera arrecta 223, 255
   dendroides 204
   hirsuta 200, 204, 223, 255
   kerstingii 235
   macrophylla 187
   paniculata 249
   pulchra 81
   simplicifolia 115, 180, 212
   spicata 20, 54, 223, 235
   suffruticosa 144, 223
   tinctoria 223, 255
   species 209, 223
Ipomoea acuminata 239
   aitonii 179, 187, 239
   alba 41, 219
   aquatica 31, 41, 235, 249
   argentaurata 239
   asarifolia 20, 80, 154, 159, 181, 199, 226, 239, 247
   batatas 29, 41, 165, 224
   cairica 239
   coptica 138, 148, 166, 239
   eriocarpa 41, 239
   hederifolia 239
   heterotricha 239
   involucrata 154, 197, 239, 255
   mauritiana 55, 170, 178, 239, 247
   nil 179, 211, 239
   ochracea 239
   pes-caprae ssp. brasiliensis 20, 149, 181, 231, 239, 247
   quamoclit 239
   rubens 239
   stolonifera 20, 239
   triloba 239
   tuba 239
   turbinata (muricata) 66, 179
   verbascoidea 239
   species 234, 239
Irenopsis aciculosa 39
Irvingia gabonensis 47, 53, 54, 80, 86, 110, 145, 187, 231
   species 145
Ischaemum longiflora 248
Isoberlinia doka 80, 228
   tomentosa (dalzielii) 42, 80, 203, 218
Isolona campanulata 130, 135, 180
   species 187
Isonema smeathmannii 42
Ixora brachypoda 47
   coccinea 239

Jacaranda mimosifolia 240
Jacobinia carnea 240
Jacquemontia ovalifolia 239
   tamnifolia 41, 160
Jardinea congoensis 56
Jasminum dichotomum 47, 195, 238
   multiflorum 239
   sambac 50, 229, 239
Jatropha curcas 8, 31, 42, 49, 191, 197, 140

INDEX TO SCIENTIFIC NAMES   301

143, 146, 148, 149, 159, 162, 164, 167, 168,
176, 179, 182, 183, 184, 190, 196, 201, 202,
211, 215, 216, 222, 223, 224, 225, 226, 228,
231, 234, Plate 11
   gossypiifolia 150, 179, 186, 194, 211, 220,
      234, 247
   multifida 31, 179, 211, 239
   podagrica 239
Jaundea pinnata 12, 213
   pubescens 106
Jussiaea suffruticosa (see Ludwigia
   octovalvis)
Justicia extensa 42, 217
   flava 144, 154, 205, 247
   insularis 41
   laxa 217

Kaempferia aethiopica 52
   species 240
Kalanchoe integra var. crenata 132, 149, 176
Keetia hispida (Canthium hispidum) 111,
   221, 237
Kerstingiella geocarpa 31
Khaya anthotheca 5, 6, 85, 91, 142, 157
   grandifoliola 6
   ivorensis 5, 6, 85, 91, 93, 105, 141, 146,
      172, 180
   senegalensis 15, 54, 85, 92, 126, 127,
      131, 134, 150, 157, 158, 173, 174, 179,
      185, 189, 195, 198, 203, 206, 208, 212,
      217, 222, 223, 233, Plate 11
   species 7, 108, 206, 223
Kigelia acutifolia 195, 203
   africana 107, 110, 127, 133, 145, 170,
      172, 176, 179, 180, 189, 197, 198, 202,
      211, 220, 226
Klainedoxa gabonensis var. oblongifolia 53,
   84, 86, 87, 88, 130
Kohautia senegalense 115, 127
Kohleria species 240
Kosteletzkya stellata 10
Kylinga erecta 52, 55, 229
   pumila 52, 55, 229, 248
   squamulata 52, 55, 229, 248
   tenuifolia 52, 55, 229
   umbellata 55

Lablab niger 55
Laccosperma opacum (Ancistrophyllum
   opacum) 80, 81, 92, 108, 111, 112
      secundiflorum (A. secundiflorum) 41,
         42, 80, 81, 92, 108, 111, 112, 128, 245
Lactuca (see Launaea)
Lagarosiphon hydrilloides 57, 235
Lagenaria breviflora (Adenopus
   breviflorus) 166, 178, 216, 222
   siceraria 39, 53, 110, 116, 117, Plate 8
Lagerstroemia indica 240
   speciosa 240
Laggera heudelotii 247
Laguncularia racemosa 54, 97, 222
Landolphia amoena (see Ancylobotrys
   amoena)
      calabarica 11
      dulcis 13, 46, 170
      heudelotii 11

hirsuta 13, 46
landolphioides 46
owariensis 11, 13, 51, 75, 129, 213
Lankesteria brevior 200
   elegans
Lannea acida 15, 42, 47, 51, 84, 106, 107, 179,
   182, 184, 189, 196, 202
   kerstingii 10, 15, 16, 47, 128, 131, 158,
      171, 195, 203, 224
   microcarpa 10, 15, 42, 47, 202
   nigritana var. nigritana 179, 183, 203
   velutina 47, 84, 158, 171, 195, 224
   welwitschii 47, 98, 223, 224
Lantana camara 13, 121, 136, 137, 140, 144,
   149, 152, 154, 167, 177, 188, 202, 208, 213,
   229, 234
   trifolia 47, 52, 137, 152, 177, 188, 195
   species 243, 245
Laportea aestuans (Fleurya aestuans) 40,
   182, 187
Lasianthera africana 115, 188
Lasiodiscus mannii var. chevalieri
   (chevalieri) 9
Lasiosiphon (see Gnidia)
Lastreopsis nigritiana (Ctenitis
   pubigera) 236
      subsimilis (C. subsimilis) 236
      vogelii (C. subcoriacea) 236
Launaea capensis (Lactuca capensis) 247
   sativa (Lactuca sativa) 41
   taraxacifolia (Lactuca taraxacifolia) 40,
      50, 170, 201, 205, 247, 251
Lawsonia inermis 139, 167, 171, 173, 183,
   187, 190, 222, 224, 229, 234
Lecaniodiscus cupanioides 11, 12, 47, 83, 88,
   97, 104, 105, 108, 156, 179, 187, 201, 229,
   233
Lecythia zabucajo 46
Leea guineensis 47, 139, 140, 144, 178, 181,
   189, 191, 193, 196, 200, 238, 248
Leersia drepanotrix 56
   hexandra 55, 249
Leguminosae 256
Lemna perpusilla (paucicostata) 57, 249
Leonotis nepetifolia var. africana 50, 137,
   154, 158, 160, 165, 183, 198
      nepetifolia var. nepetifolia 50, 137, 154,
         158, 160, 165, 198
Lepidagathis heudelotiana 153, 178
Lepidium sativum 159, 216
Lepiota species 208, 211
Lepisanthes senegalensis (Aphania
   senegalensis) 46, 98, 202, 211, 214, 228
Lepistemon owariensis 239
Leptadenia hastata 42, 153
Leptaurus daphnoides 80
Leptobyrsa decora 245
Leptochloa caerulescens 249
Leptothrium senegalense 26
Leucaena leucocephala (glauca) 8, 42, 54,
   96, 112, 114, 223, 234, 243, 248, 251, 256
Leucas deflexa 229
   martinicensis 138, 154, 158, 165
Leutothrium senegalense 56
Licania elaeosperma 231
Limnophyton obtusifolium 50

Lindernia diffusa var. diffusa 219, 247, 249
Linociera (see Chionanthus)
Liparis nervosa (guineensis & rufina) 236
Lippia multiflora 42, 49, 50, 138, 139, 142, 144, 149, 151, 152, 154, 158, 167, 168, 169, 170, 176, 177, 185, 188, 189, 200, 229
Listrostachys pertusa 237
    species 130, 185
Loa loa 153
Loeseneriella africana (Hippocratea africana) 10
    rowlandii (H. rowlandii) 10
    species 87
Lonchitis currori 248
    reducta 248
Lonchocarpus cyanescens 131, 151, 171, 183, 195, 202, 204, 223, Plate 14
    griffonianus 240
    laxiflorus 54, 127, 167, 189, 195, 223, 240
    sericeus 10, 86, 97, 104, 133, 140, 165, 179, 183, 208, 211, 225, 233, Plate 11
    species 209, 225
Lophira alata 6, 53, 78, 84, 87, 88, 108, 156, 255
    lanceolata 53, 133, 135, 139, 146, 156, 158, 160, 168, 171, 185, 192, 202, 240
Loudetia annua 56
    arundinacea 56, 82
    flavida 82
    hordeiformis 56
    kagerensis 56
    phragmitoides 106
    simplex 56, 82
    togoensis 112
Lovoa trichilioides 6, 78, 84, 85, 91, 104
Ludwigia erecta 249
    hyssopifolia 249
    leptocarpa 249
    octovalvis (Jussiaea suffruticosa) 249
    stolonifera 249
Luffa acutangula 148, 150, 179, 201, 216
    cylindrica 39, 178, 183, 191, 208, 216
Lycopersicon esculentum (lycopersicum) 35, 149, 200
Lycopodiella cernua (Lycopedium cernuum) 235, 248
Lycopodium (see Lycopodiella)
Lygodium smithianum 235
Lyngbya limnetica 58

Macaranga barteri 197
    heterophylla 15, 47, 49, 141, 197, 219
    heudelotii 147
    hurifolia 80, 98, 158, 194
Machaerium lunatum (Drepanocarpus lunatus) 47, 130, 132, 162, 166, 171, 177, 188, 192, 200, 212, 214
Macrophomina phaseolina 29
Macroptilium atropurpureus (Phaseolus atropurpureus) 32, 250, 256
    lathyroides (P. lathyroides) 32, 250
Macrosphyra longistyla 155, 229, 238
Maerua angolensis 42, 54, 126, 181, 203, 211, 214
    crassifolia 9, 42, 47, 105, 188
Maesa lanceolata 79, 149, 210, 223, 234

Maesobotrya barteri var. sparsiflora 47, 126, 130, 145, 168, 197
Maesopsis eminii 47, 53, 88, 91, 107, 148, 149, 151, 168, 174, 180, 200
Majidea fosteri 114, 240
Malacantha alnifolia 78, 80, 88, 105, 110, 152
Malaxis maclaudii 236
Mallotus oppositifolius 9, 96, 106, 108, 121, 127, 130, 147, 161, 172, 190, 195, 199, 202, Plate 11
Malpighia glabra 46, 234
Malvastrum coromandelianum 248
Mammea africana 7, 15, 47, 53, 86, 88, 92, 143, 167, 180, 184, 194, 198, 217, 229
Mangenotia eburnea 143
Mangifera indica 13, 15, 44, 146, 147, 155, 186, 192, 196, 199, 222, 225, 233
Manihot dichotoma 14
    esculenta 26, 42, 54, 138, 153, 155, 160, 165, 168, 170, 183, 187, 195, 201, 208, 210, Plate 2
    glaziovii 13, 14, 42
Manilkara multinervis ssp. lacera (lacera) 11, 92, 97, 108, 195
    obovata (multinervis) 16, 87, 98, 106, 144
Manniella gustavii 236
Manniophyton fulvum 49, 111, 135, 183, 188, 196
Manotes longiflora 152
Mansonia altissima 6, 78, 81, 85, 91, 129, 208, 212
Marantaceae 80, 81, 112
Maranthes glabra (Parinari glabra) 12, 87, 101
    polyandra (P. polyandra) 48, 49, 101, 154, 157, 222
    robusta (P. robusta) 80
Marantochloa cuspidata 81
    filipes 134
    leucantha 81, 112, 134, 189
    mannii 81
    purpurea 81, 112, 134, 189
    ramosissima 112, 189
    species 89
Marattia fraxinea 235, 248
Mareya micrantha 125, 127, 142, 143, 149, 157, 159, 171, 178, 181, 187, 206, 210, 218, 219, 220
Margaritaria discoidea (Phyllanthus discoideus) 9, 48, 49, 81, 129, 134, 143, 146, 149, 152, 156, 179, 189, 191, 201, 222
Mariscus alternifolius 82, 116, 183, 248
    flabelliformis 248
    sumatrensis (alternifolius) 31, 143
Markhamia lutea 109, 132, 154, 167, 184, 202
    tomentosa 132, 167, 184, 202, 222
Marsilea species 235
Massularia acuminata 80, 106, 129, 146, 153, 172, 217
Maytenus buchananii 121, 123, 136
    senegalensis 47, 49, 54, 102, 121, 128, 134, 139, 140, 146, 148, 153, 158, 169, 175, 188, 193, 195, 196, 203, 212
    species 136
Megaphrynium macrostachyum 81

INDEX TO SCIENTIFIC NAMES

Melanthera elliptica 55, 247
    scandens 55, 160, 247, 251
Melastomastrum capitatum 239
    theifolium 239
Melia azedarach 13, 15, 54, 113, 129, 147, 154, 165, 173, 180, 198, 210, 213, 233
Melicoccus bijugatus 46
Melinis effusa 56
    minutiflora 55, 255
    tenuissima 56
Melochia species 248
Meloidogyne javanica 45
    species 33, 35
Melosira granulata 57
Memecylon afzelii 80, 192, 200
    blakeoides (see Spathandra blakeoides)
Merremia aegyptiaca 80, 187
    dissecta 239
    hederacea 239
    kentrocaulos 239
    pinnata 239
    pterygocaulis 239
    tridentata ssp. angustifolia 197, 221, 239
    tuberosa 239
    umbellata ssp. umbellata 20, 239
Mezoneuron benthamianus 146, 199
Michelia champaca 233
Microchloa indica 56
    kunthii 56
Micrococca mercurialis 247
Microcoelia caespitosa 237
    dahomeensis 237
    macrorrhynchia 237
Microcystis aeruginosa 57
Microdesmis puberula 9, 42, 47, 104, 108, 127, 130, 135, 146, 161, 164, 167, 187, 194, 196, 199, 218
Microglossa afzelii 135, 138, 139, 148, 149, 155, 175, 181, 186, 194, 197, 218
    pyrifolia 126, 129, 139, 142, 149, 151, 154, 161, 164, 168, 169, 170, 172, 186, 197, 212, 213, 214
Microlepia speluncae 235, 248
Mikania chevalieri (cordata var. chevalieri) 42, 54, 141, 143, 152, 219, 220, 247, 255, Plate 16
Milicia excelsa (Chlorophora excelsa) 6, 14, 41, 78, 81, 83, 84, 85, 86, 88, 92, 107, 108, 121, 141, 143, 154, 162, 163, 170, 172, 179, 187, 189, 193, 194, 200
    regia (C. regia) 6, 47, 92, Plate 1
Millettia barteri 116, 184, 186, 217
    irvinei 115
    thonningii 54, 97, 105, 128, 133, 145, 179, 215, 232, 233
    zechiana 136
    species 97
Millingtonia hortensis 54, 233
Mimosa pigra 49, 139, 153, 155, 219, 249
    pudica 146, 150, 159, 210, 254
Mimusops elengi 108, 223
Mirabilis jalapa 149, 224
Mitracarpus villosus (scaber) 218, 247
Mitragyna ciliata 6, 91, 116, 137, 145, 156, 187, 197
    inermis 13, 54, 88, 92, 110, 121, 126, 134, 144, 148, 157, 158, 171, 178, 181, 197, 225, 249, 251
    stipulosa 6, 81, 83, 84, 88, 89, 92, 93, 108, 121, 127, 140, 148, 156, 202, 217
Mollugo nudicaulis 127, 141, 247
Momordica angustisepala 115
    balsamina 205
    charantia 39, 125, 131, 133, 136, 144, 150, 154, 163, 177, 187, 192, 200, 201, 203, 204, 258
    cissoides 154, 177
    foetida 141, 188, 258
Monanthotaxis foliosa (Enneastemon foliosus) 42, 47
Monechma ciliatum 138
Mongeotia scalaris 57
Monocymbium ceresiiforme 56, 81, 82, 253
Monodora brevipes 52, 179, 199, 240
    crispata 240
    myristica 52, 104, 114, 125, 159, 162, 164, 179, 192, 193, 203, 229, 240
    tenuifolia 47, 52, 80, 104, 144, 240
Mononychellus tanajua 27
Monotes kerstingii 154
Monstera species 240
Moraceae 13, 14
Morelia senegalensis 47, 105, 110, 112
Morinda lucida 80, 88, 97, 126, 133, 137, 139, 146, 154, 156, 160, 167, 173, 176, 196, 217, 224, 225, Plate 11
    morindoides 51, 121, 128, 139, 143, 156, 166, 167, 177, 178, 195, 212
Moringa oleifera 13, 15, 42, 47, 52, 53, 54, 121, 141, 147, 150, 160, 168, 184, 191, 196, 222, 228
Morus mesozygia 7, 11, 14, 47, 83, 92, 98, 108, 132, 172, 182, 198, 233
Motandra guineensis 152, 193
Mucuna flagellipes 182, 223, 226
    pruriens var. pruriens 128, 213, 226
    sloanei 226
    species 34, 250
Mundulea sericea 121, 208, 214, 217
Murraya paniculata 240
Musa paradisiaca 30, 76, 136, 147, 149, 164, 168, 173, 175, 186, 187, 195, 202, Plate 4
    paradisiaca var. sapientum 44, 132, 144
Musanga cecropioides 12, 47, 50, 51, 81, 88, 97, 108, 109, 111, 137, 166, 170, 172, 173, 189, 192, 204, 228, 254, 257
Mussaenda afzelii 8, 152, 197, 222, 238
    chippii 238
    elegans 47, 149, 150, 238
    erythrophylla 131, 142, 174, 238, Plate 15
Myrianthus arboreus 11, 12, 42, 88, 134, 137, 144, 145, 161, 186, 191, 194, 228, 254
    libericus 42, 48
    serratus 48, 110
Myristica fragrans 52, 136, 151

Naja nigricollis 153, 219
Napoleona (see Napoleonaea)
Napoleonaea leonensis (Napoleona leonensis) 9, 11, 98, 132, 142
    vogelii (Napoleona vogelii) 9, 105, 108, 156, 210

Nauclea diderrichii 6, 48, 83, 84, 86, 87, 88, 91, 107, 108, 127, 146, 167, 173, 189, 191, 197
  latifolia 16, 48, 125, 131, 139, 141, 145, 148, 150, 155, 158, 161, 173, 177, 178, 180, 188, 193, 194, 199, 202, 212, 220, 225
  popeguinii 48, 108
  vanderguchtii 109
  xanthoxylon 109
  species 168
Necepsia afzelii 11, 108, 125, 142, 170
Nelsonia canescens 50, 55, 144, 155, 169, 247, 258
Nelumbo nucifera 114, 235
Neocarya macrophylla (Parinari macrophylla) 48
Neodypsis decaryii 241
Neostachyanthus (see Stachyanthus)
Neostenanthera hamata 127
Nephelium lappaceum 46
Nephrolepis biserrata 248
  davallioides 235
Neptunia oleracea 235, 249
Nerium oleander 208, 213, 239
Nervilia adolphii 236
  fuerstenbergiana 236
  kotschyi 236
  reniformis 236
  umbrosa 236
Nesogordonia papaverifera 6, 84, 85, 86, 87, 91, 104, 105, 108, 257
Neuropeltis acuminata 10, 42
Newbouldia laevis 86, 98, 128, 129, 131, 137, 140, 146, 149, 152, 155, 161, 162, 163, 164, 169, 170, 173, 174, 176, 179, 181, 184, 186, 190, 193, 198, 203, 240
Nicolaia elatior (Phaeomaria magnifica) 240
Nicotiana rustica 9, 73, 185, 193, 212, 217, 230
  tabacum 9, 73, 122, 135, 165, 185, 193, 212, 217, 220, 230
Nostoc 257
Nymphaea lotus 31, 147, 152, 200, 235, 249
  masculata 235
  micrantha 235, 249
  species 198

Ochna afzelii 48, 104
  multiflora 98, 238
  ovata 238
  rhizomatosa 238
  schweinfurthiana 104, 178, 238
  staudtii (kibbiensis) 238
  species 238
Ochthocosmus (see Phyllocosmus)
Ocimum americanum 220
  basilicum 52, 136, 144
  canum 52
  gratissimum 131, 137, 139, 141, 144, 145, 152, 154, 166, 172, 176, 177, 182, 184, 188, 219, 231
  species 50, 52, 144, 156, 178
Octoknema borealis 80, 98, 157
Octotoma scabripennis 245
Odontonema strictum 240

Oeceoclades latifolia (Eulophidium latifolium) 236
  maculata (E. maculatum) 236
  saundersiana (E. saundersianum) 236
Oedogonium species 57
Oidium species 33
Okoubaka aubrevillei 172, 199
Olax gambecola 9
  subscorpioidea 48, 115, 133, 156, 159, 167, 189, 199, 219, 228
Oldenlandia affinis ssp. fugax 139, 163
  corymbosa (see Hedyotis corymbosa)
Olyra latifolia 106, 114, 116, 258
Omphalocarpum ahia 14, 16, 114, 179, 180, 189
  elatum 105, 107, 109, 114, 203, 204
  procerum 48, 113, 114, 204
Oncinotis glabrata 106, 238
  gracilis 238
  nitida 106, 238
Oncoba brachyanthera 114, 238
  spinosa 48, 114, 117, 133, 146, 156, 171, 199, 202, 238
Ongokea gore 48, 51, 87, 179, 190
Operculina macrocarpa 114, 239
Ophiobotrys zenkeri 109
Opilia celtidifolia 128, 148, 150, 178, 185, Plate 12
Oplismenus burmannii 56
  hirtellus 56
Opuntia species 15, 48, 51, 190, 201, 227
Orchidaceae 130
Oreodoxa (see Roystonia)
Orthopichonia barteri 13
Oryctes monoceros 68
Oryza barthii 26, 56, 82, 248
  glaberrima 23, 24, 26
  longistaminata 26, 56, 82, 248
  punctata 56, 82
  sativa 23, 24, 81, 96, 115
Osbeckia tubulosa (see Dissotis tubulosa)
Oscillatoria curviceps 58
  pimosa 58
  princeps 57
  tenuis 58
Ostryderris leucobotrya (see Aganope leucobotrya)
  stuhlmannii 87, 133, 196, 203, 211
Ostryocarpus riparius 10, 216
Ottelia ulvifolia 57
Ouratea affinis 42, 238
  calophylla 42, 238
  flava (see Campylospermum flavum)
  glaberrima (see C. glaberrimum)
  myrioneura (see Rhabdophyllum affine ssp. myrioneurum)
  reticulata (see C. reticulatum)
  schoenleiniana (see C. schoenleinianum)
  sulcata (see C. sulcatum)
  vogelii (see C. vogelii)
  species 238
Oxalis corniculata 208, 213, 235, 247
  species 240
Oxyanthus speciosus 9, 98, 155, 229, 238
  subpunctatus 238

INDEX TO SCIENTIFIC NAMES 305

tenuis 9
tubiflorus 48, 172, 182, 238
unilocularis 98, 228
Oxycaryum cubense (Scirpus cubensis) 246, 249
Oxytenanthera abyssinica 26, 41, 54, 79, 83, 84, 88, 92, 102, 105, 106, 110, 111, 112, 116, 234
Ozoroa reticulata (Heeria insignis and H. pulcherrima) 102, 127, 158

Pachycarpus lineolatus 31, 131, 158, 238
Pachypodanthium staudtii 80, 89, 137, 158, 164, 194, 212, 229
Pachystela brevipes 11, 48, 109
Palisota barteri 239
    hirsuta 126, 129, 139, 140, 149, 169, 188, 193, 196, 216, 231
    species 212
Panax species 240
Pancratium hirtum 239
    trianthum 210, 239
Panda oleosa 14, 49, 53, 195
Pandanus abbiwii 112, 116, 145, Plate 8
    utilis 240
    veitchii 112, 145, 151, 240
Panicum anabaptistum 56, 82, 113, 115
    baumannii 56
    brevifolium 56
    comorense 56
    congoense 56, 115
    dinklagei 56, 82, 115
    fluviicola 26, 115
    griffonii 56
    hochstetteri 56
    laetum 26, 56
    laxum 56
    lindleyanum 56
    maximum 56, 82, 115, 156, 160, 243, 248
    pansum 26, 56, 115
    parvifolium 56, 235
    paucinode 56, 115
    phragmitoides 56, 82, 113, 115
    porphyrrhizos 56, 115
    praealtum 56
    pubiglume 56
    repens 21, 56, 248
    subalbidum 26, 56, 106, 112, 116
    trichoides 56
    turgidum 26, 56, 82, 113
    species 25, 56
Papaver somniferum 119
Paramacrolobium coeruleum 114
Pararistolochia goldieana 238
Parinari congensis 48, 80, 170, 179
    curatellifolia 11, 79, 102, 109, 154, 157, 201, 231
    excelsa 12, 51, 92, 101, 147, 203, 204, 222, 226
    glabra (see Maranthes glabra)
    macrophylla (see Neocarya macrophylla)
    polyandra (see M. polyandra)
    robusta (see M. robusta)
    species 48
Parkia bicolor 48, 51, 97, 152, 203

    biglobosa 48, 52, 66, 122, 148, 157, 172, 182, 187, 192, 195, 203, 216, 222, 223, 226, 228
    clappertoniana 42, 48, 51, 52, 54, 104, 109, 115, 116, 160, 163, 172, 216, 222, 228, 233, 254
    species 47, 48, 49, 52
Parkinsonia aculeata 102, 234
Parogus borbonicus 244
Paropsia adenostegia (Androsiphonia adenestegia) 9, 165
Parquetina nigrescens 13, 122, 126, 133, 143, 147, 148, 153, 167, 174, 179, 182, 183, 189, 197, 201, 213, 219, 238
Paspalidium geminatum 56
Paspalum conjugatum 56, 235, 248, 258
    dilatatum 57
    polystachyum 82
    scrobiculatum (orbiculare) 26, 56, 248, 249
    vaginatum 56, 235, 248
    species 25
Passiflora edulis 46
    glabra (foetida) 20, 173, 201, 247, 255
    quadrangularis 46
    species 175
Paulinia acuminata 246
Paullinia pinnata 9, 15, 48, 122, 126, 129, 130, 132, 134, 139, 142, 145, 146, 148, 152, 157, 158, 161, 163, 168, 172, 174, 181, 182, 187, 190, 192, 193, 196, 203, 216, 222, 223, 226, 228, Plate 14
Pausinystalia lane-poolei 80, 167, 204
    yohimbe 122
Pavetta corymbosa 66
    crassipes 14, 42, 128, 160
Pavonia urens var. glabrescens 239
Peddiea fischeri 48
Pediastrum duplex 57
Pellegriniodendron diphyllum 16, 83
Pellionea species 240
Peltophorum pterocarpum 54, 233
Penianthus zenkeri 9, 129, 134, 142, 192, 199, 202
Penicillium digitatum 44
    species 29
Pennisetum americanum 23, 25, 50, 82, 89, 113, Plate 3
    hordeoides 57
    laxior (Beckeropsis laxior) 56
    pedicellatum 57, 81, 82, 113
    polystachyon 55
    purpureum 57, 79, 82, 89, 147
    subangustum 57, 190, 248
    unisetum (B. uniseta) 26, 56, 82, 89, 106, 116
    species 25
Pentaclethra macrophylla 12, 49, 53, 88, 101, 110, 114, 126, 140, 145, 167, 170, 171, 203, 208, 212, 217, 226, 233
Pentadesma butyraceum (butyracea) 9, 53, 69, 88, 110, 128, 144, 165, 231
Pentodon pentandrus 41, 152, 181, 249
Peperomia species 240
Peradinium cinctum 58
Pergularia daemia 122, 150, 170, 173, 174, 247

Pericopsis elata (Afrormosia elata) 5, 6, 84, 86, 91, 92, 93, 222
    laxiflora (A. laxiflora) 11, 79, 81, 104, 134, 143, 146, 152, 156, 161, 166, 168, 172, 174, 176, 180, 191, 199, 213, 214, 219
Perotis hildebrandtii 57, 235
    patens 57
Persea americana 13, 45, 164, 222
Petersianthus macrocarpus (Combretodendron macrocarpum) 7, 85, 87, 97, 105, 108, 135, 172, 197, 199
Petrea volubilis 240
Phaeomeria (see Nicolsia)
Phaseolus atropurpureus (see Macroptilium atropurpureus)
    aureus 32
    lathyroides (see M. lathyroides)
    lunatus 31, 32, 208, 214, 255
    mungo 32
    species 32
Phaulopsis barteri 202
    ciliata (falcisepala) 202
    imbricata 202
Phenacoccus manihoti 27
Philodendron species 240
Philoxerus vermicularis 20
Phoenix dactylifera 49
    reclinata 41, 49, 50, 79, 81, 88, 97, 107, 111, 113, 114
Phragmanthera incana 248
    nigritana 248
    species 44
Phragmites karka 79, 82, 89, 106, 111, 113, 116, 249
Phyllanthus discoideus (see Margaritaria discoidea)
    fraternus ssp. togoensis (niruri) 131, 139, 163, 180, 200
    maderaspatensis 161
    muellerianus 9, 12, 42, 48, 49, 127, 134, 135, 138, 146, 152, 154, 176, 177, 186, 197, 201, 226
    niruri var. amarus (amarus) 247
    pentandrus 149
    reticulatus var. glaber 48, 54, 147, 210, 224, 226
    reticulatus var. reticulatus 48, 54, 147, 152, 210, 224, 226, 248
Phyllocosmus africanus (Ochthocosmus africanus) 53, 80, 101
    chippii (O. chippii) 238
Physalis angulata 126, 132, 152, 153, 169, 188, 247
    micrantha 247
Physostigma venenosum 122, 124, 153, 175, 181, 183, 192, 208, 212, 215, 218, 226
Phythium species 29
Phytolacca dodecandra 125, 137, 166, 170, 191, 202, 214, 215, 228
Phytophthora colocasiae 29
    palmivora 64
    parasitica 44
    parasitica var. nicotianae 74
    species 29
Picralima nitida 104, 106, 109, 110, 122, 127, 128, 138, 156, 167, 177, 189, 199, 216
Pierreodendron kerstingii 214
Piliostigma reticulatum 10, 11, 14, 15, 48, 51, 54, 222, 228
    thonningii 10, 11, 14, 15, 41, 48, 49, 51, 54, 105, 112, 135, 137, 139, 145, 147, 155, 156, 162, 166, 169, 171, 181, 182, 185, 189, 190, 192, 195, 202, 220, 222, 223, 226, 228
Pinus 8
Piper guineense 40, 75, 122, 131, 135, 136, 137, 139, 140, 141, 158, 164, 172, 181, 188, 190, 196, 197, 202, 231
    nigrum 122, 191
    umbellatum 42, 147, 178, 180, 184, 188, 193, 197, 201, 247
    species 146, 198, 203
Piptadeniastrum africanum 6, 15, 78, 83, 84, 86, 87, 88, 101, 109, 125, 193, 212, 228
Pisonia aculeata 141, 142, 162, 234
Pistia stratiotes 50, 57, 131, 152, 155, 158, 181, 189, 245
    species 246
Pithecellobium dulce 13, 15, 55, 96, 114, 222, 234
Pityrogramma calomelanos 235
Placodiscus pseudostipularis 80, 104
Plagiohammus spinipennis 243
Planococcus njalensis 64
Platostoma africanum 142, 154, 161, 180, 186, 190, 229, 247
Platycerum elephantotis (angolense) 235
    stemaria 235
Platylepis glandulosa 236
Plectrelminthus caudatus 236, 237
Pleiocarpa bicarpellata 104
    mutica 104, 156, 168, 178, 238
    pycnantha var. tubicina 104, 110
Pleioceras barteri 115, 126, 174
Pleurotus species 40
Plumbago capensis 238, 240
    indica 125, 171, 200, 211
    zeylanica 122, 125, 128, 136, 141, 144, 155, 171, 176, 181, 182, 184, 200, 201, 211
Plumeria rubra var. acutifolia 179, 229, 240
Pneumatopteris afra (Cyclosorus afer) 236, 248
Poa species 235
Podangis dactyloceras 237
Poinsettia species 240
Pollia condensata 169
Polyalthia (see Greenwayodendron)
Polygonum lanigerum 249
    senegalense 181, 191, 198, 249
Polyspatha paniculata 258
Polystachya adansoniae 237
    affinis 237
    dolichophylla 237
    fractiflexa 237
    fusiformis 237
    galeata 237
    golungensis 237
    inconspicua 237
    laxiflora 237
    monolensis 237

INDEX TO SCIENTIFIC NAMES 307

mukandaensis 237
paniculata 237
polychaeta 237
ramulosa 237
reflexa 237
subulata 237
tessellata 237
species 130, 185
Portulaca foliosa 235
grandifolia 235
oleracea 41, 128, 134, 147, 149, 160, 162, 176, 183, 184, 187, 193, 198, 200, 201, 247
quadrifida 235, 247
Potamogeton octandrus 57, 249
schweinfurthii 57, 249
Pouzolzia guineensis 248
Pratylenchus brachyurus 29, 39
Premna hispida 122, 126, 144, 152, 154, 158, 186, 193, 238
luscens 166
quadrifolia 141, 148, 191, 238
species 156
Preussiella kamerunensis (chevalieri) 239
Priva lappulacea 247
Prosopis africana 11, 15, 49, 55, 86, 87, 91, 102, 104, 105, 109, 186, 193, 203, 222, 229
chilensis 46, 77
Protea madienensis var. elliotii (elliotii) 238
Protomegabaria stapfiana 98
Pseudarthria confertiflora 181
hookeri var. argrophylla 157, 160, 168, 198
Pseuderanthemum ludovicianum 238
tunicatum 42, 131, 138, 163, 188, 238
species 240
Pseudobrachiaria deflexa (Brachiaria deflexa) 26, 56, 248
Pseudocedrela kotschyi 83, 89, 91, 104, 107, 109, 129, 140, 146, 156, 158, 161, 172, 176, 180, 189, 195, 212, 215, 217, 223
Pseudophegopteris cruciata (Thelypteris cruciata) 236
Pseudospondias microcarpa var. microcarpa 16, 48, 51, 114, 148, 153, 168, 179
Psidium cattleianum 46
guajava 46, 102, 104, 141, 146, 179, 188, 193, 222
Psilanthus mannii 66, 204
Psophocarpus monophyllus 34
palustris 34
tetragonolobus 31, 33
Psorospermum corymbiferum var. corymbiferum 16, 143, 148, 155, 175, 184, 210, 228
corymbiferum var. kerstingii 143
febrifugum var. ferrugineum 155, 171, 183, 184
Psychotria articulata 193
obscura 49
peduncularis (Cephaelis peduncularis) 128, 155, 190, 202
psychotrioides 224
subobliqua 9
Psydrax horizontalis (Canthium horizontale) 111

parviflora (C. vulgare) 164
subcordata (C. subcordatum) 97, 122, 141, 174
venosa (C. venosum) 134, 166, 181
Pteleopsis hylodendron 111
Pteridium aquilinum 235, 248
Pteris atrovirens 248
burtonii 248
intricata 235
preussii 235
pteridiodes 235
tripartita 235
vittatas 248
Pterocarpus erinaceus 16, 55, 80, 91, 106, 107, 109, 126, 144, 155, 166, 182, 196, 200, 222, 224, 225, Plate 12
santalinoides 49, 97, 117, 126, 155, 203, 224
Pterygota macrocarpa 81, 86, 136, 188, 197
Pueraria phaseoloides 250, 255
Pulicaria crispa 50, 154, 160, 187, 191
Punica granatum 46, 122, 147, 221, 225, 240
Pupalia lappacea 134, 137, 142, 154, 171, 190, 201, 247
Pycnanthus angolensis 7, 81, 91, 92, 128, 131, 143, 149, 150, 170, 179, 190, 192, 194, 217, 231
Pycnocoma cornuta 151, 178, 210, 217, 222
macrophylla 178, 210, 222, Plate 12
Pyricularia grisea 30
Pythium aphanidermatum 38, 74

Quassia amara 122, 240
undulata (Hannoa undulata) 47, 97, 107, 156, 165, 213, 231
Quelea quelea 68
Quisqualis indica 42, 111, 136, 146, 238

Rangaeris muscicola 237
rhipsalisocia 237
Rangia grandis 42
Raphia hookeri 41, 50, 52, 53, 69, 81, 83, 88, 111, 113, 114, 115, 116, 149, 170, 212, 216, 229, 238
species 50, 79, 92, 257
Rauvolfia vomitoria 10, 114, 124, 129, 136, 140, 141, 146, 150, 157, 158, 165, 166, 168, 171, 173, 174, 178, 181, 183, 192, 196, 199, 200, 203, 204, 210, 213, 219
Ravenala madagascariensis 12, 241
Reissantia indica (Hippocratea indica) 10, 141, 157, 179, 202
Remirea maritima 20, 248
Rhabdophyllum affine ssp. myrioneurum (Ouratea myrioneura) 42, 238
Raphiostylis beninensis 128, 140, 172, 202
Rhigiocarya racemifera 14, 130, 137
Rhinacanthus virens var. virens 189, 216
Rhizobium 256
Rhizophora racemosa 97
species 87, 88, 89, 135, 143, 145, 170, 186, 209, 222, 224, Plate 7
Rhizopus nodosus 27
stolonifer 38
species 27, 38

Rhodognaphalon brevicuspe (Bombax brevicuspe) 41, 115, 134, 146, 157
    species 115
Rhoicissus revoilii 48
Rhynchelytrum repens 57, 248
    villosum 57
Rhynchosia buettneri 239
Rhytacne rottboellioides 112
    triaristata 57, 112
Richardia brasiliensis 235
Ricinodendron heudelotii 49, 107, 108, 109, 117, 126, 146, 150, 154, 178, 197, 228, 231, 255
Ricinus communis 132, 136, 143, 144, 152, 156, 159, 165, 170, 173, 177, 179, 183, 191, 206, 211, Plate 13
Rinorea dentata 110
    ilicifolia 105, 130, 140
    oblongifolia 110
    subintegrifolia 9, 152
Ritchiea reflexa 48, 134, 142, 150, 159, 238
Rothmannia hispida 226
    longiflora 126, 132, 143, 145, 150, 156, 188, 226, 227, 238
    urcelliformis 226
    whitfieldii 104, 203, 223, 227, 238
Rottboellia exaltata 55, 113, 132, 248
Rotylenchus reniformis 32
Roystonia regia (Oreodoxa regia) 233
Rubus species 46
Ruellia tuberosa 242
Ruspolia hypocrateriformis 151, 188, 232, 238
Russelia equisetiformis 240
Rutaceae 43
Rutidia glabra 48
    parviflora 200
Rytigynia canthioides 195

Saba florida 11, 140, 196, 219, 223
    senegalensis 48, 144, 172, 187, 202, 213
    thompsonii 13
Sabal mexicana 241
Sabicea africana 48, 50
    calycina 178, 181, 201, 202
    vogelii 48
Saccharum officinarum 74, 98
    spontaneum var. aegyptiacum 21, 82, 106
    spontaneum var. spontaneum 112
Sacciolepis africana 21, 26, 57, 249
    micrococca 57
Sacoglottis gabonensis 48, 79, 86, 87, 92, 97, 109, 150, 156
Salacia chlorantha (senegalensis) 11
    debilis 16
    ituriensis 11
    owariensis (pyriformis) 11
    staudtiana var. leonensis (callei and tshopoensis) 11, 16
    stuhlmanniana (lomensis) 11, 16
    togoica 11, 16
    whytei (nitida) 11
    species 11
Salvinia nymphellula 57, 449
Samanea dinklagei 152, 179, 228, 229

    saman 15, 46, 55, 114, 233, 254
Samea multiplicalis 246
Sanchezia nobilis 240
Sansevieria liberica 140, 143, 149, 152, 161, 169, 185, 195, 199
Santaloides afzelii 48
Santiria trimera 48, 204
Sapium ellipticum 109, 180, 210, 213
    grahamii 149, 159, 172, 184, 201, 210, 213, 227
Sapotaceae 13, 14
Sarcophrynium brachystachys 49, 81, 112
    prionogonium 81, 112
    species 89
Sarcostemma viminale 213, 216
Scadoxus cinnabarinus ssp. katerinae (Haemanthus cinnabarinus) 217, 239
    multiflorus (H. multiflorus) 214, 217, 239, Plate 15
Scaevola plumieri 155
Scaphopetalum amoenum 48, 80, 106, 115, 154
Scenedesmus serratus 57
Schizachyrium breviflorum var.
    breviflorum 57
    delicatum 57
    exile 57, 81, 82, 112
    maclaudii 57
    nodulosum 57
    platyphyllum 57, 82
    pulchellum 57
    ruderale 57, 82
    rupestre 57
    sanguineum 57, 82
    schweinfurthii 57
    urceolatum 57
    species 57
Schizothrix 257
Schoenefeldia gracilis 57, 80, 82
Schoenoplectus aureiglumis (Scirpus aureiglumis) 221
Schrankia leptocarpa 247
Schrebera arborea 105
Schumanniophyton magnificum 144
Schwenckia americana 138, 142, 146, 181, 192, 217
Scindapsus aureus 240
Scirpus aureiglumis (see Schoeneplectus aureiglumis)
    cubense (see Oxycaryum cubense)
Scleria boivinii 112, 113, 193
    depressa 82, 112, 113, 193
    naumanniana 112, 113, 193
Sclerocarpus africanus 197
Sclerocarya birrea 10, 48, 51, 55, 109, 145, 222, 227, 240
Sclerosperma mannii 81
Sclerotium cepivorum 38
    rolfsii 29
Scoparia dulcis 115, 133, 142, 147, 149, 152, 155, 161, 170, 186, 197, 220, 247
Scottellia klaineana var. klaineana (chevalieri and coriacea) 7, 79, 228
Scytopetalum teighemii 48, 136, 171, 177, 183
Secamone afzelii 135, 138, 179, 198, 210
    species 148

INDEX TO SCIENTIFIC NAMES

Sechium edule 39, 80, 111, Plate 3
Securidaca longepedunculata 55, 80, 106, 133, 137, 148, 154, 156, 161, 171, 178, 180, 185, 186, 188, 196, 198, 206, 209, 212, 218, 219, 220, 222, 228, 238
    welwitschii 152
Securinega virosa 48, 96, 126, 129, 134, 135, 145, 154, 175, 177, 178, 199, 215, 217, 227, 234, 248
Selaginella myosurus 235
    vogelii 235
Senecio abyssinicus 204
    biafrae (Crassocephalum biafrae) 41, 50, 201
Sericanthe chevalieri (Tricalysia chevalieri) 66
Sericostachys scandens 238
Sesamum alatum 41, 185
    indicum 40, 50, 53, 76, 125, 128, 130, 135, 138, 165, 173, 179, 185, 243
    radiatum 41, 76
Sesbania grandiflora 42, 48, 55, 145, 150, 189
    pachycarpa (bispinosa) 10, 55, 57, 255
    sericea (pubescens) 216, 255
    sesban 21, 42, 48, 55, 106, 188, 209, 255
    sudanica ssp. occidentalis (dalzielii) 255
Sesuvium portulacastrum 235, 247
Setaria anceps 82
    aurea 57
    barbata 55, 248
    chevalieri 57, 82
    longiseta 82, 157, 248
    megaphylla 82
    pallide-fusca 26, 55, 82, 248
    sphacelata 57
    verticillata 26, 57, 248
    species 25
Sherbournea bignoniiflora 10, 46, 142, 162, 187, 237
    calycina 46, 142, 237
Sida acuta 169, 248
    cordifolia 71, 132
    ovata 248
    rhombifolia 248
Simirestis welwitschii (Hippocratea welwitschii) 16
Simmondsia chinensis 77
Smeathmannia pubescens 48, 145, 151, 193
Smilax kraussiana 10, 31, 148, 152, 153, 156, 169, 182, 197, 198
Solanum aculeatissimum 179, 219
    aethiopicum (gilo) 38, 41, 136, 140, 151, 192
    americanum 38 (see also nigrum)
    anguivi 38 (see also indicum)
    anomalum 38, 41, 131, 179, 203
    dasyphyllum 9, 41, 183, 218
    erianthum (verbascifolium) 122, 248
    gilo (see aethiopicum)
    incanum 38, 134, 137, 152, 180, 186, 188, 196, 198, 213, 214
    indicum ssp. distichum 38, 131, 150, 214
    macrocarpon 9, 38, 41, Plate 3
    melongena 38
    nigrum 38, 41, 131, 148, 165, 202, 205, 214, 227, 247
    scabrum 38
    seaforthianum 240
    torvum 38, 142, 248
    tuberosum 29
    verbascifolium (see erianthum)
    wrightii 240
    species 38, 41, 209
Solenangis clavata 237
    scandens 237
Solenostemon monostachys 41, 131, 139, 143, 155, 160, 169, 176, 189, 204
    rotundifolius 31, 144, 152
Sonchus oleraceus 41
Sophora occidentalis 122, 209, 216
Sorghum arundinaceum 26, 57
    bicolor 23, 25, 50, Plate 3
    lanceolatum 57
    vogelianum 57
    species 82, 89, 113, 224
Sorindeia grandifolia 148, 186, 227
    juglandifolia 9, 48
    warneckei 48, 227
Soyauxia grandifolia 8, 79
Spathandra blakeoides (Memecylon blakeoides) 239
Spathodea campanulata 130, 132, 133, 145, 159, 175, 183, 195, 197, 202, 214, 218, 232, 233
Spermacoce verticillata (Borreria verticillata) 247
Sphaeranthus senegalensis 134, 201
Sphenocentrum jollyanum 9, 129, 131, 178, 188, 202
Sphenoclea zeylanica 249
Sphenstylis stenocarpa 31, 42
Spigelia anthelmia 128, 209, 213, 247
Spilanthes filicaulis 55, 148, 162, 184, 186, 189, 193, 219
Spirodela polyrhiza 57, 249
Spirogyra species 57
Spirulina species 35
Spondianthus preussii var. preussii 210, 215
Spondias cytherea 46
    mombin 11, 15, 46, 55, 105, 136, 141, 148, 153, 156, 171, 176, 189, 197, 204, 222, 226
Sporobolus africanus 57
    festivus 57, 82
    microprotus 57
    pyramidalis 26, 57, 248
    robustus 21, 57, 248
    sanguineus 57
    virginicus 21
    species 57
Stachyanthus occidentalis (Neostachyanthus occidentalis) 10
Stachylidium theobromae 45
Stachytarpheta cayennensis 50, 55, 127, 145, 149, 151, 152, 162, 197, 247
    indica 247
Stapelia species 240
Staurastrum paradoxum 57
Steganotaenia araliacea 140, 147, 152, 229, 234
Stenotaphrum dimidiatum 55
    secundatum 21, 55, 148, 235

Stephania abyssinica var. abyssinica 128
    dinkagei 128, 174
Stephanodiscus lantzschii 57
Sterculia foetida 53, 86, 125
    oblonga 84, 92
    rhinopetala 7, 49, 83, 91, 137, 191, 228
    setigera 12, 15, 112, 137, 171, 200
    tragacantha 10, 15, 42, 49, 79, 97, 107, 127, 134, 146, 197, 201, 219, 226, 228
Stereospermum acuminatissimum 190, 202, 240
    kunthianum 109, 134, 142, 145, 188, 199, 203, 240
Stictocardia beraviensis 201, 239
Strephonema pseudocola 73, 145, 149
Streptogyna crinita 248
Striga gesnerioides 33, 248
    hermonthica 161, 171, 223, 248, 250, Plate 16
    species 32, 248, 249, 250
Strombosia glaucescens var. lucida 7, 78, 84, 86, 87, 88, 91, 98, 163, Plate 6
Strophanthus barteri 14
    gracilis 216
    gratus 89, 125, 155, 159, 195, 196, 210, 212, 219, 238
    preussii 14, 106, 197, 212, 222, Plate 14
    sarmentosus 14, 89, 106, 152, 181, 197, 201, 212, 238
    species 112, 119, 182, 209, 212, 238
Struchium sparganophora 41, 55, 160
Strychnos aculeata 150, 159, 197, 216
    afzelii 9, 130, 156, 189, 224, 229
    innocua ssp. innocua var. pubescens 48, 211
    nux-vomica 48, 206, 209, 211
    spinosa 48, 51, 129, 143, 149, 150, 152, 156, 160, 195, 198, 211, 214, 217, 220
Stylochiton hypogaeus 225
    lancifolius 31, 225
Stylosanthes fruticosa (mucronata) 55, 130, 139, 201, 218
Suaeda monoica 257
Summerhayesia laurentii (Aerangis laurentii) 236
Swartzia fistuloides 11, 98, 109
    madagascariensis 92, 102, 104, 110, 117, 167, 209, 213, 215, 216, 217, 220
Symphonia globulifera 15, 83, 84, 92, 130, 143, 167, 203, 241
Synedra acus 57
    falciculata 57
Synedrella nodiflora 55, 162, 178, 247
Syngamia haemorrhoidalis 245
Synsepalum dulcificum 12, Plate 1
Syzygium cumini 48, 137, 140
    gineense var. guineense 48, 79, 86, 97, 106, 175, 199, 222
    guineense var. littorale 48, 51, 109
    guineense var. macrocarpum 48, 134, 145
    owariensis 48
    rowlandii 180, 226
    species 48

Tabebuia pentaphylla 233

Tabernaemontana brachyantha 113
    chippii 14
    crassa 14, 109, 110, 133, 161, 164, 171, 182, 204
    psorocarpa 125
    species 10
Tacazzea apiculata 143, 238
Tacca leontopetaloides 31, 49, 115, 130, 210, 219
Talbotiella gentii 80, 97
Talinum triangulare 40, 247, 251
Tamarindus indica 10, 11, 13, 14, 15, 42, 46, 85, 88, 91, 102, 105, 109, 135, 138, 144, 153, 156, 162, 163, 164, 171, 178, 184, 185, 187, 191, 201, 204, 217, 221, 223, 225, 226, 233
Tapinanthus bangwensis 159, 248, Plate 16
    globiferus 248
Tarenna thomasii 155, 218
Tarrietia (see Heritiera)
Teclea afzelii 48
    verdoorniana 9, 105, 122, 142, 152
Tecoma stans 234
Tecomaria capensis 55, 240
Tectaria angelicifolia 236
Tectona grandis 86, 88, 91, 95, 97, 224, 233
Teleonemia elata 245
    scrupulosa 243
Telfairia occidentalis 39, 53, 115
    species 39
Tephrosia barbigera 216
    bracteolata 55
    densiflora 216
    linearis 52, 55, 115, 180
    nana 216
    platycarpa 55, 255
    purpurea 52, 133, 139, 142, 148, 216, 225, 255
    vogelii 126, 150, 164, 177, 184, 206, 209, 212, 213, 214, 215, 216, 217, 230
    species 248
Terminalia avicennioides 97, 105, 146, 170, 182, 187, 193, 195, 202, 204
    catappa 15, 46, 53, 85, 89, 222, 226
    glaucescens 9, 13, 79, 97, 105, 106, 109, 168, 170, 198, 203
    ivorensis 6, 7, 78, 81, 84, 85, 86, 88, 91, 92, 93, 108, 109, 195, 203, 225
    laxiflora 204, 224
    macroptera 105, 106, 141, 147, 148, 154, 176, 182, 203, 226, 229
    mollis 224
    superba 7, 78, 81, 85, 87, 88, 89, 91, 107
    species 145, 192
Termitomyces species 40
Tetracera affinis 204
    alnifolia 12, 42, 122, 130, 140, 155, 182, 189, 196, 222, Plate 14
    leiocarpa 12
    podotricha 12
    potatoria 10, 12, 42, 141, 193
Tetrapleura tetraptera 16, 49, 52, 85, 150, 156, 166, 179, 182, 197, 212, 222, 228, 229
Tetrorchidium didymostemon 9, 79, 157, 179, 191, 228
Thalia welwitschii 10, 81, 89, 112, 235, 249
Thaumatococcus daniellii 12, 81, 89, 112,

INDEX TO SCIENTIFIC NAMES 311

258, Plate 1
Thelepogon elegans 57
Thelypteris (see Pseudophegopteris)
Themeda triandra 57
Theobroma cacao 59, 61, 76, 81, 98, 123, 148, 229, 239, Plate 5
Thespesia lampas 240
    populnea 10, 42, 85, 109, 110, 222, 238
Thevetia peruviana 104, 114, 150, 157, 179, 209, 210, 217, 234
Thielaviopsis paradoxa 44
    species 30
Thomandersia hensii (laurifolia) 204
Thonningia sanguinea 132, Plate 12
Thryallis glauca (Galphinia glauca) 240
Thunbergia alata 240
    chrysops 201, 238
    erecta 234
    fragrans 240
    grandiflora 240
Tieghemella heckelii 5, 6, 11, 53, 69, 81, 87, 88, 91, 93, 114, 192, 216, Plate 1
Tiliocora dielsiana 9, 10, 49
    dinklagei 10, 96, 111
    funifera (warneckei) 96
Tolypothrix 257
Torenia thouarsii 195, 204
Torulinum odoratum 249
Trachylobium verrucosum 16, 83
Trachyphrynium braunianum 49, 81, 113, 130
Trachysphaera fructigena 45, 65
Tradescantia species 240
Treculia africana var. africana 11, 51, 53, 92, 128, 142, 155, 171, 179, 191, 194
Trema orientalis (guineensis) 10, 42, 49, 101, 105, 128, 131, 132, 136, 138, 143, 147, 150, 163, 166, 176, 187, 192, 197, 214, 223, 229, 258
    species 168
Trianthema pentandra 197, 211
    portulacastrum 125, 149, 247
Triaspis odorata 115, 239
    stipulata 239
Tribulus terrestris 133, 177, 181, 187, 235, 247
Tricalysia chevalieri (see Neorosa chevalieri)
    reticulata 141
Trichilia emetica ssp. suberosa (roka) 49, 53, 55, 92, 104, 142, 150, 162, 167, 172, 178, 183, 218, 222, 231, 233, 255
    monadelpha (heudelotii) 88, 97, 126, 130, 158, 162, 172, 175, 182, 189, 195, 197, 198, 203, 204, 210, 224
    prieuriana 101, 197
    tessmannii (lanata) 79, 94
Trichodesma africanum 145, 191
Trichosanthes cucumerina var. anguina 42
Trichoscypha arborea 49, 126
    chevalieri 49
    oba 49
Triclisia dictyophylla (gilletii) 111, 123, 147, 213
    patens 51, 127
Tridactyle anthomaniaca 237
    armeniaca 237

bicaudata 237
crassifolia 237
gentilii 237
tridentata 237
Tridax procumbens 247
Trilepisium madagascariense (Bosqueia angolensis) 14, 46, 83, 86, 91, 105, 144, 224, 227
Triplochiton scleroxylon 5, 6, 81, 84, 85, 92, 93, 103, 107, 109, 175, Plate 1
Triplophyllum pilossisimum (Ctenitis pilossisima) 236
    protensum (C. protensa) 236
    securidiforme (C. securidiformis) 236
    vogelii (C. lanigera) 236
Triplotaxis stellulifera 141, 165
Tristemma coronatum 239
    hirtum 49, 239
    incompletum 49, 239
    littorale 186
Triticum aestivum 23
Triumfetta cordifolia 42, 71, 106, 145, 229, 248
    rhomboides 42, 71, 112, 169, 229, 248
    species 146
Turraea heterophylla 132, 153, 155, 161, 164, 172, 188, 239
    vogelii 153
Turraeanthus africanus 6, 91, 93, 126, 217
Tylenchus species 32
Tylophora conspicua 172, 195, 201, 204
Typha domingensis (australis) 31, 50, 81, 89, 112, 115, 235, 249

Uapaca corbisieri (esculenta) 49, 101
    guineensis 49, 86, 92, 98, 129, 142, 150, 183, 212
    heudelotii 49, 79, 86, 88, 97
    togoensis 49, 97
Umbelliferae 206
Uncaria africana 139, 238
    talbotii 155, 167, 173, 222, 238
Uraria picta 129, 156, 162, 189, 219, 251
Urelytrum annuum 82
    muricatum 82
    pallidum 82
Urena lobata 10, 55, 71, 145, 157, 203, 248
Urera cameroonensis 10, 41
    mannii 146
    oblongifolia 170
    obovata 130, 146
Urochloa mutica (Brachiaria mutica) 56, 249
Uroplata girardi 245
Usteria guineensis 142, 193, 238
Utricularia baoulensis 57
    gibba ssp. exoleta 57
    inflexa var. inflexa 57
    inflexa var. stellaris 57
    reflexa 57
    spiralis var. tortilis 57
    subulata 57
Uvaria afzelii 141, 155, 178
    angolensis ssp. guineensis 11
    chamae 11, 96, 111, 137, 141, 146, 152, 155, 160, 168, 174, 176, 178, 188, 190, 191, 202

doeringii 11, 167, 176
ovata 11, 111
Uvariastrum pierreanum 104

Vallisneria aethiopica 57, 235
Vanilla africana 237
  crenulata 116, 237
  planifolia 229
  ramosa 237
  species 237
Veitchia marrillii 241
Vernonia amygdalina 40, 96, 123, 131, 141, 146, 148, 150, 156, 158, 170, 177, 183, 184, 196, 209, 213, 214, 217, 248
  biafrae 154, 161, 186, 192, 244
  cinerea 128, 164, 247
  colorata 13, 40, 49, 141, 146, 148, 150, 157, 170, 177, 183, 184, 196, 213, 214, 217, 248
  conferta 49, 128, 130, 134, 140, 145, 153, 163, 170, 179, 180, 186, 204, 217, 229
  galamensis 77
  guineensis 129, 146, 157, 168, 193, 217, 219
  macrocyanus 216
  nigritiana 139, 146, 148, 150, 156, 157, 162, 167, 168, 174, 176, 178, 209, 239
  perrottetii 247
Verschaffeltia splendida 241
Vetieria fulvibarbis 57, 82, 111, 116
  nigritana 57, 82, 112, 116
  zizanioides 57, 229, 231
Vigna subterranea (Voandzeia subterranea) 31, 33, 256, Plate 3
  unguiculata 30, 31, 32, 33
Vismia guineensis 79, 143, 177, 186, 188, 191, 204
Vitellaria paradoxa (Butyrospermum paradoxum ssp. parkii) 11, 13, 16, 66, 97, 107, 108, 152, 161, 169, 226, 228
Vitex doniana 11, 13, 41, 51, 55, 97, 107, 116, 138, 145, 155, 171, 179, 182, 187, 188, 204, 224, 227, 228
  ferruginea 49
  fosteri 49, 185, 227
  grandifolia 49, 51, 80, 107
  micrantha 49, 107, 143
  rivularis 49
  simplicifolia 49, 183, 192
Vitis vinifera 46, 233
Voacanga africana 10, 14, 49, 124, 136, 193, Plate 12
  thouarsii 49, 79
Voandzeia subterranea (see Vigna subterranea)
Volvariella species 40
Volvox aureus 57
  tertius 57
Vossia cuspidata 21, 57, 249

Wahlenbergia perrottetii (Cephalostigma perrottetii) 181
Waltheria indica 9, 71, 126, 142, 143, 148, 152, 187, 190, 193, 199, 199, 202, 248
Washingtonia filifera 241

Wissadula amplissima 10, 71, 115, 248
Wolffia arrhiza 57, 249
Woronienella psophocarpi 34

Xanthomonas albilineans 74
  malvacerum 71
  vasculorum 74
  vesicatoria 38
Xanthosoma brasiliense 29
  mafaffa 28, 40, 216, Plate 3
Ximenia americana 49, 51, 53, 102, 123, 128, 139, 141, 143, 146, 151, 152, 155, 161, 168, 180, 185, 193, 195, 203, 209, 211, 213, 219, 234
  species 156
Xylia evansii 228, 229
Xylopia acutifolia 106
  aethiopica 106, 109, 123, 128, 133, 134, 135, 136, 141, 144, 148, 150, 160, 165, 172, 173, 175, 181, 189, 203
  parviflora 52, 109, 132
  quintasii 10, 11, 79, 88, 105, 109
  staudtii 10, 79
  villosa (see Xylopiastrum villosum)
  species 140, 146, 163, 174, 175, 180
Xylopiastrum villosum (Xylopia villosa) 125, 133, 173
Xylotripes gideon 68

Zanha golungensis 49, 88, 92, 211
Zanthoxylum chevalieri (Fagara pubescens) 161, 186
  gilletii (F. macrophylla) 98, 107, 136, 139, 141, 162, 163, 180, 185, 193, 198, 199, 212
  leprieuri (F. leprieuri) 107, 130, 131, 163, 171, 180
  viride (Fagara viridis) 91, 123, 131, 166, 170, 172, 181, 193, 196, 214, 215
  xanthoxyloides (F. zanthoxyloides) 15, 52, 54, 95, 97, 114, 123, 126, 128, 129, 136, 140, 144, 145, 153, 156, 158, 164, 169, 171, 174, 175, 176, 181, 188, 190, 191, 193, 198, 202, 214, 216, Plate 12
Zea mays 23, 50, 96, 147, 197, 200
Zeuxine elongata 236
Zingiber officinale 76, 125, 131, 136, 137, 140, 144, 153, 175, 187, 193, 217
  zerumbet 76
Ziziphus abyssinica 12, 46, 55, 79, 234
  jujuba 46
  mauritiana 10, 12, 46, 51, 52, 55, 79, 97, 127, 131, 140, 144, 150, 155, 171, 173, 178, 188, 195, 199, 203, 215, 221, 226, 234
  mucronata 46, 55, 65, 79, 83, 85, 104, 105, 106, 110, 128, 133, 134, 148, 164, 172, 178, 179, 191, 193, 194, 198, 199, 211, 221
  spina-christi var. macrophylla 46, 55, 83, 91, 105, 113
  spina-christi var. spina-christi 147, 202
Zonocerus variagatus 27
Zornia latifolia 156, 235, 247
Zygotritonia crocea 31, 49

GENERAL INDEX

See also Index to Local Names, p. 277, Common Names, p. 283 and Scientific Names, p. 288.

Ghanaian words are italicised.

Abakrampa 43, 77
Abenkwan 68
Abidjan 33
Ablemamu 23
Aboboe 30, 33
Abobya 34
Aboloo 110
Abomianu 30
Abomoso 72
Abomu 38
Abortifacient 169
Abortion 125, 192; cause 126; prevent 126; procure 125
Abotoase 100
Abralin 207
Abrine 113, 206, 207, 211
Abscesses 126, 134, 135; dental 134
Absin 120
Absorption of water 18, 21, 252
Abura Dunkwa 73
Aburi 39; Botanical Gardens 232
Abyssinian (Ethiopian) mountains 64
Accra 34, 39, 83, 96, 111, 118, 232, 246; Plains 24, 45, 245
Acetyldigoxin 118
Achimota 206
Acholis of Uganda 26
Acid 64; abric 211; alcohol 121; anacardic 45, 119; chrysophanic 120; citric 43, 77; 2, 4-dichlorophenoxyacetic 250; epoxy 77; hydrocyanic 207, 213, 214; oleic 45; oxalic 208, 213; palmitic 45; phosphoric 255; prussic 25, 207, 210; soil 76, 258; stearic 45; tannic 221; 2, 4, 5-trichlorophenoxyacetic 250
Acidic: to taste 50; water 246
Acoustic requirement 108
Acrid(ity): degree of 29; property 31
Actinomycetes 256
Active principle 206, 211, 212, 216, 218
Ada 242
Adanse Ashanti 76
Adansonin 119, 218
Adenites 191
Adjina 100
Adjuvant 223
Adomfe 64
Adoniside 118
Adoso 30
Adrubum 72
Adult male 27, 196
Adulterant 13, 14, 77, 209
Advertisement 89
Aerial tuber 28
Aescin 118
Aesthetic value xi, 95, 232, 233
Afara 87
Afforestation 99

Afram: arm 57; Plains 35, 38
Aframso 72
Africa 215, 218; Central 185; East 16, 23, 28, 30, 42, 44, 48, 50, 64, 73, 77; South 85, 96, 213, 240, 244; South-west 64; tropical 251
African: Blackwood 116; Timber and Plywood (Gh) Ltd 6, 93, 102
Afroni 72
Agbelekakro 26
Agbelima 26
Agbo infusions 141, 180
Agidi 23
Agogo 74
Agona Swedru 73
Agricultural: activities 253; chemicals 22; countries 22; land 98; output 22, 250; product 102; purpose 95; Research Station 12, 33, 43, 72, 75, 250
Agriculture: beneficial to 18; destruction of forest for 236; development of 22; harmful weed in 248; Ministry of 246; peasant 245; spreading 2
Agriculture Research Services, United States 123; Agriculturist, scientific 256
Agro-forestry operation 7
Agushie 39
Ahei 23
Ahensaw 115
Aid-agency 34
Air 18, 101, 256; hot 90; limited supply 95; movement 17
Ajmalicine 118
Akan 106, 107
Akate 64
Akatsi 73
Aketewa-foroye 39
Akla 33
Akorwu-Bana 35
Akosombo Textiles 70
Akpeteshie 51, 74; triple-distilled 51
Akple 23
Akramang Farms 44
Akroso 72
Akuafo Cheque System 59
Akuammine 122, 127
Akumadan 35, 72
Akuse/Kpong 74
Akwanseram 68
Akwapim 61, 124; Chief 18; Range 232
Akwatia 17
Akyeampong weed 244
Akyim 61
Albuminoid, digestive of 120
Alcohol: ethyl 230; extraction 96; production of 64; sugar-cane 96; Alcoholic: beverage 44, 51; extract 211; Alcoholic drink 50, 51, 69, 180; source of 45
Alertness 65, 230

313

Algal blooms 57
Aliens Compliance Order 59
Alkaline 35
Alkaloid 119, 120, 121, 122, 123, 127, 210, 211, 214, 216; active 119; crystalline 206; poisonous 210; principle 210
Allantoin 118
Amamfrom Kwahu 35
Amaranth 36
Amedzofe 29, 50
Amelioration of local climate 17
Amelondado Cocoa 61, 62
Amenity planting 232
Amenorrhoea 126, 173
America 28, 61, 124; Central 61, 240, 245; continental 244; North 240; South 26, 31, 35, 44, 73, 74, 75, 123, 240, 245; sub-tropical 77; tropical 28, 31, 45, 74, 77, 232, 239, 240, 245; American Indians 109
Ametryne 249
Ammonium chloride, local 225
Amodin 120
Ampesi 26, 27, 28, 30
Anabasine 118
Anaemia 127
Anaerobic site 256
Anaesthetic, local 120, 122, 123, 127, 206
Analgesic 126, 149, 174
Andirine 119
André Simon Memorial Fund 34
Andrographolide 118
Angeline 119
Angler, float for 110
Angola 64, 227
Animal 19, 39, 46, 66, 211, 245; aquatic 57; circulation of nitrogen through 257; feed 29, 33, 62, 68; higher 77; home for 19; laboratory 123; pet 17; products 17, 19; protection for 19, 95; remains 254; toxic to 71; wastes 254
Anisodamine 118
Anonaine 119
Ant: keep off 165; little brown 126
Antelope 1
Anthelmintic 127, 192, 200
Antibiotic 121, 122
Antidote 13, 133, 206, 218, 219; to arrow poison 218; to atropine poison 218; general cases 217, 219; to poison 150, 217; to scorpion sting 220; to snake venom 206, 218; to strychnine poisoning 218; to venom of viper 251
Antileukaemia 120
Anti-periodic 155
Antiseptic 129, 195, 202, 203, 230
Anti-spasmodic 121, 139, 160
Antracnose 38
Antraquinone 120, 121
Anuria 197
Anxiety reduction 230
Anyinam 72
Apakyi 110
Apantu 30
Apem 30
Apempa 30
Aperient 200

Aphid, green 244
Aphrodisiac 9, 30, 120, 122, 129; preparations 130
Apoplexy 130, 190
Appedite 131; loss of 230
Appetizer 130, 190
Apple 45
Aprampransa 23
Aquarium 235; domestic 57, 235
Aquatic: Biology, Institute of 246; fern 235
Arboricide 7, 242, 250
Arecoline 118
Argentina 240
Arid zone 77; semi 26, 77
Armlet 112
Aromatic: plants 229; Labiate 160, 200; scent 9
Arrow 106, 212; extraction of poisonous 251; shaft 16
Arrow poison 210, 212; antidote to 119, 206, 218
Arthritis 131, 182; rheumatoid 182
Asamankese 72
Asamoa nkwanta 258
Ascites 131, 149
Asclepin 120, 207, 211
Ashanti(s) 52, 61, 107, 113, 117; Akim District 64; Bekwae 73
Ashewa 117; music 117
Asia 76; East 36; Minor 240; South East 32, 33, 41, 43, 240; tropical 29, 30, 125, 232, 239, 240, 255; western 30; Asian origin 38; Asiatic: plant 108; species 114
Asiaticoside 171
Asparagus 42; bean 37
Asphyxia, death by 213
Assin: Apemanim District 61; Foso 68, 72, 73; Manso 72; Nsuta 61
Astec Industries 43, 44
Asthenia 132
Asthma 65, 132; spasmodic 121
Asuansi 67
Asutsuare 74
Atadwe-milk 30
Atanisatine 120
Atapkame 79
Atebubu 67; District 72
Atenteben 116; bigger types of 116
Atewa Range 29, 76
Atlantic 75
Atmosphere 1, 20, 21, 256; Atmospheric nitrogen: fixing of free 252
Atrazine 244
Atropine poisoning 118, 153, 107, 210; specific antidote to 218
Atuabo 67
Atumpan drum 107
Australia 96, 240, 244
Automobile 95, 102
Avenue 232, 233; Indian Cedar 233; Neem 233; Palm 233
Average nutritional value (ANV) 35, 36, 37, 38
Aveyime 74
Avirosan 249
Avitaminosis C 27
Avoceine 123

GENERAL INDEX 315

Avrebo 75
Awaso 17
Awiebo 67
Awuna 38
Awutu 39
Axe 104, 250; handle 104
Axim 68, 242

'Back to the land' 22
Backache 132
Backyard 32, 41, 73
Bacteria 20, 71, 256; metabolic activities of 252; nitrogen-fixing 256; partnership 256; pathogenic 252; resistant to 70; symbiotic 256
Bags 24, 34, 76, 209, 223; of various sizes 111
Bagasse 74, 98
Baker 96
Bakery 102
Baking fat, manufacture of 69
Bamboo 79, 106, 116; shoots 37
Banana 30, 44, 45, 158, 167, 184, 201; cigar-end rot 45; leaf spot of 45; Panama disease of 45; Sigatoka disease of 45; wilt 44
Banku 23, 25
Bantu tribes, semi 25
Bar: drinking 79; *fufu* 79
Barikese 72; Dam 246
Barley 25
Barrel 89; staves 107
Bark cloth 113; making 106
Basidiomycetes 40
Basin (of wood) 107
Basket 9, 68, 69, 112, 245; assorted 111; Bolga 103; industry 223; weaving 103, 111
Basketry 106, 111, 112, 209
Bassoon 116
Bast fibre 10, 71, 72, 116; plant 10, 71; source of 10; Bast Fibres Development Board 71
Bauchi 55
Bauxite 17
Bays: shelter 246; upper 246
BCCI 159
Bead 113, 114; decorative 16, 114; rattling 117; -work 113, 114
Beams 90; durable 83
Beans 32, 33, 38, 61, 65; butter 32; cocoa 62; edible 33; fermented 62; stew 33; underground 31; web blight of 32
Beast 86, 234
Bechem 72
Bedding 71, 112; products 71
Beech 103
Beer 26, 51, 88; Hausa 76; local 25; millet 163, 168, 197; native 23
Bee 13
Beetle 245; Hope 68; Rhinoceros 68
Begoro 23
Bekwai 73
Bellater 249
Belly pain 139, 159
Bengal 240
Benin Republic 27, 31, 52, 67, 68, 75
Benso 68
Bentoa 110
Berberine 118, 119, 121

Berekum 242
Beriberi 24
Beverage 25, 48, 50, 51, 52, 65, 135, 141, 146, 149, 154, 155, 162, 168, 173, 176, 180, 181, 197; alcoholic 25, 51; cocoa 62; effervescent 76; prepared from corn 23; stimulating 65; sweet 50, 51; refreshing 65, 76; variety of 62
Bihar 213
Bile 138
Billet 95, 99
Biliousness 133, 148
Binis of S. Nigeria 12, 41
Bio-control for weeds 243
Bioko 61
Biological control 242, 243, 245, 250; insects used for 243, 245; of weeds 242
Biomethane digesters, feed stock for 257
Bird 1, 14, 17, 35, 48; cage 111; —lime 13, 14, 209; weaver 68
Birefi 112
Birim valley 17
Biscuit 29
Bitter gourd 36
Black 226; —arm 71; indelible 226; pepper 52; Plum 43; rot 29, 38; shank 74
Blackpod of cocoa 64
Blacksmith('s) 61, 98; slag 199
Bladder trouble 133, 168
Bleeding: arrest 160, 190; stop 133
Blinding serum, remedy for 153; Blindness, causing 133
Blinds 112
Blood: coagulating effect on 211; grouping test 125; high pressure 163; pressure 65, 164; purifying 38, 133; spitting of 135, 190; vomiting of 159, 190
Blue 223, 227, 251; black 223, 227; dye 223, 225; —green algae 35, 256, 257; mould rot 29
Boat 90, 93, 106; building 86
Bodongo 23
Bonducin 119
Body-paint 223
Boilers 102
Boils 76, 126, 134, 135, 151; carbuncular 135; 'tumbo-fly' 134
Bokabo-Tumantu 61
Bolgatanga 73
Bolivia 240
Bonduc counters 106
Bonsa Tyre Fctory 75
Bonsaso 75
Boodoo 23, 29
Borax 108
Borer, wood 88
Botanical: garden 12, 232, 244; interest 95
Botanist, professional 232
Botianor 245
Boundary planting 233
Bow 105; string 106
Bowl 67, 107, 108, 109, 110
Bracelet 113, 114
Brain, action on 65
Bran 25, 221
Brass 225; filings 195

Brazil 26, 61, 64, 96, 234, 239, 240; South 240
Bread 23, 24; cassava 26; sweet potato 29
Breath, shortage of 132; Breathing, difficulty in 214
Bricklayer 79
Bricks 79; burnt 78, 83
Bridge: all-purpose 86; —building 86; deck of 86; hammock suspension 87; light 86; log 86
Britain 26, 33, 44, 61, 124, (235,) 236, 238, 247
Brodehene 30
Brodekokoa-apem 30
Brodewio 30
Brofoyedru 76
Bromelain 118
Bronchial troubles 135, 136, 180, 194; Bronchitis 136
Bronchus 190
Broniwawu 70
Brong Ahafo 61
Broom 68, 114, 115; household sweeping 69; making 106; sweeping 114
Brown 223, 251
Browse species 53
Brucine 118, 119, 209
Bruises 136, 187
Brush 103, 116; ceiling 115; general utility 115; handle 116; scrubbing 116; sweeping 115
Brushwood 96
Buffalo 54
Building xi, 78, 84; accessory 84; frame 9; material xi, 78; platform 9, 84; rafters 9; sample 83; temporary 68; timbers for 78, 83; wooden 83
Bulbils (aerial tuber) 27, 28
Bulbs 37, 38
Bulinus snails, plants poisonous to 215
Bull 114
Bullock ploughing 105
Bunso 12, 75
Burkina Faso (Upper Volta) 25, 32
Burma 34, 44, 88
Burns 136, 187
Burundi 64
Bus 90; stop 79
Busia weed 244
Butter 132, 140, 182, 187; dyeing 225; substitute 45
Butter Bean, web blight of 32

Cabaca 117
Cabbage: Chinese 36; Palm 69; White 36
Cabin fittings 93
Cabinet: for electronic equipment, set 92; making 99
Cadinene 120
Caffeine 65, 72, 73, 118, 120, 123, 156, 230
Cailcedrin A 208; B 208
Cake 26, 32; colouring 224; kernels beaten into 54
Calabarine 122, 208
Calabash xii, 110; broken 14; dyeing 226
Calcium 32, 34, 36, 37, 38, 39, 253; carbonate 253; oxalate 40; salts 64
Calmatambin 122

Calmness 230
Calotropin 207
Camel 54, 55, 56
Cameroun 1, 27, 50, 61, 64, 69, 73, 244
Camp 66, 79; bed, chair, table 90
Canal 235
Canbod Farms and Industries 44
Cancer 136; anti 121, 123; contributory cause 73; fight against 123; institute 123; lung 73, 230; mouth 73, 230; of the blood 136; of the breast 136; of the nose 136; of the skin 136; of the stomach 136; of the uterus 136; pancreatic 123, 136; throat 73, 230
Candle 231; -making 67, 231
Cannery 102
Canoe 103, 109; construction of 109
Capsid bug disease, of cocoa 64
Cape: Coast 38, 39, 43, 67, 77, 244; St Paul Wilt 38
Carapin 120
Carbohydrate 24, 33
Carbon 227; dioxide 67
Carbonates 253; Carbonate of soda, impure 49; local 125
Carbonization of wood 101
Carbuncle (boil) 76
Cardiac: inhibitor 206; stimulant 65; Cardiovascular damage 73, 230
Cardol 119
Caribbean 61
Carminative 136
Carotene 35, 36, 37, 53, 68; B- 34
Carpenter('s): amateur 90; local xii, 91; plane 104; professional 90; Carpentry 98; and joinery 5, 90
Carpet 113
Carrot 37
Cart, horse drawn 84
Carver 90; Carving xi, 106; suitable wood for 109; wooden 109
Cask 89
Cassaidine 121, 207; Cassaimine 207; Cassaine 121, 207
Cassava 26, 27, 30, 31, 62, 70; bread 26; dough 23, 26; dried chips 26, 109; dry rot of 27
Castanet 117
Castor oil 178, 178, 206
Cataract 123, 137, 151
Catarrh 137, 141; mucous 137
Catchment area 18
Catechin 120
Cathartic 211
Cattle 53, 54, 55, 56, 57, 129, 243; cake 54; droppings of 254; feed 54, 71; increase lactation in 54, 170; medicine for 131; poisonous to 62; rearing 221, 254; 'tumbo fly' boil in 134
Ceiling 84; batten 84; board 62
Cell culture 123
Cellulose 74
Cement 81, 102; block 79, 83; mortar 83
Cemetery 232
Central: African Republic 12, 42, 48, 64, 138; nervous system 65

GENERAL INDEX 317

Centre for Scientific: and Industrial Research xi, 83, 124; Research into Plant Medicine xii, 124, 159
Cephaeline 118
Cerambycid 243
Ceramic factory 102
Cercariae of schistosome 215
Cereals 20, 23, 24, 25, 42, 145, 179; best of the wild 26; important 248; intercrop of 33
CERES 33, 94
Chad 25, 48
Chaff, rice 24
Chair xi, 103, 111; camp 90; lazy 92
Chaksine 120, 207
Chancre 137, 198
Charcoal 7, 67, 94, 95, 99, 101, 102, 165, 197, 227; and fuel-wood compared 95, 101; domestic uses of 102; forest trees for 101; heavy 101; high yield of quality 101; industrial uses of 102; maxi-bag of 96; poor quality 96; production of 102; project 101; quality 101; Research team (Unit) 101; rot in sweet potato 29; savanna trees for 102; traders 101; types of manufacture 99, 100, 101; woodland trees for 102
Charcoal manufacture 99, 100; by-products of 101; clay kiln method 101; local method 101; portable kiln method 100; retort method 101, 102; shallow pit method 99
'Charm' to protect crops 234
Chassis 85
Chavicine 122
Cheese 32
Chemical: analyses of plants 119; compound 123, 253; composition of soil 252; dissolution of rock particles 253; fertilizer 255; preservative 83; processes in the soil 20; weathering of rock particles 253; weed control 249; Chemist, organic 123; Chemistry 124
Chest complaints 132, 137; and pains 137
Chewing: cane 74; sponge 8; stick 8, 9, 193
Chicago 123
Chicken: feed 35, 257; poisonous to 62; pox 138; unsuitable for 71
Chief 18, 22, 108
Childbirth 169; facilitate 168; flow of blood in 191; profuse bleeding after 174; to lighten pains at 169
Children: counters for 113; educate 60; general debility in 182; to revive after fainting 153; under five years 31; weak 201; worm cases in 128
Chills 156; wash for 154
China 43, 57, 58, 110, 224, 234, 240, 246, 257
Chipboard 8, 92; factory 8
Chips: cocoyam 28; plantain 30; yam 27
Chiraa 72
Chit system of payment 59
Chlorine 35
Chloropherine 121
Chocolate 62; formulations 66; manufacture of 45, 62, 69
Cholera 147
Cholagogue 138
Chrysarobin 120

Church 90, 232; building 92
Cigar 73; -end rot of banana 45
Cigarette 73, 132, 135
Cinnamon 226
Circulation of blood, depression of 206
Cirrhosa 230
Cissampeline 118
Citrus species 43: foot rot of 44; fruit rot of 44; green mould of 43; varieties 43
City 79, 89, 95, 99; centre 94; dweller 23; life 59
Clarinet 116
Classified merchantable timbers (see Table 1.2)
Clausanitine 120
Claves (musical instrument) 117
Climate: amelioration of local 17; humid 67; local 95; stabilise local 95; suitable 20, 50
Climber 235, 237; cutting of 7
Clock: table 92; wall 92
Cloth(es) 113, 224, 225, 226; bark 113; camouflaging hunter's 223; dyeing 223; funeral 223; native 113
Clothing 71; container for storing 110; secondhand 70
CO_2 67
Coach: Ghana Railway 93; luxury 85; work 85
Coagulant 14, 209
Coast: along the 24, 95; Gulf 245; Nigerian 24
Cobra: black, black-necked, spitting 219
Coca 123
Coca cola 51, 72
Cocaine 118, 121, 122, 123, 153
Cockroach, to drive away 164
Cocoa 6, 20, 59, 60, 61; Amelonado 61; bean 62, 112; black pod of 64; Board 60; brands Portem, Taksi, Wam 62, 63; brandy 62; butter 62, 63; cake 62, 63; capsid bug disease 64; export 61; farms 59; fresh juice of 64; husks 62, 64, 225; introduction of 61, 69; liquor 62, 63; mass 62; mucilage 62; pod rot of 64; powder 62, 63; processing (see Table 3.2); producer price (see Table 3.1); residue 62, 63; shells 62; sweatings of 64; swollen shoot 64; vinegar, whisky, wine 62
Cocoa Board: Monitoring and Research Dept. of 60; Policy Planning Dept. of 60
Cocoa butter substitute 67, 69; Cocoa producing nations, ascendent 61; Cocoa Products Factory: Takoradi 62; Tema 62; Cocoa Research Institute: Ghana (CRIG) 62; Nigeria (CRIN) 62; Cocoa Swollen Shoot Virus (CSSV) disease 61
Coconut 67, 228; drum 117; flour 68; frond 96; fruit 113; husk 96, 98; juice 67; oil manufacturer 67; palm 69, 110; shell 96, 98, 102, 110, 117
Cocoyam 20, 28, 29; collar rot of, leaf as spinach, root rot of 29
Codeine 118
Coffee 20, 50, 64; fruit rot of 65; growing regions 64; instant 65; leaf spot of 65; producer price (see Table 3.1); substitute 39, 50, 65, 66; wild 65
Cog-rattle 117

Coir fibre of coconut 10, 113, 115
Cola: drinks 72; fruit 159; nuts 42, 158, 186; red 73
Colchicine 118, 207, 210
Cold 138; to ward off 154
Colic 139, 159; uterine 140
College 90; of Pharmacy 123
Colour: attractive 209; black 226; blue 223; blue black 223; bright 111; brown 223; cinnamon 226; crimson 224; fixing 226, 227; green 225; grey 225; indigo blue 209, 223; purple 225; orange 225; red 68, 224; yellow 44, 225
Comb 104, 105, 106, 107, 110; hair 110; wooden 110
Combined: farms 44; operations 7
Combretannin 120
Combustion engines 96
Commerce: Bolga baskets of 111; cultivated cotton of 69
Commercial: companies 44; enterprise 79, 102; establishment 22; export potential 124; fish pond 29; *gari* manufacturer 102; timbering 2; users of fuel-wood 102
Commodity, agricultural 22
Communciation 22; means of 108
Compost(ing) 253, 254, 256; effective application of 255; period of complete 252; raw materials for 243, 252; value of as fertilizer 255; weeds for 251
Compound cooking fat, manufacture of 69
Computer technology and plant medicine 124
Conception: to ensure, to induce 187
Concrete 78, 79; foundation 83; moulding for 84; panel 83; structure 80
Conessine 12
Confectionery 64, 102; manufacture of 69
Confinement, women in their 170
Congo 42, 48, 49, 113
Conical thatch roofing, circular band of 82
Conjunctivitis 140, 151
Consciousness, to restore 154
Conservation of forests 1
Constipation 30, 39, 177, 178; stomachic for 180
Construction(al) 78, 84; heavy 5, 78; light 78; timber for 78; work 78
Consumption 140, 194
Contrabassoon 116
Contraception 27
Convalescent 62
Convulsions 140, 151, 157, 206, 120, 214
Cooker: electric 101; modern 94
Copper filings 195, 225
Coppice system of regeneration 95, 99
Copra 67, 68, 102; cake 68
Corchorin 207, 211
Cord of fuel-wood 95
Cordage 10, 71, 80, 113
Corm 29; edible 29, 30; rot of taro 29
Cormel, colour of 28
Corn 24, 41, 109, 145, 165; dough 23, 26, 29, 30, 33; milled 23
Corn cob, sheaths of 24
Cornerin 208
Cortenerin 208

Cortisone 182
Corynantheidine 120; corynantheine 120; corynanthidine 120; corynanthine 120, 127
Coryza 137, 141
Cosmetic(s) 43, 67, 68, 69; industry 45
Cost: importation 78; initial 100, 242; initial capital 101; overall 78; relative 78; relative high 99; running 94; transportation 78
Cot, baby 111
Cottage 232
Cotton 69, 70, 225, 248; absorbent, angular leaf spot, black-arm of 71; Development Board, imported, locally produced, varieties 70; wool 71, 115, 194
Cottony leak of egg plant 38
Cough 141, 201; in children 13, 141; to relieve 194; whooping 141
Coumarin 120, 207, 214
Counter-irritant 141, 160, 180, 182, 192
Counters 114; Bonduc 106
Countryside 232, 250
Courier, The 8
Cow 213; diary 62; dung 254; pea 248
Cowry bags 245
Crab 17; louse 165
Craft 103, 106; bamboo 103, 110, 111; fibre 106, 115; gourd 103, 110; indigenous 106; man xi; related 111; seed 113; sponge 115; wool 115
Craftsmanship, level of 103
Crawcraw 143
Cream: hair 67; ice 69; skin 67
Crepitin 208
Crimson 224
Crocodile 220; drive away 214; holes of 220; poisonous to 220; protect against 214; to kill 214
Crop 7, 20, 25, 26, 28, 44, 61, 67, 68, 69, 70, 71, 72, 73, 94, 98, 248, 256; alfalfa 256; backyard 35, 38; cash xi, 18, 20, 59, 76, 243; 'charm' to protect 234; cocoa 61, 64; coffee 65; cover 32, 33, 34, 250, 254, 255; developing into 9; development stage of 257; effects of weeds on 250; energy 74; food 7, 18, 26, 28, 32, 59, 62, 70, 72, 73, 76; graminaceous 243; industrial 59, 77; leguminous 32; maturation stage of 257; permanent 59; poor man's 33; post-harvest 22; proper growth of 20; protection for 95; residues 98, 254; root 29, 33; rotation 253; secondary 76; subsidiary 76; toxicity to 242; tropical rainforest 61; worth developing into 25, 30
Crops Research Institute, record cotton trials by 70
Crossopterine 120; Crossoptine 120
Crotonin 207
Crystal mills 67
CSIR xii, 83, 124
CSRPM xii, 124
Cuba 237, 238, 240
Cucumber 36; Cucurbit, leaf spot of 39
Cudgel 105
Cultivars 29 (see also variety)
Cultivated: field 243; land 248; yam 28
Cultural development 106

GENERAL INDEX

Cup 110; drinking 110
Curator, Grounds and Gardens 250
Curcin 121
Curcumin 118
Curls (in wood) 1, 92
Currency, depreciation of 59
Curry powder 52
Curtain 71
Cushion: flossy inflorescence for 245; straw for 251; weeds for 251; wool for 115
Cutlass 244, 250
Cutlery 110
Cuts 143, 201, 220
Cyclopeptides 207, 208
Cyst 252; Cystine 207; Cystisine 121, 122, 209, 211, 216

2,4-D 244, 249, 250
Daboase 8, 95, 101
Daemine 122
Dagombas 67
Dabwenya 23, 24
Dalapon herbicide 245
Dam 235; Barikese 246; Weija 246
Damongo 70, 73
Damp locations 74
Danakof Farms 44
Dancer 117
Danka 110
Danthron 118
Daturine 207
Deafness, remedy for 149
Decarine 123
Decomposing agents 20, 254
Deforestation 2, 8
Dehairing hide 221, 222
Density, variation in 85
Dentition 143, 192
Depali 23
Deserpidine 118
Design 227; and finish 103
Dessert 23, 30, 44, 45, 74, 252
Detergent 68
Developing: countries 22, 31, 33, 77, 94, 118, 119, 124, 242, 250
Dhurrine 25
Diabetes 144
Diamond mines 17
Diaphoretic 144, 191
Diarrhoea 144, 147, 210, 214; mucous 55, 147; profuse 217; to induce 199
Dictamnine 119
Diet 27; average 35; grazing animal's 53; important item of 24; important to the 68; indispensable part of 40; meatless 32; protein deficient local 34; staple 24; vegetarian 40; Dietary 230
Digestive tonic 131, 179, 189
Digitaline 123, 162
Digitoxin 118; Digotoxin 118
Dihydrodioscorine 207, 210
Dioscorine 207, 210
Discolouring, light brown 91
Disease: communicable 196; problem 29; sexually transmitted 196, 199; skin 182, 184, 185, 228; water-borne 215

Disinfecting action 38
Dislocation 147, 157
Di-sodium tetraborate 108
Dispersal, agents of: animal 242; water 242; wind 242
Display boards 89
Distillery 102
Diuretic 62, 120, 123, 133, 147, 149, 168, 175, 196, 197; important 65; principle 120; special reputation 148
Diuron 244
Divonne 8
Dizziness 133, 148, 158
Dodowa 34, 45
Dog 165; poisonous to 214; used subcutaneously on 214; Dog-bite, cure for 220
Doll 103, 106, 109
Domestic: fish pond 29; purposes 94, 98, 102
Dondo 107
Donkey 54
Door 5, 83; decorative 83; flush 83; frame 83; mat 24, 68, 103, 113; panel 83; timbers for 83
L-Dopa 118
Dormaa: Akwamu 64; District 64
Double-bass stringed instrument 116
Dough 23; corn 23, 33; fermented cassava 26; nut 29, 33; steamed cassava 26, 110
Drainage 21
Drink: non-alcoholic, non-fermented, refreshing, soft 51; strengthening 76; sweet 51; thirst-quenching 51; variety of 62
Driver ants, to disperse 165
Droppings: cattle, goat, poultry, sheep 254
Dropsy 131, 149
Drought 25; resistant 55, 56, 57; withstand 56
Drowsiness: causing 210; prevent 73
Drug 73, 123, 143, 176, 229, 231; addiction 230; dangerous 211; Development Branch 123; from natural products 119; imported 118; natural 119; of defined structure 119; plant derived (see Tables 8.1 & 9.2); research 124; source of worthwhile 123
Drum 103, 106, 107, 108, 109; Akan 107; *atumpan* 107; coconut 117; *dondo* 107; *fontomfrom* 107; single-membrane, sticks, strings, to herald the news 108
Duck: feed for 527; weeds 57
Dulcite 121
Dumas 70
Dung 98, 254; as fuel 98
Dunkwa 76
Durability, minimum standard in 78
Dwarf: abode of 1; favourite game of 114
Dyeing 222; bags 223; butter 225; calabash 226; cloth 223, 224, 225, 226; cotton 225; *edinkra* cloth 224; fabric 223, 225, 226; floor 224; fibre 224, 226; fishing tackle 223; industry 209; leather 225, 226; mats 223, 224, 225; pottery 226; raffia (craft) 223, 224; silk 225; tanned skin 225; wool 226

Dyes 13, 206, 209, 222 (see also colour); blue 251; brown 251; purplish 255; saffron 223; yellow 121, 251
Dysentery 144; amoebae 121; against 121; griping of 139
Dysmenorrhoea 149, 174
Dyspepsia 149, 189

Earache 35, 149, 175
Ear-ring 112
East Indies 44
Echitamine 119
Economic: conditions 124; constraints 70; failings 59; importance 62, 256; Economically: important xi, 40, 60; sound 66; worth while 103; Economy 91; backbone of Ghana's 61; state of 22
Ecuador 240
Edinburgh 238
Edinkra: cloth 224; signs and symbols 223
Edow 223
Educatonal background 22
Effective distribution, lack of 59
Effiduasi 72
Egg 252; high cost of 31, 35; plant 36, 38, 39; schistosome 215; stimulate laying 57
Egypt(ian) 30, 34; mummies 215
Eisam 39
Ejumako 67
Ejura 73
Electric poles 87; Electrical insulations 16; Electricity 94; Corporation of Ghana 87; generate 57, 102; pole 87; Electronic: equipment 92; products 92
Elephant 1, 27, 48, 54, 212, 213
Elephantiasis 150
Emetic 141, 150, 176, 183, 201, 219, 230; anti- 151
Emetine 118
Emeto: -cathartic 120; -purgative 126, 171, 174, 179, 188, 217
Emmenagogue 173
Emphysema 230
Enchi 75
Endemic species 97
Energy 34, 36, 37, 62; cropping 95; domestic 94; fifth of all 94; metabolizable 53; purposes 95; wood's 101; -yielding qualities 53
Engine 85, 96
Enteritis 151
Enterprise, fuel-wood as profitable 98
Entrepreneur, private 101
Enumeration survey 7
Environment: artificial working 1; beautiful 232; forest 1, 18, 20; immediate and remote xii; protective 18
Environmental Protection Council 246
Enzyme 25, 73
Ephedrine 118
Epidemic 246
Epilepsy 140, 151; Epileptic men 141
Epistaxis 151, 190
Epithelioma 136
Epsom salts 128
Equatorial Guinea 64

Erosion 20; prevent(ing) xi, 20, 243, 251
Eruptions 134, 138, 151
Erythrophlamine 207
Erythrophleine 121, 127
Erythrophleguine 206, 207
Eseranine 122; Eseridine 122, 208; Eserine 122; Esermine 208
Esiama 67; mills 67
Essarkyir 67
Essential oil 43, 53, 76, 156, 165, 229, 231
Estate 232
Ethanol 74, 230; highest yield of 74
Ethiopia 64, 77, 81, 123; Pepper 52
Ethnic group 23, 26, 67, 106
Ethno-pharmacology 124
Eto 27, 28
Euphoria, increased 230; Euphoriant 230
Europe 26, 72, 75, 124, 128, 165, 237, 238, 239; Central 103
European 26, 41, 42; countries 67, 236
Evaporation: check 20; reduce excessive 19, 254
Evil spirit: causing disease 16; drive away 16
Ewe tribe 31
Expectorant 135, 199
Exploitable girth of timber 7
Explosives, production of 71
Export 44, 61; annual 46, 54; annual production for 75; crop of importance 64; one of the first 16, 76; peak 72, 75; potential 44, 90, 124; second most important 6; trade 76; Exporting: companies 44; countries 44
Eye 133; diseases 137, 151, 167, 175, 194; drops 151, 152, 167, 168, 200; operation 123; pupil of 121; -salve 153; spat at 219; sore 151

Fabric (dyeing) 70, 71, 223, 225, 226
Face board 83
Facial marking 227
Factory 84, 90; ceramic 102; cottage 74; establishment of 75; lighting for 102; lime juice 43; paper 95; processing 62, 65; sugar 74; tyre manufacturing 75; veneer and plywood 93
Faculty of Agriculture, Dean of 26
Fagaridine 123; Fagarine 123
Fainting, revive 153
Fallow system, extended bush 253
Famine, in times of 30, 31, 210, 246
Fan 111
FAO 33, 36
Far East 41
Farm 22, 27, 35, 66, 78, 115, 255; abandoned 244; backyard 22, 243; cash 243; cocoa 59, 247, 248; commercial 22; E.A. Ackom 44; equipment 78; factory's own 74; food 59, 243, 250, 256; hands 59; hedge plant in 234; new 94; palings on 88; peasant 22, 250; shade in 94; small scale 64; traditional 22; waste 252, 254; work 59
Farmer 22, 46, 61, 68, 70, 72, 73, 94, 234, 242, 250; backyard 243; cash crop 59; Chief 59; commercial 243; crop 98; genuine 59;

GENERAL INDEX

landless 59, 60; peasant 22, 23, 44, 61, 69, 70, 72, 73, 74, 75, 243, 250, 257; traditional 22, 250, 256, 257
Farming: activities 99; backyard 23; clearing for 2; clearing forest for 94; commercial 22; illegal 2; plot 23; prepare land for 254; rice 257; soil 257; work 242; Farming land, shortage of 7; Farming methods, traditional 250; Farming system 33
Fat 24, 39, 45, 66, 73, 122, 196; accelerate separation of 67; baking 67; rich in 68; solid 32
Febrifuge 138, 157, 181
Feed 24, 27, 29, 32, 57, 62, 68
Felt 71
Fence 68, 69, 88; live 233; mat 89; of bush 234; Fencing 67, 88, 89
Fermentation 40; traditional way of 64
Fern 235; aquatic 235; decorative 235; free-floating 256
Fernando Po (Bioko) 61
Fertile topsoil 2
Fertility, to induce 169
Fertilizer 70, 255; chemical 255; commercial 6; expensive 256; imported 255; manufacture of 62, 64, 255; nigrogen 257; potash 64
Festering finger 201
Festival 73, 75
Fetish priest 118
Fever 151, 157, 173; malaria 156
Fibre 8, 10, 30, 36, 37, 44, 106, 112, 116, 224, 226; as brush 116; as sponge 30; as towel 30; bast 116; 'Bass' 116; craft 106; coir 68; flexible 116; fruit 222; low grade 71; most widely used 71; natural 70, 71; piassava 116; plant 10, 112; Products Manufacturing Co. 72; strong 116; vegetable 69
Fibreboard 8; compressed 74
Field 232; cultivated 243; football 234; games 234; rice 248, 257; sandy 243; worker 46, 250
Fig, edible 47
Filaria 153
Financial strain of importing chemical fertilizer 255
Find Your Feet 34
Fire 204, 205; bush 59, 66; ease of catching 82; fairly resistant to 78; prevalence of 66; sweeping 98
Firestone Ghana Ltd 75
Firewood 7, 8, 9, 94; gathering of 2
Firm, coach-building 85
Fish 17, 29, 102, 111, 210; breeding ground for 57; feed for 257; food for 57, 58; high cost of 31, 35; paralyse or kill 216; plant for poisoning 216; plant toxins to catch 215; poisons 215, 216, 217; pond (commercial and domestic) 29; smoking 98
Fishermen's net 224
Fishing 246; inland 57, 109; sea 109; tackle 30, 223; trap 111, 112
Fits 140, 157; fainting 153

Flatulence 157
Flavanol 121; Flavonic heteroside 122; Flavonoid 121,122
Flavour(ing) 38, 41, 43, 44, 45, 52, 61, 62, 72, 75, 76, 77
Fleas, to keep away 164
Flood(ing) 245; intolerance to 245; land liable to 74; level 245
Floor(ing) 81, 84, 85, 86, 93; dyeing 224; heavy duty 84; wooden 224Florida 238, 244, 245
Floss 115
Flour 24, 25, 29, 129; cassava 26, 33; cereal 42; groundnut 31; millet 156; peanut 32; sweet potato 29; wheat 24, 25, 26, 29, 30; winged bean 33
Flower: edible 238; holder, pot, vase 111
Flute 116
Fly 115, 153, 204; to drive away 165; sun 115
Fodder 22, 24, 25, 33, 34, 53, 54, 55, 56, 57, 62, 244, 245, 256; good source of 56, 57, 213; green 56; high quality 53; important source of 55; leaves as 32, 44; sedge as 55; useful grass for 55, 56; weed as useful 243, 251
Folinerin 208
Folklore 1, 114, 123; remedies 123
Fontomfrom drum 107
Food 22, 24, 69, 113, 126; baby 32; basic 26; chain 57; composition table 36; conventional 34; crop 7, 18, 26, 28, 32, 59, 62, 70, 72, 73, 76; demand for 22; famine 31, 67; flavouring 52, 76; give aroma to 52; human 35; important item of 74; important source of 40; legume 31, 38; plants 20, 243, 253; preparation 94; preserve 52; protein rich 32, 68; relief aid 22; requirement 25; season 52, 76; shop 33; shortage 22, 66; source of 23, 57, 58; staple 20, 23, 24, 25, 27; sweetening agent in 13; too acrid to be used as 28; value 39; wrapper 12
Foodstuff 78, 111; main source of 22
Foot rot (black shank) of *Citrus* species 44; of tobacco 74
Football 43
Forces, part salary to 24
Foreign exchange: earnings 61; earner 64, 73, 91
Forest 1, 8, 10, 11, 12, 17, 26, 49, 59, 61, 95, 96, 99, 232, 235, 247, 248, 253; as wind barrier 18; atmosphere 17; canopy 20; classification of 1; clearings 76; climatic climax of 2; climbers 8, 115; continuous band of destruction of 236; country 18; cover 19, 20; deciduous 12; degraded 7; direct benefits 2, 5, 103; destruction of 2; dry 9; edges 76; enumeration 1, 18, 20; environment 1, 18, 20; establishment of 18; floor 20, 94, 102, 253; fringe 68; high 9, 96, 257; improving stock of 7; indirect benefits 2, 17; influence on rainfall 18; maintenance of 18; major produce of 5, 99, 103; minor produce of 8, 16, 103; officer 18; poorly stocked 7; product 2, 8, 17, 103; Products

Research Institute xi, 83; proportion of 2; protective aspect 20; protective environment 18; rain 1, 109, 236; -rainfall relationship 18; region 24, 28, 30, 43, 44, 45, 52, 64; -savanna boundary 9; secondary 257; semi-deciduous 12; soil 2, 17, 20; tallest trees of 115; timber trees (see Table 1.2); transitional 73; tropical 1, 2, 68, 236, 253; unreserved 99; wet 9; zone 1, 2, 11, 18, 20, 27, 39, 40, 61, 64, 73, 76, 99, 222, 223
Forest Reserve(s) 2, 18, 99; Atewa Range 17; Asukese 18; Bia Tano 18; Bosomkese 18; Headwaters 19; in Ghana (see Table 1.1); Mpameso 18; Shelterbelt 18; Subri 8, 95, 101
Forestry, Department of xi, 2, 5, 8, 83, 99; School, Sunyani 83
Forestry Enterprises (plant exporting company) 12
Fowl feed 57
Fractures 147, 157
Framboesia 157, 204
Frame(work) 79, 85, 92
France 8, 34, 103
Frog eye 74
Fruit: aromatic 245; boiled palm 109; coconut 67; colour of 45; cultivated 11, 43; desiccated, dried 67, 68, 117; perishable 43; edible 11, 45, 47, 48, 49, 245; fermented 44; fresh 67; half-ripe 127, 133; husk 54; immature 42, 45, 51; jam 47; juice 48, 51, 52; pulp of 14, 51, 66, 76; quality 45; refreshing 43, 44, 45; ripe 12, 44, 45; -rot 35, 44, 45, 65; semi-cultivated 43; shell(ed) 133, 220, 222; sour 12; sweet 234; tart 47; tree 14, 46, 48; tropical 46; unpalatable 45; unripe 125, 128, 220; wild 11, 12, 46, 49
Fuel 8, 24, 25, 69, 74, 77, 94, 95, 102; cooking 94; dried dung as 98; dried grass as 251; for automobile 96; for domestic purpose 99; for industrial purpose 94; fossil 94, 101; less suitable material as 96; needs 99; other sources of 98; raw material for 96; sapwood as 102; savings 8; sawdust as 102; substitute 98; sugar-cane alcohol as 96; trees preferred for 94; useful material for 98; wood 1, 45, 67, 94, 256
Fuel-wood 5, 7, 8, 96, 97, 99, 101, 102; annual consumption — Ghana 7; collection 94; commercial uses of 102; consumption 8; dealer 98; first-class 97; good source of 98; head-load of 95; heavy demand for 95; high forest 97; introduced 97; investment in 98; limited availability of 96; source of 94, 95, 98, 99; plantations in Ghana (see Table 6.1); popular 94; price of 95; savanna 97; suitable trees for 95, 96; supplementing needs 95; useful as 98; useful source of 97; uses of 102; wood unsuitable as 97; woodland 97
Fufu 26, 27, 28, 30, 109, Plate 4
Fumigant 16, 141, 155, 181; Fumigation 144, 153, 154
Fungal: disease 64; thread 32; Fungi, saprophytic 252; Fungicide 64; Fungus 20, 33, 40, 44, 45
Furnishings xi, 90, 91
Furniture cheap 91; class of 90; firms 91; indoor 92; industry 93; material for 90; outdoor 92; quality 91; self-assembly 5, 93; simple 91; sophisticated 90; suitability of timbers for xii; upholstery for 71
Futa-Jallon 56

Ga 23; *-Homowo,* festival of 23
Ga-kenkey, wrapper for 24
Gabon 31
Gaffic Export and Trading 44
Galactogogue 157
Galanthamine 118
Gallic tannin 120
Gambia 24
Game xi, 19; for food 17; for sports 17; hunting ground for 95; Reserves in Ghana (see Table 1.4); and wildlife, protection of 1
Games 114; counters for 113; field 234
Gammaline 20; *Kum-akate* 64
Ganarine 122
Garden 90, 255; beauty of 234; botanical 12, 49, 232, 244; flower 243; of trees 232; vegetable 243; waste 252, 254
Garden egg 38; cottony leak of 38; grey mould of 38; *Rhipus* rot of 38
Gardener 243; backyard 250; Gardening 232
Gari 26, 32, Plate 4; commercial manufacturer 102
Garlic 38, 159, 214
Garment, fumigant for 16
Gas 23, 94, 96, 101, 102
Gastro-intestinal pains 158
Gearbox 85
Geneva 124
Genito-urinary troubles 158, 199
Germany 67, 124; East, West 61
Gesapax 249
Ghana: Broadcasting Corporation 108; Cocoa Board 61, 64; Community Farms Estates 75; Cotton Company 70; Crop Research Institute of 70; Federation of Agric. Co-op 44; International Furniture and woodworkers Exhibition (GIFEX'85) 90; Medical Schools 124; Railways 87; Rubber Estates 75; Sugar Estates Limited (GHASEL) 74; Television 108; Textiles Manufacturing Company 70
Giddiness 148, 158, 213, 214
Gift item 107
GIHOC Vegetable Oil Mills 67
Gin 74; distillation of 74; local 51, 74
Ginger 52, 76, 135, 157, 162, 163, 172, 178, 181, 193; ale 76, beer, biscuit, bread, peelings 76
Gland, inflammation of 191
Glaucoma 153
Glucoside 25, 123; phenolic 122
Glycerine, manufacture of 24
Glycerol esters of: oleic acid 45; palmitic

GENERAL INDEX

acid 45; stearic acid 45
Glyco: -alkaloid 214; protein 12; Glycoside 120, 122, 162, 210; cyanogenetic 210; 214; steroidal 122; toxic 211
Glycyrrhizin 118
Glyphosate 243, 244, 245
Goaso 72
Goat 27, 53, 54, 55, 56, 96; diarrhoea in 147; droppings of 254; lethal to 214; poisonous to 213; promote multiple birth in 54
Goldsmith 101; Gold Coast 113; Jasmine 238; Gold mines 17; slag heaps 248
Golf balls, manufacture of 16; course 234
Gomua 23
Gonococcal infection 196
Gonorrhoea 158, 196, 199
Gossypol 71
Gourd 39, 108; bitter 36; bottle 40, 110; craft 103, 110; snake-like 191; squash 40
Grafting: stock 43; propagated by 45
Grain 25, 26, 102, 112; coast 75; colour of 24, 25; fifth most important 25; quantities of 22, 24; size of 24; wild edible 25
Grains of Paradise 75, 76
Gramozone 244, 249
Gramurone 249
Grapefruit 43
Grapes 47
Grass: as arrow shaft 106; as broom 115; as whisk 115; chopped 81; dried 80, 251; for decorative wear 112; for matting 112; forest 258; introduced 235; lawn 234, 235; perennial 244, 245; stem of 80; thatch 82; with hollow stems 116; with well-developed stems 89
Grasshopper, variegated 27
Gravel (bladder trouble) 133; for building 17
Grazing 55, 56, 250; course 57
Green 225; house 237; mamba 219; manuring system 257; mould 43; 'Green revolution' 22
Grey 225; mould of pepper 38
Griping 139, 159; of dysentery 139
Groin, swellings of 191
Groundnut 30, 41, 62, 70, 73, 131, 145, 170, 179, 214, 248; flour 32; meal 62; oil 32; seller 206; tops 32
Grove, sacred 1
'Grow more food' 22
Guilandinin 119
Guinea 1, 24, 27, 48, 49, 64, 75, 76, 77; coast 75; -Congolian rainforest 1; corn 24, 25; Equatorial 64; Lower 65; savanna woodland 24, 25, 31, 33, 35, 38, 48, 66, 95, 96, 209, 221, 254; worm sores 159
Guinea corn pap 52
Guinea-pig, poisonous to 214
Guitar: box 116; strings 116
Gulf coast 245
Gum 44, 130, 138, 142, 146, 147, 153, 154, 165, 166, 177, 203, 217, 220, 227; arabic 139; commercially important 77; copal 16, 183; edible 15, 42; for anointing 15; for caulking canoes 15; good quality 15; mastic 204; of commercial quality 77; of lesser quality 1; resin 16, 131, 143, 165; scented 15; soluble 42
Gum Arabic 14; of commerce 14
Gummosis 74
Gun: maker 103; powder 203; price of 103; stock 103, 104
Gutta-percha 16, 209
Gutter 20; moulding for 84

Haematemensis 159, 190
Haematuria 160
Haemogglutinin 124
Haemoptysis 135, 160, 190
Haemorrhage: internal 190; pulmonary 190
Haemorroids 160
Haemostatic 160
Half Assini 67
Hallucinogen 230
Hamlet 79
Hani 24
Harbour construction 87
Hardboard 8, 84
Hardwoods 85, 86, 98, 101, 250; durable 86; low-classed 86; mixed tropical 101
Hare 55
Harmattan 18
Harmonium, console for 92
Harp 116
Harpsichord 116
Harvest(ing) 25, 250, 254; poor 23
Hat 103, 111, 112, 209; market mammy 111
Hausa 107; beer 76; family 225; foodstuff 41; girls 117
Havi 74
Hawaii, Government of 245
Hay 55, 56, 57
Headache 155, 160, 174, 186, 210; nervous 65
Headwaters Forest Reserves 19
Health: injurious to 73; post 118; services 118
Heart: action 123; burn 162; disease 121, 162, 176; death by paralysing 212; palpitations 162; stimulants 62, 120, 123; trouble 162
Heartwood 91, 103
Heat: of long duration 97; of the sun 17, 233; optimum conditions of 252; steady 95, 101; Heat energy, source of 102; Heater 90
Hedge 32, 232, 234, 239; attractive 40; clippings 40; dense 234; effective 40; impenetrable 234; ornamental 234; thorny 234; to exclude wild beast 234; useful 245; valuable (asset) 40
Hemp 77, 230
Hemsleyadin 118
Hentriacontane 121
Hepatitis 163, 172
Herbal: healing 118, 119; treatment 125; Herbal medicine: active parts of 119, 124; efficacy of 119; Herbalist(s') 206, 217, 221; copyright 119; illiterate 119; practising 125
Herbarium (fumigation) 77

Herbicides 243, 244, 245; application of 242; cost of 242; foliar application of 249; granular application 249; influence of 242; problems with 242
Herbs, curative 1
Hernia 163, 175; strangulated 163; umbilical 163
Herperidin methyl chalcone 118
Heterotrophic rhizosphere organisms 257
Hiccups 163
Hide 17, 221; dehairing 221, 222; for rug 19
Hi-fi system 92
Historic significance 106
Hitrine 208
Hoe 250; -handle 104
Holdings: large scale 35; small 29, 30, 32, 33, 35, 39, 41, 45, 73, 77
Holland, Rubber Estates 75
Holoside 121
Homatropine 153
Home 90, 94; private 43; Science Department 29; wealthy 251
Homophleine 121, 207
Homowo Festival 23
Honey 13, 17, 126, 128, 132, 144, 148, 150, 158, 174, 177, 179; adulterant 13
Hookworm, specific for 127
Horn 17, 116
Horse 53, 54, 55, 56, 57, 131, 137, 145, 213; mucous diarrhoea in 147
Horticulturist: commercial 236; domestic 236
Hospital 90; 118; Legon 206
Hotel 102
Household: constructional work 45; cooking 96; items 109; utility articles 103; wastes 254
Huhunya 68
Human: consumption 32, 62; dangerous to 217; failings 59; fatal to 219; illness in 218; labour 22; poison 214; settlement 2
Humus 20, 253, 254, 255; prevent accumulation of 253; raw materials for 252; rich in 20; Hunger, control 73
Huter 46, 79, 94
Hunting: ground 95; indiscriminate 19; permit regulation 19
Hut: erection of 79; flat roof of 81; poles for 10; temporary 10, 81; wall of 67
Hyacinth bean 37
Hybrid (of *Citrus* species) 43
Hydrastine 118
Hydrocarbon 121; liquid 77
Hydrocyanic acid 209; Hydrogen cyanide 32
Hygrine 121
Hyoscyamine 118, 207, 210
Hypertension 163
Hypnotic 230

IBPGR 35, 38, 40
Ice cream, manufacture of 69
Illuminant 67
Image 106
Impact: of wheels 87; resistance 85
Implement, agricultural 22

Import exchange rate, complexities of 242
Impotency 164
Improvement thinning 7
Incense 16
Incinerator, manure from 254
Incontinence of urine 164
Independence 22, 44
India 32, 33, 34, 41, 42, 213, 234, 240, 244; -Burma-Thailand Region 95; Eastern 15, 28, 77; North 240
Indian Almond 42
Indigestible 31; Indigestion 131, 164, 188, 189
Indigo 225; blue 209, 223; dyeing 223; industry 223
Indonesia 42, 52, 222, 240
Industrial: application 102; centre 232; crops 59, 77; purposes 94, 98
Industry 22, 66; agriculture dependent 59; backbone of 59; basket 223; chemical 71; cocoa 61; cosmetic 62; dyeing 209; furniture 93; important 57, 110, 111; indigo 223; inland fishing 57; leather 223; local 66; perfumery 231; pharmaceutical 62, 124; prosperous 61; regional 66; rural 66, 99, 103, 111, 206, 221, 228; small scale 103; sugar 74; tanning 209
Infants 44, 201
Infertility, major cause of 196
Inhalant 132
Ink 209, 227; Malam's 227; marking 227; purple 227; red 227; writing 227
Inland fishing 57, 109
Insect 16, 165; associated with Nut grass 243; for biological control 243, 245, 246, 250; fumitory for 165; phytophagous 243; to keep away 165
Insecticide 77, 164, 166, 168; shortage of 59
Instrument 92; musical 16, 111, 116, 117; percussion 117; price of 92; stringed 116; woodwind 116
Intercostal pains 116
Internal: decoration 90; fittings 90; tissue 67
International: Council Meeting 8; market 66; Trade Centre (ITC) 124
Intestinal parasite 127; Intestines 38; inflammation of 151; spasmodic affection of 139
Intoxicant 50, 51, 166
Intra-ocular pressure 153
Iodine 203
Iritis 151, 167, 192
Iroko 82, 85, 87
Iron 32, 34, 36, 37; box 101; cast 101; filings 162; rusty 224
Irrigated land 24; Irrigation 23, 74; project 74; sprinkler 74
Irritating hairs 34
Isochaksine 120
Isosantalene 119
Itch 166, 173
Ivory 17; tusk for 19; vegetable 114
Ivory Coast 1, 5, 15, 24, 27, 28, 33, 44, 49, 50, 61, 64, 65, 67, 68, 73, 75, 76, 77, 113, 124, 244

GENERAL INDEX

Jam 64; manufacture of 64; processed into 44
Japan 41, 110
Jaundice 167
Java 34, 41, 42, 52, 55, 234
Jelly 45, 64; colouring 224; manufacture of 64
Jiggers 165, 168
Jitteriness 230
Joinery 90
Jojoba oil 77
Jollof rice 24
Juapong 74; Textiles 70
Juaso 68
Jug (of bamboo) 110
Jute 10; Factory 72

Kaafa 23
Kaamenko 30
Kade 33, 72, 75, 250; Asuom 72
Kakaduro 76
Kakro 23, 29, 30
Kam-dye 224
Kangkong 36
Kano 69
Kaolin 167
Kapok 115
Kawain 118
Kaya-mo 24
Kenkey 23, 25, 35, 38
Kente cloth 107
Kenya 50, 64, 77, 123, 214, 226
Kenyasi 72
Kernel 66; roasted 44, 54
Kerosene 156; introduction of 69
Keta 67, 68
Keyboard 92
Khellin 118
Kibi 72
Kidney 65; disease 132, 168; stimulant 62
Kiln: clay 101; pit 101; portable 100
Kino 222, 224
Kinship patterns 106
Kiosk, wooden 79
Kisi 244
Knife sheath 209
K$_2$O 69
Koforidua 73, 245
Kokonte 26, 28, 30, 109
Kola 73, 139, 142; edible 73; nut 72, 73, 137
Kolatine 120
Koliko 28
Komenda 24
Kontomire 40
Koobi 38
Koose 33
Koranic writing boards 107
Kpekpoi 23
Kpong 24; Bawaleshie 34
Krachi 27
Krobos 51
Kum-akate 64
Kumasi 38, 72, 75, 83, 100, 111, 246; District 72; Town Plantation 99
Kuntunkuni 223
Kusi 68

Kwae 68
Kwahu 38; Scarp 30; Tafo 72
Kwamoso 68

Laboratory: medicinal plant resources 123; synthesis 119
Labour (industrial and medical): cost 250, 255; delayed 168; difficult 168; force 242, 250, 251; in weeding 250
Labour, delayed 168; pains, commencement of 169
Lactogenic 157, 169
Ladles 106, 107, 108, 110
Lagoon water 245
Lagos 69
Lake: Bosomtwe 112; Chad 24, 35, 249; Volta 35, 57, 246
Lamp 69
Lanatoside C 118
Land: fertile 1; fertility of 242; flooded 257; neighbouring 244; owner 22, 60; preparation for planting 254; Rover 96; stool 18; tenure system 22, 59, 60; uncultivated 253
Landscaping 232, 233
Lantanine 121, 208, 213
Larvicide 165
Larynx 190
Latex 13, 14, 75, 126, 133, 209; adherent 213; adhesive 213; dried 45
Lawn 232, 234; cuttings 252; suitable grass for 234, 235
Laws, customary 22
Laxative 139, 170, 177, 178, 179, 180
Leaching 253
Leaf: curl disease 74; nutrient 34; protein 34; scald 74; to tenderize meat 45
Leaf-spot 45, 65, 69, 74; angular 71; of Bambara groundnut 33
Leather 112, 221, 224, 225; dyeing 225, 226; industry 223; tanning 221; Leather goods: variety of 209
Lectern 92
Lectin, BS II 124
Legon 206, 221, 247, 250
Legume 31, 34
Lemon 43, 44, 126, 140, 149, 159, 166, 175, 183, 193, 200, 202; grass 141; juice 135, 139, 142, 162, 168, 171, 177, 182, 195, 203; oil 43; squash 43
Leper 171; Leprosy 170; Leprous: treatment 201; wounds 171
Lethal: to goat 214; to man 210; to monkey 214; to snails 215; to sheep 214; to water fleas and paramoecia 215
Lettuce 36
Leucorrhoea 172, 198
Leukaemia 136; anti- 120
Leurocristine 118, 120
Lever Brothers Ltd 68
Liane 10
Liberia 1, 24, 48, 49, 75
Lice 164, 165; fumitory for 165; to drive away 165; to keep away 165; to kill 45
Lightning 257
Lima bean 32, 37

GENERAL INDEX

Limarin 208, 214
Limb, enlargement of 150
Lime 43, 76, 77, 102, 129, 142, 166, 167, 168, 176, 177, 181, 186, 195, 196, 197, 202, 204, 255; cordial 43, 77
Linen 226
Liniment 132, 135, 140, 156, 166, 167, 172, 175, 177, 182, 183, 187, 191, 197
Lint 69; of cotton 71; stem core as 195
Lintels, moulding for 84
Linuron 249
Lipids 34
Liquers 51
Litmus substitute 223
Litters 108
Liver: inflamation of 163; problems 163, 172; stimulating 138
Liverpool 69
Livestock 24, 25, 27, 29, 32, 77, 214; feed 74; residues (as fuel) 98
Loaf (of bread) 96
$a-Lobeline 118
Local: anaesthetic 120, 122, 123, 127, 206; consumption 44
Locomotive engine, steam 102
Loganine 119
Lonchocarpin 208
London 34, 244
Lorry body 85
Loudspeakers finished in wood 92
Louvres: glass 84; wood 83
Lubricant, production of 67
Luffeine 179
Lumbago 172
Lung 190; damage, possible 230; inflammation of 177
Luteolin 121
Lye of ashes 31
Lyons 75

Maabang 61
Machinery: agricultural 22; efficient performance of 59
Madagascar 12, 244
Madhya Pradesh 213
Madness 172
Magnesium 34, 253; carbonate 253
Maize 24, 25, 62, 70, 73, 170, 214, 248; stalk 98; weed-killer 249
Malaria 119, 156, 173
Malawi 50, 73
Malaya(n) 29, 41, 75, 244; origin 28
Mali 24, 25, 76
Mallet 105
Malnutrition 27, 31
Mamba, green 219
Mambilla Plateau 50
Mammals 48
Mampong: Akwapim 124; District 72; West 72
Mandingo traders 75
Manganese 17
Mange 166, 173
Mango tree 226
Mangrove 97
Mankani: *-fitaa* or *-fufuo* 28; *-nkontia* 28; *-pa* 28; *-potowee* 28
Mankesem 74
Mansonin 208
Manure 71, 252, 253, 255; animal 70, 254; application of 253; artificial 255; farmyard 254; forest floor 253; from farm wastes 254; green 32, 33, 34, 254, 255, 256; heap 252; organic 255; pit 252; plant 254; preparation of 253; refuse dump 253, 254; weeds as 252
Maraca sticks 117; Maracas 117
Margarine 76; formulations 66, 67; incorporation into 66; manufacture of 32, 53, 69, 231
Marginal: lands 77; soils 242
Market(ing) 11, 32, 40, 42, 44, 46, 47, 48, 49, 56, 70, 111; application 102; centres 233; co-ordinate 66; cotton 69; foreign 103; international 66; local 22; mammy 111; open 232; overseas 22; procedure 59; urban 39
Marmalade 43, 77
Maryland 123
Masher 105, 106
Mask 106, 109
Mat 68, 112, 113; high quality 113; -making 24; Mat(ting) 69, 103, 106, 111, 112, 223, 224, 225; coarse 112; material for 112, 113, 245
Material: binding 80; fencing 88; house-building 83; insulating 252; locally obtained 228; man-made 90; modern plant 252; raw 96, 102, 124, 209, 252; source of 103; waterproof 67; woody 103; woven 89
Mattress 71, 112; leaves for 245; straw for 251; weeds as 251; wool for 115
Mauritania 15, 47, 77
Maytansine 121, 123, 136
MCPA 249
Meal 26, 74, 94; cocoa husk 62; cocoa pod 62; maize 146; residual 67; seeds for 77
Mealybugs 64
Measles 173
Meat: 31, 35; source of 19; to tenderize 45
Medical: doctor 118; properties 38; purposes 71; research 123
Medicinal 32, 45, 245, 251; elements 119, 123; oil 69; plants 13, 124; practice 125; prescriptions 35; properties 24, 30, 38, 44, 67, 124; purposes 76, 124, 233, 243
Medicinal elements, plant sources of (see Table 8.2)
Medicine 118, 121, 229; flavouring 52; western 118; veterinary 120
Mediterranean 30
Medulla oblongata 206
Melegueta 52, 60, 75, 76
Meliatin 165
Melon (white-green) 36
Memory damage, possible 230
Mendes of Sierra Leone 51
Menorrhagia 173, 174
Mensonso 76
Menstrual troubles 126, 149, 173; major cause of 196

GENERAL INDEX

Menstrual troubles 126, 149, 173; major cause of 196
Menstruation: absence of 173; excessive flow during 174; promoting 173
Mental activity, increased 65
Mental trouble 172, 174
Metabolic activities: of bacteria 252; of saprophytic fungi 252
Metal 78, 87; extraction 102; furniture 90; smelting 102
Methoxyl: group 64; pectins 64
Methyl salicylate 206, 209
Mexico 34, 35, 77
Mice, plants poisonous to 214
Michingoro 73
Micro-organisms 252; activities of 253; in the soil 254
Microbial action 252
Microbiologist 196
Migraine 65, 160, 174
Milk 62, 67, 127, 128, 129, 139, 140, 145, 148, 159, 170, 188, 198; breast 146, 169; in cattle 54, 170; production 62
Mill 67; private oil 67
Millet 24, 25, 131, 145, 165, 171; flour 142; bulrush 25; stalk 98
Mim-Goaso 76
Mine 88; fuel in 98
Mineral(s) 6, 17, 32, 35, 252; contents of soil 253; deposits in Ghana (see Table 1.3); elements 27; products 17; replenishing lost 253; requirement 253; salts 20, 255; second important export 17; soil 252; substantial amount of 27
Miraculin 12
Miraculous Berry 43
Miscarriage, prevent 126
Missionaries 44
Missouri Botanical Garden 77, 238
Mitraphylline 119, 121
Mitrincomine 121
Mitrinermine 121
Moisture: collection of 18; content 103; optimum conditions of 252
Molasses 74
Molluscicidal activity 251; Molluscicide 215
Monellin 12
Monkey 17, 27, 48; lethal to 214
Monocropping system of rice cultivation 257
Monomethylhydrazine 208
Monrovia 65
Mordant 223, 226; alum 225
Morindanol 121
Morphine 118, 119
Mortar 67, 83, 106, 107, 108, 109
Mosaic disease of sugarcane 74
Mosque 232
Mosquito: breeding site for 246; larvicide 165; to repel 164, 165, 244
Motif: drums as 108; traditional symbolic 110
Motor car 90
Mower, lawn 251
Mozambique 77
Mpihu 28
Mpuruman 23

Mt Gemi 50
Mt Tonkui 50
Mucunain 208
Mud 78, 79, 81; house 79; hut 83; wall 80
Mudarin 120
Mug (of bamboo) 110
Mulch(ing) 254, 255; weeds for 251
Mundibarca 75
Mung bean 37
Mupamine 120
Muscle 65, 76; stimulant 62
Muscular tissue 65
Mushroom 1, 37, 39, 40, 211; poisonous 40, 211; wild 12
Music centre 92; Musical instrument 16, 111, 116; full diversity of 117; manufacture of 103, 106; strings for 116
Muscarine 207
Muslin 35
Mutton 197
Muzzle (of animal) 245

Nankanis 42, 46
Nanomang 34
Naphthoquinone 121, 122
NAPRALERT (natural products alert) 124
Narcotic 73, 229; Narcotine 118
Nasal: douche 138; drops 161, 167, 184, 198
National: agricultural output 22, 250; benefit 66
Native: doctor 118; house 82
Natural Products Section 123
Necendrographolide 118
Necklace 113, 114
Nectar 13
Needle 113
Nematode disease 29, 32, 33, 35, 39, 44, 45, 69
Nephritis: acute 211; chronic 65
Neriin 208
Nerve: effect on 211; pain along 175; peripheral 121, 122; poisons 211; Nervous headache 65; Nervous system: central 65; paralysis of 213
Netherlangs 62
Neuralgia 175, 177
Neurologic damage 230
New Zealand 96
Ngmeda 23
Ngyiresi 111
Niacin 34, 35, 36, 37
Niangon 85
Nicosia 75
Nicotine 73, 118, 122, 165, 230
Niger 25, 32, 33; estuary 72
Nigeria 1, 23, 24, 25, 26, 28, 31, 32, 33, 38, 41, 44, 47, 50, 52, 61, 67, 68, 69, 72, 73, 75, 76, 85, 124, 218, 244; Northern 15, 24, 41, 49, 56, 57, 77, 143, 246; South-east 226; Southern 12, 41, 42, 246
Nitrogen 253, 254, 255; ability to fix 32; annual amount fixed 257; compounds 257; cycle 257; enzyme for fixing 256; fixation 256, 257; free 252, 257; in the soil 254; rich in 256; supply 257; usable 256
Nitrogen-containing compounds, organic 252, 256; Nitrogen-fixing: agents 257;

bacteria 256; property 256; total 256; Nitrogenous 25 matter 54
Nkakra 38
Nkawkaw 8
Nkenkasu 72
Nkoransa District 72
Nkulenu Industries 43
Noise breaker, hedge as 234
Nomenclature, binomial system of xii
Nor-cassaidine 121, 207
Nose-bleeding 151, 190
NRC Decree 124
Nsawam 73; Cannery 33, 35, 38, 43, 44
Nte 114 *Ntoroba foroye* 38
Nutrient 34, 35; rate of absorption 253; recycle of plant 252; rich in 40; Nutritional value 32, 38; Nutritious, highly 33
Nut, shell 45, 69
Nyankpala 70
Nyinahin 17
Nyiretia 30; *-apem* 30
Nzema 67

OANIDA in Denmark 34
Oboe 116
Obuase 17, 76, 88
Obuoho 23
Ocean Liner 93
Odurogya 116
Oedemas, general 175
Oil: coconut 52, 68; cooking 32, 53, 67; cotton seed 71; edible 52, 53, 68, 69, 231; essential 43, 53, 76, 156, 165, 229, 231; extraction of 24, 32, 68; fatty 231; fine-grade machine 77; for soap manufacture 53; groundnut 32, 52, 67; hair 231; inedible 69, 231; jojoba 77; kernel 48, 68, 69; lamp 69; lemon 43; less popular 69; linseed 231; medicinal 69, mills 67; other 231; palm (see Palm oil); palm kernel (see Palm); sesame 76; sperm whale 77; types of 68; useful 231
Oil drum 101
Oil Palm 20, 68; collar rot of 65; kernels of 98; leaf spot of 69; pericarp of 98; Research Centre 68; shells of 98; varieties (high-yielding) of 68
Ointment 171, 182, 184, 185, 192, 195, 199, 200, 202, 203, 204
Okra(o) 36, 38
Okumanin 68
Okyereko 23
Oleandrin 208
Oleoresin 204, 229
Onion 37, 38, 52, 154, 159, 161, 196, 198
Onniaba 30
'Operation feed yourself' 22
Ophthalmia 151, 175; Ophthalmic practice 153
Opon Mansi 100
Optimum time (of planting crop) 70
Orange 225; -brown 225; colour 225; fleshed variety 29; juice 43, 153, 176; oil 43; peels 43; sour, sweet 43; syrup 43; yellowish 225 (see *Citrus*)

Orchid 235, 236; climbing 237; decorative, endangered, epiphytic, export of, ground 236
Orchitis 163, 175
Ordeal, trial by 212
Organ 92; electronic 92; pipes for 116
Organic matter 20, 246, 252, 253, 254; breakdown of 20, 252; to replenish soil with 253; waste of valuable 98, 254
Ornament(al) 16, 114; coverings 112
ORSTOM 33
Osabum 30
Osakoro 30
Osameansa 30
Osameanu 30
Ostrich 54, 57, 246
Otitis 149, 175
Ouabain 118
Oven 96
Oviposition 243
Oware 106, 107, 114
Oxalates 40; Oxalic acid, salts of 213
Oxymethoxymethyl 121

Pachycarpine 118
Packing case 90
Paddle 106, 109
Padouk 183
Paint 88, 231
Pakistan 34, 243
Palanquin, Chief's 108
Palings 88
Palm 67, 11; cabbage 41, 69, 135; climbing 10, 80, 87, 89, 92, 108, 111, 112, 116, 128; collar rot of 69; decorative 241; introduced 12, 241; kernels 129, 130, 201, 208; kernel oil 52, 68, 69, 218; leaf spot of 69; nuts 42, 141, 198, 228; oil palm varieties 68; tree 50, 113, 114
Palm oil 28, 52, 68, 69, 126, 135, 138, 155, 156, 165, 166, 167, 169, 180, 183, 184, 185, 186, 187, 189, 191, 195, 204, 215, 218, 221, 225; for embrocation, for lighting, higher grade of, lower grade of 69; requirement 68
Palm wine 50, 51, 69, 110, 129, 130, 135, 137, 139, 141, 145, 148, 155, 157, 162, 164, 166, 167, 170, 172, 179, 180, 187, 189, 194, 199, 200, 210, Plate 4; *akpeteshi* distilled from 51; natural yeast in 69
Palmatine 118
Palpitations 176
Panama disease 45
Pancine 208
Panel 93; panelling 84, 93
Pap 149, 183
Papain 45, 118
Papa-yotin 120
Papaverine 118
Paper: manufacture of 8, 74; pulp 10; pulp mill 8; writing 71
Para Rubber 13, 74, 75
Paradol 119
Paraguay 244
Paralysis 176, 190
Paramoecia, plants poisonous to 215

GENERAL INDEX

Parks 90, 232; Parks and Gardens, Department 232
Parrot 17
Pastries 23, 25, 29
Pasture 55, 243
Patterns punctured on skin 227
Peanut: butter 32; flour 32
Peas, green 32
Pectin 43, 64
Pelletierine 122
Pemphisus 184
Peonol 118
Pepper: bacterial spot of 38; grey mould of 38; hot 35, 36, 38; paste 35; red 129, 164, Plate 2; sweet 36; Peppercorn 146, 166, 175
Perception, increased 230
Perfumery 16, 43; industry 231
Periplocin 122
Perspiration: promoting 144, 154
Pest: problems of 18, 29, 59; protected against 70; to cultivation 244, 248
Pestle 11
Pew 92
Pharmaceutical firm 123
Pharmacognosy, Professor of 218
Pharmacological: effect 119; purpose 124; Pharmacology 124
Phaseolunatin 208, 210, 214
PHC 119
Phellandrene 120
Phenol 208, 216; Phenolic 120, 121; resin of commerce 45
Philippines 116, 257
Phosphates 253; Phosphorus 53, 253
Photographic films 71
Photophobia 213
Photosynthesis 20
Physical performance, to facilitate 65
Physostigmine 118, 122
Physovenine 122, 208
Phytolaccotoxin 214
Phytosanitary regulations 246
Phytotoxin 113
Piano 92, 116; console for 92; manufacture 92; thumb 117
Piccolo 116
Picrotoxin 118
Picnic site 234
Pig 244; bush 27; feed 27, 57, 68, 246, 257; guinea 241; poisonous to 54, 62, 213, 214; unsuitable for 71
Pigment 227
Piles 87, 160, 176
Pillar 84
Pillow: flossy inflorescence for 245; wool for 115
Pilocarpine 118
Pineapple: black rot of 44; juice 44, 51, 176; root rot of 44; weed killer 249
Pinitol 121
Pioneer Tobacco Company (PTC) 73
Pipe 116
Piperidine 122
Pit: kiln 101; -props 88; sawer 90
Pito 23, 25
Placenta, hasten expulsion of 169

Plane, carpenter's 104
Planks, wooden 84
Plant(s): absence of 96; alkaloids (see Table 8.2); analyses 124; and soil nutrients, structure 252; antidote 206; aquatic 57; aromatic 229; as binding material 9, 10; beverages 50; wine 13; boundary 234; climbing 9, 10, 233; colours 222; cultivated 28, 242; dangerous properties of 206; decorative xi, 232, 235, 236, 237, 239, 245, 246; -derived drugs (see Tables 8.1 & 9.2); distribution 96; dried and preserved 77; dyes 223; economically useful 242; edible 30, 77; efficacious 123, 124; escapee 245, 246; exotic 49, 99, 232, 239; exporting company 12; extracts 118, 206, 209, 120, 212; exudate 16, 209; fibre yielding 10; floss-producing 115; food 20, 243, 253; for basketry 9; for decorative purposes 235; for fodder 53, 54, 55; for healing 118; for intestinal parasites 127; for poisoning fish 215, 216, 217; for round worms 127; for tapeworms 127; for threadworms 127; free-floating 245; healing properties of 206; hedge 234, 239; herbaceous 123, 125; hormones 242, 250 (see also arboricides); indigenous 239; intercropping 256; introduced 45, 232; latex producing 13, 14; leguminous 250, 255, 256; Marantaceous 112; medicinal and poisonous 13; medicinal properties of 124, 125, 206; minerals 253; mucilagenous 133; natural beauty of 232; nutrients 98, 254; nutrition 253; oil producing 95; parasites 252; poisons 207; poisonous 13, 206, 212, 214, 238; poisonous to stock 213; potted 238; protein 35; remains 254; remedial to snake venom 219; remedies 123; residue 254; root nodules of 256; screening 123, 124; stove 238; tanning-producing 221, 222; that yield sponges 115; toxix 210; toxicity to 242; toxin 212, 215; transpiraton of 18; used to relieve pain 126; waste 255; water 50, 57, 114, 235, 236, 246; wild 71, 76, 95; wild rubber producing 75; woody 123, 125
Plantain 20, 26, 30, 60, 167, 192; crown rot of 30; fruit rot of 30; peelings 30, 60; pitting 30; red 29, 33; speckle 30; stalk rot of 30 Plantation 8, 12, 61, 67, 69, 72, 74, 77, 88, 99, 248; cassava 244; cocoa 61; fuel-wood 95, 97, 99; Kumasi Town 99; oil palm 68, 244; rubber 75; World Bank Project 68
Plastic 74, 77
Plate 107, 108, 110, 245
Platter 107, 109
Pleurisy 177
Plough, wooden 105
Plumbagin 121, 122
Plumbagol 122
Plywood 5, 79, 83, 84, 93, 116; core of 93; factory 93; manufacture of 6, 93
Pneumonia 177
Poachers 19
Pod: ashes 229; rot of cocoa 64

GENERAL INDEX

Poison(ous) 13, 45, 121, 206, 211, 214, 242; antidote to 217; arrow 210, 212, 251; bites 219; cumulative 250; deadly 210; delay action of 217; extracts 212; fish 215, 216, 217; general 210; heart 210; minimise effect of 206; neutralise 206; plants 13, 206, 212, 214, 238; principle 206; property 212; suspected to be 120; to stock 213, 214; to vomit 219; toxins 211; wild yam 27
Poisoning: accidental 206, 217; arrow 177, 206; general 177, 206
Pokuase/Media 74
Poles 11, 78, 79, 84, 87, 105
Polyagglutinable red cells 124; type III 124
Polyphenolic substance 71
Polysaccharides, pectin 64
Pomade 16, 67
Pond 235
Population: centres 99; concentration of 99; in developing country 118; increase in 22; majority of 95; of Ghana 8; urban 27
Porridge 23, 24, 25, 26, 27, 28, 130
Port 59; cotton 69
Portuguese 43, 44, 75; traders 73
Post(s) 79; and Telecommunication Dept 87
Pot: cooking 227; flower 110; -herb 29, 32, 42, 251
Potash 30, 132, 198, 209, 255; source of 228
Potassium 62, 253; nitrate 129, 203; oxide 69
Potato 28, 31
Potency, increase the 51
Potions 118
Pottery 226
Poultry 25; droppings 254; feed preparation 24
Powdery mildews 33, 45
Predators 28
Pregnancy: later stages of 169; quick delivery in 169; pregnant woman 126, 178, 189; oedemas in 175
Premnine 122
Prescriptions 75, 119, 125, 126, 127, 130, 197
Preservation: of wood 88; problems of food 22
Preservative 108; impregnated with 87
Preserves, fruits used for 45
Prestea 68
Price: 59; cheap 90, 103; high(ly) 75, 92, 95; world 62, 75, 76
Primagram 249
Primary: Health Care (PHC) 119; products 93; requirement 23, 78
Private sector 44
Producer price (see Table 3.1)
Producing country 69, 74; important 28, 64; largest 64, 75; main 67; major 31, 32, 33, 68, 73; principal 76; third major 61
Production: 22, 24, 59, 61, 66, 67, 70, 72, 74, 75, 76, 77; kola nut 72; of antibodies 40; palm kernel oil 68; palm oil 68; sugarcane 74; world's 27, 61, 64
Propeller, aircraft 85
Protected timberland 99, 253
Protein 12, 25, 27, 33, 34, 36, 37, 39, 45, 55; concentrate 34; crude 24; deficiency 31; good quality 35; high content 68; high quality 40; higher in 53; rich in 32, 33, 40, 68; rich source of 35; shortage of 31; value 24
Proteolytic (enzyme) 120
Protoveratrines A and B 118
Pseudo-curanine 208
Pseudoephedrine 118
Psychological addiction 230
Psychosis, toxic 230
Pterygospermine 121
Public building 232
Pulpit 92
Pulse 65
Pumpkin 36, 40, 110
Puppet 106
Purchasing power 94, 100
Purgative 41, 122, 139, 149, 158, 168, 170, 172, 173, 177, 179, 188, 191, 196, 197, 210
Purging property 48, 219
Purple 225, 227
Purse 111
Pusa 27
Pwalagu Tomato Factory 35
Pyrethrin spray 77
Pyrethrum 77

Quassin 122
Quassinoid 122
Quercetin 118; Quercitoside 122
Quinidine 118; Quinine 118, 121; substitute 155, 156
B-quinovine 120

Rabbit 55; food for 239; harmful to 214
Radio 92; cabinet 92; gram 92
Raffia 223, 224; bast 116; craft 223; fibre 116
Rafters 67, 83
Rails 84, 86, 87, 95, 103
Railway: coachwork 85; coaches 86, 93; lines, construction of 87
Rain 20; intercepted 20; rainfall 12, 18, 20, 25, 65, 70, 252, 254
Raphides 31; of calcium oxalate 40
Rat, plants poisonous to 214
Rattlers 114, 117
Raw materials 59, 69, 70, 72
Rayon, manufacture of 71
Recorder 116
Recreational: area 233; ground 234
Rectal injection 147, 200
Red 224, 225, 227; -brown 224; indelible 227; Sea 15, 77
Reed screens 89
Refuse dumps, manure from 254
Regeneration: coppice system of 95, 99; from seed 242; from stolon 242; from vegetative parts 242
Regions (in Ghana): Ashanti 17, 27, 35, 61, 64, 68, 72, 73, 76, 88; Brong Ahafo 23, 24, 27, 31, 61, 64, 69, 70, 72, 73, 76, 244; Central 30, 43, 61, 64, 67, 68, 72, 74, 76, 77, 245; Eastern 17, 23, 30, 35, 61, 64, 68, 72, 74, 245; Northern 31, 69, 73; Upper 31, 35; Upper East 69, 72, 73; Upper West 69, 72, 73; Volta 30, 38,

GENERAL INDEX

50, 61, 64, 69, 70, 73, 107; Western 6, 17, 23, 24, 61, 64, 67, 68, 73, 75, 101, 102
Rehabilitation programme 59
Reinforcement, structural 80
Relative humidity 18
Relaxant 230
Religious: beliefs 106; ceremonies 16; occasion 113
Reptiles 220
Rescinnamine 118
Research: purposes 72; station 242; Unit 101
Reserpine 118
Reservoir 246
Residential areas 232
Resin(ous) 16, 123, 126, 143, 144, 145, 167, 229; acrid 179, 211; body 119; bonducin 16; gum 16, 131; oleo- 204, 229; phenolic 16; red 222, 224; root- 131
Resistance to: abrasion 84; marine borers 87; salt water 86; water 87, 88
Resonance box 92, 117; with cane tongue 117; with metal tongue 117
Resonator 117
Resource: indigenous 119; traditional 119; underexploited 66
Respiration 67; respiratory troubles 180
Rest House 43
Restaurant 79, 102
Revenue, source of 73
Rheumatic pain 76; rheumatism 131, 180; articular 131, 182
Rhizome 52; edible 30
Rhizopus rot 38
Riboflavin 36, 37
Rice 24, 25, 32, 41, 52, 109, 131, 140, 163, 173, 187, 215; chaff 24, 98; colouring boiled 223; consumption of 24; -eating people 215; farming 257; field 248; growers 257; producing country 24; production 24, 257; substitute 26; upland 248; water 24; weed-killer 249; wet 257
Ricin 206; ricinine 206
Rickets 182
Rifle: air 103; bolt action 103
Ringworm 182, 183; of the scalp 182, 183
Ritual occasion 113
River 20, 86; Afram 19, 246; Bia 19; Dayi 246; Fum 19; Fure 19; Klemu 19; Niger 246; Ochi 19; Offin 19; Pawnpawn 246; Pompo 19; Pra 8; Tain 19
Rock particles: chemical weathering of 253; dissolution of 253
Rome 75
Roof(ing) (material) 81, 82, 85; frame 83; most popular 82; rafters 83
Root(s) 1, 26, 53; as chewing stick 9; cheapest of 26; crop 29, 33; edible 30; most costly of 27; nodules 256; peelings 54; resin 131; rot 44; stilt 12
Rope 9, 10, 68
Rorifone 118
Rosagenin 208
Rosaries 113; seeds for 16, 113
Rotenone 121, 208, 217
Rottlerin 121
Rotundine 118

Roundwood 96, 98, 99
Roundworms 127
Royal Botanic Garden(s): Edinburgh 238; Kew 236, 238
Rubber 8, 13, 14; adulterant 14; crude 75; Growers' Association Estates 75; Para 13, 74, 75; pure 13, 75; wild 75
Rubefacient 182
Rum 138, 142, 169, 191
Ruminant, feed for 257
Rural: area 78, 94, 95, 118; charcoal trader 101; community 31, 117, 118; dweller 103; industry 103
Rural life, essential item in 94
Rust 204; preventing 102
Rutin 118, 208
Rwanda 64

Sack, jute 72
Sacred grove 253
Saffron dyes 223
Sahara 57, 75, 95
Sahel 25, 26, 77, 96, 98; savanna 248
Sailing-ship era 27
Salad 35, 38, 40, 41, 71; mixed fruit 45
Saligenin 118
Salmento-cymerin 209
Salt(s) 34, 41, 125, 129, 130, 131, 141, 142, 145, 147, 151, 154, 159, 161, 162, 163, 167, 173, 174, 177, 178, 179, 186, 191, 195, 196, 197, 204, 221, 222; -bush tree 174; common 49, 227; concentrations of 67; of oxalic acid 208; pickling with 34; soluble 253; substitute 45, 49, 50; vegetable 130
Saltpond 206
Sambunigrin 123
Samreboi 6, 93, 102
Sand 17; binders 20
Sansas 117; -Santonin 118
Saponin 31, 206, 209, 214, 215, 220; concentrations of 207, 208, 209; -producing plants 228
Sapowpa 115
Saprophytic fungi 252
Sapwood 98; as fuel 102
Sasabonsam 1
Satsuma, *Citrus* variety 43
Saturation point 18
Sauce 24, 31, 35, 38
Savanna: coastal 30, 44; country 18, 20; derived 73; northern 18, 23, 66, 72; south-eastern 45; southern 23, 27, 30, 38, 39, 40, 44, 99; zone 11, 23, 26, 32, 95, 99, 223, 224
Saw: band 102; circular 102; mill 90, 98, 102
Sawdust 98, 204, 252; as fuel 102; as snuff 139
Sawe 8
Scabies 184
Scaffold(ing) 84
Scarf ring 114
Scenery 232, 233
Schistosome eggs 215; Schistosomiasis 184, 215; -carrying 215; means of controlling 215
School 90; yard 233
Scientific: research 22; training 22
Scillarens A and B 118

Sclerotium rot 29
Scopalamine 118, 207
Scorpion sting 219; antidote to 184, 206, 220, 221; immunity against 221; preventive of 220, 221, 251
Screen 112, 234
Scrub 95
Sculptor 90
Scurvy 27, 184
Sea 20, 103; level 232; open 18; shore 20, 248; weed 58, 224
Season: dry 18, 56, 244, 245; growing 70; lean 35; rainy 18, 25, 251; short 25, 60; wet 246
Seasonal dry-up, prevent 19
Seasoner 131
Seasoning 90; food 52
Secondary host 215
Secret knowledge, wealth of 119
Sedative 151, 174, 230
Sedge, biological control of 243
Seed: beni 77; cake 71, 255; coloured 214; craft 106; decorative 113, 114; edible 39, 42, 46, 62, 66, 73, 76, 125; food value of 39; hollowed-out 114, 116; improvement of 22; lethal 210; mature 244; of improved variety 70; poisonous 113; pulp around 49, 51, 66, 76; sap around 62; shaving of 54; shell 114
Seedling 54, 56
Sefwi Bekwai 23
Sefwi Wiawso 93; District 61, 64
Sekodumasi 72
Sekyedumase 23
Selection by screening 124
Self: medication 125, 206, 217; -sufficiency 22
Semi-: arid 77; Bantu tribes (see Bantu tribes); cultivated 31, 40; desert regions 252; parasite 67
Senegal 24, 25, 31, 32, 77
Sennosides A and B 118
Senya Bereku 30
Sense alteration 230
Serwaa 28
Sesame 77
Sese 68
Settlement 66
Sewing machine 92
Sexual damage, possible 230
Shaft guides 88
Shallot 38, 52; black rot of 38; white rot of 38
Shea butter 52, 66, 67, 126, 135, 138, 145, 151, 156, 160, 161, 162, 166, 172, 180, 181, 182, 187, 192, 195, 202, 224, 225; colouring 225; producer price of (see Table 3.1)
Shea Nuts Marketing Board 66
Shed 78, 79, 234; poles for 10; temporary 10, 79
Sheep 27, 54, 55, 96; diarrhoea in 147; droppings of 254; fatalities in 213; harmful to 214; lethal to 214; poisonous to 62; promote multiple birth in 54; rearing 221; to kill 213
Sheep-dips, insecticide for 166

Sheet, roofing 81
Shell-oil 45
Shelterbelt Forest Reserve: Aboniyere 18; Amama 18; Angoben 18; Aparapi 18; Bia 18; Pru 18; Totua 18
Shifting cultivation 253
Shimba Hills, Kenya 123
Shingles 81; split wood as 81; the life of 81
Ship 90, 93; building 86; bulkheads of 86; deck 86; rail 86
Shoe polish 62
Shotgun: double barrel 103; single barrel 103
Shrimp 216
Shrine 234
Shrub 235, 237; branching 40; fast growing 40; ornamental 234
Shuttle 110
Sierra Leone 1, 24, 39, 42, 48, 51, 65, 69, 72, 73, 75, 76
Sieve 110
Sigatoka disease 45
Sign boards 89
Silica ashes 56
Silicates 253
Silk 225
Silver 211
Silybin 118
Sinusitis 184
Skimmianin 123
Skin: diseases (and affections) 182, 184, 185, 228; eruptions 134; puncturing of 227; tanned 225; to stain 223
Slasher, mechanical 251
Slave trade 28, 75
Sleep(iness) 210; prevent 73
Sleepers 87; cross-ties of, metal, timber for, wooden 87
Smallpox 185
Smithsonian Institution, Washington 2, 40, 243, 256
Smoke 94, 95, 101
Smooth wear 84
Smuggling produce 59
Snail 1, 17, 143; *Bulinus* 215; lethal to 215; plants poisonous to 215; soap to kill 215; water 57, 215
Snake 218, 219; holes of 220; in Africa 218; poisonous 219, 220; to keep away 220; to paralyse 220; to repel 220
Snake-bite 185; antidote to 220; general cases 219; immunity to 220; local treatment of 219; preventive measures against 220, 221; snake-venom: antidote to 206, 218; plants remedial to 219
Sneezing 185; powder 154
Snuff 73, 137, 139, 160, 161, 162, 184, 185, 186, 217; -box 114
Soap 172, 215; antiseptic 228; detergent 228; introduction of 69; local 155; -making 30, 67, 209, 228, 231; manufacture of 24, 68, 69, 71, 77; manufacturing 62; manufacturing company 68; manufacturing industry 69; native 125, 143; soft 62, 231
Sociability 230
Social: background 22; structure, development

trends in 106
Socrates 206
Soda 29, 224; water 132
Sodium: arsenite 7, 250; chlorate 250; sulphamate 250
Soft drink, non-alcoholic 51
Soft rot, bacterial 35
Soil 19, 20, 67, 76, 98, 254, 257; acidic 76, 258; aeration 20, 255, 256; application of manure to 253; bind the 20, 243; chemical composition 252; clay 255; constituents 253; effect of manure on 255; erosion 95, 254; exhausted 258; farming 257; fertility 249, 253, 254; good 258; improvement 255; indicator 244, 257; marginal 242; matrix 253; mineral 252; moisture 242; nitrogen in 256, 257; physical breakup of 20; phytotoxicity 242; poor 33, 242, 258; porous 20; quality 253; return of minerals to 253; rocky 258; saline 258; spongy 20; structure 252, 255; sub-tropical 253; supply of organic residue to 253; texture 20; tropical 253; upper layer of 253; water 253; with low humus content 252; weed as binder 251; yellowish 253
Solanine 122, 209, 214
Solvent 230
Somaliland 48
Sonora desert 77
Soot 101, 227
Sooty mould, parasitic 39
Sore(s) 186, 201; eyes 151, 186; guinea worm 159; gums 186, 190; mouth 186, 190; on the ear 203; syphilitic 137, 198; throat 186; venereal 199; water 202
Soumara 122
Sound-proofed doors 83
Soup: groundnut 32; okro 39; palm 48, 68, 132, 170, 191
Sparteine 118
Spear: hunting 212; shaft 105
Specific gravity (of trees) 96
Spice 52, 75, 76, 126, 135, 140, 141, 142, 145, 146, 147, 160, 168, 176, 177, 180, 187, 189, 192, 193, 194, 201
Spider bite 221; antidote to 206
Spigeline 209
Spinach 40, 42; most popular 40; young leaves as 26, 29, 33, 28, 77; weeds as 251
Spindle 107
Spine, tuberculosis of 194
Spine-like excrescence 116
Spinning game 114
Spirit (drink) 130, 148, 176, 199; manufacture of 62
Spleen, inflammation of 163
Sponge 39; bathing 115; chewing 8, 115; craft 115; fibre as 30; preparation 106; washing 115
Spoon 107, 108, 109, 110
Sprains 136
Spraying machine 59
Sprouts, bulbs, tubers, etc. 37
Sri Lanka 34, 244
Stairways 84
Stand (of trees): dense 245; in a courtyard 232; large 95; mixed 233; pure 233
Staple (food) 23, 26, 27, 28, 30
Star Apple 43
Starch 54, 73; cassava 26; sweet potato 29
State: Farms Corporation 72, 75; functions 90
Stem-borer 243
Steps 84
Sterility 187; in men 164
Steroidal sapogenins 27
Stew 24, 35, 38, 39
Stilts 83
Stimulants 72, 76, 138, 154, 230
Stock 32, 55, 56, 87; fodder for 24; losses 213; plants poisonous to 213, 214; poisonous to 25, 210, 214; survey 7; toxic to 56
Stomach: ache 149, 164, 188, 189; disorders 188; pains 188; ulceration of 196
Stomachic 130, 131, 139, 140, 144, 189, 190, 210
Stomatitis 186, 190
Stones 17
Stool(s) 67, 99, 103, 106, 107, 108; Akan 106; context and function of 106; kitchen 90; land 18; native 107
Storage, problems of 22
Storage facilities, lack of 22
Storm 97
Stove: fuel-saving 8; sawdust 98
Straw 112, 113, 252, 254
Stream 74, 86
Strength: and durability 90; and energy 62; loss of 132; minimum standard in 78; restoring failing 32
Stroke 130, 176, 190
Strophanthidin 122; Strophanthin 209, 210, 211, 212
Structure: permanent 79; temporary 78
Strychnine poisoning 118, 206, 209; specific antidote to 218
Stye 152
Styptic 133, 151, 159, 160, 190, 244
Sub-tropical country 43; Sub-tropics 232, 245
Subri Forest Reserve 95
Sucrose 12
Sudan 26, 31, 33, 34, 41, 47, 77, 206, 246; savanna (woodland) 25, 31, 33, 35, 38, 48, 77, 96, 221, 248
Sudd 246
Sudorific 144, 191
Sugar 30, 51, 62, 64, 73, 131, 158, 177; annual need 74; cane 74, 95; crystals 74; extraction of 74; industry 74; manufacture 74, 98; substitute 29, 30
Sugarcane 248; alcohol extraction from 96; gummosis of 74; juice 51; leaf scald of 74; mosaic disease of 74; weed-killer 249
Suhum 64
'Sulphur oil' 128, 181, 194
Sun 31, 80, 112, 155; cracking in 81; hot afternoon 82; light 17, 20, 67, 244; rays of 17; scorching 20, 233; spray dried in 35; warming effect 17

Sunyani 72, 83
Superbine 207
Superphosphate 255
Surf boat 86
Surgical purposes 71, 115
Swamp(y) 97; areas 21, 74, 257; land 24, 26
Sweet potato: black rot of, blue mould of, charcoal rot of, dry rot of, flour 29
Sweetening agent 13, 189; Sweetner, possible sources 12; Sweetness, destroy taste for 12; Sweets 34
Swellings 191, 201, 227; of syphilitic origin 198
Switzerland 44
Swollen shoot of cocoa 64
Symbiosis 257
Symbol 227
Sympathicostenic 122
Sympatholitic action 120
Sympathomemetic 122
Syphilis 192, 198, 199; bony, chronic, heredo-, immunity to, serious cases of, spawns of 199
Syringe 110
Syringine 122

2, 4, 5-T 250
Table 111; camp 90
Tackle, fishing 30, 223
Taenifuge 192
Tafo 72
Takoradi 73, 93, 96; Veneer and Lumber Company 6, 93
Tamale 23, 27, 67, 70, 73
Tan 221
Tannin 44, 45, 206, 221, 222; gallic 120; high proportion of 209
Tanoso 72
Tanzania 50, 77, 116
Tape recorder 92
Tapeworm, expulsion of 122; specific for 127
Tapioca 27
Tapori 105, 106
Tar 101
Tarkwa 75
Taro 37; corm rot of 29; leaf blight 29; sclerotium rot 29; soft rot 29
Tatare 30, 33
Tattoo-like marks 227; Tatooing 209, 227
Taungya system 7
Tea 50, 142; -like infusion 144, 161, 164, 177; scenting 50, 229; substitute 50
Techiman 23, 72
Techimantia 72
Tecleanone 122
Teeth: carious 193; cleaners 8, 9; stain 9; Teething 143, 192; easy 144
Tekyiman District 72
Telegraph: cross-arms 87; poles 78, 87
Television set 92
Tema 68, 70; Textiles 70
Temperature 12, 17, 96, 101; favourable 17, 20; high 210
Tephrosine 206, 209, 215, 216
Teppa 72; District 61
Termite: attack 83, 254; excreta 254; keep away 165; -proof 45, 67, 78, 81, 86; protect house from 102
Terpene 120
Tetanus 192
Tetradine 118
Tetrahydrocannabinol 118
Tetteh Quarshie, Opanyin 61
Texas 245
Textile: companies 69, 70; manufacture 26
Thailand 34
Thatch(ing) 81, 245; palm fronds as 68, 69, 81; weeds as material 245, 251
Thaumatin 12
Theatre 90
Theobromine 62, 73, 118, 120, 120, 123, 148
Theophylline 118
Thevetin 209
Thiamin 35, 36, 37
Thickets 95; dense 245
Third World Countries 8; development in 22
Thirst, control 73
Thorns of hedge plant 234
Threadworm 127
Throat: inflammation of 191; syphilis of 198; troubles and affections 194
Thrush 194
Tichiri 72
Ticks, to keep away 164
Tietie 9, 10, 80
Tiger nut 129, 183
Tikobo No. 1 67
Tile, floor 62
Tiller shoots 25
Timber 1, 6, 8, 67, 83, 85, 86, 87, 88, 89, 90, 91, 98, 99, 103; as plank 84; choicest 92; classified merchantable (see Table 1.2); destruction of forest for 236; durable 78, 86, 88, 92, 105; durable under water 87; excess water content in 90; export of 5, 93; exportable 5, 99; first class 91; for building 78; for fencing 88; for gun stock 104; for panelling 93; hard 78, 86, 88; heavy 78, 86, 88, 105; high-class 84; industry 6; international value of 5; less economically important 93; local 90, 99; lower grade 91; market 90; Marketing Board 83; merchantable 5, 91, 93; of even grain 107; medium hardness 78; medium weight 78, 107; off-cuts 90; processed 93; production 5; quality of 90, 93; resistant to abrasion 84; resistant to fire 78; resistant to marine borer 87; resistant to water 78, 87; sawn 79, 86, 87, 88, 98; seasoned 90; selected 83, 90; suitability of 83; termite-proof 78; unmerchantable 85, 91
Timboin 122
Tisane 137, 138, 147, 182
Tobacco 73, 102, 132, 139, 141, 166, 186, 193; air-cured, fire-cured, flue-cured 73; foot rot of, frog eye, leaf spot of 74; processed, raw 73
Togo 27, 31, 33, 46, 64, 67, 68, 113
'Tom-Brown' 23
Tomato 35, 36, 38; paste 35

GENERAL INDEX 335

Tongu 23; Irrigation Site 35
Tonic 130, 156, 162, 163, 167, 188; astringent 139; digestive 131, 179, 189; drink 34; for horses 131
Tonkui, Mt 50
Tonsils, inflammation of 191
Tool 103, 104; handle 103, 111
Toothache 76, 192
Tortoise 17
Totem poles 109
Tourism 19
Towel, fibre as 30
Towns 79, 95, 99; major 232; outskirts of 89
Townships 95; lighting for 102
Toxalbumen 120, 121; Toxalbumin 206
Toxic(icities) 214, 216, 217, 242; psychosis 230; Toxin 206, 211, 216; plant 212
Toy: hoops 112; vehicles 111
Trachea 190
Trachoma 151, 194
Trade: Fair 83; names 51; secret 225
Trade Winds, north-east 18
Tradition 106; Traditional: medicine 124; symbolic motifs 110
Transitional zones 97
Transpiration 18, 21, 22, 253
Transport: lack of 59; water 109, 246
Transportation cost 255; wastage on 103
Trap 103; fishing 111, 112
Tray, serving 110
Tree(s): aesthetic values of 232; association of 2; boles of 5; burnt 253; charcoal 101; cultivation of 95; dead 7; decorative 233, 235, 240; dominant 1; economic 7; emergent 1, 94; evergreen 233; for mortar 108; fork of 92; gum copal 16, 209; gum yielding 14, 15, 209; height of 45; high forest 101; huge forest 103; inaccessible 66; indigenous 43; introduced 16, 46, 54, 233; leaf of 18; middle storey 94; of medium hardness 97; parent 45; pioneer 257; plantation 95; preferred for fuel 94, 95; protection of 232; resin yielding 15, 209; rich distribution of 96; savanna 97; shade-bearing 232; shade (of) 232, 233, 234, 240; shelter of 232; small 94; timber 94; uneconomic 7; understorey 79; upper canopy 94; wide range of 96; with high specific gravity 96; with low specific gravity 96; with required resonance 107; with required vibration 107
Trestles, concrete 86
Treviso 75
Trial by ordeal 212
Tribal: land 18; markings 227; Tribe 31, 42, 107
Tricliseine 123; Triclisine 123
Trigonelline 209
Tripartite system 60
Tripoli 75
Trolley, serving 110
Tropical: areas 244; countries 43, 232, 243, 244; rainforest 18; rains 253;

shelterwood 7; species 5; temperatures 20
Tropics 2, 27, 65, 67, 69, 71, 88, 232, 233, 245
Trotro 85
Trough 107
Truck: mammy 85; wooden 85
Trumpet 116
Trypanosomiasis 185, 194
Tsetsefly 185
Tsito Avenorpeme/Torve 74
Tuber 12, 13, 26, 28, 29, 33, 37, 53; aerial 28; cheapest of 26; edible 30; most costly of 27
Tuberculosis 140; of the spine 194
Tubocurarine 118
Tuei 23
Tulukunin 120
Tumours 194
Turnip 37
Turpenoid nature 119
Twifo Praso 72
Twifu 68
Twine 9, 68, 71
Tyre factory 75

Uganda 50, 64, 243; Acholis of 26
Ulcers 171, 194; syphilitic 198
Umbrella: ribs of 108; state 108; tree 82
UNDP 159
Uniform: availability 38; price level 35
Uniformity (of wood) 85
United Kingdom 75
United States 23, 26, 61, 70, 87, 92, 224, 234, (235,)236; Agriculture Research Services 123
University of 90; Cape Coast 124, 232; Ghana, Legon 22, 26, 29, 72, 75, 124, 232, 244, 250; Ife 218; Illinois, Chicago 123; Science and Technology, Kumasi 124, 232
Upholstery 71, 115
Urban: centre 9, 41, 59, 68, 79, 94, 95, 99, 118; dweller 23, 94, 95, 99; life-style 94
Urbanization 99
Urethral: complaints 199; discharges 196, 199; stricture 199; troubles 196, 199
Urinary: complaints 196, 199; troubles 199
Urinary passage, diseases of 35
Urination, promoting 147
Urine: blood passed in 160; excessive flow of 217; incontinence of 164
Urino-genitary complaints 197
USA 34, 77, 124, 215
USAID 159
USSR 62
Utensil 103, 106, 107, 108, 109, 110; domestic 110; drinking 110
Uterus troubles 187
Utility articles 110, 114; Utilization Branch xi, 83, 99

Vapour bath 132
Variety: A 63440 71; Allen 26J 70; Allen 333 70; Allen 444 70; B 7-1 71; B 7-2 71; B 71-3 71; B 72-2 71; BJA 592 70; Brondal 29; Centennial 29; Clementine 43; Cuba 108 71; Cuba 2032 71; dry ground 48; Dura 58; dwarf type 67; early maturing 29;

Eland 29; Ex-Mokwa 71; G 45 71; Gem 29; GT-3 71; GT-7 71; HAR 444 70; high yielding 70; improved 70; in height 74; Jewel 29; King de Semis 43; Kwesi Nyarko 43; Lake Tangelo 43; Late Velencia 43; Miguela Tangelo 43; Nemagold 29; Nyankpala white 71; Orange fleshed 29; Ortenique 43; Otumi 43; Ovelleto 43; Pisifera 68; Ponkan 43; red stemmed 74; Sanguino 43; Satsuma 43; Shama 43; seedless 46; Sekkan 43; susceptible 61; Tenera 68; Thai Red 71; Washington 43; white fleshed 29; white-stemmed 74; wild 44; world 74
Varnishes 231
Vase 110
Vegetable 23, 35, 39, 41, 77, 111, 243; ANV of 36; cultivation of 41; eaten green as 44; fleshy 35, 36; gelatinous 39; immature fruits as 39;45; in emergency 41; introduced 41; ivory 114; leaf(y) 36, 40; leaf(y) shoots as 39; matter 252; remains 20; semi-cultivated 40; wastes 254; wild 12, 41; weeds as 251; young pods as 32, 33, 42
Vegetarian 39
Vegetation 96; aquatic 57; forest 97; scanty 96; sparsely distributed 96, 99; types of 96
Vegetational zones 25
Vegetative: cover 20; matter 253
Vehicle 84, 96; toy 111; utility 84
Velvet Tamarind 43
Veneer 5, 91, 92, 93; manufacture of 6; mill 6; and Plywood Factory 93; timber for 102
Venereal diseases 137, 158, 172, 192, 196, 200; cankers of 199
Venom, blinding 219
Vermifuge 200; purging 178
Vermin 165; poisonus to 215
Vernonine 123, 162, 209, 214
Vertigo 200
Vesicant 200
Veterinary medicine 120
Video recorder 92
Viennese 123
Vietnam 257
Village 66, 71, 79, 115; hedge plant in 234; neighbouring 76
Vincaleukoblastine 118
Vincamine 118
Vines, strong 87
Viola 116
Violin 116; bow frog of 116; fingerboard of 116; pegs of 116; tailpiece of 116
Viper 219, 251; carpet 219; Gaboon 219
Viral infection 246
Virus disease 27, 64, 74, 246
Vitamin 32, 35, 40; antiscorbutic 184; B 24, 25, 27, 44, 45; C 27, 32, 35, 36, 37, 38, 44, 45; E 34, 45; value 25; Vitamin A 24, 34, 44, 45, 53, 68; precursors of 68
Volta 25; Basin Research Project 71; River Authority 246

Vomit(ing) 131, 206, 213; arrest 151; causing 210, 211, 214, 217, 218, 219; induce 150, 199, 206, 217, 218, 219; of blood 159, 190; prevent 151; Vomitive 201

Wa 27; District 72
Wagons 84
Wales 75
Walk early, enable children to 201; Walking stick 209
Wall hangings 112
Walnut 103
War 75; Second World 24
Warehouse 84
Wart-hogs 35, 57
Washington: state 2, 256
Waste disposal, efficient means of 255
Water: absorbing compounds from 57; absorption of 20, 21; acidic 246; conserve 20; container for 110; courses 74; drinking 12; holding capacity 225; indicator of underground 12; melon 36; oxygenating 57; perennial flow of 19; pervious to 255; plant 235, 236; plant as source of drinking 12; rain 20; regulate flow of 18; resistant to 78; retentive properties 20; run-off 19, 20; shed 18; snails 215; surface 257; table 19; transport 109, 246; vapour 18; weed, notorious 235
Water fleas, plants poisonous to 215
Water snails 57, 215; lethal to 215; plants to control 215
Wax 17; liquid vegetable 77; print 70
Weakness, restorative for 132
Weather: desiccating effect on 18; -proofing 102; protect from 88; unfavourable 22; wet 32
Weaver birds 68
Web blight of bean 32
Webley and Scott 103
Weed 32, 242, 252, 254; abundant 245, 250; advantages of 242, 243, 251; Akyeampong 244; aquatic 246, 249; as fodder 251; as preventive of erosion 251; as soil binder 251; as spinach 251; as straw 251; as thatching material 251; as vegetable 251; biological control of 242, 244, 245, 246, 250; burning of 250; Busia 244; cheapest means of control 243; common 243, 246; cost of control 242; decorative 235, 247, 248; definition 242; disadvantages of 242; disposal of 252; drought-tolerant 245; effect on crops 250; ferns 248; for compost and mulch 251; grasses 248, 249; harmful 248; herbaceous 247; infestation 242; irradication 246; light demanding 250; mechanical control 243, 244; notorious 82, 244, 246; noxious 243, 245; parasitic 32, 248; pest 244, 245, 248; physical control 243; Research Organization 242; seeds of 252; sedges 248, 249; semi-aquatic 249; serious 246, 248; shrubby 247; suppression 255; troublesom 244; woody 247; world's

GENERAL INDEX 337

worst 243
Weed control 249; biological 243, 244, 246, 249; by burning 250; by flooding 245; by grazing 250; chemicals used in 249; cost of 250; difficulty in 244; manual 243, 250; mechanical 243, 251; physical 243; traditional method of 250; widely used means of 243
Weed-killers: chemical contact 249; pineapple 249; maize 249; rice 249; sugarcane 249; translocated 249; Weedicides: post-emergent 250; pre-emergent 249
Weevil 244; cola 165
Weija 23; Dam 243
Wenchi 44, 73
West: coast 26; Indies 28, 240, 244
West African: Black Pepper 52; Mills Ltd, Takoradi 62
Wharf piles 87
Wheat 25; flour 26, 29, 30
Wheel 85, 87; barrow 84; -making 85
Whisk 115
Whistle 116
White rot of onion 38
White ants, to drive away 165
Whitlow 210
WHO 34, 118, 123, 159; collaboratory centre 123
Whooping cough 141, 201
Wig stand 111
Wild: but edible grains 25; coffee 65; Mango 43
Wind 18; break 234; desiccating 18, 95
Window 83; frame 83; timber for 83
Wine 50, 69, 137; cocoa 62; coconut 67, 68; manufacture of 62; palm 68, 69; production 64
Winneba 39, 67
Witch 212
Wodur 109; *abe* 109
Woggles 114
Wood 8, 78, 79, 83, 84, 88, 89, 90, 92, 93, 95, 101, 107, 116, 117; as source of fuel 94; -ashes 49, 175, 222, 228, 229; beams of 79; borer 88; choice of 103; consumption of 8; curls 1, 92; dead 5, 94; dealers 90; felled 95; for comb carving 110; for mortar 108; figured 1, 92; impact resistance 95; gas 102; land 46; -lot 95; moisture content of 90; 103; off-cut 102; planks of 79; product 116; pulp 8; quality of 92; rack of 92; round 88; seasoned 103; shelf of 92; shavings 96, 98; soft 88, 97; solid 85; strips of 84; treatment of 103; uniformity of 85; waste 96; white 91, 107; wool 98;

worker 90
Wooden: buildings 82, 83; carving 109; floor 224; truck 85
Wooden house 82, 83; timbers suitable for 83
Woodland 232; savanna 8, 95, 99
Wool 115, 193, 224, 226; craft 115
Worker: average 94; temporary 251
Workmanship 91; degree of 90; high standard of 111
World 23, 40; demand 77; important nuts of 45; Indicative Plan 24; market price 61; most poisonous snakes in 219; New 235; Wide Fund for Nature (WWF) xi, 8
Worm 122, 127, 158
Wound(s) 143, 186, 201, 218; circumcision 138, 204; leprous 171; snake-bite 220
Wristlet 113

Xanthotoxin 118
X-ray films 71
Xylophone 117

Yacht 90, 93
Yakayaka 26
Yam 12, 73; antiscorbutic properties of 27; boiled 27, 215; internal brown spot of 27; peelings as feed for livestock 27; replacement of 27; soft rot of 27; tuber rot of 27; wild 27; wild edible 27; wild poisonous 27; zone of West Africa 27
Yamfo 72
Yang 85
Yapei 72
Yarn 71
Yaws 157, 204
Yeast 69
Yellow 225, 251
Yellowish: -brown 186, 204; -orange 217, 225; -red 225
Yendi 27, 70
Yield: annual 76; average 70; double the 74; herbicidal application on 242; higher 255, 256; increase 23, 76; loss of 243; low 22; of charcoal 101; reduce 242, considerably 243; reduction in 242, 243; total 73
Yohimbine 118, 120, 122; Yohimbinine 122
Yoke 105
Yoruba 141; Agbo infusions 141, 155, 180

Zaire 26, 50
Zambia 34, 50
Zebella 72
Zimbabwe 73, 77